GEOMETRIC INTEGRATION THEORY

PRINCETON MATHEMATICAL SERIES

Editors: MARSTON MORSE and A. W. TUCKER

1. The Classical Groups, Their Invariants and Representation. By HERMANN WEYL.
2. Topological Groups. By L. PONTRJAGIN. Translated by EMMA LEHMER.
3. An Introduction to Differential Geometry with Use of the Tensor Calculus. By LUTHER PFAHLER EISENHART.
4. Dimension Theory. By WITOLD HUREWICZ and HENRY WALLMAN.
5. The Analytic Foundations of Celestial Mechanics. By AUREL WINTNER.
6. The Laplace Transform. By DAVID VERNON WIDDER.
7. Integration. By EDWARD JAMES McSHANE.
8. Theory of Lie Groups: I. By CLAUDE CHEVALLEY.
9. Mathematical Methods of Statistics. By HARALD CRAMÉR.
10. Several Complex Variables. By SALOMON BOCHNER and WILLIAM TED MARTIN.
11. Introduction to Topology. By SOLOMON LEFSCHETZ.
12. Algebraic Geometry and Topology. Edited by R. H. FOX, D. C. SPENCER, and A. W. TUCKER.
14. The Topology of Fibre Bundles. By NORMAN STEENROD.
15. Foundations of Algebraic Topology. By SAMUEL EILENBERG and NORMAN STEENROD.
16. Functionals of Finite Riemann Surfaces. By MENAHEM SCHIFFER and DONALD C. SPENCER.
17. Introduction to Mathematical Logic, Vol. I. By ALONZO CHURCH.
18. Algebraic Geometry. By SOLOMON LEFSCHETZ.
19. Homological Algebra. By HENRI CARTAN and SAMUEL EILENBERG.
20. The Convolution Transform. By I. I. HIRSCHMAN and D. V. WIDDER.
21. Geometric Integration Theory. By HASSLER WHITNEY.
22. Qualitative Theory of Differential Equations. By V. V. NEMICKII and V. V. STEPANOV. Translated under the direction of SOLOMON LEFSCHETZ.
23. Topological Analysis. By GORDON T. WHYBURN.
24. Analytic Functions. By NEVANLINNA, BEHNKE and GRAUERT, et al.
25. Continuous Geometry. By JOHN VON NEUMANN. Foreword by ISRAEL HALPERIN.
26. Riemann Surfaces. By L. AHLFORS and L. SARIO.
27. Differential and Combinatorial Topology. Edited by STEWART S. CAIRNS.

GEOMETRIC INTEGRATION THEORY

By

Hassler Whitney

PRINCETON, NEW JERSEY
PRINCETON UNIVERSITY PRESS
1957

Published, 1957, by Princeton University Press
London: Oxford University Press

L. C. Card 57-5463

Second Printing 1962
Third Printing 1966

PRINTED IN THE UNITED STATES OF AMERICA

Preface

In various branches of mathematics and its applications, in particular, in differential geometry and in physics, one has often to integrate a quantity over an r-dimensional manifold M in n-space E^n, for instance, over a surface in ordinary 3-space. Considering M (or a portion of M) as the image of part of r-space E^r, integration over M is reduced to integration in E^r, where standard theory applies. However, it is important to know in what manner the integral over M depends on the position of M in E^n, assuming the quantity to be integrated is defined throughout a region R containing M. Thus we must consider the integral over M as a function of the position of M in E^n. The main purpose of this book is to study this function, in a broad geometric and analytic setting.

Starting with Chapter V, we use a postulational approach. Assuming the simplest properties of what r-dimensional integration in n-space should be like, we are led to a theory which turns out to be precisely the integration of differential forms, which may be of a very general character. Hence the role of differential forms in integration theory is more firmly fixed, and at the same time the scope of the theory is considerably increased.

The subject requires an understanding of the geometric properties of the "direction" of an r-dimensional element in n-space, and of course the fundamentals of calculus in several dimensions. The classical treatment, using coordinate systems, results in sometimes lengthy formulas, which do not make the underlying geometric ideas clear, and whose parts depend on the coordinate system employed. Hence in the first part of the book we give a full exposition of this material, in an elementary manner. The geometric approach is gradually coming into use at present; it is hoped that the early chapters may help in making the methods accessible to the general reader.

An overall picture of what the book is about may be obtained from the introductory chapter; we show how the simplest hypotheses lead to the basic tools employed, and we illustrate these tools particularly in the 3-dimensional case. For a more complete outline of results, one may read the introductory pages to the different chapters. Preliminary material that is somewhat outside the scope of the study but is needed in various parts of the book is collected in the appendices.

The body of the book falls into three Parts. The first Part, Classical theory, leads up to the theory of the Riemann integral; we include also a study of smooth (i.e. differentiable) manifolds. The early chapters should be accessible to the beginning graduate student. The second Part, General theory, gives a postulational approach. More maturity on the part of the reader is assumed here. In the last Part, we continue the general study, using Lebesgue theory.

Except where we consider smooth manifolds, we remain always in Euclidean n-dimensional space E^n. Since we use normed spaces as a tool, the metric of E^n is employed; however, the fundamental geometric ideas and theorems are independent of the metric. The study of smooth manifolds in Chapter IV lies outside the main stream of thought of the book; it is included because of its wide interest and application. We prove de Rham's Theorem by elementary means (much like de Rham's original proof). In this manner, the theorem lies close to a theorem deriving the cohomology properties of a complex from abstract integration theory; see § 12 of Chapter VII.

For other expositions bearing some relation to our Part I, see the books mentioned below of Bourbaki (for Chapter I), Lichnerowicz (for Chapters I, II and III), and de Rham (for Chapter IV).

We now give a brief description of our approach to the general problem of r-dimensional integration in n-space. An "integral" is something defined over oriented r-dimensional cells, and over linear combinations of cells; it becomes a function of polyhedral r-chains. This function is linear, and hence we call it a "cochain." We define two norms in the linear space of polyhedral chains, the "flat" norm $|A|^\flat$ and the "sharp" norm $|A|^\sharp$. The cochains which are bounded functions in one of these norms are "flat" or "sharp" correspondingly. A straightforward proof shows that a sharp cochain corresponds to a differential form, in that the value of the cochain on any polyhedral chain equals the integral of the form over the chain. Thus the theory of a certain class of differential forms is derived from the simplest assumptions about the integral. The similar theorem in the flat case is due to Wolfe.

In the case of n-dimensional integration in n-space, the "flat" theory is equivalent to the Lebesgue theory of bounded measurable functions. To obtain all locally summable functions, one should define a space of cochains more general than is possible through a norm. We know of no conditions expressible simply in the r-dimensional case which lead to differential forms in all cases, and to all measurable locally summable functions in the n-dimensional case. In the introductory pages of Chapter V we give conditions leading to arbitrary continuous forms. See also the introductory pages of Chapter VIII.

The domains of integration are always oriented, for $r > 0$; we have $\int_{-A} \omega = -\int_A \omega$, or in the terminology of cochains, $X \cdot (-A) = -X \cdot A$. In theories where orientation properties play no role, we prefer to consider the integration as 0-dimensional; see the last section of the Introduction. Since the spaces of polyhedral chains have been given norms, we may "integrate" over any element of the completions of the spaces, that is, over any "flat chain" or "sharp chain". One type of flat chain is given by an oriented (curved) piece of an r-dimensional manifold in E^n. We show this in very general circumstances in Chapter X. Another type of flat chain is determined by a continuous summable function in E^n whose values are r-vectors; see § 7 of Chapter VI (also § 25 of the Introduction). In this case, an apparently n-dimensional integral may be interpreted as an r-dimensional integral.

The differential forms ω coming from flat cochains are "flat" forms. They are measurable functions satisfying two boundedness conditions. Using properties of flat cochains, the standard properties of forms are derived; the exterior differential $d\omega$ exists (though it may not be definable through differentiation), as does the image $f^*\omega$ of ω if f is a Lipschitz mapping. Hence also cohomology with real coefficients in polyhedra (and in Lipschitz spaces, see below) may be studied through flat forms (as in de Rham's Theorem). In particular, the cup product of flat forms is anti-commutative, and obeys the other standard relations for products of cochains.

The "mass" of flat and sharp chains is definable. In the last chapter, we study the structure of sharp chains A of finite mass. One may find the "part A_Q of A" in any Borel set Q. Generalizing the notion of the r-vector of an oriented r-cell, one may define the r-vector $\{B\}$ of any r-chain B. Now with A given, $\Phi(Q) = \{A_Q\}$ is an additive function of Borel sets in E^n, whose values are r-vectors. It is shown that this set function characterizes the chain A. The theory of these chains now becomes the theory of these set functions.

In the theory of distributions and currents in a manifold (due to L. Schwartz and G. de Rham, see the book of de Rham quoted below), one starts with a simple space of forms (cochains), and obtains the currents (which include smooth singular chains) as linear functions on the forms with some continuity conditions. We do just the reverse: starting with chains, cochains are obtained as linear functions. Because of this, the spaces of cochains and chains obtained are quite different from the above spaces of forms and currents; the explicit study of our chains and cochains has little relation to the standard theory of currents.

On the other hand, starting with smooth singular chains in a Riemannian manifold and introducing sequences of semi-norms, one may obtain directly by a limiting process various spaces of currents and distributions. This procedure, offering certain advantages over the usual one, has been given by James Eells, Jr., in Geometric aspects of currents and distributions, *Proc. Nat. Ac. of Sci. 41* (1955), 493–496.

This book had its beginnings in a study of integration in "Lipschitz spaces"; see Algebraic topology and integration theory, *Proc. Nat. Ac. of Sci. 33* (1947), 1–6. The theory in Euclidean space was at that time restricted principally to the case of sharp cochains (there called "tensor cochains"). The discovery by J. H. Wolfe, in *Tensor Fields Associated with Lipschitz Cochains*, Harvard Thesis, 1948, that flat cochains (then called Lipschitz cochains) correspond to differential forms, caused a fundamental change in the point of view. The study in Euclidean space now became primary; having found that the theory of cochains could be built on norms of chains, these norms became the basic tool. For reasons of space, Lipschitz spaces were finally dropped out of the book; the author expects to give the theory for this case in a separate memoir. An account of the more recent state of the work (including results on Lipschitz spaces) may be found in *r-dimensional integration in n-space*, Proc. International Congress of Mathematicians, Providence, 1952. A study of the Riemann integral in a geometric manner similar to the early chapters of the book was made by Paul Olum in a Senior Thesis at Harvard in 1940.

A choice of notations was sometimes difficult, owing to the overlapping of the fields of integration theory and algebraic topology. Since the operations of exterior differentiation of forms and of taking the coboundary of cochains coalesce in the present work, a single symbol should be used; we finally chose the d of analysis rather than the δ of topology. We use ∇ in place of d for the ordinary differential to avoid confusion with the above d. The symbols $\vee$ and $\wedge$ chosen for the products in Grassmann algebra correspond exactly to the usual symbols $\smile$ and $\frown$ in topology; see Chapter IX.

We shall have occasion to refer to the following texts. This will be done by mention of the author's name.

Banach, S. *Théorie des operations linéaires*, Subwencji funduszu kultury narodowej, Warszawa, 1932.

Bourbaki, N. *Éléments de mathématique*, Livre II, Algèbre, Chapitre III, Hermann, Paris, 1948.

Halmos, P. *Measure Theory*, Van Nostrand, 1950.

Lichnerowicz, A. *Algèbre et analyse linéaires*, Masson, Paris, 1947.

de Rham, G. *Variétés différentiables*, Hermann, Paris, 1955.

Saks, S. *Theory of the Integral*, Subwencji funduszu kultury narodowej, Warszawa, 1937.

A reference such as (V, 10.4) applies to equation (4) in § 10 of Chapter V; (App. I, 7) means § 7 of Appendix I.

The author wishes to acknowledge the permission of J. H. Wolfe to incorporate the results of his thesis into the book. He is indebted to Norman Z. Wolfsohn for much help in the manuscript, and to James Eells for collaboration in revising Chapter XI, for other help in the manuscript, and for aid in proofreading.

HASSLER WHITNEY

May, 1956
Institute for Advanced Study
Princeton, New Jersey

Table of Contents

Part I: Classical theory

Chapter I. Grassmann algebra

Chapter II. Differential forms

Chapter III. Riemann integration theory

GEOMETRIC INTEGRATION THEORY

Introduction

The purpose of this preliminary chapter is first of all to provide motivation for the methods and tools appearing in the book, and secondly to illustrate some of the general considerations through a study of special cases, particularly in three dimensions. The Introduction and the body of the book are independent of each other; however, the meaning of the full theory will become clearer if the Introduction is read in conjunction with the rest of the book.

In Part A (which is rather abstract in character), we ask what a theory of r-dimensional integration in n-space should look like. An "integral" $X \cdot \sigma$ is defined over an oriented r-cell σ for instance, and changes sign if the orientation is reversed. We may now define $X \cdot (a\sigma) = a(X \cdot \sigma)$ for real numbers a, and $X \cdot (A + B) = X \cdot A + X \cdot B$; assuming that a subdivision of cells does not affect the integral, we now have a linear function defined over "polyhedral r-chains." To give this linear function analytical properties, we introduce some continuity hypotheses. Next we study the local nature of integration. Near any point, in any r-dimensional direction, the integral over a cell is approximately proportional to the r-dimensional volume of the cell (with the strongest continuity hypothesis); this fact leads to the construction of a point function $D_X(p)$ which acts on "r-vectors" $\{\sigma\}$ of oriented r-cells σ. These r-vectors must have certain simple properties, which in turn lead directly to the construction of Grassmann algebra. Finally, D_X becomes a differential r-form, whose integral $\int_\sigma D_X$ over any σ equals the original $X \cdot \sigma$.

In Part B, we start with the elements of Grassmann algebra as derived above, and work out from a geometric point of view some of the fundamentals of calculus. We consider vector analysis in three dimensions, differentials, Jacobians, transformation of "multiple integrals," manifolds, and the Theorems of Stokes and de Rham.

The purpose of Part C is to introduce the reader to some of the general methods in the later parts of the book. In the last two sections, we touch on the way some particular modes of integration may be considered as r-dimensional for different r.

A. THE GENERAL PROBLEM OF INTEGRATION

1. The integral as a function of the domain. For an integration theory there must certainly be various possible "domains of integration."

Whatever kind of process integration (with real values) is, a definite "integrand" X will give a real number when applied to a permissible domain A. Thus, for a fixed X, we have a real valued function of domains A; we denote the value on A by $X \cdot A$. We consider integration in Euclidean space.

If we are to call the integration r-dimensional, we must certainly include among the permissible domains the simplest r-dimensional figures. An r-cell σ, consisting of a closed bounded part of an r-plane, bounded by a finite number of pieces of $(r-1)$-planes, is such a figure. We make our first hypothesis:

HYPOTHESIS (H_1). The integral over σ depends on the orientation of σ; a reversal of orientation reverses the sign of the integral.

We discuss the meaning of and reason for this hypothesis. A line segment σ^1 has two end points p and q; the two orientations of σ^1 are the two directions along σ^1, from p to q and from q to p; we may denote the two oriented cells by $pq = -qp$ and by $qp = -pq$ respectively. They may be defined by a choice of the vector $q - p$ (from p to q) or $p - q$ (from q to p). We may orient a triangle $p_0p_1p_2$ by choosing an ordered pair of independent vectors in it; for instance, the pair $(p_1 - p_0, p_2 - p_0)$. Interchanging these or reversing the direction of either would reverse the orientation. Similarly, an r-cell σ is oriented by the choice of an ordered set of r independent vectors in it. A 0-cell, i.e. a single point, has no orientation properties.

The triangle $\sigma = p_0p_1p_2$, oriented as above, has a boundary $\partial\sigma$, consisting of the oriented segments p_0p_1, p_1p_2, and p_2p_0. The boundary $\partial(pq)$ of the oriented segment pq consists of the point q counted positively, and the point p counted negatively. The boundary $\partial\sigma$ of an r-cell σ contains its $(r-1)$-faces, properly oriented (App. II, 5).

For any oriented cell σ, let $-\sigma$ denote the oppositely oriented cell; then (H_1) may be written in the form

(1) $$X \cdot (-\sigma) = -X \cdot \sigma.$$

We consider some examples. Let ϕ be a real valued function defined in 3-space E^3, and let C be an oriented curve, from the point p to the point q. If we integrate the rate of change of ϕ along C, we obtain $\phi(q) - \phi(p)$. If the orientation of C were reversed, we would obtain $\phi(p) - \phi(q)$. Next, consider any 1-dimensional integral $\int_A \omega$. Let $\sigma = p_0p_1p_2$ be an oriented triangle, cut into the two oriented triangles $\sigma' = p_0p'p_2$, $\sigma'' = p'p_1p_2$, by the segment $p'p_2$, with p' in p_0p_1. Then

(2) $$\int_{\partial\sigma'} \omega + \int_{\partial\sigma''} \omega = \int_{\partial\sigma} \omega;$$

for

$$\int_{p_0p'} \omega + \int_{p'p_1} \omega = \int_{p_0p_1} \omega, \qquad \int_{p'p_2} \omega + \int_{p_2p'} \omega = 0,$$

the latter since $p'p_2$ and p_2p' are oppositely oriented. This type of relation is fundamental for geometric properties of the integral; it would be impossible if orientation properties were disregarded.

Of course integration over non-oriented domains is possible and sometimes of importance; in this case, much more general types of domains are permissible, but the geometric properties are largely lost. We prefer to think of such integration as 0-dimensional; see § 26 below.

The most typical requirement of integration theory is additivity, or in the present exposition, *invariance under subdivisions* (used in the example above):

HYPOTHESIS (H_2). If the oriented r-cell σ is cut into similarly oriented r-cells $\sigma_1, \cdots, \sigma_m$, then

$$X\cdot\sigma = X\cdot\sigma_1 + \cdots + X\cdot\sigma_m. \tag{3}$$

2. Polyhedral chains. We wish to write the boundary $\partial\sigma$ of the triangle $\sigma = p_0p_1p_2$ as a domain of integration. With the point p' in p_0p_1 as in § 1, we would like to write $\partial\sigma$ in various ways, such as

$$\partial\sigma = p_0p_1 + p_1p_2 + p_2p_0 = p_0p' + p'p_2 - p_0p_2 + p'p_1 + p_1p_2 - p'p_2$$

etc. This suggests the definition of a *polyhedral r-chain A* as being a linear combination of oriented r-cells, with real numbers as coefficients, together with the properties

$$1\sigma = \sigma, \qquad 0\sigma = 0, \qquad a(-\sigma) = (-a)\sigma = -(a\sigma), \tag{1}$$

and invariance under subdivision: if the oriented cell σ is cut into $\sigma_1, \cdots, \sigma_m$, then σ and $\sigma_1 + \cdots + \sigma_m$ are the same polyhedral chain. The definitions of aA (for real numbers a) and $A + B$ are obvious; the set of polyhedral r-chains now forms a linear space.

Because of (H_1) and (H_2), it is possible to define $X\cdot A$ for any polyhedral r-chain $A = \sum a_i\sigma_i$, by the relation

$$X\cdot\sum a_i\sigma_i = \sum a_i(X\cdot\sigma_i); \tag{2}$$

X is now a *linear function of polyhedral r-chains*. For this reason, we call X an *r-cochain*. That X is a cochain is equivalent to assuming that $X\cdot\sigma$ is defined, with the properties (H_1) and (H_2).

The boundary of $A = \sum a_i\sigma_i$ is defined to be $\partial A = \sum a_i\,\partial\sigma_i$; this is easily seen to be a well defined polyhedral $(r - 1)$-chain. A polyhedral 0-chain is an expression $A^0 = \sum a_ip_i$, the p_i being points; we set $\partial A^0 = 0$.

3. Two continuity hypotheses. For a satisfactory integration theory, the permissible domains must include oriented curved r-cells for instance. One should be able to obtain these as limits of polyhedral r-chains, and the integral should be definable as the limit of the integrals over the approximating polyhedral chains. This requires some continuity hypotheses on the integral. We give two hypotheses in this section with which a satisfactory general theory may be obtained; if we include also the hypothesis of the next section, the integral has simpler analytical properties.

Let σ be an oriented triangle, of area $|\sigma|$. If we cut the plane containing σ into small rectangles and let $\tau_1, \cdots, \tau_m$ be those contained in σ, then these τ_i fill up most of σ, and it is natural to require that $\sum X \cdot \tau_i$ be near $X \cdot \sigma$. This will follow from:

HYPOTHESIS (H_1'). Given the r-cochain X, there is a number N_1 such that

(1) $$|X \cdot \sigma| \leqq N_1 |\sigma|, \qquad \text{all oriented } r\text{-cells } \sigma,$$

where $|\sigma|$ is the r-dimensional volume of σ.

This is of course a stronger hypothesis than needed for the above requirement. We assume it largely for the sake of the analytical methods described in § 6 below.

Take the case $r = 0$. We may consider the "0-dimensional volume" of a point p to be 1. Any 0-cochain X corresponds to a real function ϕ: $\phi(p) = X \cdot p$ for all points p. Now (H_1') says that $|\phi(p)| \leqq N_1$, all p; that is, ϕ is bounded. These functions are too general; we must restrict them further. We look for a hypothesis suggested by the case $r = 1$.

Take the above triangle σ again; let τ be the union of the τ_i. We may consider the boundary $\partial\tau$ of τ as an approximation to $\partial\sigma$, even though it is made up of segments not parallel to the sides of σ in general. Taking $r = 1$, we may require that $X \cdot \partial\tau$ be near $X \cdot \partial\sigma$, i.e. that $X \cdot \partial(\sigma - \tau)$ be small, as a result of the area of $\sigma - \tau$ being small.

HYPOTHESIS (H_2'). Given the r-cochain X, there is a number N_2 such that

(2) $$|X \cdot \partial\sigma^{r+1}| \leqq N_2 |\sigma^{r+1}|, \qquad \text{all oriented } (r+1)\text{-cells } \sigma^{r+1}.$$

Note that (H_2') is trivially satisfied if $r = n$.

For $r = 0$, (H_2') says that for any oriented segment pq,

(3) $$|X \cdot q - X \cdot p| = |X \cdot \partial(pq)| \leqq N_2 |q - p|,$$

$|q - p|$ being the length of pq; that is, the function $\phi(p) = X \cdot p$ satisfies a Lipschitz condition.

Any cochain satisfying (H_1') and (H_2') we call a *flat* cochain.

4. A further continuity hypothesis. If the cell σ is moved into a nearby position σ', we may assume that $X \cdot \sigma'$ is near $X \cdot \sigma$. We shall consider rigid motions without turning; that is, translations by means of vectors v. Let $T_v\sigma$ denote the new cell. Our hypothesis is that $X \cdot T_v\sigma$ differs from $X \cdot \sigma$ by at most some fixed multiple of the r-volume $|\sigma|$ times the distance $|v|$ of translation.

Hypothesis (H'_3). Given the r-cochain X, there is a number N_3 such that for any oriented r-cell σ and vector v,

(1) $$|X \cdot T_v\sigma - X \cdot \sigma| \leqq N_3 |\sigma| |v|.$$

In the case $r = 0$, this hypothesis is equivalent to (H'_2). For $r = n$, it is non-trivial, whereas (H'_2) is trivial.

Any cochain satisfying all three hypotheses we call *sharp*. We shall see that sharp cochains correspond to differential forms. (This holds also in the flat case; see Chapter IX.)

5. Some examples. We help elucidate some of the hypotheses through the study of a steady flow of fluid in oriented 3-space E^3. Take any oriented 2-cell σ. Let (v_1, v_2) define its orientation, and choose a vector v_3 so that (v_1, v_2, v_3) (or equivalently, (v_3, v_1, v_2)) defines the given orientation of E^3; then the *positive* direction through σ is the direction in the sense of v_3. Let $X \cdot \sigma$ be the quantity (positive or negative) of fluid flowing through σ in the positive direction in unit time; this is the *flux* across σ. Clearly (H_1) and (H_2) hold; hence X is a 2-cochain. Of course $X \cdot S$ for any oriented surface S is definable.

If the density of fluid and the velocity of flow are bounded, then clearly (H'_1) holds. Now take any 3-cell τ. If fluid is being neither created nor destroyed, then (we assume the density constant in time) the total rate of flow out of τ, which equals $X \cdot \partial\tau$, must be 0. In general, $X \cdot \partial\tau$ equals the total rate of creation of fluid in τ. Hence (H'_2) is equivalent to assuming that the total rate of creation per unit volume is bounded.

With the same flow of fluid, consider the *circulation* along an oriented curve C. At a point p of C, if $u(p)$ is the unit tangent vector at p in the positive direction along C, $v(p)$ is the velocity vector of the fluid at p, and $\rho(p)$ is the density at p, then the circulation is

(1) $$Y \cdot C = \int_C \rho v \cdot u$$

(compare (18.3) below). Again (H_1) and (H_2) hold.

Hypothesis (H'_1) will follow from the boundedness of v and ρ. Given an oriented 2-cell σ, $Y \cdot \partial\sigma$ is the circulation around the boundary $\partial\sigma$. Taking arbitrarily small cells σ near a point p, we see that (H'_2) will follow if curl $(v) = \nabla \times v$ exists and is finite; compare (21.4).

Suppose the flow is through a pipe. Then the above cochains X and Y are not defined throughout E^3, but only in the region of flow; the hypotheses need be assumed only in this region.

6. The case $r = n$. For an n-cochain X which satisfies (H'_1) in oriented E^n ((H'_2) is satisfied trivially), it is standard Lebesgue theory that there is a bounded measurable function Φ such that (using the Lebesgue integral)

(1) $$X \cdot \sigma = \int_\sigma \Phi, \qquad \text{all } n\text{-cells } \sigma^n \text{ oriented like } E^n.$$

We consider briefly the simpler case when (H_3') is also satisfied. Given the point p, let $\sigma_1, \sigma_2, \cdots$ be a sequence of n-cells oriented like E^n, in smaller and smaller neighborhoods of p. Set

$$\Phi(p) = \lim_{i\to\infty} \frac{X\cdot\sigma_i}{|\sigma_i|}. \tag{2}$$

We indicate the proof of existence and uniqueness of the limit. Let $\tau_1, \tau_2, \cdots$ be a similar sequence of cubes. Suppose i_0 and k_0 are such that for $i \geqq i_0$ and $k \geqq k_0$, σ_i and τ_k are in a neighborhood of p of diameter $<\epsilon$. We may take k so large that translations of τ_k nearly fill up σ_i:

$$\sigma_i = \sigma' + R, \qquad \sigma' = T_{v_1}\tau_k + \cdots + T_{v_s}\tau_k, \qquad |R| \text{ small.}$$

By (H_3'),

$$\left|\frac{X\cdot\sigma'}{|\sigma'|} - \frac{X\cdot\tau_k}{|\tau_k|}\right| \leqq \frac{1}{s}\sum_{j=1}^{s}\left|\frac{X\cdot T_{v_j}\tau_k - X\cdot\tau_k}{|\tau_k|}\right| \leqq N_3\epsilon,$$

and the statement follows, using (H_1').

Using (H_3') again, it is clear that Φ satisfies

$$|\Phi(p+v) - \Phi(p)| \leqq N_3|v|, \tag{3}$$

and that (1) holds (using the Riemann integral). Moreover,

$$|X\cdot\sigma - |\sigma|\Phi(p_0)| = \left|\int_\sigma [\Phi(p) - \Phi(p_0)]\,dp\right| \leqq N_3\zeta|\sigma| \tag{4}$$

if all points of σ are within ζ of p_0.

7. The r-vector of an oriented r-cell. Let X be a sharp r-cochain in E^n. Then for each oriented r-plane P in E^n, we may consider $X\cdot\sigma$ for r-cells in P, and hence find a function Φ_P in P as in § 6. For any point p, the values of $\Phi_P(p)$ for the various oriented r-planes P through p are of interest.

We shall use a closely related function. Given σ and p, let P be the r-plane through p parallel to σ and oriented like σ, and set

$$D_X(p)\cdot\{\sigma\} = |\sigma|\,\Phi_P(p). \tag{1}$$

We must give meaning to $D_X(p)$, to $\{\sigma\}$, and to their combination. With the σ_i as in § 6, (1) and (6.2) give

$$D_X(p)\cdot\{\sigma\} = \lim_{i\to\infty}\frac{|\sigma|}{|\sigma_i|}X\cdot\sigma_i. \tag{2}$$

As a function of σ, the right hand side is known as soon as we know the set of r-planes parallel with σ, the orientation of σ, and $|\sigma|$; we call this triple the *r-vector* $\{\sigma\}$ of σ. We may now define $D_X(p)$ to be that real valued function defined on all r-vectors of oriented r-cells which is given by (1). (Later $D_X(p)$ will be taken to be defined on more special spaces $\mathbf{T}_r$.)

We remark that $D_X(p)\cdot\{\sigma\}$ is unchanged if we alter the metric in E^n; for using (2), we see that $|\sigma|/|\sigma_i|$ is unchanged, as is $X\cdot\sigma_i$.

Given $\{\sigma\}$ and the real number $a \neq 0$, let $a\{\sigma\}$ denote $\{\sigma'\}$ for any oriented σ' parallel with σ, oriented like or opposite to σ according as $a > 0$ or $a < 0$, and such that $|\sigma'| = |a||\sigma|$. If σ_1 and σ_2 are parallel, say $\{\sigma_2\} = a\{\sigma_1\}$; set

$$\{\sigma_1\} + \{\sigma_2\} = (1 + a)\{\sigma_1\},$$

for $a \neq -1$. If we include a "zero r-vector" 0, then with these definitions the r-vectors associated with a fixed set of parallel r-planes form a linear space isomorphic with the real numbers. In any such space of r-vectors, we see easily from (2) that $D_X(p)$ is linear:

$$(3) \qquad D_X(p)\cdot(a\{\sigma\} + b\{\sigma'\}) = aD_X(p)\cdot\{\sigma\} + bD_X(p)\cdot\{\sigma'\}.$$

8. On r-vectors and boundaries of $(r + 1)$-cells. Let us immerse the set of all r-vectors of oriented r-cells in a linear space $\mathbf{S}_r$ (of infinite dimension for $0 < r < n$) as follows. Take a fixed point p_0. An element α of $\mathbf{S}_r$ is a finite set of distinct r-planes $P_1, \cdots, P_m$ through p_0, together with an r-vector α_i associated with each P_i; we may include extra planes P_j, if we associate the zero r-vector with them. We form $a\alpha$ by replacing each α_i by $a\alpha_i$. Given α and β, we may take enough planes $P_1, \cdots, P_m$ so that α and β are defined by α_i and β_i in P_i respectively $(i = 1, \cdots, m)$; let $\alpha + \beta$ be defined by $\alpha_i + \beta_i$ in P_i. Clearly $\mathbf{S}_r$ is independent of the choice of p_0. The linear spaces described in the last section are linear subspaces of $\mathbf{S}_r$.

We may extend $D_X(p)$ to be a linear function in $\mathbf{S}_r$ by defining

$$(1) \quad D_X(p)\cdot\alpha = \sum_{i=1}^{m} D_X(p)\cdot\alpha_i \quad \text{if} \quad \alpha \text{ is defined by } \alpha_i \text{ in } P_i\ (i = 1, \cdots, m).$$

Now take any oriented $(r + 1)$-cell τ. Its boundary is an r-chain $\sigma_1 + \cdots + \sigma_m$. We wish the sum of the corresponding r-vectors to be 0:

$$(2) \quad \{\sigma_1\} + \cdots + \{\sigma_m\} = 0 \quad \text{if} \quad \sigma_1 + \cdots + \sigma_m = \partial\tau \text{ for some } \tau.$$

Requiring all such relations to hold turns $\mathbf{S}_r$ into a linear space $\mathbf{T}_r$. We shall call any element of $\mathbf{T}_r$ an *r-vector*. (Strictly speaking, let $\mathbf{S}_r'$ denote the linear subspace of $\mathbf{S}_r$ generated by all such elements $\{\sigma_1\} + \cdots + \{\sigma_m\}$; then $\mathbf{T}_r$ is the quotient space.) We now turn $D_X(p)$ into a linear function in the space $\mathbf{T}_r$, by letting its value on an element of $\mathbf{T}_r$ be its value on any corresponding element of $\mathbf{S}_r$. To show that this is possible, we must prove the following relation:

$$(3) \quad D_X(p)\cdot\{\sigma_1\} + \cdots + D_X(p)\cdot\{\sigma_m\} = 0 \quad \text{if} \quad \sigma_1 + \cdots + \sigma_m = \partial\tau.$$

We may suppose p is in τ. Given $\lambda > 0$, let us contract E^n towards p by the factor λ; then τ becomes τ_λ and σ_i becomes $\sigma_{\lambda i}$, and we have

$$\partial\tau_\lambda = \sum \sigma_{\lambda i}, \qquad |\tau_\lambda| = \lambda^{r+1}|\tau|, \qquad |\sigma_{\lambda i}| = \lambda^r |\sigma_i| .$$

Take a sequence $\lambda_1, \lambda_2, \cdots \to 0$. We may use $\sigma_{\lambda_1 i}, \sigma_{\lambda_2 i}, \cdots$ in (7.2); with the help of (7.1), we find

$$D_X(p)\cdot\{\sigma_i\} = \lim_{k\to\infty} \frac{|\sigma_i|}{|\sigma_{\lambda_k i}|} X\cdot\sigma_{\lambda_k i} = \lim_{k\to\infty} \frac{1}{\lambda_k^r} X\cdot\sigma_{\lambda_k i}.$$

Also, by (H_2'),

$$\left| \frac{1}{\lambda_k^r} \sum_i X\cdot\sigma_{\lambda_k i} \right| = \frac{1}{\lambda_k^r} |X\cdot\partial\tau_{\lambda_k}| \leq \frac{1}{\lambda_k^r} N_2 |\tau_{\lambda_k}| = \lambda_k N_2 |\tau| .$$

These relations give (3).

9. Grassmann algebra. We shall find a special manner of writing elements of $\mathbf{T}_r$. Take any ordered set $(v_1, \cdots, v_r)$ of independent vectors. Let σ be the parallelepiped with a point p as vertex and with these vectors along the edges from p, oriented by the v_i. We define the symbol $v_1 \vee \cdots \vee v_r$ by

(1) $$v_1 \vee \cdots \vee v_r = \{\sigma\}.$$

Now any element of $\mathbf{T}_r$ can be written as a sum of such elements.

From the properties of § 7, we see that the "product" $v_1 \vee \cdots \vee v_r$ is skew symmetric:

(2) $$v_2 \vee v_1 = -v_1 \vee v_2, \quad \text{and hence} \quad v \vee v = 0.$$

Also

(3) $$v_1 \vee \cdots \vee (av_i) \vee \cdots \vee v_r = a(v_1 \vee \cdots \vee v_i \vee \cdots \vee v_r).$$

We shall prove that it is linear in v_1:

(4) $$(v_1 + v_1') \vee v_2 \vee \cdots \vee v_r = v_1 \vee v_2 \vee \cdots \vee v_r + v_1' \vee v_2 \vee \cdots \vee v_r;$$

hence it is linear in all the v_i.

If v_1' is in the r-plane determined by $v_1, \cdots, v_r$, this is a simple geometric fact about addition of volumes. We assume this is not the case, and consider a few values of r.

For $r = 1$, a 1-vector is now represented by a vector; the 1-vector $\{pq\}$ of the oriented segment pq is represented by the vector $q - p$. Relation (4) says that addition of 1-vectors (appearing on the right) is equivalent to addition of vectors (appearing on the left). We show this as follows. Choose a point p_0, and define the points and triangle

$$p_1 = p_0 + v_1, \qquad p_2 = p_1 + v_1' = p_0 + (v_1 + v_1'), \qquad \sigma = p_0p_1p_2.$$

By (8.2),

$$\{p_0p_1\} + \{p_1p_2\} + \{p_2p_0\} = 0, \quad \text{hence} \quad \{p_0p_2\} = \{p_0p_1\} + \{p_1p_2\};$$

this is the required relation.

For $r = 2$, take the above points p_i, and also the points $q_i = p_i + v_2$ $(i = 0, 1, 2)$. Set $\sigma' = q_0q_1q_2$. The p_i and q_i are vertices of a 3-cell τ whose faces are σ and σ', and also three parallelograms σ_{01}, σ_{12}, σ_{02}, where σ_{ij} contains p_ip_j. We have clearly

$$\{\sigma\} = \{\sigma'\} = \tfrac{1}{2}v_1 \vee v_1',$$

$$\{\sigma_{01}\} = v_1 \vee v_2, \qquad \{\sigma_{12}\} = v_1' \vee v_2, \qquad \{\sigma_{02}\} = (v_1 + v_1') \vee v_2.$$

With proper regard to orientations, we see that

$$\partial\tau = \sigma' - \sigma + \sigma_{01} + \sigma_{12} - \sigma_{02};$$

applying (8.2) gives (4) for $r = 2$. The general case is similar.

The set of all vectors in E^n forms a vector space $V = V(E^n)$. Let $e_1, \cdots, e_n$ be a base in V. Then any vector v can be written uniquely as $\sum v^i e_i$; the v^i are the components of v. Since we consider 1-vectors and vectors as the same, the e_i form a base in $\mathbf{T}_1$. For $r = 2$, using (2), (3) and (4) gives

$$(5) \qquad v_1 \vee v_2 = \sum_{i,j=1}^{n} v_1^i v_2^j (e_i \vee e_j) = \sum_{i<j} \begin{vmatrix} v_1^i & v_2^i \\ v_1^j & v_2^j \end{vmatrix} e_i \vee e_j.$$

It may be shown that the $e_{ij} = e_i \vee e_j$ $(i < j)$ are independent in $\mathbf{T}_2$; by (5), they form a base in $\mathbf{T}_2$. Similarly, the $e_{\lambda_1 \cdots \lambda_r} = e_{\lambda_1} \vee \cdots \vee e_{\lambda_r}$ $(\lambda_1 < \cdots < \lambda_r)$ form a base in $\mathbf{T}_r$, and any r-vector α can be written uniquely in the form

$$(6) \qquad \alpha = \sum_{\lambda_1 < \cdots < \lambda_r} \alpha^{\lambda_1 \cdots \lambda_r} e_{\lambda_1 \cdots \lambda_r}.$$

The $\alpha^{\lambda_1 \cdots \lambda_r}$ are the "components" of α.

It follows that $\mathbf{T}_r$ is of dimension $\binom{n}{r}$. In particular, $\mathbf{T}_n$ is of dimension 1, with base element $e_{1 \cdots n}$, and each $\mathbf{T}_k$ for $k > n$ contains the zero element only.

Through the definition

$$(7) \qquad (v_1 \vee \cdots \vee v_r) \vee (v_{r+1} \vee \cdots \vee v_{r+s}) = v_1 \vee \cdots \vee v_{r+s},$$

we have a bilinear multiplication between $\mathbf{T}_r$ and $\mathbf{T}_s$, with product in $\mathbf{T}_{r+s}$. If we include the space $\mathbf{T}_0$ = the real numbers, the system of the $\mathbf{T}_r$ with these operations is the *Grassmann algebra* of V. Let us denote $\mathbf{T}_r$ by $V_{[r]}$.

For $n \leqq 3$, any r-vector α (for any $r > 0$) equals $\{\sigma\}$ for some oriented r-cell σ. This is not the case for $n \geqq 4$; for instance, $e_{12} + e_{34}$ cannot be written in this form. Any α of the form $\{\sigma\}$ is a *simple* r-vector.

10. The dual algebra. The set of linear functions f in a vector space V forms a vector space, with the following definitions: the function af has the value $af(v)$ at v; the function $f+g$ has the value $f(v)+g(v)$ at v. This space is the *conjugate space* $\bar{V}$ of V.

Since $D_X(p)$ is a linear function in the vector space $\mathbf{T}_r = V_{[r]}$, we may consider it as being an element of the conjugate space of $V_{[r]}$; we denote this conjugate space by $V^{[r]}$. We call its elements *r-covectors*; the elements of $\bar{V} = V^{[1]}$ are *covectors*. We shall find a special manner of representing elements of $V^{[r]}$.

Let $e_1, \cdots, e_n$ be a base in V. Setting

$$e^i \cdot v = e^i \cdot (v^1 e_1 + \cdots + v^n e_n) = v^i \tag{1}$$

defines a linear function e^i in V, i.e. an element of $\bar{V}$. The elements $e^1, \cdots, e^n$ are easily seen to form a base in the base *dual* to the e_i. Now any element f of $\bar{V}$ may be written uniquely as $\sum f_i e^i$; the f_i are the components of f. Hence also $\bar{V}$ and V are of the same dimension; since $V^{[r]}$ and $V_{[r]}$ are conjugate, they are also of the same dimension.

The e^i may be defined by the relation

$$e^i \cdot e_j = \delta^i_j = 1 \quad \text{if} \quad i = j, \qquad = 0 \quad \text{if} \quad i \neq j. \tag{2}$$

Let us define base elements $e^{\lambda_1 \cdots \lambda_r}$ in $V^{[r]}$ by the same formula:

$$e^{\lambda_1 \cdots \lambda_r} \cdot e_{\lambda_1 \cdots \lambda_r} = 1 \qquad (\lambda_1 < \cdots < \lambda_r), \tag{3}$$

and $e^{\lambda_1 \cdots \lambda_r} \cdot e_{\mu_1 \cdots \mu_r} = 0$ for other μ_i $(\mu_1 < \cdots < \mu_r)$.

As a consequence of (2), we have

$$f(v) = f \cdot v = \sum_{i,j=1}^{n} f_i v^j e^i \cdot e_j = \sum_{i=1}^{n} f_i v^i; \tag{4}$$

similarly, writing

$$\xi = \sum_{\lambda_1 < \cdots < \lambda_r} \xi_{\lambda_1 \cdots \lambda_r} e^{\lambda_1 \cdots \lambda_r}, \tag{5}$$

the $\xi_{\lambda_1 \cdots \lambda_r}$ are the components of ξ, and we have, using (9.6),

$$\xi(\alpha) = \xi \cdot \alpha = \sum_{\lambda_1 < \cdots < \lambda_r} \xi_{\lambda_1 \cdots \lambda_r} \alpha^{\lambda_1 \cdots \lambda_r}. \tag{6}$$

Of course a change of base results in a change in components. For $n = 3$, $r = 2$, (5), (9.6) and (6) read

$$\xi = \xi_{12} e^{12} + \xi_{13} e^{13} + \xi_{23} e^{23}, \qquad \alpha = \alpha^{12} e_{12} + \alpha^{13} e_{13} + \alpha^{23} e_{23}, \tag{7}$$

$$\xi \cdot \alpha = \xi_{12} \alpha^{12} + \xi_{13} \alpha^{13} + \xi_{23} \alpha^{23}. \tag{8}$$

We wish to define expressions like $f^1 \vee \cdots \vee f^r$, the f^i being in $\bar{V}$ and the result being in $V^{[r]}$. We wish this multiplication to be skew symmetric and linear in each variable, and we wish to have $e^{\lambda_1}\vee\cdots\vee e^{\lambda_r} = e^{\lambda_1\cdots\lambda_r}$. This determines the multiplication. For instance, for $r = 2$,

$$(9) \qquad f\vee g = \sum_{i,j=1}^{n} f_i g_j e^i \vee e^j = \sum_{i<j} \begin{vmatrix} f_i & f_j \\ g_i & g_j \end{vmatrix} e^{ij}.$$

The value of $(f^1\vee\cdots\vee f^r)\cdot(v_1\vee\cdots\vee v_r)$ is the determinant with the elements $f^i\cdot v_j$. For instance, for $r = 2$,

$$(10) \qquad (f\vee g)\cdot(v\vee w) = \begin{vmatrix} f\cdot v & f\cdot w \\ g\cdot v & g\cdot w \end{vmatrix}.$$

To show this, we note that both sides of (10) are skew symmetric and linear in the quantities v, w; hence it is sufficient to prove this in the particular case $v = e_k$, $w = e_l$, $k < l$. But this follows at once from (9) and (3). Note that if we use (9) and (9.5) to obtain $(f\vee g)\cdot(v\vee w)$, comparing with (10) gives the Lagrange identity

$$(11) \qquad \begin{vmatrix} \sum f_i v^i & \sum f_i w^i \\ \sum g_i v^i & \sum g_i w^i \end{vmatrix} = \sum_{i<j} \begin{vmatrix} f_i & f_j \\ g_i & g_j \end{vmatrix} \begin{vmatrix} v^i & v^j \\ w^i & w^j \end{vmatrix}.$$

11. Integration of differential forms. A *differential r-form* ω in E^n is a function whose values $\omega(p)$ are r-covectors. Hence, for any p and r-vector α, $\omega(p)\cdot\alpha$ is a real number. If ω is continuous, we may define its integral over any oriented r-cell σ in E^n as follows. Take a fine subdivision of σ into oriented r-cells $\sigma_1, \cdots, \sigma_s$; choose a point p_i in each σ_i; form the sum

$$(1) \qquad \sum \omega(p_i)\cdot\{\sigma_i\};$$

take the limit of this, using a sequence of subdivisions with diameters of cells approaching 0. *It is not necessary that E^n be metric or oriented.*

For a sharp r-cochain, D_X is a continuous differential form, and

$$(2) \qquad X\cdot\sigma = \int_\sigma D_X = \lim \sum D_X(p_i)\cdot\{\sigma_i\}.$$

For if P is the r-plane of σ, oriented like σ, and we choose a metric in E^n, we may apply (7.2). See (V, 10).

B. Some Classical Topics

12. Grassmann algebra in metric oriented n-space. In metric E^n, scalar products $u\cdot v$ of vectors are defined. Take any vector u. The function $\phi_u(v) = u\cdot v$ (we write also $\phi_u\cdot v$) of vectors v is linear; hence ϕ_u is a definite element of the conjugate space $\bar{V}$ of V. This is easily shown to define an isomorphism between V and $\bar{V}$. Thus any linear function ψ in V can be written in the form $\psi(v) = u\cdot v$ for a unique vector u. Since $V_{[r]}$

and $V^{[r]}$ are conjugate for each r, a definite choice of metric in $V_{[r]}$ (see (I, 12.1)) establishes a definite isomorphism between $V_{[r]}$ and $V^{[r]}$.

Now take E^n to be oriented also. Then there is a definite "unit n-vector" α_0, which equals $\{\sigma\}$ for any n-cell σ of unit volume, oriented like E^n. For any orthonormal base $(e_1, \cdots, e_n)$ (consisting of perpendicular unit vectors defining the given orientation of E^n), we clearly have

(1) $$\alpha_0 = e_{1\cdots n} = e_1 \vee \cdots \vee e_n.$$

There is also a definite "unit n-covector" ω_0 such that

(2) $$\omega_0 = e^{1\cdots n} = e^1 \vee \cdots \vee e^n, \qquad \omega_0 \cdot \alpha_0 = 1.$$

Given any r-vector α, set

(3) $$\Phi(\beta) = \omega_0 \cdot [\alpha \vee \beta], \qquad \text{all } (n-r)\text{-vectors } \beta.$$

Now Φ is a linear function in $V_{[n-r]}$ and hence is a definite element of $V^{[n-r]}$. Thus we have a definite isomorphism between $V_{[r]}$ and $V^{[n-r]}$.

13. The same, $n = 3$. Applying the results of the last section shows that in metric oriented E^3, all spaces $V_{[r]}$ and $V^{[r]}$ are isomorphic in a definite way either to $V_{[1]} = V$ or to $V_{[0]} =$ the reals. The isomorphism between $V = V_{[1]}$ and $\bar{V} = V^{[1]}$ is given by the scalar product; we consider the isomorphisms between V and $V^{[2]}$ and between V and $V_{[2]}$.

Take any vector v; set

(1) $$\Psi_v(\alpha) = \omega_0 \cdot (v \vee \alpha), \qquad \text{all 2-vectors } \alpha.$$

Since Ψ_v is a linear function of α, it is a 2-covector.

Next, take any 2-vector α; set

(2) $$\Theta_\alpha \cdot w = \omega_0 \cdot (\alpha \vee w), \qquad \text{all vectors } w.$$

Since this function of w is linear, it is given by the scalar product of w by a definite vector Θ_α.

We use this to define the *vector product* of vectors:

(3) $$u \times v = \Theta_{u \vee v};$$

thus $u \times v$ is defined by

(4) $$(u \times v) \cdot w = \omega_0 \cdot (u \vee v \vee w), \qquad \text{all vectors } w.$$

Since all operations on the right of (4) are linear, $u \times v$ is a bilinear function of u and v. Since $v \vee u = -u \vee v$ and $u \vee u = 0$, we have

(5) $$v \times u = -u \times v, \qquad u \times u = 0.$$

Let (e_1, e_2, e_3) be an orthonormal base giving the orientation chosen in E^3. Let each of u, v, w be each of e_1, e_2, e_3 in (4) in turn. As one example, we have

(6) $$(e_1 \times e_3) \cdot e_2 = \omega_0 \cdot (e_1 \vee e_3 \vee e_2) = -\omega_0 \cdot \alpha_0 = -1.$$

In this manner we see that

(7) $$e_1 \times e_2 = e_3, \qquad e_1 \times e_3 = -e_2, \qquad e_2 \times e_3 = e_1.$$

Working out $u \times v = \sum_{i,j} u^i v^j e_i \times e_j$ gives

(8) $$u \times v = \begin{vmatrix} u_2 & u_3 \\ v_2 & v_3 \end{vmatrix} e_1 - \begin{vmatrix} u_1 & u_3 \\ v_1 & v_3 \end{vmatrix} e_2 + \begin{vmatrix} u_1 & u_2 \\ v_1 & v_2 \end{vmatrix} e_3.$$

We shall find a geometric interpretation of the vector product. Given u and v, choose an orthonormal base (e_1, e_2, e_3) orienting E^3 properly so that u is in the e_1-direction and v is in the plane of e_1 and e_2, on the same side of e_1 that e_2 is. Say

$$u = ae_1, \qquad v = b'e_1 + be_2; \qquad \text{then} \quad a \geqq 0, \qquad b \geqq 0.$$

Then applying (7) gives $u \times v = abe_3$. Thus, if u and v are independent, in which case a and b are $\neq 0$, then $u \times v$ is a vector perpendicular to both, and oriented so that $(u, v, u \times v)$ gives the orientation of E^3. Otherwise, $u \times v = 0$. Let σ be the parallelogram with a point p as vertex and sides along u and v. Then $|\sigma| = ab$, and hence the length of $u \times v$ is

(9) $$|u \times v| = |\sigma|.$$

We shall find the components of the 2-covector Ψ_v in (1) in an orthonormal coordinate system. Using (10.5) and (10.3), we have for instance

$$(\Psi_v)_{13} = \Psi_v(e_{13}) = \omega_0 \cdot \left(\sum v^i e_i \vee e_{13} \right) = \omega_0 \cdot v^2 e_2 \vee e_1 \vee e_3 = -v^2.$$

We find

(10) $$\Psi_v = v^3 e_{12} - v^2 e_{13} + v^1 e_{23}.$$

Similarly, for the vector Θ_α in (2),

(11) $$\Theta_\alpha = \alpha^{23} e_1 - \alpha^{13} e_2 + \alpha^{12} e_3.$$

14. The differential of a mapping. Let f be a mapping of E^n (or of an open set in E^n) into E^m which is smooth; that is, with coordinate systems in the spaces, the first partial derivatives of each component of the mapping function exist and are continuous. The *differential* ∇f of f is a concept which may be used in place of these partial derivatives, as follows. Take any point p and any vector v in E^n; then for each real number $t > 0$, $p + tv$ is a point in E^n. We set

(1) $$\nabla_v f(p) = \nabla f(p, v) = \lim_{t \to 0+} \frac{1}{t} [f(p + tv) - f(p)].$$

This is a vector in E^m, tangent to the curve which is the image under f of the line through p in the direction of v. For each p, we have a function $\nabla f(p)$ mapping vectors in E^n into vectors in E^m; it is elementary to show that this function is linear.

If $m = 1$ and E^m is the space $\mathfrak{A}$ of real numbers, we have a real valued function ϕ in E^n; with coordinates $(x^1, \cdots, x^n)$ in E^n and corresponding vectors $e_1, \cdots, e_n$ along the axes at the point p, we clearly have

$$(2) \qquad \nabla_{e_i}\phi(p) = \frac{\partial\phi(p)}{\partial x^i}, \qquad i = 1, \cdots, n.$$

In this case, $\nabla_v\phi(p)$ is a real linear function of vectors v, and thus $\nabla\phi(p)$ is a covector. Hence $\nabla\phi$ is a differential 1-form in E^n, called the *gradient* of ϕ.

We return to the general case. Given p and vectors $v_1, \cdots, v_r$ in E^n, set

$$(3) \qquad \nabla f(p, v_1\vee\cdots\vee v_r) = \nabla f(p, v_1)\vee\cdots\vee\nabla f(p, v_r);$$

this defines a linear transformation of r-vectors in E^n into r-vectors in E^m, which we also call $\nabla f(p)$.

Now let ω be any r-form in E^m. Take any point $q = f(p)$ in E^m. Then $\omega(q)\cdot\alpha'$ is a linear function of r-vectors α' in E^m; hence $\omega(q)\cdot\nabla f(p, \alpha)$ is a linear function of r-vectors α in E^n, and is thus an r-covector in E^n, which we call $(f^*\omega)(p)$. Now $f^*\omega$ is an r-form in E^n. The definition is given by

$$(4) \qquad (f^*\omega)(p)\cdot\alpha = \omega(f(p))\cdot\nabla f(p, \alpha), \qquad \text{all } r\text{-vectors } \alpha \text{ in } E^n.$$

Because of (3), we find, for differential forms ω and ξ in E^m,

$$(5) \qquad f^*(\omega\vee\xi) = f^*\omega\vee f^*\xi.$$

15. Jacobians. Let f be a smooth mapping of E^n (assumed metric and oriented) into E^m. Then the image under $\nabla f(p)$ of the unit n-vector α_0 of E^n is an n-vector in E^m, which we call the *Jacobian* $J_f(p)$ of f at p. Take for instance $n = 2$, $m = 3$, and let (e_1, e_2) be an orthonormal base in E^2. The images from p of these vectors are

$$(1) \qquad w_1(p) = \nabla f(p, e_1), \qquad w_2(p) = \nabla f(p, e_2),$$

and the Jacobian at p is the 2-vector

$$(2) \qquad J_f(p) = \nabla f(p, \alpha_0) = \nabla f(p, e_1\vee e_2) = w_1(p)\vee w_2(p).$$

If $J_f(p)$ is $\neq 0$, then $w_1(p)$ and $w_2(p)$ are independent, which clearly implies that f is one-one in a neighborhood of p; in this case the image of a neighborhood of p is a smooth piece S of surface in E^3.

Suppose f maps E^n into itself. Then $J_f(p)$ is a multiple $a\alpha_0$ of α_0; we call this number a the *algebraic Jacobian* $\bar{J}_f(p)$. Thus, with the unit n-covector ω_0 of E^n,

$$(3) \qquad J_f(p) = \bar{J}_f(p)\alpha_0, \qquad \omega_0\cdot J_f(p) = \bar{J}_f(p).$$

The term "Jacobian" is commonly used to denote $\bar{J}_f(p)$; note that $\bar{J}_f(p)$ is independent of the metric or orientation of E^n.

Take the case $n = 2$, $m = 3$ again. Using orthonormal coordinates (s, t) in E^2 and (x, y, z) in E^3, write the components of $\nabla_{e_1} f(p) = \partial f(p)/\partial s$ etc. as a set of three numbers. We have

$$\text{(4)} \qquad \frac{\partial f(p)}{\partial s} = \left(\frac{\partial x}{\partial s}, \frac{\partial y}{\partial s}, \frac{\partial z}{\partial s}\right), \qquad \frac{\partial f(p)}{\partial t} = \left(\frac{\partial x}{\partial t}, \frac{\partial y}{\partial t}, \frac{\partial z}{\partial t}\right).$$

Define the "Jacobian determinants"

$$\frac{\partial(x, y)}{\partial(s, t)} = \begin{vmatrix} \partial x/\partial s & \partial y/\partial s \\ \partial x/\partial t & \partial y/\partial t \end{vmatrix}, \text{ etc.}$$

Then by (9.5),

$$\text{(5)} \qquad J_f(p) = \frac{\partial(x, y)}{\partial(s, t)} e_{12} + \frac{\partial(x, z)}{\partial(s, t)} e_{13} + \frac{\partial(y, z)}{\partial(s, t)} e_{23}.$$

Define the "Jacobian vector" $\tilde{J}_f(p)$ at p as that vector corresponding to the 2-vector $J_f(p)$; see (13.2). By (2) and (13.3), we may write it in the form (using orthonormal coordinates)

$$\text{(6)} \qquad \tilde{J}_f(p) = \frac{\partial f(p)}{\partial s} \times \frac{\partial f(p)}{\partial t} = w_1(p) \times w_2(p).$$

By (13.11) or (13.8),

$$\text{(7)} \qquad \tilde{J}_f(p) = \left(\frac{\partial(y, z)}{\partial(s, t)}, -\frac{\partial(x, z)}{\partial(s, t)}, \frac{\partial(x, y)}{\partial(s, t)}\right).$$

16. Transformation of the integral. Let ω be a uniformly continuous differential 2-form in the bounded open set R of the oriented plane E^2. We consider a smooth mapping f of the bounded open set R_0 of the space E'^2 onto R, with Jacobian $J_f \neq 0$ at all points. We wish to express $\int_R \omega$ as an integral over R_0.

Cut E'^2 into small rectangles, and let $\sigma_1, \cdots, \sigma_s$ be those lying in R_0. The image $\tau_i = f(\sigma_i)$ of σ_i is a small "curvilinear parallelogram" in R. Let p_i be a corner of σ_i, and let v_{i1}, v_{i2} be vectors on adjacent sides of σ_i, so that

$$\{\sigma_i\} = v_{i1} \vee v_{i2}.$$

Now

$$w_{i1} = \nabla f(p_i, v_{i1}), \qquad w_{i2} = \nabla f(p_i, v_{i2})$$

are side vectors of a parallelogram τ_i' which is a good approximation to τ_i if σ_i is small. By § 11, clearly

$$\text{(1)} \qquad \sum \omega(q_i) \cdot \{\tau_i'\}, \qquad q_i = f(p_i),$$

is a good approximation to $\int_R \omega$. By (14.3),

$$\{\tau_i'\} = w_{i1} \vee w_{i2} = \nabla f(p_i, v_{i1} \vee v_{i2}) = \nabla f(p_i, \{\sigma_i\}),$$

and by (14.4),

$$\sum \omega(q_i)\cdot\{\tau_i'\} = \sum \omega(f(p_i))\cdot\nabla f(p_i, \{\sigma_i\}) = \sum (f^*\omega)(p_i)\cdot\{\sigma_i\},$$

which is a good approximation to $\int_{R_0} f^*\omega$. Thus we see that

$$\int_R \omega = \int_{R_0} f^*\omega. \tag{2}$$

(The detailed proof is given in (III, 8).)

From this we shall derive the usual formula, using Jacobians, and taking $E'^2 = E^2$. With the unit 2-vector α_0 in E^2,

$$\{\tau_i'\} = \nabla f(p_i, \{\sigma_i\}) = |\sigma_i|\, \nabla f(p_i, \alpha_0) = |\sigma_i|\, J_f(p_i).$$

Let $\bar\omega$ be the real function corresponding to ω; that is, with the unit 2-covector ω_0 in E^2,

$$\omega(q) = \bar\omega(q)\omega_0, \qquad \bar\omega(q) = \omega(q)\cdot\alpha_0. \tag{3}$$

Now by (15.3),

$$\begin{aligned}\bar\omega(q_i)\,|\tau_i'| &= \omega(q_i)\cdot\{\tau_i'\} \\ &= \bar\omega(q_i)\omega_0\cdot|\sigma_i|\, J_f(p_i) = \bar\omega(q_i)\bar J_f(p_i)\,|\sigma_i|\,.\end{aligned}$$

Summing and taking limits, we have

$$\int_R \bar\omega = \int_{R_0} \bar\omega(f(p))\bar J_f(p)\,dp, \tag{4}$$

using the Riemann integral in both cases.

The same formulas hold in any number of dimensions. Note that neither side of (2) depends on a choice of metric.

17. Smooth manifolds. As a typical example, we take a piece S of smooth surface in E^3. At each point p of S there is a tangent plane $T(p)$. Some neighborhood U of p in $T(p)$ projects in a one-one way into S. A coordinate system in $T(p)$ projects into a coordinate system in S; this coordinate system is a smooth mapping of part of the space $\mathfrak{A}^2$ of pairs of real numbers into S. Where two such coordinate systems overlap, they are related by a smooth mapping of part of $\mathfrak{A}^2$ into itself, with non-vanishing Jacobian. This suggests the general definition of a smooth manifold, using such coordinate systems, without reference to any containing space, as in (II, 10).

With S as above, let v be a vector in $T(p)$. The points $p_t = p + tv$ in $T(p)$ project into points q_t in S (for t not too large). These points q_t form a "parametrized curve" C in S, which we may take as defining a corresponding vector "of S" at p. In place of the points p_t, we could use any function p_t' in $T(p)$ with $p_0' = p$ which has the same tangent vector at $t = 0$: $(\partial p_t'/\partial t)_{t=0} = v$; this would project into a parametrized curve

C' with points q'_t in S equivalent to C. We may use the definitions of av and of $v + w$ in $T(p)$ to give corresponding definitions in S at p. The vectors in S at p now form a vector space $V(p)$. Using this vector space, we may define r-vectors and r-covectors in S at p; hence we may define differential forms in S. In a general smooth manifold (not in Euclidean space), we may define vectors at a point by means of parametrized curves, and define addition of vectors with the help of coordinate systems.

To define r-dimensional integration in an n-dimensional manifold M, one must first choose simple r-dimensional domains of integration; (rectilinear) r-cells are not defined here. One may choose oriented pieces of r-dimensional submanifolds of M for such domains. We have now the problem of defining the integral of an r-form ω over an oriented piece of an r-manifold.

Take first $n = 1$; then we have an oriented curve C, abstractly defined. It is of course equivalent to an interval of the real numbers. Now $\int_C \omega$ is defined for a 1-form ω. Suppose C is in E^3 and ω is defined in a neighborhood of C. Let $p_0, p_1, \cdots, p_m$ be a division of C into short arcs. Then the vectors $p_{i+1} - p_i$ are nearly tangent to C, and

$$\sum_{i=0}^{m-1} \omega(p_i)\cdot\{p_i p_{i+1}\} = \sum_{i=0}^{m-1} \omega(p_i)\cdot(p_{i+1} - p_i) \tag{1}$$

is an approximation to the integral $\int_C \omega$.

Take an oriented piece S of surface in E^3 again. Supposing S is a small piece, we may cut it up into small curvilinear pieces $\tau_1, \cdots, \tau_s$, for instance images $f(\sigma_1), \cdots, f(\sigma_s)$ of rectangles in a coordinate system (see the τ_i in § 16). With tangent vectors w_{i1}, w_{i2} at the vertex q_i of τ_i, we may form

$$\sum \omega(q_i)\cdot(w_{i1} \vee w_{i2}) \tag{2}$$

and use this as an approximation to $\int_S \omega$; compare (16.1). Because of (16.2), we see that we may equivalently define $\int_S \omega$ as being $\int_{S_0} f^*\omega$, if $S = f(S_0)$, S_0 being part of the Euclidean plane.

The latter definition is independent of the fact that S is in E^3. With S in E^3, if ω is defined throughout a neighborhood of S in E^3, we could use polyhedral approximations to define the integral; see Chapter X.

18. Particular forms of integrals in 3-space. Here we have r-dimensional integration for $r = 0$, 1, 2 and 3. Letting E^3 be metric and oriented, an r-covector in E^3 corresponds either to a real number or to a vector (see § 13); this gives us special forms of the integrals, which we discuss.

$r = 0$. A 0-cell is a point p (without orientation properties). A 0-covector is a real number; hence a 0-form is a real function. The integral of the 0-form ϕ over the 0-cell p is the number $\phi(p)$. A polyhedral 0-chain is an expression $A = \sum a_i p_i$, and the integral of ϕ over A is

(1) $$\int_A \phi = \sum a_i \phi(p_i).$$

In particular if σ is the oriented 1-cell pq, then

(2) $$\partial(pq) = 1q + (-1)p, \qquad \int_{\partial(pq)} \phi = \phi(q) - \phi(p).$$

$r = 1$. With an oriented curve C and a 1-form ω^1, an approximation to $\int_C \omega^1$ was given in (17.1). We replace ω^1 by a vector function v in E^3, as follows. For each p, $v(p)$ is the vector such that the scalar product $v(p)\cdot w$ equals $\omega^1(p)\cdot w$ for all vectors w; see § 12. With a common notation, a tangent vector along C is called the "line element" $d\mathbf{s}$, and the integral is written (using (17.1))

(3) $$\int_S \omega^1 = \int_S v\cdot d\mathbf{s} = \lim \sum v(p_i)\cdot(p_{i+1} - p_i).$$

This is the "circulation" of v along C.

$r = 2$. In this case, an approximation to $\int_S \omega^2$ was given in (17.2). The 2-form ω^2 corresponds to a vector function v, as follows. For each p, $\omega^2(p) = \Psi_{v(p)}$, as in (13.1). This relation gives

$$\omega^2(p)\cdot\alpha = \omega_0\cdot[v(p)\vee\alpha], \qquad \text{all 2-vectors } \alpha.$$

Hence, by (13.4),

(4) $$\omega^2(q_i)\cdot(w_{i1}\vee w_{i2}) = \omega_0\cdot[w_{i1}\vee w_{i2}\vee v(q_i)] = (w_{i1}\times w_{i2})\cdot v(q_i);$$

with the notations of (17.2), the sum of these quantities is an approximation to $\int_S \omega^2$. The vector

(5) $$\mathbf{N}_i = w_{i1}\times w_{i2}$$

is called the "surface element" at q_i, and in analogy with (3), the integral is written

(6) $$\int_S \omega^2 = \int_S v\cdot d\mathbf{N} = \lim \sum v(q_i)\cdot\mathbf{N}_i.$$

This is the "flux" of v across S (see § 5).

$r = 3$. The integral $\int_R \omega^3$ is defined for regions R in oriented E^3 and a 3-form ω^3. With the unit 3-covector ω_0, $\omega^3(p) = \bar{\omega}(p)\omega_0$ for a real function $\bar{\omega}$ as in (16.3), and

(7) $$\int_R \omega^3 = \int_R \bar{\omega},$$

the last integral being the Riemann integral. If we cut R into small cells σ_i oriented like E^3 and choose a point p_i in each σ_i, then

$$\sum \omega^3(p_i)\cdot\{\sigma_i\} = \sum \bar{\omega}(p_i)\,|\,\sigma_i\,| \tag{8}$$

is an approximation to the integral.

19. The Theorem of Stokes. Let ω be a smooth r-form in E^n. Then the general Theorem of Stokes says that there is an $(r+1)$-form $\omega' = d\omega$ in E^n such that for any oriented piece M of a smooth manifold of dimension $r+1$ in E^n with boundary ∂M,

$$\int_M d\omega = \int_{\partial M} \omega. \tag{1}$$

From general considerations we will see that ω' must exist; we deduce its form in the next section. The usual formulas in E^3 are given in § 21.

Define the r-cochain X in E^n by

$$X\cdot\sigma = \int_\sigma \omega, \qquad \text{all oriented } r\text{-cells } \sigma. \tag{2}$$

We define an $(r+1)$-cochain $Y = dX$ by the formula

$$Y\cdot\tau = X\cdot\partial\tau = \int_{\partial\tau} \omega, \qquad \text{all oriented } (r+1)\text{-cells } \tau; \tag{3}$$

compare (1.2). It may be proved that X satisfies (H_2'), or equivalently, that Y satisfies (H_1'); see (V, 10). That Y satisfies (H_2') is trivial: For any oriented $(r+2)$-cell τ',

$$Y\cdot\partial\tau' = X\cdot\partial\partial\tau' = X\cdot 0 = 0. \tag{4}$$

Thus Y is a flat $(r+1)$-cochain. We call it the *coboundary* dX of X. If we assume that the partial derivatives of ω satisfy a Lipschitz condition, then we may show that Y is sharp. There is then (§ 11) a corresponding $(r+1)$-form $\omega' = D_Y$ such that

$$\int_\tau \omega' = Y\cdot\tau = \int_{\partial\tau} \omega,$$

giving (1) for cells and hence for M. (Of course ω' exists under much more general conditions; see for instance (III, 16) and (IX, 12).)

20. The exterior differential. Given the smooth r-form ω in E^n, the $(r+1)$-form $\omega' = d\omega$ of § 19 is called the *exterior differential* of ω. Using an affine coordinate system $(x^1, \cdots, x^n)$ in E^n, we wish to find the components of ω' in terms of those of ω.

Take first $r = 0$. Given the point p, take any i, set $p_t = p + te_i$, and consider the segment pp_t $(t > 0)$. Since

$$\int_0^t \omega'_i\, dx^i = \int_{pp_t} \omega' = \int_{\partial(pp_t)} \omega$$
$$= \omega(p_t) - \omega(p) = \int_0^t \frac{\partial \omega}{\partial x^i}\, dx^i,$$

we must have

(1) $$r = 0: \qquad \omega'_i = \frac{\partial \omega}{\partial x^i}.$$

Take next $r = 1$. Given p and i, j, let us write x, y for x^i, x^j, and omit the other coordinates. Take any $h, k > 0$, and let σ be the oriented rectangle with corners

$$p = (x, y), \qquad p' = (x + h, y), \qquad p'' = (x + h, y + k), \qquad p''' = (x, y + k).$$

Since $\int_{p'}^{p''} \omega = \int_y^{y+k} \omega_j(x + h, t)\, dt$ etc., we have

(2)
$$\int_{\partial\sigma} \omega = \int_y^{y+k} [\omega_j(x + h, t) - \omega_j(x, t)]\, dt - \int_x^{x+h} [\omega_i(s, y + k) - \omega_i(s, y)]\, ds.$$

Now

$$\omega_j(x + h, t) - \omega_j(x, t) = \int_x^{x+h} \frac{\partial}{\partial x} \omega_j(s, t)\, ds,$$

and similarly for the other integrand. Hence we must have

(3) $$\int_\sigma \omega' = \int_{\partial\sigma} \omega = \int_x^{x+h} \int_y^{y+k} \left[\frac{\partial}{\partial x}\omega_j(s, t) - \frac{\partial}{\partial y}\omega_i(s, t)\right] dt\, ds.$$

Taking h and k arbitrarily small, this shows that

(4) $$r = 1: \qquad \omega'_{ij} = \frac{\partial \omega_j}{\partial x^i} - \frac{\partial \omega_i}{\partial x^j}.$$

The same method may be used in the general case. In particular,

(5) $$r = 2: \qquad \omega'_{ijk} = \frac{\partial \omega_{jk}}{\partial x^i} - \frac{\partial \omega_{ik}}{\partial x^j} + \frac{\partial \omega_{ij}}{\partial x^k}.$$

There are similar formulas relating ω and ω' directly. For instance,

(6) $$r = 1: \qquad d\omega(p)\cdot(v \vee w) = \nabla_v[\omega(p)\cdot w] - \nabla_w[\omega(p)\cdot v].$$

We mention some general properties. For any r-form ω and s-form ξ,

(7) $$dd\omega = 0, \qquad d(\omega \vee \xi) = d\omega \vee \xi + (-1)^r \omega \vee d\xi.$$

If f is a smooth mapping of one Euclidean space into another and ω is an r-form in the second, then

(8) $$df^*\omega = f^*\, d\omega.$$

If ω is an r-form defined in a manifold, we may define $d\omega$ with the help of a coordinate system; because of (8), the definition is independent of the coordinate system employed. The above properties continue to hold, except for (6).

21. Some special formulas in metric oriented E^3. We give examples of the exterior differential and Stokes' Theorem.

$r = 0$. Here, the exterior differential is simply the differential, i.e. the gradient: $d\phi = \nabla\phi$. (This formula fails for $r > 0$.) With the "del" operator

$$\nabla = e_1 \frac{\partial}{\partial x} + e_2 \frac{\partial}{\partial y} + e_3 \frac{\partial}{\partial z},$$

the vector function corresponding to $d\phi$ (see § 18) is

$$\nabla\phi = \text{grad}\,(\phi) = \frac{\partial\phi}{\partial x} e_1 + \frac{\partial\phi}{\partial y} e_2 + \frac{\partial\phi}{\partial z} e_3. \tag{1}$$

For an oriented curve C from p to q, (19.1) gives, with the notations of § 18,

$$\int_C \nabla\phi\cdot d\mathbf{s} = \int_C \text{grad}\,(\phi)\cdot d\mathbf{s} = \int_C d\phi = \phi(q) - \phi(p). \tag{2}$$

If C is an interval of the real axis, this is the "fundamental theorem of the integral calculus."

$r = 1$. Given the smooth vector function v in E^3, there is a corresponding 1-form ω, this has an exterior differential $\omega' = d\omega$, and this 2-covector valued function corresponds to a vector function v', called the *curl* of v. Because of (10.2), $\omega_i = v^i$, and because of (13.10),

$$\omega'_{12} = v'^3, \qquad \omega'_{13} = -v'^2, \qquad \omega'_{23} = v'^1. \tag{3}$$

Hence, by (20.4),

$$\begin{aligned} \nabla\times v &= \text{curl}\,(v) \\ &= \left(\frac{\partial v^3}{\partial y} - \frac{\partial v^2}{\partial z}\right) e_1 + \left(-\frac{\partial v^3}{\partial x} + \frac{\partial v^1}{\partial z}\right) e_2 + \left(\frac{\partial v^2}{\partial x} - \frac{\partial v^1}{\partial y}\right) e_3. \end{aligned} \tag{4}$$

For an oriented piece S of surface with boundary curve C, (19.1) gives

$$\int_S (\nabla\times v)\cdot d\mathbf{N} = \int_S \text{curl}\,(v)\cdot d\mathbf{N} = \int_C v\cdot d\mathbf{s}. \tag{5}$$

This is the commonly called "Theorem of Stokes."

$r = 2$. Given the smooth vector function v, there is a corresponding 2-form ω, and $d\omega = \omega'$ is a 3-form corresponding to a real function $\bar{\omega}'$

as in (16.3), called the *divergence* of v. The relation between v and ω is given by (3), and $\bar{\omega}' = \omega'_{123}$. Hence, by (20.5),

$$\nabla \cdot v = \operatorname{div}(v) = \frac{\partial v^1}{\partial x} + \frac{\partial v^2}{\partial y} + \frac{\partial v^3}{\partial z}. \tag{6}$$

For a region R (oriented like E^3), with boundary S, (19.1) gives, using the volume element dR,

$$\int_R \nabla \cdot v\, dR = \int_R \operatorname{div}(v)\, dR = \int_S v \cdot d\mathbf{N}. \tag{7}$$

By taking particular vector functions in (5) or (7), we obtain special standard formulas. Recall that they are all cases of (19.1), which formula holds independently of a metric in the space.

The formula $dd\omega = 0$ gives, in the cases $r = 0$ and $r = 1$,

$$\nabla \times \nabla \phi = \operatorname{curl}(\operatorname{grad}(\phi)) = 0, \tag{8}$$

$$\nabla \cdot (\nabla \times \phi) = \operatorname{div}(\operatorname{curl}(\phi)) = 0; \tag{9}$$

these are easily verified directly.

We may start with a real function ϕ, find its gradient, interpret this as a vector function, and find its divergence, giving a real function again; this is the *Laplacian* of ϕ. By (1) and (6), this is

$$\nabla \cdot \nabla \phi = \Delta \phi = \frac{\partial^2 \phi}{\partial x^2} + \frac{\partial^2 \phi}{\partial y^2} + \frac{\partial^2 \phi}{\partial z^2}. \tag{10}$$

Because of its construction, the right hand side is independent of the choice of orthonormal coordinates.

22. An existence theorem. In a convex subset of E^3, any vector function is a gradient provided its curl is 0 (it is then commonly called an exact differential), any vector function is a curl provided its divergence is 0, and any real function is a divergence. (The functions are assumed smooth.) These are cases of the general theorem, that a smooth r-form ω ($r \geqq 1$) in a convex subset of E^n equals $d\xi$ for some $(r-1)$-form ξ if and only if $d\omega = 0$. Note that for a 0-form (real function) ϕ, $d\phi = 0$ if and only if ϕ is constant. The theorem fails for more general subsets of E^n or for manifolds; see the next section.

We give a formula denoted by $(n;\ r)$, for the components of a possible ξ in terms of those of the r-form ω in E^n, for various values of n and r, in a cube containing the origin. For $n = r = 1$, set

$$\xi(x) = \int_0^x \omega_1(t)\, dt; \tag{1;1}$$

then $(d\xi)_1(x) = d\xi(x)/dx = \omega_1(x)$, and hence $d\xi = \omega$.

For $n = 2$, $r = 1$, we carry out the idea above by integrating along the path from $(0, 0)$ to $(0, y)$ and thence to (x, y):

$$\xi(x, y) = \int_0^y \omega_2(0, t)\, dt + \int_0^x \omega_1(s, y)\, ds. \tag{2;1}$$

Then $\partial\xi(x, y)/\partial x = \omega_1(x, y)$, and since $\partial\omega_2/\partial x - \partial\omega_1/\partial y = 0$,

$$\begin{aligned}\frac{\partial\xi(x, y)}{\partial y} &= \omega_2(0, y) + \int_0^x \frac{\partial\omega_1(s, y)}{\partial y}\, ds \\ &= \omega_2(0, y) + [\omega_2(x, y) - \omega_2(0, y)] = \omega_2(x, y).\end{aligned}$$

We write down some further cases:

$$\xi_1(x, y) = 0, \qquad \xi_2(x, y) = \int_0^x \omega_{12}(s, y)\, ds; \tag{2;2}$$

$$\xi(x, y, z) = \int_0^z \omega_3(0, 0, u)\, du + \int_0^y \omega_2(0, t, z)\, dt + \int_0^x \omega_1(s, y, z)\, ds; \tag{3;1}$$

$$\begin{cases} \xi_1(x, y, z) = 0, \qquad \xi_2(x, y, z) = \int_0^x \omega_{12}(s, y, z)\, ds, \\ \xi_3(x, y, z) = \int_0^y \omega_{23}(0, t, z)\, dt + \int_0^x \omega_{13}(s, y, z)\, ds; \end{cases} \tag{3;2}$$

$$\xi_{12}(x, y, z) = \xi_{13}(x, y, z) = 0, \qquad \xi_{23}(x, y, z) = \int_0^x \omega_{123}(s, y, z)\, ds. \tag{3;3}$$

A general proof of existence of ξ will be given in (IV, 25).

23. De Rham's Theorem. The Theorem of de Rham relates differential forms in a manifold to homology and cohomology properties of the manifold; see (IV, Theorem 29A). We shall discuss the theorem briefly for the case of a torus M, in the dimension 1.

The points of M are described by naming a pair of angular coordinates (θ, ψ), $0 \leqq \theta \leqq 2\pi$, $0 \leqq \psi \leqq 2\pi$, where $\theta = 0$ is identified with $\theta = 2\pi$, and similarly for ψ. Thus we have

$$(0, \psi) = (2\pi, \psi), \qquad (\theta, 0) = (\theta, 2\pi). \tag{1}$$

The points $(\theta, 0)$ as θ goes from 0 to 2π form an oriented closed curve C_1; similarly, the points $(0, \psi)$ form C_2. Now let C be any oriented closed curve. By an elementary argument we may show that C can be deformed into an oriented closed curve C' which follows C_1 a number μ_1 of times, and then C_2 μ_2 times; μ_1 and μ_2 are any integers. For $\mu_1 = \mu_2 = 0$, C' reduces to a point.

We call a differential form ω *closed* if $d\omega = 0$, and *derived* if $\omega = d\xi$ for some ξ. Since $dd\xi = 0$, any derived form is closed. Call ω and ω' *cohomologous* if $\omega - \omega'$ is derived. A closed r-form, together with all cohomologous forms, forms an r-dimensional *differential cohomology class* in M (with real coefficients); the set of these classes form the r-dimensional *differential cohomology space* $\mathbf{H}^r$. By the Theorem of de Rham, this is isomorphic to the r-dimensional algebraic cohomology space with real coefficients.

The *fundamental periods* of the closed 1-form ω in the torus are the numbers

$$P_1(\omega) = \int_{C_1} \omega, \qquad P_2(\omega) = \int_{C_2} \omega. \tag{2}$$

For any cohomologous ω', Stokes' Theorem gives

$$P_i(\omega') - P_i(\omega) = \int_{C_i} (\omega' - \omega) = \int_{C_i} d\xi = \int_{\partial C_i} \xi = 0,$$

since C_i is closed. Hence we may speak of the fundamental periods of a differential cohomology class. We shall show that there is a unique class with any given pair of numbers as periods. This is the original formulation of the theorem in the present situation.

Define the closed 1-forms

$$\omega_1^* = d\theta, \qquad \omega_2^* = d\psi. \tag{3}$$

Though θ and ψ are not single valued in M, because of (1), clearly ω_1^* and ω_2^* are. The fundamental periods of ω_1^* and of ω_2^* are $(2\pi, 0)$ and $(0, 2\pi)$ respectively; hence

$$P_1(a\omega_1^* + b\omega_2^*) = 2\pi a, \qquad P_2(a\omega_1^* + b\omega_2^*) = 2\pi b. \tag{4}$$

Thus there are closed forms with any given fundamental periods.

We next note that if the oriented closed curve C is deformed into C' as above, then for any closed 1-form ω,

$$\int_C \omega = \int_{C'} \omega = \mu_1 P_1(\omega) + \mu_2 P_2(\omega). \tag{5}$$

isomorphic to the r-dimensional algebraic cohomology space.
For we may find a succession of curves $C_0 = C, C_1, \cdots, C_m = C'$, as follows. For each i, there is an oriented 2-cell σ_i with boundary $\partial\sigma_i = B_i - A_i$, such that C_{i+1} contains B_i, C_i contains A_i, and C_i and C_{i+1} are the same otherwise. Now

$$\int_{C_{i+1}} \omega - \int_{C_i} \omega = \int_{\partial\sigma_i} \omega = \int_{\sigma_i} d\omega = 0,$$

and the first relation in (5) follows; the second is clear.

We must show still that if ω and ω' are closed 1-forms with the same fundamental periods, then their difference $\eta = \omega' - \omega$ is derived: $\eta = d\phi$. We may define ϕ by the formula (2;1) in § 22:

$$\phi(\theta, \psi) = \int_0^\psi \eta_2(0, t)\, dt + \int_0^\theta \eta_1(s, \psi)\, ds. \tag{6}$$

To show that ϕ is single-valued, apply (6) to give

$$\phi(\theta, 2\pi) - \phi(\theta, 0) = \int_0^{2\pi} \eta_2(0, t)\, dt = \int_{C_2} \eta = 0.$$

Similarly $\phi(2\pi, \psi) = \phi(0, \psi)$. That $d\phi = \eta$ follows as in § 22.

We remark that for any oriented closed curve C, we can find the associated numbers μ_1, μ_2 through ω_1^*, ω_2^*. With C' as before,

$$(7) \qquad \int_C \omega_i^* = \int_{C'} \omega_i^* = \mu_1 P_1(\omega_i^*) + \mu_2 P_2(\omega_i^*) = 2\pi\mu_i.$$

The Theorem of de Rham also relates products of forms to products of algebraic cohomology classes or intersections of cycles. We give one formula. Clearly $\omega_1^* \vee \omega_2^*$ is the unit 2-covector function in M; hence

$$(8) \qquad \int_M \omega_1^* \vee \omega_2^* = 4\pi^2.$$

It follows easily, for any closed ω_1, ω_2, that

$$(9) \qquad \int_M \omega_1 \vee \omega_2 = 4\pi^2[P_1(\omega_1)P_2(\omega_2) - P_2(\omega_1)P_1(\omega_2)] \quad ;$$

we use the fact that ω_i is cohomologous to a form $\omega_i' = a_i\omega_1^* + b_i\omega_2^*$.

C. Indications of General Theory

24. Normed spaces of chains and cochains. Given the flat r-cochain X in E^n (§ 3), let $|X|$ and $|dX|$ be the smallest N_1 and N_2 satisfying the hypotheses (H_1') and (H_2') respectively. Then the function

$$(1) \qquad |X|^\flat = \sup\{|X|, |dX|\}$$

of flat cochains is a norm, the *flat norm*, and the linear space of flat r-cochains with this norm is a Banach space $\mathbf{C}^{\flat r}$.

Let the *mass* of the polyhedral r-chain $A = \sum a_i\sigma_i$ be

$$(2) \qquad |A| = \sum |a_i| |\sigma_i| \qquad \text{(if the } \sigma_i \text{ are non-overlapping)};$$

then the definition of $|X|$ gives $|X \cdot A| \leqq |X| |A|$, and

$$|X \cdot \partial D| \leqq |dX| |D|, \qquad \text{all polyhedral } (r+1)\text{-chains } D.$$

Given X and A, take any such D. Then

$$|X \cdot A| = |X \cdot (A - \partial D) + X \cdot \partial D|$$
$$\leqq |X| |A - \partial D| + |dX| |D| \leqq |X|^\flat(|A - \partial D| + |D|).$$

This suggests defining the *flat norm* of the polyhedral r-chain A by

$$(3) \quad |A|^\flat = \inf\{|A - \partial D| + |D|\}, \qquad \text{all polyhedral } (r+1)\text{-chains } D.$$

This is a norm in the space of polyhedral r-chains, and is the smallest norm such that, for all flat X,

$$(4) \qquad |X \cdot A| \leqq |X|^\flat |A|^\flat.$$

For an example, take an oriented 1-cell σ, and let $\sigma' = T_v\sigma$ denote σ translated by the vector v. Now σ and σ' are two sides (of length $|\sigma|$) of a parallelogram τ; the other two sides A, B are of length $|v|$, and $|\tau| \leqq |\sigma||v|$. Hence

$$(5) \qquad |T_v\sigma - \sigma|^\flat \leqq |T_v\sigma - \sigma - \partial\tau| + |\tau| \leqq (2 + |\sigma|)|v|.$$

Since

$$|X\cdot(T_v\sigma - \sigma)| \leqq |X|^\flat |T_v\sigma - \sigma|^\flat \leqq (2 + |\sigma|)|X|^\flat |v|,$$

$X\cdot T_v\sigma$ is near $X\cdot\sigma$ if v is small.

Each flat r-cochain X has a *coboundary* dX, which is a flat $(r + 1)$-cochain, defined by

$$(6) \qquad dX\cdot B = X\cdot\partial B, \qquad \text{all polyhedral } (r+1)\text{-chains } B.$$

This relation is the abstract counterpart of Stokes' Theorem (19.1).

The space of polyhedral r-chains in the flat norm is not complete; if we complete it, we obtain the space $\mathbf{C}_r^\flat$ of *flat* r-chains. Now $\mathbf{C}^{\flat r}$ is the conjugate space of $\mathbf{C}_r^\flat$.

We may define the *sharp norms* $|X|^\#$ of cochains and $|A|^\#$ of polyhedral r-chains similarly. Completing the space of r-chains in the sharp norm gives the space $\mathbf{C}_r^\#$ of sharp r-chains, which includes $\mathbf{C}_r^\flat$.

Now integration theory may be set up as follows: Define a norm (for instance the sharp or flat norm) in the linear space of polyhedral r-chains; then an "integrand" (cochain) X is a bounded linear function in this space; it satisfies continuity conditions, depending on the norm chosen. Now $X\cdot A = \int_A D_X$ is defined not only over polyhedral r-chains, but also over all elements of the completed space of chains; with properly chosen norms, this space will include curved pieces of r-dimensional manifolds, etc. This is the program we carry out starting with Chapter V.

25. Continuous chains. We return for the moment to the torus M of § 23. Let $C_{1\psi}$ be the oriented closed curve of points (θ, ψ) as θ runs from 0 to 2π. Since $C_{1\psi}$ may be deformed into C_1, we have, for any closed 1-form ω in M,

$$(1) \qquad \int_0^{2\pi} [\omega(\theta, \psi)\cdot e_1]\, d\theta = \int_0^{2\pi} \omega_1(\theta, \psi)\, d\theta = \int_{C_{1\psi}} \omega = P_1(\omega),$$

where e_1 is the unit vector in the θ-direction. Integrating with respect to ψ and dividing by 2π gives

$$(2) \qquad \frac{1}{2\pi}\int_M \omega\cdot e_1 = \frac{1}{2\pi}\int_0^{2\pi}\int_0^{2\pi} [\omega(\theta, \psi)\cdot e_1]\, d\theta\, d\psi = P_1(\omega).$$

This expresses $P_1(\omega)$ by means of the Riemann integral of $\omega\cdot e_1$ over the whole of M. We have here a 1-dimensional integral $P_1(\omega)$ in the guise of a 2-dimensional integral.

We ask now, *dropping the restriction that ω be closed*, can the left hand side of (2) be interpreted as a 1-dimensional integral? We may do this as follows. First, the curves $C_{1\psi}$ may be considered as simple 1-dimensional figures, and

$$X \cdot C_{1\psi} = \int_{C_{1\psi}} \omega = F(\psi)$$

is defined. Now take a fine subdivision $\mathfrak{S}$ of $(0, 2\pi)$ by the points $\psi_0 = 0$, $\psi_1, \cdots, \psi_m = 2\pi$. Corresponding to $\mathfrak{S}$, define the 1-chain (similar to a polyhedral 1-chain)

$$(3) \qquad A_{\mathfrak{S}} = \sum_{i=0}^{m-1} (\psi_{i+1} - \psi_i) C_{1\psi_i}.$$

Then

$$X \cdot A_{\mathfrak{S}} = \sum (\psi_{i+1} - \psi_i) X \cdot C_{1\psi_i} = \sum F(\psi_i)(\psi_{i+1} - \psi_i).$$

Taking a sequence $\mathfrak{S}_1, \mathfrak{S}_2, \cdots$ of such subdivisions, with mesh $\to 0$, this shows that

$$(4) \qquad \lim_{k \to \infty} X \cdot A_{\mathfrak{S}_k} = \int_0^{2\pi} F(\psi)\, d\psi,$$

which is the desired integral over M. Moreover, we show below that in the flat norm (§ 24), the sequence $A_{\mathfrak{S}_1}, A_{\mathfrak{S}_2}, \cdots$ is a Cauchy sequence; hence it has a limit A:

$$(5) \qquad \lim_{k \to \infty}{}^{\flat} A_{\mathfrak{S}_1} = A.$$

Now $X \cdot A = \int_0^{2\pi} F(\psi)\, d\psi$, so that the integral in (2) is simply the value of X on the flat 1-chain A.

To illustrate the convergence in (5), take the square $0 \leqq \theta \leqq 2\pi$, $0 \leqq \psi \leqq 2\pi$, in $\mathfrak{A}^2$ (see § 17) instead of in M; the $C_{1\psi_i}$ are now segments σ_i. Suppose $\mathfrak{S}_l$ is a refinement of $\mathfrak{S}_k$. Then each term $(\psi_{i+1} - \psi_i)\sigma_i$ in $\mathfrak{S}_k$ is replaced in $\mathfrak{S}_l$ by a sum of terms $(\psi_{i,j+1} - \psi_{ij})\sigma_{ij}$; if $\psi_{i+1} - \psi_i < \epsilon$ (all i), then (24.5) gives

$$\left| \sum_j (\psi_{i,j+1} - \psi_{ij})\sigma_{ij} - (\psi_{i+1} - \psi_i)\sigma_i \right|^{\flat} = \left| \sum_j (\psi_{i,j+1} - \psi_{ij})(\sigma_{ij} - \sigma_i) \right|^{\flat}$$
$$\leqq \sum_j (\psi_{i,j+1} - \psi_{ij}) \left| \sigma_{ij} - \sigma_i \right|^{\flat} \leqq (\psi_{i+1} - \psi_i)(2 + 2\pi)\epsilon.$$

Therefore

$$(6) \qquad \left| A_{\mathfrak{S}_l} - A_{\mathfrak{S}_k} \right|^{\flat} \leqq \sum_i (\psi_{i+1} - \psi_i)(2 + 2\pi)\epsilon = 2\pi(2 + 2\pi)\epsilon,$$

and (5) follows for this case (and hence for the above case).

Note that it is the e_1 in (2) which determined the flat chain A. In general, in an n-dimensional Riemannian manifold M, we may have a continuous point function $\alpha(p)$, whose values are r-vectors; the function α defines a flat r-chain A, called a *continuous r-chain*. An r-form $\omega = D_X$ corresponds to a flat (in fact, sharp) cochain X (see (11.2)), and using the Riemann integral,

$$X\cdot A = \int_M \omega\cdot\alpha. \tag{7}$$

Suppose the function α is smooth. Then it may be shown (VI, 9) that the boundary ∂A of the corresponding chain A is defined by a function α' formed from the partial derivatives of α.

26. On 0-dimensional integration. We may divide kinds of integration into two types: (a) those in which orientation properties of the domains are used, and (b) those where these properties are not used. Under (a), there is of necessity a certain dimension number r attached to the integral. This might even be the case under (b), but the essential geometric properties are missing; we show below how to set up such an integral as a 0-dimensional integral.

Whenever a domain of integration can be turned in space (thus moving in part in directions not parallel to its original position), orientation properties are probably needed; witness the discussion of flux in § 5. Where this type of motion is not used, orientation can in general be disregarded. Thus, as in (18.7), the integral $\int_R \omega$ of a 3-form ω in oriented E^3 over a domain R oriented like E^3 can also be written as the Riemann integral $\int_R \bar{\omega}$ of the corresponding real function $\bar{\omega}$ (assuming E^3 is metric); in the latter, orientation properties are not used. In the transformation formula (16.2) or (16.4) however, one must pay attention to orientation.

A typical example of an integral of type (b) is the evaluation of the total mass of a mass distribution. We might have mass spread out over curves or over surfaces in E^3; but these figures need not be oriented in determining the mass. A more general integral of the same sort is the integral of a real function ϕ in a metric space S which has a (positive) measure function μ attached. We may consider ϕ as a 0-form; then $X\cdot p = \phi(p)$ is a 0-cochain in S. We show how to consider any compact measurable subset Q of S as a "flat" 0-chain A_Q, such that

$$\int_Q \phi\, d\mu = X\cdot A_Q; \tag{1}$$

we assume ϕ is continuous, so a Riemann type definition may be used.

For each subdivision $\mathfrak{S}$ of Q into measurable sets $Q_1, \cdots, Q_m$, together with a point p_i chosen in each Q_i, we have a 0-chain

$$A_{\mathfrak{S}} = \sum \mu(Q_i) p_i; \tag{2}$$

for a sequence of such sets with mesh $\to 0$, the limit of the corresponding chains is a "flat" chain A. The proof rests on the following fact: Suppose each Q_i is of diameter $<\epsilon$, and $\mathfrak{S}'$ is a refinement of $\mathfrak{S}$. Say Q_i is divided into sets Q_{ij}, and p_{ij} is in Q_{ij}. Then considering $p_i p_{ij}$ as an abstract 1-cell and noting from (24.3) that $| \partial(pq) |^\flat \leqq$ distance from p to q,

$$| A_{\mathfrak{S}'} - A_{\mathfrak{S}} |^\flat = \Big| \sum_{i,j} \mu(Q_{ij})(p_{ij} - p_i) \Big|^\flat = \Big| \sum_{i,j} \mu(Q_{ij})\, \partial(p_i p_{ij}) \Big|^\flat$$

$$\leqq \epsilon \sum_{i,j} \mu(Q_{ij}) = \epsilon\mu(Q).$$

The usual proof now shows that the flat 0-chain A exists. Since

$$X \cdot A_{\mathfrak{S}} = \sum_i \mu(Q_i) X \cdot p_i = \sum_i \phi(p_i)\mu(Q_i),$$

it is clear that (1) holds.

We remark that if we set

$$\phi A_{\mathfrak{S}} = \sum \phi(p_i)\mu(Q_i)p_i,$$

the flat limit gives a 0-chain ϕA_Q; there is a "unit" 0-cochain X_0, defined by $X_0 \cdot p = 1$ (all p), with corresponding real function $D_{X_0}(p) = 1$, and $X_0 \cdot \phi A = \int_Q \phi \, d\mu$. This type of operation is studied in (VII, 1).

PART I

CLASSICAL THEORY

I. Grassmann Algebra

We give here the algebraic basis for the theory of r-dimensional integration in n-space; it will be used in Chapter III to set up the integral of a differential form. Why this algebra appears in integration theory will be seen in Chapter V; see also Part A of the Introduction. We use a geometric approach; with the introduction of coordinate systems, the usual formulas involving components are derived.

The Riemann integral, being an n-dimensional integral in n-space, does not require the use of Grassmann algebra. However, the transformation theorem, and still more, the Theorem of Stokes, are best understood with its help; hence we employ the algebra in Chapter III from the start. For the purposes of that chapter, the important sections of the present chapter are §§ 1, 2, 3, 5, 8, 9, and parts of 10 and 12. In the general theory (starting with Chapter V), the norms of § 13 are basic. The products of §§ 6 and 7, though clearly a fundamental part of the subject, are needed only in special parts of the theory.

For other accounts of Grassmann algebra, see for instance Bourbaki and Lichnerowicz.

The theory is concerned with a vector space V. One defines, abstractly, "products" of vectors (§ 1), thus forming r-vectors of V. Similar products in the conjugate space $\bar{V}$ of V (App. I, 3) give r-covectors, which act on r-vectors to give numbers. These objects are known in tensor analysis as contravariant and covariant alternating tensors. The actual appearance in integration theory (V, 10) of r-covectors is as alternating multilinear functions; these are discussed in § 4. In the study of coordinate systems in § 5, we find the properties of components often used to characterize the objects.

The exterior and interior products correspond to products in algebraic topology; see (IV, 29), (IX, 14), (IX, 16), and (X, 11).

The r-vectors with immediate geometric interpretation are the "simple" ones; see § 9 and also § 7 of the Introduction. It is this interpretation we make use of in the definition of the integral in Chapter III. Properties of linear transformations, duality, and magnitude occupy the latter part of the Chapter.

1. Multivectors. Let V be a vector space. We define first *2-vectors*, or *bivectors*, of V. A *2-vector* is an expression of the form

$$\alpha = a_1(u_1 \vee v_1) + \cdots + a_k(u_k \vee v_k), \tag{1}$$

with any number of summands, written in any order; the a_i are real numbers, and the u_i and v_i, vectors of V. We form $\alpha + \beta$ by writing one expression after the other, joined by a plus sign; we form $a\alpha$ by replacing each a_i by aa_i. The expressions are to satisfy the relations

(2) $$a(u\vee v) + b(u\vee v) = (a + b)(u\vee v), \qquad 1(u\vee v) = u\vee v,$$

(3) $$(u + u')\vee v = u\vee v + u'\vee v, \qquad u\vee(v + v') = u\vee v + u\vee v',$$

(4) $$(au)\vee v = u\vee(av) = a(u\vee v),$$

(5) $$u\vee u = 0.$$

Using these rules gives

$$u\vee v + v\vee u = u\vee u + u\vee v + v\vee u + v\vee v = (u + v)\vee(u + v) = 0;$$

hence

(6) $$v\vee u = -u\vee v.$$

An *r-vector* is an expression of the form

(7) $$a_1(v_{11}\vee\cdots\vee v_{1r}) + \cdots + a_k(v_{k1}\vee\cdots\vee v_{kr}),$$

with the same properties. The distributive law applies to each variable; a term with any vector repeated is zero. For $r = 0$, we have real numbers (scalars); for $r = 1$, vectors. The r-vectors from a vector space, which we denote by $V_{[r]}$.

We say r is the *degree* of the r-vector, though perhaps "dimension" is a better term.

We can describe $V_{[r]}$ in accurate terms, as follows. Let L^r be the linear space of real valued functions F, defined on the r-fold Cartesian product $V^{(r)} = V \times \cdots \times V$, which vanish except at a finite set of points of $V^{(r)}$. If F has the value a_i at $v_{i1} \times \cdots \times v_{ir}$ $(i = 1, \cdots, k)$, and is zero elsewhere, we write

(8) $$F = a_1(v_{11} \cdots v_{1r}) + \cdots + a_k(v_{k1} \cdots v_{kr});$$

the terms may then be commuted, and a rule like (2) may be used. We write 0 for $0(v_1 \cdots v_r)$, and $v_1 \cdots v_r$ for $1(v_1 \cdots v_r)$.

With these notations, let L_0^r be the subset of all functions F which may be written in any of the following forms (letting A, B, and C denote any expressions $v_1 \cdots v_s$, possibly vacuous):

(9) $$A(v + v')B - AvB - Av'B,$$

(10) $$A(av)B - a(AvB),$$

(11) $$AvBvC.$$

Let L_1^r be the subspace of L^r spanned by L_0^r, i.e. the set of all linear combinations of such functions. Then $V_{[r]}$ is the quotient space (App. I, 5)

(12) $$V_{[r]} = L^r \bmod L_1^r;$$

let $a(v_1 \vee \cdots \vee v_r)$ be the element of $V_{[r]}$ corresponding to $a(v_1 \cdots v_r) \in L^r$.

The proof of (6) shows that interchanging two vectors in any expression $v_1 \vee \cdots \vee v_r$ changes the sign of the term; applying this any number of times, we find, if $(\lambda_1, \cdots, \lambda_r)$ is a permutation of $(1, \cdots, r)$, that (see the notations in (App. I))

$$(13) \qquad v_{\lambda_1} \vee \cdots \vee v_{\lambda_r} = \epsilon_{\lambda_1 \cdots \lambda_r} v_1 \vee \cdots \vee v_r.$$

If the λ_i are not distinct, both sides are 0.

2. Multicovectors. Using the conjugate space $\bar{V}$ of V (App. I, 3), form the space

$$(1) \qquad V^{[r]} = \bar{V}_{[r]}$$

of multicovectors in terms of $\bar{V}$ exactly as $V_{[r]}$ was formed in terms of V. We say an r-covector is of *degree* r. Let expressions $\sum a_i f^{i1} \vee \cdots \vee f^{ir}$ denote r-covectors.

Define the *scalar product* of r-covectors and r-vectors as the bilinear operation such that

$$(2) \qquad \begin{aligned} &(f^1 \vee \cdots \vee f^r) \cdot (v_1 \vee \cdots \vee v_r) \\ &= \sum_\lambda \epsilon^{\lambda_1 \cdots \lambda_r} f^1(v_{\lambda_1}) \cdots f^r(v_{\lambda_r}) = \begin{vmatrix} f^1(v_1) \cdots f^1(v_r) \\ \cdot \ \cdot \ \cdot \ \cdot \ \cdot \ \cdot \ \cdot \\ f^r(v_1) \cdots f^r(v_r) \end{vmatrix}. \end{aligned}$$

For the case $r = 0$, we have multiplication of real numbers; for $r = 1$, the operation of a covector on a vector; for $r = 2$, $(f \vee g) \cdot (u \vee v) = f(u)g(v) - f(v)g(u)$.

Relation (2) certainly defines all scalar products. Since elements of $V^{[r]}$ and of $V_{[r]}$ may be written in many different ways, we must show that the definition is independent of the manner of writing the elements. We may prove this as follows. Say $V^{[r]} = M^r \bmod M_1^r$, as in (1.12). Let M^r operate on L^r by the rule

$$\Big(\sum_i a^i f^{i1} \cdots f^{ir}\Big) \cdot \Big(\sum_j b_j v_{j1} \cdots v_{jr}\Big) = \sum_{i,j} a^i b_j \sum_\lambda \epsilon^\lambda f^{i1}(v_{j\lambda_1}) \cdots f^{ir}(v_{j\lambda_r});$$

clearly this is well defined and bilinear. If we show that an element of M_0^r times an element of L^r is zero, as is an element of M^r times an element of L_0^r, it will follow that a bilinear operation is defined correspondingly between $V^{[r]}$ and $V_{[r]}$, which is exactly the scalar product. We sketch the proof of the first fact in the case $r = 2$; the general proof will then be clear. First,

$$\begin{aligned} (f + f')g \cdot uv &= (f + f')(u)g(v) - (f + f')(v)g(u) \\ &= [f(u) + f'(u)]g(v) - [f(v) + f'(v)]g(u), \end{aligned}$$

and we find $[(f+f')g-fg-f'g]\cdot uv=0$. Use of (1.10) is similar, since $(af)(u)=af(u)$. Finally,

$$ff\cdot uv=f(u)f(v)-f(v)f(u)=0=0\cdot uv.$$

Let $e_1, \cdots, e_n$ and $e^1, \cdots, e^n$ be dual bases in V and in $\bar{V}$ respectively (App. I, 3). Write, for any $\lambda=(\lambda_1, \cdots, \lambda_r)$,

$$e_\lambda = e_{\lambda_1\cdots\lambda_r} = e_{\lambda_1}\vee\cdots\vee e_{\lambda_r}, \qquad e^\lambda = e^{\lambda_1\cdots\lambda_r} = e^{\lambda_1}\vee\cdots\vee e^{\lambda_r}. \tag{3}$$

Then for any $\lambda=(\lambda_1, \cdots, \lambda_r)$ and $\mu=(\mu_1, \cdots, \mu_r)$,

$$e^\lambda\cdot e_\mu = \delta^\lambda_\mu. \tag{4}$$

For if the λ_i are not distinct, then $e^\lambda=0$; similarly for e_μ. If μ is not a permutation of λ, then some λ_i does not appear in μ, $e^{\lambda_i}(e_{\mu_j})=0$ for all j, and $e^\lambda\cdot e_\mu=0$. It is immediate from (2) that $e^\lambda\cdot e_\lambda=1$. If μ is a permutation of λ, we reduce the proof to the case $\mu=\lambda$ by using (1.13).

In general, we use only the e_λ *and* e^λ *with* $\lambda_1<\cdots<\lambda_r$.

Using (4) gives (see the definition of $\sum_{(\mu)}$ in (App. I))

$$\left.\begin{aligned} &e^\lambda\cdot\sum_{(\mu)}\alpha^\mu e_\mu = \alpha^\lambda \qquad (5)\\ &\sum_{(\mu)}\omega_\mu e^\mu\cdot e_\lambda = \omega_\lambda \qquad (6)\end{aligned}\right\} \quad (\lambda_1<\cdots<\lambda_r).$$

Clearly *the scalar product is characterized* as that skew-symmetric bilinear operation satisfying (4). Compare § 10 of the Introduction.

3. Properties of $V_{[r]}$ **and** $V^{[r]}$. Note that through (2.2), each r-covector ξ defines a linear function H_ξ in $V_{[r]}$. We prove

THEOREM 3A. *Let* V *be of dimension* n. *Then for* $0\leqq r\leqq n$, $V_{[r]}$ *and* $V^{[r]}$ *are of dimension* $\binom{n}{r}$; *for* $r>n$, *all* r*-vectors and* r*-covectors vanish. With the operation of scalar products,* $V^{[r]}$ *may be identified with the conjugate space of* $V_{[r]}$. *With dual bases as in* § 2, *the elements*

$$e_\lambda,\ e^\lambda \qquad (\lambda_1<\cdots<\lambda_r) \tag{1}$$

form dual bases in $V_{[r]}$ *and in* $V^{[r]}$ *respectively.*

First, given any r-vector $\alpha=\sum_i a_i v_{i1}\vee\cdots\vee v_{ir}$, writing $v_{ij}=\sum_k v^k_{ij}e_k$, multiplying out, and using (1.13) shows that α is a linear combination of the e^λ in (1); also, if $r>n$, then $\alpha=0$. To show that the e_λ are independent, suppose there were a relation $\sum_{(\lambda)}a^\lambda e_\lambda=0$. Applying (2.5) shows that $a^\lambda=0$ for each λ. Hence the e_λ in (1) form a base in $V_{[r]}$. Similarly, the e^λ in (1) form a base in $V^{[r]}$. Hence also the statement about dimensions is correct. If the e_λ and $\bar{e}_\lambda$ are dual bases in $V_{[r]}$ and $\overline{V_{[r]}}$, then letting e^λ

correspond to $\bar{e}_\lambda$ obviously gives an isomorphism between $V^{[r]}$ and $\overline{V_{[r]}}$, which is defined by the above function H.

4. Alternating r-linear functions. The theorem below will be used in (II, 8) and in (V, 9). Let $L^r_{\text{alt}}(V)$ be the vector space of alternating r-linear functions F in V; that is, F is a function of r variables, linear in each, such that

$$(1) \qquad F(v_{\lambda_1}, \cdots, v_{\lambda_r}) = \epsilon_{\lambda_1 \cdots \lambda_r} F(v_1, \cdots, v_r).$$

For r-linear functions, (1) holds if and only if F vanishes whenever a vector is repeated; compare § 1.

THEOREM 4A. *For each such F, let ΨF be the corresponding linear function in $V_{[r]}$, defined by*

$$(2) \qquad (\Psi F)\Big(\sum_i a_i v_{i1} \vee \cdots \vee v_{ir}\Big) = \sum_i a_i F(v_{i1}, \cdots, v_{ir}).$$

Then Ψ is an isomorphism of $L^r_{\text{alt}}(V)$ onto $\overline{V_{[r]}}$. Hence (see Theorem 3A)

$$(3) \qquad V^{[r]} = \bar{V}_{[r]} \approx \overline{V_{[r]}} \approx L^r_{\text{alt}}(V).$$

First, given F, let ΦF be the linear function in L^r (see § 1) defined by (2), with Φ and $\sum a_i v_{i1} \cdots v_{ir}$ on the left; this is clearly well defined. We see at once that ΦF vanishes on all elements of L^r_0; hence it defines a linear function ΨF in $V_{[r]}$.

Using a base $e_1, \cdots, e_n$ in V, set

$$(4) \qquad f^\lambda(e_{\mu_1}, \cdots, e_{\mu_r}) = \delta^\lambda_\mu \qquad (\lambda_1 < \cdots < \lambda_r);$$

there is a unique element f^λ of $L^r_{\text{alt}}(V)$ which has these values. Given $F \in L^r_{\text{alt}}(V)$, its *components* are the numbers $F_\lambda = F(e_{\lambda_1}, \cdots, e_{\lambda_r})$ $(\lambda_1 < \cdots < \lambda_r)$. Now

$$(5) \qquad F = \sum_{(\lambda)} F_\lambda f^\lambda.$$

For take any μ, $\mu_1 < \cdots < \mu_r$. Then

$$\Big(\sum_{(\lambda)} F_\lambda f^\lambda\Big)(e_{\mu_1}, \cdots) = \sum_{(\lambda)} F_\lambda \, \delta^\lambda_\mu = F_\mu = F(e_{\mu_1}, \cdots, e_{\mu_r}),$$

and since both $\sum_{(\lambda)} F_\lambda f^\lambda$ and F are r-linear and alternating, (5) follows.

Hence the f^λ span $L^r_{\text{alt}}(V)$. To show that they are independent, suppose $G = \sum_{(\lambda)} a_\lambda f_\lambda = 0$. Then

$$0 = G(e_{\mu_1}, \cdots) = \sum_{(\lambda)} a_\lambda \, \delta^\lambda_\mu = a_\mu \qquad (\mu_1 < \cdots < \mu_r).$$

Therefore the f^λ form a base in $L^r_{\text{alt}}(V)$.

Now with the dual base element, $\bar{e}_\lambda$ (§ 3),

$$(\Psi f^\lambda)(e_\mu) = f^\lambda(e_{\mu_1}, \cdots) = \delta^\lambda_\mu = e^\lambda \cdot e_\mu = \bar{e}_\lambda(e_\mu);$$

hence Ψ carries the base elements f^λ of $L^r_{\text{alt}}(V)$ into the base elements $\bar{e}_\lambda$ of $\overline{V_{[r]}}$, proving (3).

5. Use of coordinate systems. Given the base $e_1, \cdots, e_n$ in V, and corresponding base elements e_λ, e^λ in $V_{[r]}$ and $V^{[r]}$ (§ 2), define the *components* α^λ of r-vectors α and ω_λ of r-covectors ω by

(1) $$\alpha = \sum_{(\lambda)} \alpha^\lambda e_\lambda, \qquad \omega = \sum_{(\lambda)} \omega_\lambda e^\lambda$$

(see Theorem 3A). Then α^λ and ω_λ may be found from (2.5) and (2.6):

(2) $$\alpha^\lambda = e^\lambda\cdot\alpha, \quad \omega_\lambda = \omega\cdot e_\lambda \qquad (\lambda_1 < \cdots < \lambda_r).$$

We may use (2) to define α^λ and ω_λ for all λ, not requiring $\lambda_1 < \cdots < \lambda_r$. Using (1) and (2.4) gives

(3) $$\omega\cdot\alpha = \sum_{(\lambda)} \omega_\lambda \alpha^\lambda.$$

Given vectors $v_1, \cdots, v_r$ and covectors $f^1, \cdots, f^r$, if $v_i = \sum_j v_i^j e_j$, etc., the components of the v_i and f^i form matrices, from which we form the determinants

(4) $$D^\lambda(v_1, \cdots, v_r) = \begin{vmatrix} v_1^{\lambda_1} & \cdots & v_1^{\lambda_r} \\ \cdot & \cdots & \cdot \\ v_r^{\lambda_1} & \cdots & v_r^{\lambda_r} \end{vmatrix}, \qquad D_\lambda(f^1, \cdots, f^r) = \begin{vmatrix} f^1_{\lambda_1} & \cdots & f^1_{\lambda_r} \\ \cdot & \cdots & \cdot \\ f^r_{\lambda_1} & \cdots & f^r_{\lambda_r} \end{vmatrix}.$$

Since $e^{\lambda_i}(v_j) = v_j^{\lambda_i}$, $f^i(e_{\lambda_j}) = f^i_{\lambda_j}$, (2) and (2.2) give

(5) $$(v_1\vee\cdots\vee v_r)^\lambda = D^\lambda(v_1, \cdots, v_r), \qquad (f^1\vee\cdots\vee f^r)_\lambda = D_\lambda(f^1, \cdots, f^r).$$

Hence, by (3),

(6) $$(f^1\vee\cdots\vee f^r)\cdot(v_1\vee\cdots\vee v_r) = \sum_{(\lambda)} D_\lambda(f^1, \cdots, f^r) D^\lambda(v_1, \cdots, v_r);$$

the equality of the right hand sides of (6) and of (2.2) is called the *general Lagrange identity* (recall that $f(v) = f\cdot v = \sum_i f^i v_i$).

We now consider a *change of coordinates*. Let $e_1', \cdots, e_n'$ be a new base, related to the old one by (App. I, 1.1); the dual bases are related by (App. I, 3.6). Define the determinants

(7) $$D^\lambda_\mu = \sum_\nu \delta^\lambda_\nu a^{\nu_1}_{\mu_1} \cdots a^{\nu_r}_{\mu_r} = \begin{vmatrix} a^{\lambda_1}_{\mu_1} & \cdots & a^{\lambda_1}_{\mu_r} \\ \cdot & \cdots & \cdot \\ a^{\lambda_r}_{\mu_1} & \cdots & a^{\lambda_r}_{\mu_r} \end{vmatrix},$$

with a similar definition for D'^λ_μ, in terms of the a'^i_j. The corresponding transformation of components of r-vectors and r-covectors is given by

(8) $$\alpha'^\lambda = \sum_{(\mu)} D^\lambda_\mu \alpha^\mu, \qquad \omega'_\lambda = \sum_{(\mu)} D'^\mu_\lambda \omega_\mu.$$

We prove the first relation. Let e'_ν denote $e'_{\nu_1}\vee\cdots\vee e'_{\nu_r}$. Given $\alpha = \sum_{(\mu)}\alpha^\mu e_\mu$, use (App. I, 1.1) to find

$$\alpha = \sum_{\nu}\sum_{(\mu)} a^{\nu_1}_{\mu_1}\cdots a^{\nu_r}_{\mu_r}\alpha^\mu e'_\nu.$$

In the first sum, collect together those terms in which the indices are a permutation of some set $\lambda_1 < \cdots < \lambda_r$; this sum then becomes a double sum $\sum_{(\lambda)}\sum_\nu^{[\lambda]}$, where $\sum_\nu^{[\lambda]}$ is the sum over all permutations ν of this λ. Recalling the definition of δ^λ_ν and using (1.13) gives

$$\alpha = \sum_{(\lambda)}\sum_{\nu}\sum_{(\mu)} \delta^\lambda_\nu a^{\nu_1}_{\mu_1}\cdots a^{\nu_r}_{\mu_r}\alpha^\mu e'_\lambda = \sum_{(\lambda)}\sum_{(\mu)} D^\lambda_\mu \alpha^\mu e'_\lambda,$$

from which the relation follows.

6. Exterior products. We define the *exterior product* of r-vectors and s-vectors by the formula

$$(1)\qquad (v_1\vee\cdots\vee v_r)\vee(w_1\vee\cdots\vee w_s) = v_1\vee\cdots\vee v_r\vee w_1\vee\cdots\vee w_s,$$

extending it to all r-vectors and s-vectors by linearity. This must be shown to be independent of the manner of writing elements of $V_{[r]}$ and $V_{[s]}$. The proof is like the corresponding proof in § 2, and will be omitted.

The associative law $\alpha\vee(\beta\vee\gamma) = (\alpha\vee\beta)\vee\gamma$ is obvious. Using (1.13), we find the commutative law

$$(2)\qquad \beta\vee\alpha = (-1)^{rs}\alpha\vee\beta \qquad (\alpha\in V_{[r]},\ \beta\in V_{[s]}).$$

As a consequence, $\alpha\vee\alpha = 0$ if r is odd. But this need not be the case if r is even. For instance, with the notations of § 2,

$$(e_{12}+e_{34})\vee(e_{12}+e_{34}) = 2e_{1234}.$$

Define the *exterior product* of multicovectors in the same way; the same laws hold.

As special cases, $v\vee w$ and $f\vee g$ have the same meaning as formerly; and $v_1\vee\cdots\vee v_r$ and $f^1\vee\cdots\vee f^r$ may now be considered as exterior products. If a is a number, define also

$$(3)\qquad a\vee\alpha = \alpha\vee a = a\alpha, \qquad a\vee\omega = \omega\vee a = a\omega.$$

We now study components, with a base $e_1, \cdots, e_n$ in V. Suppose λ and μ are distinct sets of distinct numbers (taken from 1 to n); let ν be the union of these sets, in some permutation. Then (1.13) gives at once

$$(4)\qquad e_\lambda\vee e_\mu = \delta^\nu_{\lambda\mu}e_\nu.$$

For instance, $e_{24}\vee e_{13} = -e_{1234}$. We prove

$$(5)\qquad (\alpha\vee\beta)^\nu = \sum_{(\lambda)(\mu)} \delta^\nu_{\lambda\mu}\alpha^\lambda\beta^\mu.$$

For if $\alpha \vee \beta = \gamma$, then

$$\alpha \vee \beta = \sum_{(\lambda)} \alpha^\lambda e_\lambda \vee \sum_{(\mu)} \beta^\mu e_\mu = \sum_{(\lambda)(\mu)} \alpha^\lambda \beta^\mu e_\lambda \vee e_\mu = \sum_{(\nu)} \gamma^\nu e_\nu;$$

collecting together those terms in $\sum_{(\lambda)(\mu)}$ in which λ and μ together form a permutation of a single ν, (5) follows from (4). Similarly, for multicovectors,

$$(6) \qquad (\omega \vee \xi)_\nu = \sum_{(\lambda)(\mu)} \delta^{\lambda\mu}_\nu \omega_\lambda \xi_\mu.$$

For instance, if ω and ξ are 2-covectors,

$$(\omega \vee \xi)_{1234} = \omega_{12}\xi_{34} - \omega_{13}\xi_{24} + \omega_{14}\xi_{23} + \omega_{23}\xi_{14} - \omega_{24}\xi_{13} + \omega_{34}\xi_{12}.$$

If ω is a covector and ξ an r-covector, we find

$$(7) \qquad (\omega \vee \xi)_{\lambda_1 \cdots \lambda_{r+1}} = \sum_{k,(\mu)} \delta^{k\mu_1\cdots\mu_r}_{\lambda_1\lambda_2\cdots\lambda_{r+1}} \omega_k \xi_{\mu_1\cdots\mu_r}$$

$$= \sum_{i=1}^{r+1} (-1)^{i-1} \omega_{\lambda_i} \xi_{\lambda_1 \cdots \hat{\lambda}_i \cdots \lambda_{r+1}}.$$

Consider the direct sum $\tilde{V}$ of the spaces $V_{[0]}$ (the reals), $V_{[1]}$ (i.e. V), $V_{[2]}, \cdots, V_{[n]}$; this is a vector space of dimension 2^n, with base elements e_λ together with the number 1. With the exterior products, $\tilde{V}$ is a ring, with the number 1 as unit; it is commonly called the *Grassmann algebra* of V. The elements $e_1, \cdots, e_n$, with the number 1, generate the algebra; that is, any element is a linear combination of products of these elements.

A more general system is formed by including also the spaces $V^{[r]}$ and the interior products of the next section.

7. Interior products. We define two products (which will be used rarely in later chapters):

$$(r\text{-covector}) \wedge (s\text{-vector}) = (s - r)\text{-vector} \quad \text{if} \quad r \leqq s,$$

$$(r\text{-covector}) \wedge (s\text{-vector}) = (r - s)\text{-covector} \quad \text{if} \quad r \geqq s;$$

for $r = s$, both reduce to the scalar product.

Letting superscripts denote degrees, the products are defined by the relations

$$(1) \qquad \xi^{s-r} \cdot (\omega^r \wedge \alpha^s) = (\xi^{s-r} \vee \omega^r) \cdot \alpha^s \quad \text{if} \quad r \leqq s,$$

$$(2) \qquad (\omega^r \wedge \alpha^s) \cdot \beta^{r-s} = \omega^r \cdot (\alpha^s \vee \beta^{r-s}) \quad \text{if} \quad r \geqq s.$$

The right hand side of (1), for fixed ω^r and α^s, is a linear function of ξ^{s-r}; by Theorem 3A and (App. I, Lemma 3a), there is therefore a unique $(s - r)$-vector β^{s-r} such that $(\xi^{s-r} \vee \omega^r) \cdot \alpha^s = \xi^{s-r} \cdot \beta^{s-r}$ for all ξ^{s-r}. Set

$\omega^r \wedge \alpha^s = \beta^{s-r}$; then (1) holds. In the same fashion, (2) defines $\omega^r \wedge \alpha^s$ for $r \geqq s$. The two products are clearly bilinear.

If $r = s$, setting $\xi^0 = 1$ in (1) gives

$$(3) \qquad \omega^r \wedge \alpha^r = 1 \cdot (\omega^r \wedge \alpha^r) = (1 \vee \omega^r) \cdot \alpha^r = \omega^r \cdot \alpha^r;$$

the same relation follows from (2).

Take $r = 0$ in (1). Then ω^0 is a number a, and

$$\xi^s \cdot (a \wedge \alpha^s) = (\xi^s \vee a) \cdot \alpha^s = (a\xi^s) \cdot \alpha^s = \xi^s \cdot (a\alpha^s);$$

with a similar use of (2), we have

$$(4) \qquad a \wedge \alpha = a\alpha, \qquad \omega \wedge a = a\omega.$$

Using (1) several times gives

$$\eta^{t-r-s} \cdot [\xi^r \wedge (\omega^s \wedge \alpha^t)] = (\eta \vee \xi) \cdot (\omega \wedge \alpha) = [(\eta \vee \xi) \vee \omega] \cdot \alpha$$
$$= [\eta \vee (\xi \vee \omega)] \cdot \alpha = \eta \cdot [(\xi \vee \omega) \wedge \alpha];$$

a similar relation follows from (2). Hence

$$(5) \qquad \xi^r \wedge (\omega^s \wedge \alpha^t) = (\xi^r \vee \omega^s) \wedge \alpha^t \qquad (r + s \leqq t),$$

$$(6) \qquad (\omega^r \wedge \alpha^s) \wedge \beta^t = \omega^r \wedge (\alpha^s \vee \beta^t) \qquad (s + t \leqq r).$$

We give relations for the components of $\omega \wedge \alpha$, if $\omega \in V^{[r]}$, $\alpha \in V_{[s]}$:

$$(7) \qquad (\omega \wedge \alpha)^\mu = \sum_{(\lambda)(\nu)} \delta_\lambda^{\mu\nu} \omega_\nu \alpha^\lambda = \sum_{(\nu)} \omega_\nu \alpha^{\mu\nu} \qquad (r \leqq s),$$

$$(8) \qquad (\omega \wedge \alpha)_\nu = \sum_{(\lambda)(\mu)} \delta_{\mu\nu}^\lambda \omega_\lambda \alpha^\mu = \sum_{(\mu)} \omega_{\mu\nu} \alpha^\mu \qquad (r \geqq s).$$

We shall prove the first part of (7). Take any $\xi \in V^{[s-r]}$. Using (5.3) and (6.6) gives

$$\xi \cdot (\omega \wedge \alpha) = (\xi \vee \omega) \cdot \alpha = \sum_{(\lambda)} (\xi \vee \omega)_\lambda \alpha^\lambda = \sum_{(\lambda)(\mu)(\nu)} \delta_\lambda^{\mu\nu} \xi_\mu \omega_\nu \alpha^\lambda.$$

Take a fixed μ, and set $\xi = e^\mu$. Then this relation, with (2.5), gives the first part of (7). The second part is immediate, and the proof of (8) is similar.

For an r-covector ω and an s-vector α,

$$(9) \quad \omega \wedge (v_1 \vee \cdots \vee v_s) = \sum_{(\lambda)(\mu)} \epsilon_{\lambda\mu} [\omega \cdot (v_{\mu_1} \vee \cdots \vee v_{\mu_r})] v_{\lambda_1} \vee \cdots \vee v_{\lambda_{s-r}} \qquad (r \leqq s),$$

$$(10) \quad (f^1 \vee \cdots \vee f^r) \wedge \alpha = \sum_{(\lambda)(\mu)} \epsilon_{\mu\lambda} [(f^{\mu_1} \vee \cdots \vee f^{\mu_s}) \cdot \alpha] f^{\lambda_1} \vee \cdots \vee f^{\lambda_{r-s}} \qquad (r \geqq s).$$

Note that for each λ, just one μ occurs. For instance, for $r = 2$,

$$\omega \wedge (v_1 \vee v_2 \vee v_3) = [\omega \cdot (v_2 \vee v_3)] v_1 - [\omega \cdot (v_1 \vee v_3)] v_2 + [\omega \cdot (v_1 \vee v_2)] v_3.$$

To prove (9), suppose $\omega = f^{s-r+1} \vee \cdots \vee f^s$, and choose any $f^1, \cdots, f^{s-r}$; set $\xi = f^1 \vee \cdots \vee f^{s-r}$. Take the scalar product of ξ with both sides of (9).

The left hand side equals $(f^1 \vee \cdots \vee f^s)\cdot(v_1 \vee \cdots \vee v_s)$, which is the determinant $D = |f^i(v_j)|$, and the right hand side is clearly a Laplace's expansion of D. Since the f^i were arbitrary, (9) follows. The proof of (10) is similar.

8. n-vectors in n-space. If V is of dimension n, then, by Theorem 3A, $V_{[n]}$ is of dimension 1. Hence if $\alpha_0 \neq 0$ is an n-vector, then any n-vector equals $a\alpha_0$ for some real number a. The sets of all n-vectors $a\alpha_0$ with $a > 0$ and with $a < 0$ we call the two *orientations* of V.

With a base in V, formulas (5.1) become

(1) $$\alpha = \alpha^{1\cdots n} e_{1\cdots n}, \qquad \omega = \omega_{1\cdots n} e^{1\cdots n};$$

by Theorem 3A, $e_{1\cdots n} \neq 0$, $e^{1\cdots n} \neq 0$. The transformation formulas (5.8) become

(2) $$\alpha'^{1\cdots n} = D\alpha^{1\cdots n}, \qquad \omega'_{1\cdots n} = D'\omega_{1\cdots n}, \qquad D = |a^i_j|, \quad D' = 1/D.$$

Suppose $(u_1, \cdots, u_n)$ and $(v_1, \cdots, v_n)$ are independent sets. Say $v_i = \sum_j v^j_i u_j$. Then working out $v_1 \vee \cdots \vee v_n$ and using (1.13) gives

(3) $$v_1 \vee \cdots \vee v_n = D u_1 \vee \cdots \vee u_n, \quad \text{if} \quad v_i = \textstyle\sum_j v^j_i u_j, \qquad D = |v^j_i|.$$

(This also follows directly from (2), if we let the v_i and the u_i be the old and new bases respectively, and use $\alpha = v_1 \vee \cdots \vee v_n$.) Hence the two sets determine the same or opposite orientation of V according as $D > 0$ or $D < 0$.

It is an elementary fact that if $D > 0$, then the u_i may be deformed into the v_i, keeping them independent. That is, there are continuous functions $u_1(t), \cdots, u_n(t)$, such that $u_i(0) = u_i$, $u_i(1) = v_i$, and the $u_i(t)$ are independent for each t. For instance, using a metric in V (App. I, 9), we may first deform both sets into orthonormal sets. Then, by rotations, we may carry u_1 into v_1, then u_2 into v_2, etc. Finally, when u_{n-1} is carried into v_{n-1}, we must have $u_n = \pm v_n$; since the determinant D in (3) remains positive, we have $u_n = v_n$.

9. Simple multivectors. An r-vector α is called *simple*, or *decomposable*, if it can be written in the form $\alpha = v_1 \vee \cdots \vee v_r$; similarly for r-covectors. For $r = 0$, 1, or n, any r-vector in n-space is clearly simple; from §§ 11 and 12 below, we see that this is true also for $r = n - 1$, and for r-covectors, for the same values of r. In $\mathfrak{A}^4$, $e_{12} + e_{34}$ for instance is not simple, as follows from (13.9), (13.13) and (12.7).

We shall give a geometric interpretation of simple r-vectors.

LEMMA 9a. $v_1 \vee \cdots \vee v_r = 0$ *if and only if the* v_i *are dependent.*

If the v_i are dependent, then some v_j is a linear combination of the others; say $v_r = a_1 v_1 + \cdots + a_{r-1} v_{r-1}$. Substituting into $v_1 \vee \cdots \vee v_r$ and using $v_i \vee v_i = 0$ gives $v_1 \vee \cdots \vee v_r = 0$. Now suppose the v_i are independent. Then they form part of a base $v_1, \cdots, v_n$. By Theorem 3A, $v_1 \vee \cdots \vee v_r$ is now a base element of $V_{[r]}$, and hence is $\neq 0$.

Given an r-vector α, let $V(\alpha)$ be the set of all vectors u such that $\alpha \vee u = 0$; this is a subspace of V.

LEMMA 9b. *If* $\alpha = v_1 \vee \cdots \vee v_r \neq 0$, *then* $V(\alpha)$ *is the subspace of* V *spanned by* $v_1, \cdots, v_r$.

By the last lemma, $v_1, \cdots, v_r$ are independent; also, $u \in V(\alpha)$ if and only if $v_1, \cdots, v_r, u$ are dependent, i.e. u is a linear combination of the v_i.

If there is no metric in V, we cannot define volumes; however, we may define the *ratio* $\rho(R, R')$ of the "volumes" of the bounded open sets R, R', as the number with the following property: Given the base $e_1, \cdots, e_n$, and $\epsilon > 0$, there is a $\delta > 0$ for which the following is true. Using the e_i as base vectors of a rectangular coordinate system, cut V into equal cubes of side h, where $h \leqq \delta$; if N and N' are the numbers of cubes in R and in R' respectively, then $|\rho(R, R') - N/N'| < \epsilon$. In particular, we can tell whether or not two open sets R, R' have "equal volume".

By an *oriented volume* $\tilde{R}$ in V, we mean the pair consisting of a class of open sets of equal volume in V (all sets of this volume being in the class), and an orientation of V. If $\tilde{R}$ and $\tilde{R}'$ are oriented volumes in V, let $\tilde{\rho}(\tilde{R}, \tilde{R}')$ denote the ratio of the corresponding volumes, with the plus or minus sign, according as the two orientations agree or disagree. See also (VII, 11.7).

If $v_1, \cdots, v_n$ is a base in V, let $\Lambda(v_1, \cdots, v_n)$ denote the oriented volume determined by the open set of all vectors $\sum a_i v_i$, with $0 < a_i < 1$, and the orientation of V given by $v_1 \vee \cdots \vee v_n$ (§ 8).

LEMMA 9c. *If* $u_1, \cdots, u_n$ *and* $v_1, \cdots, v_n$ *are bases in* V, *then*

$$v_1 \vee \cdots \vee v_n = \tilde{\rho}[\Lambda(v_1, \cdots, v_n), \Lambda(u_1, \cdots, u_n)] u_1 \vee \cdots \vee u_n. \tag{1}$$

Using some elementary geometry, we may prove this as follows. Some u_k is not in the plane of $v_2, \cdots, v_n$; there are numbers $c_2, \cdots, c_n$ such that

$$v_1' = v_1 + c_2 v_2 + \cdots + c_n v_n = a_1 u_k, \qquad a_1 \neq 0.$$

If we replace v_1 by v_1', neither side of (1) is changed. Continuing in this manner, we may suppose $v_i = a_i u_{\lambda_i}$ for each i. A permutation of the v_i leaves (1) unchanged or changes the sign of both sides; hence we may suppose $\lambda_i = i$. Replacing v_i by $b_i v_i$ multiplies both sides of (1) by b_i; hence we may suppose $v_i = u_i$. But (1) is trivial in this case.

By an *oriented r-volume* in V, we mean the object consisting of an r-dimensional subspace W of V, and an oriented volume in W.

THEOREM 9A. *There is a one-one correspondence between simple r-vectors* $\neq 0$ *and oriented r-volumes in* V, *defined as follows. Given* $\alpha = v_1 \vee \cdots \vee v_r \neq 0$, *the ordered set* $(v_1, \cdots, v_r)$ *determines an r-dimensional subspace* W *of* V, *an orientation of* W, *and a parallelepiped in* W, *and hence an oriented volume in* W, *and an oriented r-volume in* V.

The oriented r-volume is determined by $v_1, \cdots, v_r$. To show that it is uniquely determined by α, suppose $\alpha = u_1 \vee \cdots \vee u_r$ also. By Lemma 9b, W is spanned by the u_i also. By (1) (used in W), $\tilde{\rho} = 1$; hence the same oriented volume in W is determined, and hence the same oriented r-volume in V.

Given any oriented r-volume in V, it is obviously determined by some $\alpha = v_1 \vee \cdots \vee v_r$. Suppose it is also determined by $\beta = u_1 \vee \cdots \vee u_r$. Then both the u_i and the v_i form bases in the subspace W, and determine oriented volumes in W, which are the same; that is, $\tilde{\rho} = 1$ in (1), and hence $\beta = \alpha$.

10. Linear mappings of vector spaces. Let ϕ be a linear mapping of V into V'. There is a unique corresponding linear mapping of $V_{[r]}$ into $V'_{[r]}$, such that

$$\phi(v_1 \vee \cdots \vee v_r) = \phi v_1 \vee \cdots \vee \phi v_r. \tag{1}$$

The proof that $\phi\alpha$ is independent of the expression of α as a sum of simple r-vectors is like a corresponding proof in § 2.

An immediate consequence of (1) is

$$\phi(\alpha \vee \beta) = \phi\alpha \vee \phi\beta. \tag{2}$$

If ϕ^* is the dual mapping of $\overline{V}'$ into $\overline{V}$ (App. I, 3), it defines a linear mapping of $V'^{[r]}$ into $V^{[r]}$, by the formula

$$\phi^*(f^1 \vee \cdots \vee f^r) = \phi^* f^1 \vee \cdots \vee \phi^* f^r. \tag{3}$$

Hence

$$\phi^*(\xi \vee \omega) = \phi^*\xi \vee \phi^*\omega. \tag{4}$$

The pair ϕ, ϕ^* preserves scalar products:

$$\phi^*\omega \cdot \alpha = \omega \cdot \phi\alpha \qquad (\omega \in V'^{[r]}, \quad \alpha \in V_{[r]}); \tag{5}$$

hence, considering $V^{[r]}$ as the conjugate space of $V_{[r]}$ (Theorem 3A), ϕ^* is the linear mapping dual to ϕ. The proof of (5) is immediate from (3), (2.2) and (App. I, 3.8):

$$\begin{aligned} \phi^*(f^1 \vee \cdots \vee f^r) \cdot (v_1 \vee \cdots \vee v_r) &= (\phi^* f^1 \vee \cdots \vee \phi^* f^r) \cdot (v_1 \vee \cdots \vee v_r) \\ &= \sum_\lambda \epsilon^\lambda \phi^* f^1(v_{\lambda_1}) \cdots \phi^* f^r(v_{\lambda_r}) = \sum_\lambda \epsilon^\lambda f^1(\phi v_{\lambda_1}) \cdots f^r(\phi v_{\lambda_r}) \\ &= (f^1 \vee \cdots \vee f^r) \cdot (\phi v_1 \vee \cdots \vee \phi v_r) = (f^1 \vee \cdots \vee f^r) \cdot \phi(v_1 \vee \cdots \vee v_r). \end{aligned}$$

There are two formulas relating interior products:

$$\omega \wedge \phi\alpha = \phi(\phi^*\omega \wedge \alpha) \qquad (\omega \in V'^{[r]}, \quad \alpha \in V_{[s]}, \quad r \leqq s), \tag{6}$$

$$\phi^*(\omega \wedge \phi\alpha) = \phi^*\omega \wedge \alpha \qquad (\omega \in V'^{[r]}, \quad \alpha \in V_{[s]}, \quad r \geqq s). \tag{7}$$

For instance, (6) follows from the relations

$$\begin{aligned} \xi \cdot \phi(\phi^*\omega \wedge \alpha) &= \phi^*\xi \cdot (\phi^*\omega \wedge \alpha) = (\phi^*\xi \vee \phi^*\omega) \cdot \alpha \\ &= \phi^*(\xi \vee \omega) \cdot \alpha = (\xi \vee \omega) \cdot \phi\alpha = \xi \cdot (\omega \wedge \phi\alpha). \end{aligned}$$

Let ϕ and ψ be linear mappings of V into V' and of V' into V'' respectively. Then $\psi\phi$ and $\phi^*\psi^*$ map V into V'' and $\overline{V''}$ into $\bar{V}$ respectively; we let these define mappings of $V_{[r]}$ into $V''_{[r]}$ and of $V''^{[r]}$ into $V^{[r]}$ by the same relations:

$$(\psi\phi)\alpha = \psi(\phi\alpha), \qquad (\phi^*\psi^*)\omega = \phi^*(\psi^*\omega). \tag{8}$$

Working out $(\phi^*\psi^*)\omega\cdot\alpha$ as in the proof of (App. I, 3.10) gives

$$(\psi\phi)^* = \phi^*\psi^*. \tag{9}$$

Let ϕ be a linear mapping of V into itself. Since all n-vectors α are multiples of a single one α_0 (§ 3), there is a unique number D_ϕ such that $\phi\alpha = D_\phi\alpha$; D_ϕ is called the *determinant* of ϕ. If $e_1, \cdots, e_n$ is a base in V, and $\phi e_i = \sum_j \phi_i^j e_j$, we find at once, using $\alpha = e_{1\cdots n}$, that $D_\phi = |\phi_i^j|$.

11. Duality. Suppose V is of dimension n. Since $V_{[r]}$, $V^{[r]}$, $V_{[n-r]}$ and $V^{[n-r]}$ are of the same dimension, they are isomorphic. We give here an isomorphism between $V_{[r]}$ and $V^{[n-r]}$, determined solely by the choice of an n-vector or n-covector. (For $V_{[r]}$ and $V^{[r]}$, see the next section.)

Choose an n-vector α_0 and an n-covector ω_0 such that $\omega_0\cdot\alpha_0 = 1$. Set

$$\mathscr{D}\alpha = \omega_0\wedge\alpha, \qquad \mathscr{D}'\omega = \omega\wedge\alpha_0. \tag{1}$$

We shall prove

$$\mathscr{D}'\mathscr{D}\alpha = \alpha, \qquad \mathscr{D}\mathscr{D}'\omega = \omega, \tag{2}$$

which shows that each of $\mathscr{D}$, $\mathscr{D}'$ is an isomorphism onto.

We may choose a base $e_1, \cdots, e_n$ so that $\alpha_0 = e_{1\cdots n}$; then $\omega_0 = e^{1\cdots n}$. Using (7.10), (7.9), (2.5), and (2.6), we find

$$\mathscr{D}\alpha = \sum_{(\lambda)(\mu)} \epsilon_{\mu\lambda}\alpha^\mu e^\lambda, \tag{3}$$

$$\mathscr{D}'\omega = \sum_{(\lambda)(\mu)} \epsilon_{\lambda\mu}\omega_\mu e_\lambda; \tag{4}$$

in each sum, for a given λ, just one μ occurs. In particular (always if $\alpha_0 = e_{1\cdots n}$),

$$\mathscr{D}e_\lambda = \sum_{(\mu)} \epsilon_{\lambda\mu}e^\mu, \qquad \mathscr{D}'e^\lambda = \sum_{(\mu)} \epsilon_{\mu\lambda}e_\mu, \tag{5}$$

with only one non-zero term in each sum. For instance, with $n = 5$, $\mathscr{D}e_{12} = e^{345}$, $\mathscr{D}e_{13} = -e^{245}$. Hence

$$\mathscr{D}e_\lambda = \epsilon_{\lambda\mu}e^\mu, \qquad \mathscr{D}'e^\mu = \epsilon_{\lambda\mu}e_\lambda, \quad \text{if} \quad \epsilon_{\lambda\mu} \neq 0; \tag{6}$$

hence also $\mathscr{D}'\mathscr{D}e_\lambda = e_\lambda$, $\mathscr{D}\mathscr{D}'e^\mu = e^\mu$, from which (2) follows.

Applying (7.5) and (7.6) gives (if the degrees are correct)

$$\omega\wedge\mathscr{D}'\xi = \omega\wedge(\xi\wedge\alpha_0) = (\omega\vee\xi)\wedge\alpha_0 = \mathscr{D}'(\omega\vee\xi),$$

$$\mathscr{D}\beta\wedge\alpha = (\omega_0\wedge\beta)\wedge\alpha = \omega_0\wedge(\beta\vee\alpha) = \mathscr{D}(\beta\vee\alpha);$$

setting $\alpha = \mathscr{D}'\xi$ in the first and $\omega = \mathscr{D}\beta$ in the second gives

(7) $$\omega\wedge\alpha = \mathscr{D}'(\omega\vee\mathscr{D}\alpha) \qquad (\deg(\omega) \leqq \deg(\alpha)),$$

(8) $$\omega\wedge\alpha = \mathscr{D}(\mathscr{D}'\omega\vee\alpha) \qquad (\deg(\omega) \geqq \deg(\alpha)),$$

thus expressing the interior products in terms of the exterior products.

Applying (7.3), (7), (1), and (7.3) again gives

(9) $$(\omega\vee\mathscr{D}\alpha)\cdot\alpha_0 = \omega\cdot\alpha \qquad (\deg(\omega) = \deg(\alpha));$$

a similar formula follows from (8).

Lemma 11a. *A multivector or multicovector is simple if and only if its dual is.*

Suppose for instance $\alpha = v_1\vee\cdots\vee v_r \neq 0$ is simple, and $r < n$. Choose $v_{r+1}, \cdots, v_n$ so that $v_1, \cdots, v_n$ form a base in V (see Lemma 9a), and so that $\alpha_0 = v_1\vee\cdots\vee v_n$; let $f^1, \cdots, f^n$ be the dual base. Then, by (6), $\mathscr{D}\alpha = f^{r+1}\vee\cdots\vee f^n$, which is simple.

12. Euclidean vector spaces. Let V be Euclidean, and let Φ be the isomorphism of V onto $\bar{V}$, given by $(\Phi u)(v) = u\cdot v$ (App. I, Lemma 9a). Setting $\Phi u\cdot\Phi v = u\cdot v$ defines a scalar product in $\bar{V}$; now $\bar{V}$ is Euclidean. If $e_1, \cdots, e_n$ is an orthonormal base in V, then $\Phi e_1, \cdots, \Phi e_n$ is the dual base in $\bar{V}$; thus $\Phi e_i = e^i$.

Define scalar products in $V_{[r]}$ by the relation

(1) $$(u_1\vee\cdots\vee u_r)\cdot(v_1\vee\cdots\vee v_r) = (\Phi u_1\vee\cdots\vee\Phi u_r)\cdot(v_1\vee\cdots\vee v_r);$$

we will prove the required properties (App. I, 9) below.

By (2.2),

(2) $$(u_1\vee\cdots\vee u_r)\cdot(v_1\vee\cdots\vee v_r) = \begin{vmatrix} u_1\cdot v_1 & \cdots & u_1\cdot v_r \\ \cdot & \cdot & \cdot \\ u_r\cdot v_1 & \cdots & u_r\cdot v_r \end{vmatrix}.$$

Using

(3) $$\Phi(v_1\vee\cdots\vee v_r) = \Phi v_1\vee\cdots\vee\Phi v_r$$

and linearity, Φ becomes a linear mapping of $V_{[r]}$ into $V^{[r]}$. Setting $\Phi\alpha\cdot\Phi\beta = \alpha\cdot\beta$, we now have all but the last relation in

(4) $$\Phi e_\lambda = e^\lambda, \qquad \Phi\alpha\cdot\Phi\beta = \Phi\alpha\cdot\beta = \alpha\cdot\beta, \qquad |\Phi\alpha| = |\alpha|.$$

Because of the first relation, Φ is an isomorphism of $V_{[r]}$ onto $V^{[r]}$.

With an orthonormal base,

$$\Phi \sum_{(\lambda)} \alpha^\lambda e_\lambda = \sum_{(\lambda)} \alpha^\lambda \Phi e_\lambda = \sum_{(\lambda)} \alpha^\lambda e^\lambda; \tag{5}$$

thus components are preserved by $\Phi : (\Phi\alpha)_\lambda = \alpha^\lambda$. In terms of components,

$$\alpha\cdot\beta = \sum_{(\lambda)} \alpha^\lambda \beta^\lambda = \beta\cdot\alpha, \qquad \omega\cdot\xi = \sum_{(\lambda)} \omega_\lambda \xi_\lambda = \xi\cdot\omega. \tag{6}$$

For instance, $\alpha\cdot\beta = \Phi\alpha\cdot\beta = \sum_{(\lambda)}(\Phi\alpha)_\lambda\beta^\lambda = \sum_{(\lambda)}\alpha^\lambda\beta^\lambda$. In particular,

$$|\alpha| = (\alpha\cdot\alpha)^{1/2} = \Big[\sum_{(\lambda)} (\alpha^\lambda)^2\Big]^{1/2}, \quad |\omega| = (\omega\cdot\omega)^{1/2} = \Big[\sum_{(\lambda)} (\omega_\lambda)^2\Big]^{1/2}, \tag{7}$$

which proves that $|\alpha| > 0$ if $\alpha \neq 0$, $|\omega| > 0$ if $\omega \neq 0$; hence these are norms, and the last part of (4) holds.

THEOREM 12A. *$|\alpha|$ and $|\omega|$ are conjugate norms in $V_{[r]}$ and $V^{[r]}$ respectively.*

This follows from (4) and (App. I, Lemma 9a).

Note that $\Phi\alpha$ is simple if and only if α is.

The case $r = 2$ gives

$$0 \leqq (u\vee v)\cdot(u\vee v) = \begin{vmatrix} u\cdot u & u\cdot v \\ v\cdot u & v\cdot v \end{vmatrix} = |u|^2\,|v|^2 - (u\cdot v)^2;$$

also (Lemma 9a) $u\vee v = 0$ if and only if u, v are dependent. These facts give the *Schwarz inequality:*

$$|u\cdot v| \leqq |u|\,|v|, \tag{8}$$

$$|u\cdot v| = |u|\,|v| \quad \text{if and only if} \quad u \text{ and } v \text{ are dependent.} \tag{9}$$

Since this is true in any Euclidean vector space, it is true in $V_{[r]}$ and $V^{[r]}$; hence

$$|\alpha\cdot\beta| \leqq |\alpha|\,|\beta|, \qquad |\omega\cdot\xi| \leqq |\omega|\,|\xi|, \tag{10}$$

equality holding if and only if α and β, respectively ω and ξ, are dependent. Hence also (App. I, 9)

$$|\alpha+\beta| \leqq |\alpha| + |\beta|, \qquad |\omega+\xi| \leqq |\omega| + |\xi|. \tag{11}$$

If α has just one component, there is a simple expression for $\alpha\vee\beta$:

$$\text{If } \alpha = \alpha^{1\cdots r}e_{1\cdots r}, \quad \text{then} \quad \alpha\vee\beta = \alpha^{1\cdots r} \sum_{(\lambda), r<\lambda_1} \beta^\lambda e_{1\cdots r\lambda}. \tag{12}$$

For the right hand side of (6.5), using $\nu_1 < \cdots < \nu_r$, vanishes unless $(\nu_1, \cdots, \nu_r) = (1, \cdots, r)$; if this is so, there is one term, with $\lambda = (1, \cdots, r)$ and $\mu = (\nu_{r+1}, \cdots)$; this term is $\alpha^{1\cdots r}\beta^{\nu_{r+1}\cdots}$, and (12) follows.

We now prove

$$|\alpha\vee\beta| \leqq |\alpha|\,|\beta| \quad \text{if} \quad \alpha \text{ or } \beta \text{ is simple.} \tag{13}$$

Say $\alpha = v_1 \vee \cdots \vee v_r$. Choose an orthonormal coordinate system so that $v_1, \cdots, v_r$ are in the plane of $e_1, \cdots, e_r$. Then $\alpha = \alpha^{1\cdots r} e_{1\cdots r}$, and (12) gives

$$| \alpha \vee \beta |^2 = \sum_{(\nu)} [(\alpha \vee \beta)^\nu]^2 = (\alpha^{1\cdots r})^2 \sum_{(\lambda), r < \lambda_1} (\beta^\lambda)^2 \leqq | \alpha |^2 \, | \beta |^2.$$

REMARK. (13) fails if the assumption is omitted, as has been shown in an example by J. H. Wolfe, with $n = 6$: take $\alpha = \beta = \sum_{i<j} a_{ij} e_{ij}$, where all a_{ij} are 1 except that $a_{13} = a_{46} = -1$; then $| \alpha |^2 = 15$, $| \alpha \vee \alpha |^2 = 252$. He has shown that (13) holds if the factor $\binom{r+s}{r}$ is inserted, where $\deg(\alpha) = r$, $\deg(\beta) = s$.

As a special case of (13), and of a corresponding inequality for multicovectors, we have

$$(14) \qquad | u_1 \vee \cdots \vee u_r \vee v_1 \vee \cdots \vee v_s | \leqq | u_1 \vee \cdots \vee u_r | \, | v_1 \vee \cdots \vee v_s |,$$

$$(15) \qquad | f^1 \vee \cdots \vee f^r \vee g^1 \vee \cdots \vee g^s | \leqq | f^1 \vee \cdots \vee f^r | \, | g^1 \vee \cdots \vee g^s |,$$

and hence

$$(16) \quad | v_1 \vee \cdots \vee v_r | \leqq | v_1 | \cdots | v_r |, \qquad | f^1 \vee \cdots \vee f^r | \leqq | f^1 | \cdots | f^r |.$$

Since

$$v_1 \vee \cdots \vee v_r - u_1 \vee \cdots \vee u_r = (v_1 - u_1) \vee u_2 \vee \cdots \vee u_r + v_1 \vee (v_2 - u_2) \vee u_3 \vee \cdots \vee u_r + \cdots + v_1 \vee \cdots \vee v_{r-1} \vee (v_r - u_r),$$

we find

$$(17) \qquad | v_1 \vee \cdots \vee v_r - u_1 \vee \cdots \vee u_r | \leqq r M^{r-1} \epsilon \quad \text{if } | v_i - u_i | \leqq \epsilon, \quad | u_i | \leqq M, \quad | v_i | \leqq M.$$

A bound for $| \alpha |$ in terms of a bound for its components follows from (7):

$$(18) \qquad | \alpha | \leqq \binom{n}{r}^{1/2} M \quad \text{if} \quad | \alpha^\lambda | \leqq M, \quad \text{all } \lambda \qquad (\deg(\alpha) = r).$$

To prove a converse relation, choose an orthonormal coordinate system, and set $\beta^\lambda = \alpha^\lambda / | \alpha^\lambda |$ if $\alpha^\lambda \neq 0$, and $\beta^\lambda = 0$ otherwise. Then $\alpha^\lambda \beta^\lambda = | \alpha^\lambda |$, and (10) gives

$$\sum_{(\lambda)} | \alpha^\lambda | = \sum_{(\lambda)} \alpha^\lambda \beta^\lambda = \alpha \cdot \beta \leqq | \alpha | \, | \beta |.$$

Since $| \beta | = \left[\sum_{(\lambda)} (\beta^\lambda)^2 \right]^{1/2} \leqq \binom{n}{r}^{1/2}$, this gives

$$(19) \qquad \sum_{(\lambda)} | \alpha^\lambda | \leqq \binom{n}{r}^{1/2} | \alpha | \qquad (\deg(\alpha) = r).$$

We now prove

$$(20) \quad | v_1 \vee v_2 \vee \cdots \vee v_r | = | v_1 | \, | v_2 \vee \cdots \vee v_r | \quad \text{if} \quad v_1 \cdot v_i = 0 \quad \text{for} \quad i > 1.$$

For if we use (2), the top row and left hand column of the determinant consists of zeros, except for the term $v_1 \cdot v_1 = |v_1|^2$; the rest of the determinant gives $|v_2 \vee \cdots \vee v_r|^2$.

If $v_1, \cdots, v_r$ are independent, they determine a subspace W of V, and r-volumes are defined in W. Now

$$(21) \qquad |v_1 \vee \cdots \vee v_r| = r\text{-volume of parallelepiped of } v_1, \cdots, v_r.$$

For this holds obviously if the v_i are orthogonal (use (7), with the first r axes along the v_i); in the general case, we may choose orthogonal vectors $u_1, \cdots, u_r$ so that $v_1 \vee \cdots \vee v_r = u_1 \vee \cdots \vee u_r$, and apply (9.1). (A proof of (21) by induction, using (20), is easy to give; compare the proof of (9.1)).

Let W be any oriented r-dimensional subspace of V. Associating W with a unit volume in W gives an oriented r-volume, determining uniquely an r-vector $\alpha(W)$ (Theorem 9A), which we call the *unit r-vector* or the *r-direction* of W. If $e_1, \cdots, e_r$ is an orthonormal set in W, determining its positive orientation, then $\alpha(W) = e_{1 \cdots r}$. In particular, if V is oriented, its n-direction $\alpha(V) = e_{1 \cdots n}$ is defined. Its *unit n-covector* is $e^{1 \cdots n}$.

If V is oriented, we may use $\alpha(V)$ to define the duality operators of § 11. In particular, for $n = 3$, given the vectors u, v, $\mathscr{D}(u \vee v)$ is a covector, and equals Φw for a unique vector w; w is the *vector product* of u and v in ordinary vector analysis. Note that if α is an r-direction, then there is a dual $(n - r)$-direction α', containing those vectors orthogonal to the vectors of α. The set of r-directions in E^n is called a *Grassmann manifold* (see (IV, 9)), and the correspondence $\alpha \to \alpha'$ is a well known duality in projective geometry.

Let V be oriented, with corresponding unit n-covector $\omega_0 = e^{1 \cdots n}$. Given any oriented subspace W, define β by

$$(22) \qquad \omega_0 \wedge \alpha(W) = \Phi \beta;$$

then it is easily seen (by a suitable choice of a base) that $\beta = \alpha(W')$, W' being the oriented orthogonal complement of W.

Let ϕ be a linear mapping of V into V' (both Euclidean). We may define the *magnitude* of ϕ by

$$(23) \qquad |\phi| = \sup \{ |\phi(v)| : \ |v| = 1 \}.$$

Then $|\phi|$ is the smallest number with the property

$$(24) \qquad |\phi v| \leq |\phi| \, |v|.$$

Given ϕ, choose coordinate systems in V and in V' as follows. Choose e_1 so that $|e_1| = 1$ and $|\phi e_1|$ is a maximum. Next choose e_2 so that $|e_2| = 1$, $e_1 \cdot e_2 = 0$, and $|\phi e_2|$ is a maximum, etc. Set $\phi e_i = v'_i$. If $i < j$, a simple calculation (first putting ϕ inside, then expanding, then differentiating, then setting $\theta = 0$) gives

$$\frac{d}{d\theta} |\phi(\cos \theta e_i + \sin \theta e_j)|^2_{\theta = 0} = 2 v'_i \cdot v'_j;$$

because of the maximum property of e_i, $v_i'\cdot v_j' = 0$. Hence we may choose an orthonormal base $e_1', \cdots, e_m'$ in V' so that, setting $n' = \inf\{n, m\}$,

$$\phi e_i = c_i e_i' \qquad (i = 1, \cdots, n'), \qquad \phi e_i = 0 \quad (i > n'), \tag{25}$$

$$|\phi| = c_1 \geqq c_2 \geqq \cdots \geqq c_{n'} \geqq 0. \tag{26}$$

Given any r-vector α of V, using (25) gives the components in V' of $\phi\alpha$:

$$\phi\alpha = \sum_{(\lambda)} \alpha^\lambda \phi e_\lambda = \sum_{(\lambda), \lambda_r \leqq n'} (\alpha^\lambda c_{\lambda_1} \cdots c_{\lambda_r}) e_\lambda'. \tag{27}$$

Hence, using (26),

$$|\phi\alpha|^2 = \sum_{(\lambda), \lambda_r \leqq n'} (\alpha^\lambda c_{\lambda_1} \cdots c_{\lambda_r})^2 \leqq |\phi|^{2r} \sum_{(\lambda)} (\alpha^\lambda)^2,$$

and thus

$$|\phi\alpha| \leqq c_1 \cdots c_r |\alpha| \leqq |\phi|^r |\alpha| \qquad (\deg(\alpha) = r). \tag{28}$$

Take any r-covector ω in V'. Then, for any r-vector α of V,

$$|\phi^*\omega\cdot\alpha| = |\omega\cdot\phi\alpha| \leqq |\omega|\,|\phi\alpha| \leqq |\omega|\,|\phi|^r |\alpha|;$$

hence, by (App. I, 8.7),

$$|\phi^*\omega| \leqq |\phi|^r |\omega| \qquad (\deg(\omega) = r). \tag{29}$$

13. Mass and comass. Let V be Euclidean; then we have norms $|\alpha|$ and $|\omega|$ in $V_{[r]}$ and in $V^{[r]}$ respectively. In the general theory of integration (starting with Chapter V), we shall need new norms, defined as follows.

The *mass* of an r-vector α is

$$|\alpha|_0 = \inf\left\{\sum_i |\alpha_i| : \alpha = \sum \alpha_i, \text{ the } \alpha_i \text{ simple}\right\}. \tag{1}$$

We do not know if there exist simple r-vectors α_i with $\alpha = \sum \alpha_i$, $|\alpha|_0 = \sum |\alpha_i|$; we know no bound for the number of terms needed in (1).

The *comass* of an r-covector ω is

$$|\omega|_0 = \sup\{|\omega\cdot\alpha| : \alpha \text{ simple}, |\alpha| = 1\}. \tag{2}$$

These are related to the earlier norms as follows:

$$|\alpha|_0 \geqq |\alpha|, \qquad |\omega|_0 \leqq |\omega|. \tag{3}$$

See also (9), (10) below. The first is immediate from (1); the second holds, since $|\omega|$ is conjugate to $|\alpha|$ (Theorem 12A), and hence is given by (2), omitting the restriction that α is simple (App. I, 8.7).

For any ω, and for any α, expressed as $\sum\alpha_i$ with the α_i simple, (2) gives

$$|\omega\cdot\alpha| \leqq \sum |\omega\cdot\alpha_i| \leqq |\omega|_0 \sum |\alpha_i|;$$

hence, using (1),

$$|\omega\cdot\alpha| \leqq |\omega|_0 |\alpha|_0. \tag{4}$$

THEOREM 13A. *Mass and comass are conjugate norms in* $V_{[r]}$ *and in* $V^{[r]}$ *respectively; hence*

(5) $$|\omega|_0 = \sup\{|\omega\cdot\alpha|: |\alpha|_0 = 1\},$$

(6) $$|\alpha|_0 = \sup\{|\omega\cdot\alpha|: |\omega|_0 = 1\}.$$

Clearly $|a\alpha|_0 = |a||\alpha|_0$, $|\alpha+\beta|_0 \leqq |\alpha|_0 + |\beta|_0$. If $\alpha \neq 0$, then $|\alpha|_0 \geqq |\alpha| > 0$. Hence mass is a norm. If $\omega \neq 0$, then in an orthonormal coordinate system, $\omega_\lambda = \omega\cdot e_\lambda \neq 0$ for some λ; since e_λ is simple, $|\omega|_0 \geqq |\omega_\lambda| > 0$. Thus comass is a norm. To prove the theorem, it is sufficient to prove (5), because of (App. I, Lemma 8e). Let w denote the right hand side. Because of (4), $w \leqq |\omega|_0$. Since (clearly) $|\alpha|_0 = |\alpha|$ if α is simple, the set of α appearing in (2) is a subset of those appearing in (5); therefore $|\omega|_0 \leqq w$. Hence (5) holds.

If α is simple and $|\alpha| = 1$, we may choose an orthonormal base such that $\alpha = e_{1\cdots r}$. Then $\omega\cdot\alpha = \omega_{1\cdots r}$; thus (2) shows that

(7) $$|\omega|_0 = \sup\{\omega_{1\cdots r}: \text{orthonormal bases}\}.$$

Since the set of all orthonormal bases forms a compact space in an obvious fashion, and $\omega_{1\cdots r}$ (with ω fixed) is continuous in this space, the bound is attained. Hence also

LEMMA 13a. *For each r-covector* ω *there is an r-vector* β *such that*

(8) $$\omega\cdot\beta = |\omega|_0, \qquad \beta \text{ simple}, \qquad |\beta| = 1.$$

(This follows also from (App. I, Lemma 8b).)

We now give further relations between the old and the new norms:

(9) $$|\alpha|_0 = |\alpha| \text{ if } \alpha \text{ is simple}, \ |\alpha|_0 > |\alpha| \text{ otherwise};$$

(10) $$|\omega|_0 = |\omega| \text{ if } \omega \text{ is simple}, \ |\omega|_0 < |\omega| \text{ otherwise}.$$

The first part of (9) is clear. To prove the first part of (10), say $\omega = \Phi\alpha$ (§ 12). Then α is simple, and using (12.4),

$$|\omega|^2 = \omega\cdot\alpha \leqq |\omega|_0|\alpha|_0 = |\omega|_0|\alpha| = |\omega|_0|\omega|,$$

which proves $|\omega|_0 \geqq |\omega|$. Using (3) gives the result.

Suppose now that $\omega = \Phi\alpha$ is not simple; then α is not simple. Choose β by Lemma 13a; then neither α nor β is a multiple of the other, and the remark following (12.10) gives

$$|\omega|_0 = \omega\cdot\beta = \alpha\cdot\beta < |\alpha||\beta| = |\alpha| = |\omega|.$$

To prove the last part of (9), suppose α is not simple; we may suppose $|\alpha| = 1$. Set $\omega = \Phi\alpha$, and choose β as before. Again we have $|\omega|_0 < |\alpha||\beta| = 1$. Set $\xi = \omega/|\omega|_0$. Then $|\xi|_0 = 1$, and

$$|\xi\cdot\alpha| = |\omega\cdot\alpha|/|\omega|_0 > |\omega\cdot\alpha| = |\alpha|^2 = 1,$$

and (6) gives $|\alpha|_0 \geqq |\xi\cdot\alpha| > 1$, as required.

Let ϕ be a linear mapping of V into V'. Then, corresponding to (12.28) and (12.29),

$$(11) \qquad |\phi\alpha|_0 \leqq |\phi|^r|\alpha|_0, \qquad |\phi^*\omega|_0 \leqq |\phi|^r|\omega|_0.$$

To prove the first relation, given $\epsilon > 0$, write

$$\alpha = \sum \alpha_i, \qquad \text{the } \alpha_i \text{ simple}, \qquad \sum |\alpha_i| \leqq |\alpha|_0 + \epsilon.$$

Then

$$\phi\alpha = \sum \phi\alpha_i, \qquad \sum |\phi\alpha_i| \leqq |\phi|^r \sum |\alpha_i| \leqq |\phi|^r(|\alpha|_0 + \epsilon),$$

and since the $\phi\alpha_i$ are simple, the relation follows. To prove the second, take any simple α with $|\alpha| = 1$. Then using the relation just proved, (4), and (9),

$$|\phi^*\omega\cdot\alpha| = |\omega\cdot\phi\alpha| \leqq |\omega|_0 |\phi\alpha|_0 \leqq |\phi|^r|\omega|_0,$$

and (2) gives the relation.

Examples. Using an orthonormal base, we prove

$$(12) \qquad |ae^{12} + be^{34}|_0 = \sup\{|a|, |b|\}.$$

Supposing $\xi = e^{12} + ce^{34}$, $|c| \leqq 1$, we shall prove $|\xi|_0 \leqq 1$; then, using (7), (12) will follow. To find $|\xi|_0$ from (2), we need clearly use only those simple α lying in the space of $e_1, \cdots, e_4$. Now $\alpha = v_1 \vee v_2$, with v_1 and v_2 being orthogonal unit vectors in this space. Since rotations in the spaces of e_1, e_2 and of e_3, e_4 do not affect the form of ξ, we may suppose $v_1 = a_1e_1 + a_3e_3$. If $v_2 = \sum b_ie_i$, then (5.5) gives

$$\alpha^{12} = (v_1 \vee v_2)^{12} = a_1b_2, \qquad \alpha^{34} = a_3b_4.$$

Hence

$$\xi\cdot\alpha = \sum_{(\lambda)} \xi_\lambda\alpha^\lambda = \alpha^{12} + c\alpha^{34} = a_1b_2 + ca_3b_4.$$

Set

$$u_1 = a_1e_1 + ca_3e_2, \qquad u_2 = b_2e_1 + b_4e_2;$$

then

$$(\xi\cdot\alpha)^2 = (u_1\cdot u_2)^2 \leqq |u_1|^2|u_2|^2 = (a_1^2 + c^2a_3^2)(b_2^2 + b_4^2) \leqq |v_1|^2|v_2|^2 = 1,$$

proving $|\xi|_0 \leqq 1$.

We now prove

$$(13) \qquad |ae_{12} + be_{34}|_0 = |a| + |b|.$$

We may suppose $a > 0$, $b \geqq 0$. Set $\omega = e^{12} + e^{34}$, $\alpha = ae_{12} + be_{34}$. Then $|\omega|_0 = 1$, $\omega\cdot\alpha = a + b$; hence, by (6), $|\alpha|_0 \geqq a + b$. The converse is clear.

Remark. In a Euclidean vector space, for each vector $v \neq 0$ there is a unique covector f such that $|f| = 1$, $f(v) = |v|$, given by $f = \Phi u$, $u = v/|v|$. In a normed vector space, a covector with these properties exists (App. I, Lemma 8b), but need not be unique. This happens in the present case, with $V_{[r]}$ and its conjugate space $V^{[r]}$. For

$$|e^{12}|_0 = |e^{12} + e^{34}|_0 = 1,$$

$$e^{12}\cdot e_{12} = (e^{12} + e^{34})\cdot e_{12} = 1 = |e_{12}|_0.$$

14. Mass and comass of products. We shall give some inequalities here; they are not all the best possible. First, for exterior products:

(1) $$|\alpha\vee\beta|_0 \leqq |\alpha|_0 |\beta|_0,$$

(2) $$|\omega\vee\xi|_0 \leqq \binom{r+s}{r} |\omega|_0 |\xi|_0 \qquad (\omega \in V^{[r]}, \quad \xi \in V^{[s]}),$$

(3) $$|\omega\vee\xi|_0 \leqq |\omega|_0 |\xi|_0 \quad \text{if} \quad \omega \text{ or } \xi \text{ is simple.}$$

To prove (1), take any $\epsilon > 0$, and write $\alpha = \sum \alpha_i$, the α_i simple, $\sum |\alpha_i| < |\alpha|_0 + \epsilon$, and similarly for β. Then using (13.9) and (12.14) gives

$$|\alpha\vee\beta|_0 \leqq \sum_{i,j} |\alpha_i\vee\beta_j|_0 \leqq \sum_{i,j} |\alpha_i| |\beta_j| < (|\alpha|_0 + \epsilon)(|\beta|_0 + \epsilon),$$

giving (1). To prove (2), choose an orthonormal coordinate system for $\omega\vee\xi$ by the remark following (13.7), and use (6.6), giving

$$|\omega\vee\xi|_0 = (\omega\vee\xi)_{1\cdots r+s} = \sum_{(\lambda)(\mu)} \delta^{\lambda\mu}_{1\cdots r+s}\omega_\lambda\xi_\mu.$$

Since $|\omega_\lambda| \leqq |\omega|_0$ and $|\xi_\mu| \leqq |\xi|_0$, by (13.7), and the sum has at most $\binom{r+s}{r}$ terms, (2) follows.

Example. We cannot omit the numerical factor in (2). For take $r = s = 2$, and $\omega = \xi = e^{12} + e^{34}$. Then $|\omega|_0 = 1$, by (13.12), and hence

$$|\omega\vee\xi|_0 = |2e^{1234}|_0 = 2 = 2|\omega|_0|\xi|_0.$$

To prove (3), suppose ω is simple. Using the same coordinate system as before, write $\omega = f^1\vee\cdots\vee f^r$, and $f^i = g^i + h^i$, g^i and h^i being linear combinations of $e^1, \cdots, e^{r+s}$ and $e^{r+s+1}, \cdots, e^n$ respectively. Then

$$\omega = \omega' + \sum_k \omega_k, \qquad \omega' = g^1\vee\cdots\vee g^r,$$

each ω_k containing some e^j with $j > r + s$. By a rotation in the space of $e_1, \cdots, e_{r+s}$, we may suppose $\omega' = ae^{1\cdots r}$ for some a. We now follow the proof of (2), giving

$$|\omega\vee\xi|_0 = (\omega\vee\xi)_{1\cdots r+s} = \omega_{1\cdots r}\xi_{r+1\cdots r+s} \leqq |\omega|_0 |\xi|_0.$$

We now prove similar inequalities for interior products:

(4) $$|\omega\wedge\alpha|_0 \leqq |\omega|_0 |\alpha|_0 \qquad (\omega \in V^{[r]},\ \alpha \in V_{[s]},\ r \geqq s),$$

(5) $$|\omega\wedge\alpha|_0 \leqq \binom{s}{r} |\omega|_0 |\alpha|_0 \qquad (\omega \in V^{[r]},\ \alpha \in V_{[s]},\ r \leqq s),$$

(6) $$|\omega\wedge\alpha|_0 \leqq |\omega|_0 |\alpha|_0 \qquad \text{if } \omega \text{ is simple.}$$

To prove (4) and (5), take any $\beta \in V_{[r-s]}$ in the first case, and any $\xi \in V^{[s-r]}$ in the second. Then, in the two cases,

$$|(\omega\wedge\alpha)\cdot\beta| = |\omega\cdot(\alpha\vee\beta)| \leqq |\omega|_0 |\alpha\vee\beta|_0 \leqq |\omega|_0 |\alpha|_0 |\beta|_0,$$

$$|\xi\cdot(\omega\wedge\alpha)| = |(\xi\vee\omega)\cdot\alpha| \leqq |\xi\vee\omega|_0 |\alpha|_0 \leqq \binom{s}{r} |\xi|_0 |\omega|_0 |\alpha|_0.$$

Taking $|\beta|_0 = 1$, $|\xi|_0 = 1$, and using (13.5) and (13.6) gives (4) and (5). Relation (6) follows similarly, using (3).

EXAMPLE. We cannot omit the numerical factor in (5), even if α is simple. For take $\omega = e^{12} + e^{34}$, $\alpha = e_{1234}$; then, using (13.12), $|\omega|_0 = |\alpha|_0 = 1$. By (7.7) and (13.13),

$$|\omega\wedge\alpha|_0 = |e_{34} + e_{12}|_0 = 2 = 2|\omega|_0 |\alpha|_0.$$

15. On projections. We shall give some inequalities which will be used in Chapter IX; the last one occurs in Chapter IV. Let V be Euclidean, and let W be any subspace. Then any vector v may be written uniquely in the form

(1) $$v = \pi v + \pi' v, \qquad \pi v \in W, \qquad \pi' v \text{ orthogonal to } W;$$

πv is the *orthogonal projection* of v into W. It is a linear mapping of V into W; hence $\pi\alpha$ is defined for r-vectors α.

Let W_1 and W_2 be oriented r-dimensional subspaces of V, with r-directions $\alpha(W_1)$, $\alpha(W_2)$ (§ 12). We define the *angle* θ between them and the *distance* $|W_2 - W_1|$ between their r-directions by the formulas

(2) $$\cos\theta = \alpha(W_1)\cdot\alpha(W_2), \qquad |W_2 - W_1| = |\alpha(W_2) - \alpha(W_1)|,$$

taking $0 \leqq \theta \leqq \pi$. Because of (12.10), θ is defined. Note that

(3) $$|W_2 - W_1|^2 = [\alpha(W_2) - \alpha(W_1)]\cdot[\alpha(W_2) - \alpha(W_1)] = 2 - 2\cos\theta,$$

and hence

(4) $$|W_2 - W_1| = 2\sin\tfrac{1}{2}\theta.$$

Letting π denote the orthogonal projection into W_1, we prove

(5) $$\pi\alpha(W_2) = \cos\theta\,\alpha(W_1), \qquad \pi\alpha(W_2)\cdot\alpha(W_1) = \cos\theta.$$

For we may suppose

$$\alpha(W_1) = e_{1\ldots r}, \qquad \alpha(W_2) = v_1\vee\cdots\vee v_r, \qquad v_i = \pi v_i + v_i',$$

and $v_i'\cdot e_j = 0$ for $j \leqq r$; hence

$$\alpha(W_2)\cdot\alpha(W_1) = (\pi v_1\vee\cdots\vee\pi v_r)\cdot e_{1\ldots r} = \pi\alpha(W_2)\cdot\alpha(W_1),$$

proving the second part of (5). Since $\pi\alpha(W_2)$ is a multiple of $\alpha(W_1)$, the first part follows from this.

Letting $|Q|$ denote the (r-dimensional) volume of the polyhedral set Q in W_2, we prove

$$|\pi(Q)| = |\cos\theta|\,|Q| \qquad (Q \subset W_2). \tag{6}$$

From elementary properties of volumes it follows that we may suppose that Q is the parallelepiped determined by $v_1, \cdots, v_r$, with $v_1\vee\cdots\vee v_r = \alpha(W_2)$. Then $|Q| = 1$, and $|\pi(Q)| = |\pi\alpha(W_2)| = |\cos\theta|$, as required.

We end with some inequalities about vectors in W_2:

$$|\cos\theta|\,|v| \leqq |\pi v| \leqq |v|, \qquad |v - \pi v| \leqq |W_2 - W_1|\,|v| \qquad (v \in W_2). \tag{7}$$

We may suppose $|v| = 1$. Choose $v_2, \cdots, v_r$ in W_2 so that the set v, $v_2, \cdots, v_r$ is orthonormal. Clearly $|\pi v| \leqq |v|$; also $|\pi v_i| \leqq |v_i|$, and (5) and (12.16) give

$$|\cos\theta| = |\pi\alpha(W_2)| \leqq |\pi v|\,|\pi v_2|\cdots|\pi v_r| \leqq |\pi v|,$$

proving the first inequality. Also, by Lemma 9a and (12.14), setting $\alpha_i = \alpha(W_i)$,

$$|\alpha_1\vee v| = |(\alpha_1 - \alpha_2)\vee v + \alpha_2\vee v| = |(\alpha_1 - \alpha_2)\vee v| \leqq |\alpha_1 - \alpha_2|.$$

Finally, using (12.20),

$$|\alpha_1\vee v| = |\alpha_1\vee(v - \pi v)| = |v - \pi v|,$$

giving the rest of (7).

II. *Differential Forms*

In this chapter we give some of the fundamentals of calculus in Euclidean space E^n from a geometric point of view. In particular, differential forms are treated. Smooth manifolds are defined, and it is shown how to handle differential forms on these manifolds. Rather little of Chapter I is needed in most of the present chapter. The metric character of E^n is used for convenience, but most of the formulas clearly hold in an affine space.

To start with, the general properties of a smooth (i.e. continuously differentiable) mapping of one Euclidean space E^n into another one, E^m, are given. In the special case $m = 1$, we have real functions defined in E^n. We prove some elementary theorems, partly to illustrate the general methods. The study is then continued to include the properties of differential forms. From the usual theory of partial differentiation, no large amount is needed in the present book. Most of that is given here, in a manner which does not use coordinate systems, and keeps the full meaning always clear. The usual properties of coordinate systems follow easily. The geometric meaning of Jacobians is given, and a theorem on inverse functions is proved.

The basic properties of the operation of exterior differentiation of differential forms are given next. This operation forms a part of the general Stokes' Theorem; its abstract counterpart, the coboundary operation on cochains, is at the heart of the later general theory.

In smooth manifolds one loses not only the metric character, but also the affine character of Euclidean spaces. Vectors and covectors must be defined at individual points. This is done in a direct manner, through the use of parametrized curves and real functions in the manifold. The theory of differential forms can now be carried over. In studying the exterior differential, it is assumed that the manifold is 2-smooth and the differential form is smooth. These assumptions will be weakened in (III, 17); see also (IX, 12) and (X, 9).

1. The differential of a smooth mapping. Let f be a mapping of an open subset R of E^n into E^m. For any point $p \in R$ and any vector v of E^n, we define the *derivative of f at p with respect to v* by the formula (the notation $t \to 0+$ meaning that we keep $t > 0$)

$$(1) \qquad \nabla_v f(p) = \nabla f(p, v) = \lim_{t \to 0+} \frac{1}{t} [f(p + tv) - f(p)],$$

if this exists; this is a vector in E^m. We say that f is *smooth* if this exists and is continuous in p, for each v; we will see below (§ 5) that this is true if and only if f is continuously differentiable, using a rectangular coordinate system.

If ϕ is a smooth mapping of the interval $(0, t)$ into E^m, and e is the number 1, considered as a vector in 1-space, then $\nabla_e \phi(s) = d\phi(s)/ds$ is the tangent vector to the arc at $\phi(s)$; we have

$$\phi(t) - \phi(0) = \int_0^t \nabla_e \phi(s)\, ds. \tag{2}$$

We may prove this by the ordinary methods of calculus (compare Chapter III), or introduce a coordinate system in E^m and obtain it from the corresponding formula for each coordinate.

Given f, p and v as above, setting $\phi(t) = f(p + tv)$ shows that (if $p + sv \in R$ for $0 \leqq s \leqq t$)

$$f(p + tv) - f(p) = \int_0^t \nabla_v f(p + sv)\, ds. \tag{3}$$

THEOREM 1A. *If f is smooth, then for each p, $\nabla_v f(p)$ is a linear function of v, mapping $V(E^n)$ into $V(E^m)$.*

We shall let $\nabla f(p)$ denote this function; $\nabla f(p)$ is the *differential* of f at p.

Clearly $\nabla_{av} f(p) = a\nabla_v f(p) (a \geqq 0)$; we must prove

$$\nabla_{v_1+v_2} f(p) = \nabla_{v_1} f(p) + \nabla_{v_2} f(p). \tag{4}$$

Set

$$p_t = p + tv_1, \qquad q_t = p_t + tv_2 = p + t(v_1 + v_2).$$

Given $\epsilon > 0$, choose $\zeta > 0$ so that

$$|\nabla_{v_i} f(p') - \nabla_{v_i} f(p)| < \epsilon \quad \text{if} \quad |p' - p| < \zeta \qquad (i = 1, 2).$$

Taking t so small that the segments pp_t and $p_t q_t$ are in $U_\zeta(p) \cap R$, applying (3) to each of these gives

$$|f(p_t) - f(p) - t\nabla_{v_1} f(p)| = \left|\int_0^t [\nabla_{v_1} f(p_s) - \nabla_{v_1} f(p)]\, ds\right| \leqq \epsilon t,$$

$$|f(q_t) - f(p_t) - t\nabla_{v_2} f(p)| \leqq \epsilon t.$$

Hence

$$\left|\frac{f(q_t) - f(p)}{t} - [\nabla_{v_1} f(p) + \nabla_{v_2} f(p)]\right| \leqq 2\epsilon,$$

from which (4) follows.

In the case $m = 1$, we have a mapping of E^n into the reals, i.e. a real function ϕ in E^m. Now $\nabla_v \phi(p)$ is a real valued function, linear in v at each p; hence $\nabla \phi(p)$ is a covector at p. The differential $\nabla \phi$ is also called the *gradient* of ϕ at p. By definition,

$$\nabla \phi(p) \cdot v = \nabla \phi(p, v). \tag{5}$$

2. Some properties of differentials. We prove first three lemmas.

LEMMA 2a. *Let f be smooth in R. Then for each compact set $Q \subset R$ and each $\epsilon > 0$ there is a $\zeta > 0$ such that for any p, q, and vector v,*

$$(1) \qquad |\nabla_v f(q) - \nabla_v f(p)| \leqq \epsilon |v| \quad \text{if} \quad p, q \in Q, |q - p| < \zeta.$$

Take an orthonormal base in E^n, and choose ζ so that this holds for each of the vectors $e_1, \cdots, e_n$, using ϵ/n in place of ϵ. Now take any $v = \sum v^i e_i$; then $|v^i| \leqq |v|$. By Theorem 1A,

$$|\nabla_v f(q) - \nabla_v f(p)| = |\sum_i v^i[\nabla_{e_i} f(q) - \nabla_{e_i} f(p)]|$$
$$\leqq \sum_i |v^i| \epsilon/n \leqq \epsilon |v|.$$

LEMMA 2b. *Let f be smooth in the convex open set R', and suppose that for some $p_0 \in R'$,*

$$(2) \qquad |\nabla_v f(p) - \nabla_v f(p_0)| \leqq \epsilon |v| \qquad (p \in R', \text{any vector } v).$$

Then (3) *below holds for any p, $q \in R'$.*

Setting $p_t = (1 - t)p + tq$ and using (1.3) gives, if $u = q - p$,

$$f(q) - f(p) - \nabla_u f(p_0) = \int_0^1 [\nabla_u f(p_t) - \nabla_u f(p_0)]\, dt;$$

(3) follows at once from this.

LEMMA 2c. *Let f be smooth in R, and let $Q \subset R$ be compact. Then for each $\epsilon > 0$ there is a $\zeta > 0$ such that*

$$(3) \qquad |f(q) - f(p) - \nabla f(p_0, q - p)| \leqq \epsilon |q - p|$$
$$\text{if } p_0 \in Q \text{ and } p, q \in U_\zeta(p_0).$$

Say $Q' = \bar{U}_{\zeta_0}(Q) \subset R$. Using Q', find $\zeta \leqq \zeta_0$ by Lemma 2a; then (3) follows by applying Lemma 2b to $U_\zeta(p_0)$.

Given the smooth mapping f, define the quantities

$$(4) \qquad |\nabla f(p)| = \sup\{|\nabla_v f(p)| : |v| = 1\},$$

$$(5) \qquad |\nabla f|_Q = \sup\{|\nabla f(p)| : p \in Q\};$$

if R is the domain of f, let $|\nabla f|$ denote $|\nabla f|_R$.

Since $\nabla_v f(p)$ is linear and hence continuous in v, $|\nabla f(p)|$ is finite; if f is real valued, it is simply the magnitude of the covector $\nabla f(p)$ (App. I, 8.7). Also $|\nabla f(p)|$ is continuous in R; hence $|\nabla f|_Q$ is finite if Q is compact. Clearly (using (1.3))

$$(6) \qquad |\nabla_v f(p)| \leqq |\nabla f(p)|\, |v| \leqq |\nabla f|\, |v|,$$

$$(7) \qquad |f(q) - f(p)| \leqq |\nabla f|_{pq} |q - p| \quad \text{if} \quad \text{segment } pq \subset R.$$

Let f and g be smooth mappings of open sets $R \subset E^n$ and $S \subset E^m$ into S and E^l respectively; then $gf = g \circ f$ is a smooth mapping of R into E^l, and

$$\nabla (g \circ f)(p, v) = \nabla g[f(p), \nabla f(p, v)]; \tag{8}$$

thus if ∇f (at p) carries v into v', and ∇g (at $f(p)$) carries v' into v'', then $\nabla (g \circ f)$ (at p) carries v into v''.

To prove this, say $|\nabla f(p')| \leqq M$ in some neighborhood of p. Set

$$p_t = p + tv, \qquad q = f(p), \qquad q_t = f(p_t).$$

Take any $\epsilon > 0$. By Lemma 2c, we may choose $t_0 > 0$ so that

$$|q_t - q - t\nabla f(p, v)| \leqq \epsilon t |v|,$$

$$|g(q_t) - g(q) - \nabla g(q, q_t - q)| \leqq \epsilon |q_t - q|,$$

and $|\nabla f(p_t)| \leqq M$, if $0 < t \leqq t_0$. Then

$$|\nabla g(q, q_t - q) - t\nabla g(q, \nabla f(p, v))| = |\nabla g(q, q_t - q - t\nabla f(p, v))|$$
$$\leqq |\nabla g(q)| \, |q_t - q - t\nabla f(p, v)| \leqq \epsilon t |v| \, |\nabla g(q)|.$$

Also, by (7), $|q_t - q| \leqq Mt|v|$. Hence

$$|g(q_t) - g(q) - t\nabla g(q, \nabla f(p, v))| \leqq \epsilon t |v| [|\nabla g(q)| + M]$$

for $0 < t \leqq t_0$; this gives (8).

If we use the notation $\nabla f(p)\cdot v$ in place of $\nabla f(p, v)$, as in (1.5), then (8) reads

$$\nabla (g \circ f)(p)\cdot v = \nabla g(f(p))\cdot[\nabla f(p)\cdot v]. \tag{8'}$$

We may write this in the form

$$\nabla (g \circ f)(p) = \nabla g(f(p))\circ\nabla f(p), \tag{8''}$$

or $\nabla (g \circ f) = \nabla g \circ \nabla f$ simply.

If ϕ and ψ are smooth real valued functions in R, and $\phi\cdot\psi$ is their product, the usual proof gives $\nabla_v(\phi\cdot\psi) = \psi\nabla_v\phi + \phi\nabla_v\psi$; hence

$$\nabla (\phi\cdot\psi) = \psi\nabla\phi + \phi\nabla\psi. \tag{9}$$

3. Differential forms. A *differential r-form* ω, or *r-form* for short, in the set $Q \subset E^n$, is a function defined in Q, whose values are r-covectors; r is the *degree* (or dimension) of ω. We say ω is *s-smooth* if, for each r-vector α, $\omega(p)\cdot\alpha$ is s-smooth; see § 5 below.

A 0-form in Q is a real valued function in Q. If E^n is oriented, and ω_0 is the unit n-covector of E^n (I, 12), an n-form is uniquely expressible in the form

$$\omega(p) = \bar{\omega}(p)\omega_0, \qquad \bar{\omega}(p) \text{ a real function}; \tag{1}$$

see (I, 8).

Supposing ω is defined in Q, define the *magnitude* $|\omega|$ and the *comass* $|\omega|_0$ of ω by

(2) $$|\omega| = \sup \{|\omega(p)|: p \in Q\}, \qquad |\omega|_0 = \sup \{|\omega(p)|_0: p \in Q\},$$

if these are finite. (See also (App. III, 5.1).) Then, by (I, 13.4),

(3) $$|\omega(p)\cdot\alpha| \leqq |\omega(p)|_0 |\alpha|_0 \leqq |\omega|_0 |\alpha|_0;$$

the same relation holds with $|\omega|$, $|\alpha|$.

4. Smooth mappings. Let f be a smooth mapping of the open set $R \subset E^n$ into E^m. By Theorem 1A, for $p \in R$, $\nabla f(p)$ is a linear mapping of $V(E^n)$ into $V(E^m)$. By (I, 10.1), a corresponding linear mapping $\nabla f(p)$ of $V_{[r]}(E^n)$ into $V_{[r]}(E^m)$ is defined by

(1) $$\nabla f(p, v_1 \vee \cdots \vee v_r) = \nabla f(p, v_1) \vee \cdots \vee \nabla f(p, v_r).$$

By (I, 10.2),

(2) $$\nabla f(p, \alpha \vee \beta) = \nabla f(p, \alpha) \vee \nabla f(p, \beta).$$

The mapping $\nabla f(p)$ of $V(E^n)$ into $V(E^m)$ has a dual (App. I, 3), which we denote by f^*_p. If ω is an r-form in $S \subset E^m$, and $f(R) \subset S$, then for any $p \in R$, $\omega(f(p))$ is an r-covector, and by (I, 10.3), $f^*_p(\omega(f(p)))$ is an r-covector in E^n, which we denote by $(f^*\omega)(p)$. Thus an r-form $f^*\omega$ is defined in R:

(3) $$f^*\omega(p) = f^*_p(\omega(q)), \qquad q = f(p).$$

By (I, 10.5), for any r-vector α in E^n,

(4) $$f^*\omega(p)\cdot\alpha = f^*_p(\omega(q))\cdot\alpha = \omega(q)\cdot\nabla f(p, \alpha), \qquad q = f(p).$$

In particular, for $r = 0$, we have a real function ϕ, and

(5) $$(f^*\phi)(p) = \phi(f(p)) = (\phi \circ f)(p); \qquad \text{thus } f^*\phi = \phi f \text{ (}\phi\text{ real valued).}$$

We prove (see (8.8) for a generalization)

(6) $$\nabla(f^*\phi) = f^*(\nabla\phi) \qquad (\phi \text{ real valued}).$$

For let u be a vector in E^n. Using (1.5), (5), (2.8) (ϕ maps $S \subset E^n$ into the reals), and (4) gives

$$\begin{aligned}\nabla(f^*\phi)(p)\cdot u &= \nabla(\phi f)(p, u) = \nabla\phi[f(p), \nabla f(p, u)]\\ &= \nabla\phi(f(p))\cdot\nabla f(p, u) = f^*(\nabla\phi)(p)\cdot u.\end{aligned}$$

Given the differential forms ω and ξ, their (exterior) *product* $\omega\vee\xi$ is defined by $(\omega\vee\xi)(q) = \omega(q)\vee\xi(q)$. By (I, 10.4),

(7) $$f^*(\omega\vee\xi) = f^*\omega \vee f^*\xi.$$

With f and g as in (2.8), that relation gives

$$\nabla(g \circ f)(p, \alpha) = \nabla g[f(p), \nabla f(p, \alpha)]. \tag{8}$$

Hence, if ω is defined in $g(f(R))$,

$$[(gf)^*\omega]\cdot\alpha = \omega[(gf)(p)]\cdot\nabla(gf)(p, \alpha) = \omega[gf(p)]\cdot\nabla g[f(p), \nabla f(p, \alpha)]$$
$$= (g^*\omega)(f(p))\cdot\nabla f(p, \alpha) = [f^*(g^*\omega)](p)\cdot\alpha;$$

therefore

$$(g \circ f)^*\omega = f^*g^*\omega. \tag{9}$$

By (I, 12.28), (I, 12.29) and (I, 13.11), for r-vectors α and r-covectors ω,

$$|\nabla f(p, \alpha)| \leqq |\nabla f(p)|^r |\alpha| \leqq |\nabla f|^r |\alpha|, \tag{10}$$

$$|\nabla f(p, \alpha)|_0 \leqq |\nabla f(p)|^r |\alpha|_0 \leqq |\nabla f|^r |\alpha|_0, \tag{11}$$

$$|f^*\omega(p)| \leqq |\nabla f(p)|^r |\omega(f(p))| \leqq |\nabla f|^r |\omega|, \tag{12}$$

$$|f^*\omega(p)|_0 \leqq |\nabla f(p)|^r |\omega(f(p))|_0 \leqq |\nabla f|^r |\omega|_0. \tag{13}$$

The *Lipschitz constant* of a mapping f into E^m (or mapping into any normed linear space) is

$$\mathfrak{L}_f = \mathfrak{L}(f) = \sup\left\{\frac{|f(q) - f(p)|}{|q - p|} : p \neq q\right\}; \tag{14}$$

define $\mathfrak{L}_{f|Q}$, requiring the above p and q to lie in Q. Then

$$|\nabla f(p)| \leqq \mathfrak{L}_f \text{ if } f \text{ is smooth}, \tag{15}$$

$$|\nabla f| = \mathfrak{L}_f \text{ if } f \text{ is smooth and } R \text{ is convex}. \tag{16}$$

The first relation is obvious; the second follows from the first and (1.3).

We say f is *Lipschitz* if $\mathfrak{L}_f$ is finite.

5. Use of coordinate systems. Recall (App. I, 1) that arithmetic n-space $\mathfrak{A}^n$ is the set of all n-tuples $x = (x^1, \cdots, x^n)$ of real numbers; it has natural base vectors $\bar{e}_1, \cdots, \bar{e}_n$, and a natural scalar product.

A mapping f of an open set $R \subset E^n$ into E^m is *regular at* p if f is smooth in a neighborhood of p and $\nabla_v f(p) \neq 0$ for all $v \neq 0$; equivalently, if $\nabla_{v_1} f(p), \cdots, \nabla_{v_n} f(p)$ are independent, for some set of vectors $v_1, \cdots, v_n$ (and hence for all independent sets). If this holds, then $m \geqq n$, and $\nabla f(p)$ is one-one in $V(E^n)$. f is *regular in* Q if it is regular at all points of Q; f is *regular* if it is regular in its domain of definition.

A *smooth* or *curvilinear coordinate system* in E^n is a one-one regular mapping χ of an open set $O \subset \mathfrak{A}^n$ into E^n. The image $\chi(O)$ is open (Theorem 7A). If $p = \chi(x)$, the numbers $x^1, \cdots, x^n$ are the *coordinates* of p. The *coordinate vectors* at $p = \chi(x)$ are the vectors

$$e_i(p) = \nabla\chi(x, \bar{e}_i) = \partial p/\partial x^i \qquad (i = 1, \cdots, n). \tag{1}$$

Since χ is regular, these are independent, and hence form a base in $V(E^n)$.

For any $\lambda = (\lambda_1, \cdots, \lambda_r)$, set

(2) $$e_\lambda(p) = e_{\lambda_1}(p)\vee\cdots\vee e_{\lambda_r}(p).$$

Using (1) and (4.1) gives

(3) $$e_\lambda(p) = \nabla\chi(x, \bar{e}_\lambda) \qquad (p = \chi(x)).$$

Let $\bar{e}^1, \cdots, \bar{e}^n$ and $e^1(p), \cdots, e^n(p)$ be the bases dual to the $\bar{e}_i$ and the $e_i(p)$ respectively. Now e^i is a 1-form in E^n, and hence χ^*e^i is a 1-form in $\mathfrak{A}^n$. By (4.4) and (1),

$$\chi^*e^i(x)\cdot\bar{e}_j = e^i(p)\cdot\nabla\chi(x, \bar{e}_j) = e^i(p)\cdot e_j(p) = \delta^i_j.$$

Hence $\chi^*e^i(x) = \bar{e}^i$. Also, by (4.7), $\chi^*e^\lambda = \chi^*e^{\lambda_1}\vee\cdots\vee\chi^*e^{\lambda_r}$. Now

(4) $$\chi^*e^i(x) = \bar{e}^i, \quad \chi^*e^\lambda(x) = \bar{e}^\lambda, \qquad \text{all } x \text{ in the domain of } \chi.$$

For any p in E^n and r-vector α, using (3) shows that

(5) $$\text{if} \quad \alpha = \sum_{(\lambda)} \alpha^\lambda e_\lambda(p) \quad \text{and} \quad \bar{\alpha} = \sum_{(\lambda)} \alpha^\lambda \bar{e}_\lambda, \quad \text{then} \quad \alpha = \nabla\chi(x, \bar{\alpha}),$$

where $p = \chi(x)$. Thus $\nabla\chi$ preserves components of r-vectors, and in particular, of vectors.

By (4.4) and (3), $\chi^*\omega(x)\cdot\bar{e}_\lambda = \omega(p)\cdot e_\lambda(p)$. Hence

(6) $$(\chi^*\omega)_\lambda(x) = \omega_\lambda(p) \qquad (p = \chi(x)).$$

Thus χ^* preserves components of r-forms. See also (13) below.

For affine coordinate systems, (App. I, 12.6) gives

(7) $$\chi(x^1, \cdots, x^n) = \chi(0, \cdots, 0) + \sum_i x^i e_i \qquad (\chi \text{ affine}).$$

If further, $e_i\cdot e_j = \delta^j_i$, the coordinate system is *Cartesian*, or *orthonormal*.

The differential $\nabla\phi$ of a real function ϕ in R is a 1-form, with components

(8) $$[\nabla\phi(p)]_i = \nabla\phi(p, e_i(p)) = \partial\phi(p)/\partial x^i.$$

In particular, the coordinates x^i of p have differentials, with components

(9) $$[\nabla x^i(p)]_j = \partial x^i/\partial x^j = \delta^i_j;$$

hence

(10) $$\nabla x^i(p) = e^i(p), \qquad \text{commonly written } dx^i.$$

With this notation, an r-form ω becomes

(11) $$\omega(p) = \sum_{(\lambda)} \omega_\lambda(p)e^\lambda(p) = \sum_{(\lambda)} \omega_\lambda(p)\, dx^{\lambda_1}\vee\cdots\vee dx^{\lambda_r}.$$

Given the real function ϕ in E^n, set $\phi^*(x) = \phi(\chi(x))$ in O. By (6),

$$\nabla\phi^*(x)\cdot\bar{e}_i = \nabla\phi(p)\cdot e_i(p) \qquad (p = \chi(x)). \tag{12}$$

Thus ϕ and ϕ^* have the same partial derivatives:

$$\partial\phi^*(x)/\partial x^i = \partial\phi(p)/\partial x^i \qquad (\phi^* = \chi^*\phi,\ p = \chi(x)). \tag{13}$$

Let f be a smooth mapping of an open set $R \subset E^n$ into E^m, and let χ be a coordinate system in E^n. Then, generalizing (8),

$$\partial f(p)/\partial x^i = \nabla f(p, e_i(p)). \tag{14}$$

Let us translate (2.8) into the usual notation, using coordinate systems in the three spaces. Let $e_i = e_i(p)$, $e'_j = e'_j(f(p))$ denote the coordinate vectors in E^n and E^m at p and at $f(p)$ respectively. If $\nabla f^j(p, v)$ $(j = 1, \cdots, m)$ are the components of $\nabla f(p, v)$ etc., (2.8) gives

$$\begin{aligned}\nabla(gf)^j(p, e_i) &= \nabla g^j[f(p), \sum_k \nabla f^k(p, e_i)e'_k] \\ &= \sum_k \nabla g^j[f(p), e'_k]\nabla f^k(p, e_i).\end{aligned}$$

If x^i, y^k, z^j are the coordinates of p, $f(p)$, and $g(f(p))$ respectively, this reads

$$\frac{\partial z^j}{\partial x^i} = \sum_k \frac{\partial z^j}{\partial y^k}\frac{\partial y^k}{\partial x^i}.$$

Consider a *transformation of coordinates*. Let χ and χ' be coordinate systems; at any point p, let the old and the new base vectors be related as in (App. I, 1.1):

$$e_i(p) = \sum_j a^j_i(p)e'_j(p), \qquad e'_i(p) = \sum_j a'^j_i(p)e_j(p), \tag{15}$$

and similarly for $e^i(p)$, $e'^i(p)$. By (9),

$$\frac{\partial x^i}{\partial x'^j} = \nabla x^i(p, e'_j(p)) = \sum_k a'^k_j(p)\nabla x^i(p, e_k(p)) = a'^i_j(p),$$

etc.; hence

$$a^i_j(p) = \frac{\partial x'^i}{\partial x^j}, \qquad a'^i_j(p) = \frac{\partial x^i}{\partial x'^j}. \tag{16}$$

By (I, 5.8) and (16), the formulas for the transformation of components of r-vector functions and of r-forms are

$$\alpha'^\lambda(p) = \sum_{(\mu)} D^\lambda_\mu(p)\alpha^\mu(p), \qquad \omega'_\lambda(p) = \sum_{(\mu)} D'^\mu_\lambda(p)\omega_\mu(p), \tag{17}$$

where

$$D^\lambda_\mu(p) = \left|\frac{\partial x'^{\lambda_i}}{\partial x^{\mu_j}}\right|, \qquad D'^\lambda_\mu(p) = \left|\frac{\partial x^{\lambda_i}}{\partial x'^{\mu_j}}\right|. \tag{18}$$

Let f map $R \subset E^n$ into E^m. With an affine coordinate system χ in E^n as in (7), $\nabla_{e_i} f(p) = \partial f(p)/\partial x^i$; clearly $\nabla_v f(p)$ is continuous in p for each v if and only if the $\partial f(p)/\partial x^i$ are continuous. If $f(p) = (y^1, \cdots, y^m)$, this holds if and only if the $\partial y^j/\partial x^i$ are continuous. This is the usual definition of f being *smooth* (or *continuously differentiable*, or *of class* C^1). An easy proof by induction shows that f is *k-smooth* (or *of class* C^k), i.e. the partial derivatives of order $\leqq k$ exist and are continuous, if and only if, for each set of vectors $v_1, \cdots, v_k$ in E^m, $\nabla_{v_k} \cdots \nabla_{v_1} f(p)$ exists and is continuous.

Let the coordinate system χ be k-smooth. Then it is easily seen that a real function ϕ is k-smooth if and only if all partial derivatives of ϕ of order $\leqq k$ exist and are continuous; a similar theorem holds for r-forms ω.

Let f be smooth. Then for any continuous r-vector function $\alpha(p)$ and any coordinate system χ, (14) gives

$$(19)\quad \nabla f(p, \alpha(p)) = \sum_{(\lambda)} \alpha^\lambda(p) \nabla f(p, e_\lambda(p)) = \sum_{(\lambda)} \alpha^\lambda(p) \frac{\partial f(p)}{\partial x^{\lambda_1}} \vee \cdots \vee \frac{\partial f(p)}{\partial x^{\lambda_r}}.$$

LEMMA 5a. *Let f be a k-smooth mapping of the open set $R \subset E^n$ into E^m, and let $\alpha(p)$ be a $(k-1)$-smooth r-vector function in R. Then $\nabla f(p, \alpha(p))$ is $(k-1)$-smooth in R.*

This follows at once from (19).

LEMMA 5b. *With f as above, let ω be a $(k-1)$-smooth r-form in $S \supset f(R)$. Then $f^*\omega$ is $(k-1)$-smooth in R.*

Using an affine coordinate system in E^n, the last lemma shows that each $[f^*\omega(p)]_\lambda = \omega(f(p)) \cdot \nabla f(p, e_\lambda)$ is $(k-1)$-smooth; hence so is $f^*\omega$.

Let f be as above, and let the e_i and the e^i be dual bases in E^m. With a corresponding affine coordinate system in E^m, any $y \in E^m$ equals $q_0 + \sum y^i e_i$ (q_0 fixed). We prove:

$$(20)\qquad \text{if} \quad f(p) = q_0 + \sum_i f^i(p) e_i, \qquad \text{then} \quad f^* e^i = \nabla f^i.$$

For (4.4) gives

$$(f^* e^i)(p) \cdot v = e^i \cdot \nabla_v f(p) = e^i \cdot \sum_j \nabla_v f^j(p) e_j = \nabla_v f^i(p).$$

6. Jacobians. Let E^n be oriented, with unit n-vector α_0. Then given the smooth mapping f of $R \subset E^n$ into E^m, the *Jacobian*, or *Jacobian n-vector*, of f at $p \in R$ is

$$(1)\qquad J_f(p) = \nabla f(p, \alpha_0).$$

Using an expression $\alpha_0 = v_1 \vee \cdots \vee v_n$ shows that f is regular at p (see § 5) if and only if $J_f(p) \neq 0$; see (4.1) and (I, Lemma 9a).

If $J_f(p) \neq 0$, then $\nabla f(p, v) \neq 0$ if $v \neq 0$. We prove

LEMMA 6a. *If f is regular at p, then for any vector v,*

$$|\nabla f(p, v)| \geqq |J_f(p)| |v| / |\nabla f(p)|^{n-1}. \tag{2}$$

We may suppose $|v| = 1$. Choose an orthonormal base $e_1, \cdots, e_n$, with $e_1 = v$; set $w_i = \nabla f(p, e_i)$. Then, by (I, 12.16),

$$\begin{aligned} |J_f(p)| &= |\nabla f(p, e_1) \vee \cdots \vee \nabla f(p, e_n)| \\ &\leqq |\nabla f(p, e_1)| \cdots |\nabla f(p, e_n)| \leqq |\nabla f(p, v)| |\nabla f(p)|^{n-1}, \end{aligned}$$

which gives (2).

Suppose $E^m = E'^n$ is oriented, with unit n-vector α_0'. Then (compare (3.1)) there is a unique real valued function $\bar{J}_f(p)$, the *algebraic Jacobian* of f at p, defined by

$$J_f(p) = \bar{J}_f(p)\alpha_0'. \tag{3}$$

This is the Jacobian in the usual sense of the word.

LEMMA 6b. *Let E^n and E'^n be oriented, and let f and g be smooth mappings of $R \subset E^n$ into $S \subset E'^n$ and of S into E^m respectively. Then*

$$J_{gf}(p) = \bar{J}_f(p) J_g(f(p)). \tag{4}$$

If, further, $E^m = E''^n$ is oriented, then

$$\bar{J}_{gf}(p) = \bar{J}_f(p) \bar{J}_g(f(p)). \tag{5}$$

With α_0, α_0' as before, we have, by (4.8),

$$\begin{aligned} J_{gf}(p) &= \nabla gf(p, \alpha_0) = \nabla g[f(p), \nabla f(p, \alpha_0)] \\ &= \bar{J}_f(p) \nabla g[f(p), \alpha_0'] = \bar{J}_f(p) J_g(f(p)); \end{aligned}$$

relation (5) follows from this.

With f as in the beginning of the section, let ω be an n-form defined in a neighborhood of $f(R)$. Then

$$f^*\omega(p) \cdot \alpha_0 = \omega(f(p)) \cdot \nabla f(p, \alpha_0) = \omega(f(p)) \cdot J_f(p). \tag{6}$$

Let $E^m = E'^n$, and let ω_0, ω_0' be the unit n-covectors of E^n and E'^n respectively. Define $\bar{\omega}$, $\bar{\omega}^*$ as in (3.1):

$$\omega(q) = \bar{\omega}(q)\omega_0', \qquad f^*\omega(p) = \bar{\omega}^*(p)\omega_0. \tag{7}$$

By (6),

$$\bar{\omega}^*(p)\omega_0 \cdot \alpha_0 = f^*\omega(p) \cdot \alpha_0 = \omega(f(p)) \cdot J_f(p) = \bar{\omega}(f(p))\omega_0' \cdot J_f(p);$$

since $\omega_0 \cdot \alpha_0 = \omega_0' \cdot \alpha_0' = 1$, using (3) gives

$$\bar{\omega}^*(p) = \bar{J}_f(p) \bar{\omega}(f(p)). \tag{8}$$

We discuss briefly the Jacobian of a transformation of coordinates. Let χ, χ' be overlapping coordinate systems; then $\psi = \chi'^{-1}\chi$, where defined, is regular, mapping an open set in $\mathfrak{A}^n$ into $\mathfrak{A}^n$. Say $\nabla\psi(p,\bar{e}_i) = \sum_j \psi_i^j(p)\bar{e}_j$. By (I, 1.13), $\bar{e}_\mu = \epsilon_\mu \bar{e}_{1\cdots n}$ for $\mu = (\mu_1, \cdots, \mu_n)$; hence

$$J_\psi(p) = \nabla\psi(p, \bar{e}_{1\cdots n}) = \sum_{\mu_1} \psi_1^{\mu_1}(p)\bar{e}_{\mu_1} \vee \cdots \vee \sum_{\mu_n} \psi_n^{\mu_n}(p)\bar{e}_{\mu_n}$$
$$= \sum_\mu \epsilon_\mu \psi_1^{\mu_1}(p) \cdots \psi_n^{\mu_n}(p)\bar{e}_{1\cdots n}.$$

Let x^i and y^i be the coordinates of p under χ and χ' respectively. Then comparing with (5.15) and (5.16) shows that $\psi_i^j(p) = \partial y^j/\partial x^i$. Therefore

$$(9) \qquad J_\psi(p) = \begin{vmatrix} \psi_1^{\mu_1}(p) & \cdots & \psi_n^{\mu_1}(p) \\ \cdot & \cdots & \cdot \\ \psi_1^{\mu_n}(p) & \cdots & \psi_n^{\mu_n}(p) \end{vmatrix} = \begin{vmatrix} \partial y^1/\partial x^1 & \cdots & \partial y^1/\partial x^n \\ \cdot & \cdots & \cdot \\ \partial y^n/\partial x^1 & \cdots & \partial y^n/\partial x^n \end{vmatrix}.$$

7. The inverse and implicit function theorems. The theorem on inverse functions is

THEOREM 7A. *Let f be an s-smooth mapping ($s \geqq 1$) of the open set $R \subset E^n$ into E'^n, and suppose $J_f(p_0) \neq 0$. Then there are neighborhoods U of p_0 and U' of $q_0 = f(p_0)$ such that f is a one-one mapping of U onto U', and f^{-1}, considered in U' only, is s-smooth, with $J_{f^{-1}}(q_0) \neq 0$.*

Since f is regular at p_0, $F_{p_0}(v) = \nabla f(p_0, v)$ is a one-one linear mapping of $V(E^n)$ into $V(E'^n)$; its inverse $F_{p_0}^{-1}$ has the same property. Take

$$N = | F_{p_0}^{-1} |, \qquad \eta \leqq 1/2N.$$

Choose $\zeta_0 > 0$ so that $\bar{U}_{\zeta_0}(p_0) \subset R$. Letting Q consist of the point p_0 only, choose $\zeta \leqq \zeta_0$ by Lemma 2c (with η in place of ϵ), requiring also that $J_f \neq 0$ in $U_\zeta(p_0)$. Set

$$\rho = \zeta/4N, \qquad U' = U_\rho(q_0), \qquad U = f^{-1}(U') \cap U_\zeta(p_0).$$

Now U and U' are neighborhoods of p_0 and q_0 respectively, and $f(U) \subset U'$.

To show that f is one-one in U, take any $p, p' \in U$, with $f(p) = f(p')$. Then by the choice of ζ,

$$|p' - p| \leqq |F_{p_0}^{-1}|\,|F_{p_0}(p' - p)| \leqq N\,|f(p') - f(p) - \nabla f(p_0, p' - p)|$$
$$\leqq N\eta\,|p' - p| \leqq |p' - p|/2,$$

proving $p = p'$.

To show that f maps U onto U', take any $q \in U'$. We shall define in succession vectors and points $w_1, v_1, p_1, q_1, w_2, \cdots$, and will have, for $i = 1, 2, \cdots$,

$$(1) \qquad w_i = q - q_{i-1}, \quad v_i = F_{p_0}^{-1}(w_i), \quad p_i = p_{i-1} + v_i, \quad q_i = f(p_i),$$

$$(2) \qquad |w_i| \leqq (N\eta)^{i-1}|w_1| < \rho/2^{i-1},$$

$$(3) \qquad |v_i| \leqq N(N\eta)^{i-1}|w_1| < N\rho/2^{i-1},$$

$$(4) \qquad |p_i - p_0| \leqq N|w_1|[1 + N\eta + \cdots + (N\eta)^{i-1}] < 2N|w_1| < \zeta/2.$$

First define w_1 and v_1 by (1). Since $q \in U_\rho(q_0)$,

$$(5) \qquad |w_1| < \rho, \qquad |v_1| \leqq N|w_1| < N\rho,$$

giving (2) and (3) for $i = 1$. Define p_1 and q_1 by (1); then (4) holds for $i = 1$.

We now use induction. Define w_{i+1} and v_{i+1} by (1). By (4), p_i and p_{i-1} are in $U_\zeta(p_0)$; hence

$$\begin{aligned} |w_{i+1}| = |w_i + q_{i-1} - q_i| &= |f(p_{i-1}) - f(p_i) - \nabla f(p_0, -v_i)| \\ &\leqq \eta |v_i| \leqq (N\eta)^i |w_1|, \end{aligned}$$

proving (2); (3) follows also. Define p_{i+1} and q_{i+1} by (1); since $p_{i+1} - p_0 = (p_i - p_0) + v_{i+1}$, (4) follows by induction, using (3).

By (2), $\lim q_i = q$. Using (1) and (3) shows that, for $i < j$,

$$|p_j - p_i| = |v_{i+1} + \cdots + v_j| \to 0 \qquad \text{as} \quad i, j \to \infty;$$

hence we may set $p = \lim p_i$. By (4), $|p - p_0| \leqq \zeta/2$; hence $p \in U_\zeta(p_0)$. Also $f(p) = \lim f(p_i) = q$, and $p \in U$, as required.

We show next that f^{-1} is smooth in U'. Since $J_f(p) \neq 0$ in U, F_p is one-one in U. Take any $q^* = f(p^*) \in U'$ ($p^* \in U$) and any vector w; we shall prove

$$(6) \qquad \nabla f^{-1}(q^*, w) = F_{p^*}^{-1}(w).$$

Since F_{p^*} and hence $F_{p^*}^{-1}$ is a continuous function of p^*, the statement will be proved.

Given $\epsilon > 0$, take (supposing $w \neq 0$)

$$N_1 = |F_{p^*}^{-1}|, \qquad \eta \leqq \epsilon/2N_1^2|w|, \qquad \eta \leqq 1/2N_1.$$

Using p^* in place of p_0, find $\zeta_1 > 0$ as we found ζ in the proof above, requiring $U_{\zeta_1}(p^*) \subset U$. Set

$$\rho_1 = \zeta_1/4N_1, \qquad U_1' = U_{\rho_1}(q^*), \qquad U_1 = f^{-1}(U_1') \cap U.$$

Now take any t such that $q = q^* + tw \in U_1'$; say $q = f(p)$, $p \in U$. Starting with p^* and q^* in place of p_0 and q_0, use the proof above to define the w_i, v_i, p_i, q_i ($i \geqq 1$). Since $tw = q - q^* = w_1$, we have

$$\begin{aligned} |f^{-1}(q^* + tw) - f^{-1}(q^*) - F_{p^*}^{-1}(tw)| &= |p - p^* - v_1| = |p - p_1| \\ &\leqq |v_2| + |v_3| + \cdots \leqq 2N_1^2\eta t |w| \leqq t\epsilon, \end{aligned}$$

which gives (6).

Introduce coordinate systems $(x^1, \cdots, x^n)$ and $(y^1, \cdots, y^n)$ into E^n and E'^n respectively. Since f and f^{-1} are smooth and $f^{-1} \circ f$ is the identity in U, we have

$$\sum_j \frac{\partial x^i}{\partial y^j} \frac{\partial y^j}{\partial x^k} = \delta_k^i \qquad (i, k = 1, \cdots, n).$$

Since the determinant $|\partial y^j/\partial x^k|$ equals $J_f(p)$, which is $\neq 0$, we can solve these equations for the $\partial x^i/\partial y^j$ at each $q \subset U'$. Also (6.5) with $g = f^{-1}$ shows that $J_{f^{-1}}(q_0) \neq 0$. Since f is s-smooth, the $\partial x^i/\partial y^j$ are $(s-1)$-smooth; hence f^{-1} is s-smooth, and the proof is complete.

THEOREM 7B. *Let f be a regular mapping (§ 5) of the open set $R \subset E^n$ into E^m. Then f is locally one-one.*

That is, each $p_0 \in R$ is in a neighborhood U such that f is one-one in U. Since f is regular at p_0, $\nabla f(p_0)$ maps $V(E^n)$ onto a vector space V_p of dimension n in $V(E^m)$; the points $f(p_0) + w$ $(w \in V_{p_0})$ form the *tangent plane* T_{p_0} to $f(R)$ at $f(p_0)$. Let π_{p_0} be the projection of E^m onto T_{p_0}; set $g_{p_0}(p) = \pi_{p_0}(f(p))$, $p \in R$. Clearly $\nabla g_{p_0}(p_0, v) = \nabla f(p_0, v)$; hence g_{p_0} is regular at p_0. By Theorem 7A, g_{p_0} is one-one in a neighborhood of p_0; hence so is f.

Using the inverse function theorem, we shall sketch a proof of the implicit function theorem. (Usually the former is derived from the latter.)

Let the s-smooth real functions

$$(7) \qquad F_i(u_1, \cdots, u_n; x_1, \cdots, x_m) \qquad (i = 1, \cdots, n)$$

vanish at $(0, \cdots, 0)$, and suppose the determinant $D = |\partial F_i/\partial u_j|$ is $\neq 0$ there. Set $F_{n+i}(u_1, \cdots) = x_i$ for $i = 1, \cdots, m$. The complete set of F_i now defines an s-smooth mapping F of a neighborhood of 0 in $\mathfrak{A}^n$ into $\mathfrak{A}^n$; since $D \neq 0$, $J_F(0, \cdots, 0) \neq 0$. Hence there is an s-smooth inverse function Φ near $(0, \cdots, 0)$. Writing $\Phi = (\Phi_1, \cdots, \Phi_{n+m})$, the relation $F\Phi =$ identity gives

$$(8) \; F_i(\Phi_1(t_1, \cdots; x_1, \cdots), \cdots; \Phi_{n+1}(t_1, \cdots; x_1, \cdots), \cdots) = \begin{Bmatrix} t_i & (i \leq n), \\ x_{i-n} & (i > n). \end{Bmatrix}$$

Using the definition of F_{n+i} shows that

$$\Phi_{n+i}(t_1, \cdots; x_1, \cdots) = x_i \qquad (i = 1, \cdots, m).$$

Set

$$(9) \quad \phi_i(x_1, \cdots, x_m) = \Phi_i(0, \cdots, 0; x_1, \cdots, x_m) \qquad (i = 1, \cdots, n).$$

Setting $t_i = 0$ in (8) now gives

$$(10) \qquad F_i(\phi_1(x_1, \cdots, x_m), \cdots, \phi_n(x_1, \cdots, x_m); x_1, \cdots, x_m) = 0$$

for $i = 1, \cdots, n$. Thus $u_i = \phi_i(x_1, \cdots, x_m)$ is a solution (the only one) of (7) near $(0, \cdots, 0)$; the ϕ_i are s-smooth.

8. The exterior differential. To each smooth r-form ω defined in an open set $R \subset E^n$ corresponds an $(r+1)$-form $d\omega$, its *exterior differential*, defined as follows. For each $p \in R$, it is that linear function of $(r+1)$-vectors (see (I, Theorem 3A)) such that

$$(1) \quad d\omega(p)\cdot(v_1 \vee \cdots \vee v_{r+1}) = \sum_{i=1}^{r+1} (-1)^{i-1} \nabla_{v_i} \omega(p) \cdot (v_1 \vee \cdots \hat{v}_i \cdots \vee v_{r+1}).$$

Note that, since the v_i do not depend on p,

$$\nabla_{v_i}\omega(p)\cdot(v_1\vee\cdots) = \nabla_{v_i}[\omega(p)\cdot(v_1\vee\cdots)].$$

(For the reason for this definition, see (III, 11) or §§ 19 and 20 of the Introduction.)

That the result does not depend on the representation of an $(r+1)$-vector as a sum of products may be seen as follows. Given ω, set

$$F(p; v_1, \cdots, v_r) = \omega(p)\cdot(v_1\vee\cdots\vee v_r);$$

then this is an element of $L^r_{\text{alt}}(V(E^n))$, for each $p \in R$ (I, Theorem 4A). Set

$$\bar{d}F(p; v_1, \cdots, v_{r+1}) = \sum_{i=1}^{r+1}(-1)^{i-1}\nabla_{v_i}F(p; v_1, \cdots \hat{v}_i \cdots, v_{r+1}).$$

This is clearly a well defined element of $L^{r+1}_{\text{alt}}(V(E^n))$, for $p \in R$. The isomorphism of (I, Theorem 4A) carries $\bar{d}$ into d, which is therefore well defined.

Note that the differential $\nabla\omega$ and the exterior differential $d\omega$ of an r-form ω are not the same if $r \geqq 1$; $\nabla\omega(p)$ is a linear transformation Φ^r_p of $V = V(E^n)$ into $V^{[r]}$, not an element of $V^{[r+1]}$. For $r = 0$, Φ^r_p transforms V into $V^{[0]} = \mathfrak{A}$, and defines an element of $\bar{V} = V^{[1]}$; here, $\nabla\omega(p) = d\omega(p)$.

An intrinsic characterization of the operator d in smooth manifolds will be given in § 13 below.

Some special cases of (1) are:

$$(2) \qquad r = 0: \qquad d\omega = \nabla\omega; \quad \text{thus} \quad d\omega(p)\cdot v = \nabla_v\omega(p).$$

$$(3) \qquad r = 1: \qquad d\omega(p)\cdot(u\vee v) = \nabla_u\omega(p)\cdot v - \nabla_v\omega(p)\cdot u.$$

Of course $d\omega = 0$ if $r \geqq n$.

We shall prove the following fundamental properties:

$$(4) \qquad (d\omega)_\lambda(p) = \sum_{i=1}^{r+1}(-1)^{i-1}\frac{\partial}{\partial x^{\lambda_i}}\omega_{\lambda_1\cdots\hat{\lambda}_i\cdots\lambda_{r+1}}(p).$$

This relation, in the cases $r = 0$ and $r = 1$, reads

$$(d\omega)_i = \frac{\partial\omega}{\partial x^i}, \qquad (d\omega)_{ij} = \frac{\partial\omega_j}{\partial x^i} - \frac{\partial\omega_i}{\partial x^j}.$$

Noting that (2) applies to the 0-form $\omega_\lambda(p)$,

$$(5) \qquad d\sum_{(\lambda)}\omega_\lambda(p)e^\lambda(p) = \sum_{(\lambda)}\nabla\omega_\lambda(p)\vee e^\lambda(p) = \sum_{(\lambda)}d(\omega_\lambda)(p)\vee e^\lambda(p),$$

if the coordinate system is 2-smooth. In the notation of (5.11),

$$(5') \qquad d \sum_{(\lambda)} \omega_\lambda(p)\, dx^{\lambda_1} \vee \cdots \vee dx^{\lambda_r} = \sum_{(\lambda)} \nabla\, \omega_\lambda(p) \vee dx^{\lambda_1} \vee \cdots \vee dx^{\lambda_r}.$$

For a smooth r-form ω and a smooth s-form ξ,

$$(6) \qquad d(\omega \vee \xi) = d\omega \vee \xi + (-1)^r \omega \vee d\xi.$$

For any 2-smooth ω,

$$(7) \qquad dd\omega = 0.$$

If f is a 2-smooth mapping of $R \subset E^n$ into $S \subset E^m$, and ω is a smooth r-form in S, then

$$(8) \qquad df^*\omega = f^*d\omega.$$

First we prove (4) for an affine coordinate system (see (5.7)). Setting $\lambda(i) = (\lambda_1, \cdots \hat{\lambda}_i \cdots, \lambda_{r+1})$, we find

$$\begin{aligned}(d\omega)_\lambda(p) = (d\omega)(p) \cdot e_\lambda &= \sum (-1)^{i-1} \nabla_{e_{\lambda_i}} \omega(p) \cdot e_{\lambda(i)} \\ &= \sum (-1)^{i-1} [\partial \omega(p)/\partial x^{\lambda_i}] \cdot e_{\lambda(i)} = \sum (-1)^{i-1} \partial \omega_{\lambda(i)} p)/\partial x^{\lambda_i}.\end{aligned}$$

(The last step fails in the general case.)

Next we prove (5) for an affine coordinate system. Then $\omega = \sum \omega_\lambda(p) e^\lambda$; since d is linear, it is sufficient to consider the case $\omega = w\xi$, with w a smooth function and $\xi = e^\mu$. Now

$$\nabla_{e_j}(w\xi) \cdot e_\nu = \nabla_{e_j}[w(\xi \cdot e_\nu)] = (\nabla_{e_j} w)\xi_\nu = (\nabla w)_j \xi_\nu.$$

Hence, with $\lambda(i)$ as before, using (I, 6.7) gives

$$\begin{aligned}[d(w\xi)]_\lambda = d(w\xi) \cdot e_\lambda &= \sum_{i=1}^{r+1} (-1)^{i-1} \nabla_{e_{\lambda_i}} (w\xi) \cdot e_{\lambda(i)} \\ &= \sum_{i=1}^{r+1} (-1)^{i-1} (\nabla w)_{\lambda_i} \xi_{\lambda(i)} = (\nabla w \vee \xi)_\lambda,\end{aligned}$$

giving $d(w\xi) = \nabla w \vee \xi$, and hence (5), for this case.

Next we prove (6). Using an affine coordinate system, and the result just proved and (2.9),

$$\begin{aligned} d(\omega \vee \xi) = d \sum_{(\lambda)(\mu)} \omega_\lambda \xi_\mu e^\lambda \vee e^\mu &= \sum \nabla (\omega_\lambda \xi_\mu) \vee e^\lambda \vee e^\mu \\ &= \sum (\nabla\, \omega_\lambda \xi_\mu + \omega_\lambda \nabla\, \xi_\mu) \vee e^\lambda \vee e^\mu \\ &= \sum \nabla\, \omega_\lambda \vee e^\lambda \vee \xi_\mu e^\mu + (-1)^r \sum \omega_\lambda e^\lambda \vee \nabla\, \xi_\mu \vee e^\mu \\ &= \sum_{(\lambda)} \nabla\, \omega_\lambda \vee e^\lambda \vee \sum_{(\mu)} \xi_\mu e^\mu + (-1)^r \sum_{(\lambda)} \omega_\lambda e^\lambda \vee \sum_{(\mu)} \nabla\, \xi_\mu \vee e^\mu, \end{aligned}$$

which gives (6).

We prove (7) first for a 0-form ϕ. Take an affine coordinate system, and set $\xi = d\phi$. Now $\xi_k = \partial\phi/\partial x^k$, and hence (4) gives

$$(dd\phi)_{ij} = (d\xi)_{ij} = \frac{\partial}{\partial x^i}\left(\frac{\partial\phi}{\partial x^j}\right) - \frac{\partial}{\partial x^j}\left(\frac{\partial\phi}{\partial x^i}\right) = 0.$$

In the general case, using an affine coordinate system, and using (5), (6), the result just proved, and the fact that $de^\lambda = 0$ (since e_λ is constant),

$$dd\omega = d\sum_{(\lambda)} d\omega_\lambda \vee e^\lambda = \sum_{(\lambda)} (dd\omega_\lambda \vee e^\lambda \pm d\omega_\lambda \vee de^\lambda) = 0.$$

A direct proof that $dd\omega\cdot e_{\lambda_1\cdots\lambda_{r+2}} = 0$ is also simple.

We prove (8) first in the case that $\omega = e^i$, a dual base element, with an affine coordinate system in E^m. Then $f^*de^i = f^*0 = 0$. Also, by (5.20), $df^*e^i = ddf^i = 0$.

Next we show that if (8) holds for both ω and ξ, then it holds for $\omega\vee\xi$: using (4.7) and (6) gives

$$\begin{aligned} df^*(\omega\vee\xi) &= d(f^*\omega\vee f^*\xi) = df^*\omega\vee f^*\xi \pm f^*\omega\vee df^*\xi \\ &= f^*d\omega\vee f^*\xi \pm f^*\omega\vee f^*d\xi = f^*(d\omega\vee\xi \pm \omega\vee d\xi) = f^*d(\omega\vee\xi). \end{aligned}$$

To prove (8) for any ω, choose an affine coordinate system in S, and write $\omega = \sum_{(\lambda)}\omega_\lambda\vee e^{\lambda_1}\vee\cdots\vee e^{\lambda_r}$. We need now merely prove (8) for ω_λ and for each e^{λ_i}; we have done this for e^{λ_i}, and (4.6) gives it for ω_λ.

To prove (5) in any coordinate system χ, use (5.10), giving

$$de^i(p) = ddx^i(p) = 0; \tag{9}$$

hence applying (6) to the left hand side of (5) gives the right hand side.

Finally, to prove (4) in the general case, set $\tilde\omega = \chi^*\omega$. Since the coordinates in $\mathfrak{A}^n$ are affine, (4) holds for $\tilde\omega$. By (5.5) (applied to $d\omega$) and (8), if $p = \chi(x)$,

$$(d\omega)_\lambda(p) = (\chi^*d\omega)_\lambda(x) = (d\chi^*\omega)_\lambda(x) = (d\tilde\omega)_\lambda(x).$$

By (5.5), $\tilde\omega_\lambda(x) = \omega_\lambda(p)$; hence, by (5.13), $\partial\tilde\omega_\lambda(x)/\partial x^k = \partial\omega_\lambda(p)/\partial x^k$. Therefore, applying (4) to $\tilde\omega$ gives

$$(d\omega)_\lambda(p) = \sum (-1)^{i-1}\frac{\partial}{\partial x^{\lambda_i}}\tilde\omega_{\lambda(i)}(x) = \sum (-1)^{i-1}\frac{\partial}{\partial x^{\lambda_i}}\omega_{\lambda(i)}(p).$$

We give one more formula for $d\omega$, in a smooth coordinate system:

$$d\omega(p) = \sum_i e^i(p)\vee\frac{\partial\omega(p)}{\partial x^i}. \tag{10}$$

If we replace the coordinate system by an affine system with the same coordinate vectors at p, the quantities on the right are unchanged; hence we may assume the coordinate system is affine. Using (5) and (5.8) (applied to each ω_λ) gives

$$d\omega(p) = \sum_{(\lambda)} \sum_i \frac{\partial \omega_\lambda(p)}{\partial x^i}\, e^i(p) \vee e^\lambda(p) = \sum_i e^i(p) \vee \frac{\partial}{\partial x^i} \sum_{(\lambda)} \omega_\lambda(p) e^\lambda(p),$$

which is (10).

9. A representation of vectors and covectors. We show how vectors and covectors in E^n are represented by parametrized curves F and real functions ϕ respectively. Let e be the unit vector (the number 1) in the reals.

By a *p-curve* F in E^n we mean a smooth mapping F of a neighborhood of 0 in the reals into E^n, with $F(0) = p$. It defines a vector in E^n:

$$(1) \qquad W_F = dF(t)/dt\big|_{t=0} = \nabla F(0, e).$$

By a *p-function* ϕ in E^n we mean a smooth real function ϕ, defined in a neighborhood of p, with $\phi(p) = 0$. It defines a covector in E^n, namely, the gradient $\nabla \phi(p)$. (The condition $\phi(p) = 0$ is assumed merely for convenience.) Clearly all covectors are obtained in this manner. If $\bar{e} = 1$ is the unit covector in the reals, then $\nabla \phi(p) = \phi_p^* \bar{e}$.

Given F and ϕ, we have, by (1.5),

$$\nabla \phi(p) \cdot W_F = \nabla \phi(p, \nabla F(0, e)) = \nabla (\phi \circ F)(0, e);$$

hence

$$(2) \qquad \nabla \phi(p) \cdot W_F = d(\phi \circ F)/dt\big|_{t=0} = \lim_{t \to 0+} \phi(F(t))/t.$$

LEMMA 9a. *Let f be a smooth mapping of the open set $R \subset E^n$ into the open set $S \subset E^m$. Then for any p-curve F in R, if $F'(t) = f(F(t))$, then $\nabla f(p)$ carries W_F into $W_{F'}$. Also, for any q-function ϕ in S, if $\phi^*(p') = \phi(f(p'))$ $(p' \in R)$ and $f(p) = q$, then f_p^* carries $\nabla \phi(q)$ into $\nabla \phi^*(p)$.*

Since $\nabla (f \circ F)(0, e) = \nabla f(p, \nabla F(0, e))$, we have

$$(3) \qquad W_{F\circ} = \nabla f(p, W_F),$$

giving the first part. The second part follows from (4.5), (4.6) and (4.3).

Addition of vectors and covectors may clearly be carried out as follows: For p-curves F and G,

$$(4) \qquad W_F + W_G = W_H \quad \text{if} \quad H(t) = p + [F(t) - p] + [G(t) - p];$$

for p-functions ϕ and ψ,

$$(5) \qquad \nabla \phi + \nabla \psi = \nabla \gamma \quad \text{if} \quad \gamma(q) = \phi(q) + \psi(q).$$

Also if $F'(t) = F(at)$, then $W_{F'} = aW_F$; $\nabla (a\phi) = a \nabla \phi$.

10. Smooth manifolds. In various places in mathematics and its applications, one has to deal with spaces that, locally, are like Euclidean spaces, but are not so in the large. Such a space, a "manifold", may appear as a subset of a Euclidean space; for instance, as a surface in 3-space (a sphere, a torus, etc., or a connected open subset of one of these). Or it may be defined abstractly, as the phase space in a dynamical system. The manifold M_0 in (IV, 9) appears in both ways. A manifold is a topological space; that is, open and closed sets are defined, with the usual properties.

A *smooth manifold*, or *differentiable manifold*, M, of dimension n, is a connected topological space, which we also call M, and a set of *coordinate systems*, as follows. Each coordinate system is a homeomorphism χ_i of an open set $O_i \subset \mathfrak{A}^n$ into M. (We could use only a single O.) A finite or denumerable number of the $U_i = \chi_i(O_i)$ cover M. If $U_i \cap U_j \neq 0$, then $\psi_{ij} = \chi_j^{-1}\chi_i$, where defined, is smooth and regular (see § 5). Any homeomorphism χ of an open set in $\mathfrak{A}^n$ into M which satisfies the above conditions with these coordinate systems is itself a coordinate system.

If the ψ_{ij} are required to be s-smooth, then M is s-smooth; s may be ∞. If the ψ_{ij} are analytic, so is M. A manifold may be s-smooth in terms of a set of coordinate systems, and s'-smooth ($s' > s$) in terms of a subset of these; if it is s'-smooth, then it is s-smooth (add the omitted coordinate systems). Euclidean space E^n is an analytic manifold, using all analytic regular homeomorphisms of open subsets of $\mathfrak{A}^n$ into E^n.

If a subset of the coordinate systems are picked out, such that the U_i cover M, and all corresponding ψ_{ij} have positive algebraic Jacobians $J_{\psi_{ij}}$ (§ 6), these are said to *orient* M. M is *orientable* if there exists such a set. The projective plane, for instance, is not orientable.

A mapping f of one k-smooth manifold, M, into another one, M', is *s-smooth* ($s \leqq k$) if the following holds. For any $p \in M$, say $p \in U_i$, $f(p) \in U'_j$; then $\chi_j'^{-1} f \chi_i$ (where defined in $\mathfrak{A}^n$) is s-smooth. For $M' = \mathfrak{A}$, we have s-smooth real functions in M; for $M = \mathfrak{A}$, we have s-smooth parametrized curves in M'.

11. The tangent space of a smooth manifold. Let M be a smooth manifold. For any $p \in M$, p-curves and p-functions may be defined, as in § 9. We call two p-curves F and G *equivalent* if, in some coordinate system about p, $W_{\chi^{-1}F} = W_{\chi^{-1}G}$. If χ' is another coordinate system about p, and $\psi = \chi'^{-1}\chi$, then by Lemma 9a, $W_{\chi'^{-1}F}$ and $W_{\chi'^{-1}G}$ are both the image of $W_{\chi^{-1}F}$ under ψ; hence the definition of equivalence is independent of the coordinate system used.

By a *vector in M at p* we mean a class of equivalent p-curves. We may add vectors u and v at p in M as follows. Using the coordinate system χ, form H from $\chi^{-1}F$ and $\chi^{-1}G$ as in (9.4), and let the p-curve χH define

$u + v$. By Lemma 9a and the linearity of $\nabla\psi$ at $\chi^{-1}(p)$ (with ψ as above), the definition is independent of the coordinate system employed. If v is defined by F, then av is defined by $F'(t) = F(at)$. Now the vectors in M at p form a vector space $V(M, p)$, of dimension n, the *tangent space of M at p*. The set of all $V(M, p)$ $(p \in M)$ is the *tangent space* of M.

Two p-functions ϕ, θ in M are *equivalent* (at p) if $\nabla(\chi^*\phi) = \nabla(\chi^*\theta)$ at $\chi^{-1}(p)$, in some coordinate system. By Lemma 9a, this is independent of the coordinate system used. A *covector* in M at p is a class of equivalent p-functions. Two covectors in M at p are added by adding corresponding p-functions. The covectors at p form a vector space $\bar{V}(M, p)$. Using the last expression in (9.2), $\bar{V}(M, p)$ becomes the conjugate space of $V(M, p)$.

With a coordinate system χ about p, $\nabla\chi(q)$ $(p = \chi(q))$ is an isomorphism of $V(E^n)$ onto $V(M, p)$, and χ^*_q is an isomorphism of $\bar{V}(M, p)$ onto $\bar{V}(E^n)$.

Let f be a smooth mapping of the smooth manifold M into the smooth manifold M'. Given the vector v at p in M, defined by $F(t)$, set $F'(t) = f(F(t))$; this defines a vector v' in M', which we call $\nabla f(p, v)$. Let χ and χ' be coordinate systems about p and $f(p)$ respectively; set $G(t) = \chi^{-1}(F(t))$, $G'(t) = \chi'^{-1}(F'(t))$; then $G'(t) = \gamma(G(t))$, where $\gamma = \chi'^{-1}f\chi$. If $\bar{F}(t)$ had been used to define v in place of $F(t)$, and $\bar{F}'$, $\bar{G}$, $\bar{G}'$ are defined as above, then $\bar{G}$ and G would be equivalent; hence (Lemma 9a) G' and $\bar{G}'$ would be equivalent, and hence F' and $\bar{F}'$ also. Thus $\nabla f(p, v)$ is well defined; it is clearly linear. By Lemma 9a, it agrees with the previous use of ∇f if M and M' are Euclidean. Also, it agrees with the above defined $\nabla\chi$. The same considerations hold for f^*_p, a linear mapping of $\bar{V}(M', f(p))$ into $\bar{V}(M, p)$; moreover, (4.4) holds for vectors and covectors. If f and g are smooth mappings of M into M' and of M' into M'', then (2.8) and (4.9) (at the moment for covectors) hold.

12. Differential forms in smooth manifolds. For each $p \in M$, let $\omega(p)$ be an r-covector in M at p (defined, using $V(M, p)$). Then ω is an *r-form* in M. Similarly r-vector functions $\alpha(p)$ in M are defined. If f is a smooth mapping of M into M', then $\nabla f(p, \alpha(p))$ and $f^*_p\omega(f(p))$ are defined, as in § 4; in particular $\nabla\chi$ and χ^*_q are defined. Relations (4.1) through (4.8) and the analogue of (4.9) hold, as follows from the discussion in § 11.

We say ω is *continuous* if each $\chi^*\omega$ is continuous. More generally, suppose M is s-smooth, and let $s' \leqq s - 1$ be an integer. Then ω is *s'-smooth* if each $\chi^*\omega$ is. If this holds near p with χ, then it holds also with χ'. For $\chi'^*\omega = (\chi\chi^{-1}\chi')^*\omega = \psi^*(\chi^*\omega)$ $(\psi = \chi^{-1}\chi')$, and since ψ is s-smooth, ψ^* is s'-smooth (see (5.16) and (5.17)). (For $r = 0$, we may use $s' = s$.) Similarly we have continuous and s'-smooth r-vector functions; in particular, vector functions.

We now study the exterior differential $d\omega$ of r-forms ω in M. For $r = 0$, we may set $d\omega = \nabla\omega$, as in (8.2). If $r > 0$, we cannot define

$\nabla_v\omega$ directly; for if $q \neq p$, then $\omega(q)$ and $\omega(p)$ lie in different vector spaces, $V^{[r]}(M, q)$ and $V^{[r]}(M, p)$; and if a coordinate system is used to define $\nabla_v\omega$, the result in general depends on the coordinate system used. However, the particular combination of derivatives in (8.1) is independent of the coordinate system, as we now show.

Let M be 2-smooth, and let ω be smooth in M. Given $p \in M$, choose a coordinate system χ about p, and set

$$(1) \qquad d\omega(p) = \chi^{-1*}d\chi^*\omega(p).$$

Suppose we had used χ_1 instead of χ. Then $\psi = \chi_1^{-1}\chi$ is 2-smooth, and by (8.8), $\psi^{-1*}d = d\psi^{-1*}$. Applying this to the form $\psi^*\chi_1^*\omega$ gives

$$\begin{aligned}\chi_1^{-1*}d\chi_1^*\omega &= \chi_1^{-1*}d\psi^{-1*}\psi^*\chi_1^*\omega\\ &= \chi_1^{-1*}\psi^{-1*}d\psi^*\chi_1^*\omega = \chi^{-1*}d\chi^*\omega.\end{aligned}$$

If f is a 2-smooth mapping of M into M', then $df^*\omega = f^*d\omega$ (ω in M'), as follows at once on using coordinate systems. The other properties of § 8 follow also.

If M is not 2-smooth, the definition of smoothness of an r-form $\omega (r \geqq 1)$ in M fails; $\chi^*\omega$ might be smooth, but $\chi_1^*\omega$ might not be. Hence $d\omega$ cannot be defined by (1). It can also happen that $d\omega$ exists but has different values as defined with the help of χ_1 and χ_2. We give an example, with M an open set in E^2, and with $\psi = \chi_1^{-1}\chi_2$ given as follows. Letting (x, y) denote points of $\mathfrak{A}^2$, set

$$f(x, y) = \frac{x^2y}{(x^2 + y^2)^{1/2}}, \qquad \psi(x, y) = (x + f(x, y), y);$$

let $f(0, 0) = 0$. Then $\partial f/\partial x \to 0$ and $\partial f/\partial y \to 0$ as $(x, y) \to (0, 0)$, and hence we see that f and ψ are smooth. Clearly $J_\psi \neq 0$ near $(0, 0)$. Note that the second cross partial derivatives of f at the origin are 1 and 0, and hence

$$\left(\frac{\partial^2\psi}{\partial x\partial y} - \frac{\partial^2\psi}{\partial y\partial x}\right)_{(0,0)} = (1, 0).$$

Let ξ be a 1-form in $\mathfrak{A}^2$. Set $q = (0, 0)$. A direct computation of the single component of $d\psi^*\xi$ and of $\psi^*d\xi$ gives

$$(d\psi^*\xi)_{12}(q) - (\psi^*d\xi)_{12}(q) = \xi(q)\cdot\left(\frac{\partial^2\psi}{\partial x\partial y} - \frac{\partial^2\psi}{\partial y\partial x}\right)_q = \xi_1(q).$$

This shows that if ω in M is such that $\omega_1(\chi(q)) \neq 0$ (in the coordinate system χ), then $d\omega$ has different values at $\chi(q)$ in the two coordinate systems.

13. A characterization of the exterior differential. Let M be a 2-smooth manifold. Consider the class of all smooth forms defined in open subsets of M. We shall show that the operator d on these forms is characterized by the following properties.†

(a) $d(\omega_1 + \omega_2) = d\omega_1 + d\omega_2$ where ω_1 and ω_2 are defined.

(b) $d(\phi\omega) = \nabla\phi\vee\omega + \phi\, d\omega$ where ϕ and ω are defined.

(c) If $\omega = \nabla\phi_1\vee\cdots\vee\nabla\phi_r$ in an open set, then $d\omega = 0$ there.

These properties are known to hold. Now let d' be any operator with these properties. Take any ω, and any p in the domain of ω. Using a coordinate system in a neighborhood U of p, write $\omega = \sum_{(\lambda)}\omega_\lambda(p)e^\lambda(p)$ in U. Applying the above properties shows that $d'\omega$ is given by (8.5); hence $d'\omega = d\omega$.

† See Lichnerowicz, p. 166.

III. Riemann Integration Theory

The purpose of this chapter is to present the fundamental properties of the integral of Riemann type in any number of variables, in a manner that brings out clearly the geometric content. Beyond the elements of the subject, we prove two basic theorems, the transformation formula (Theorem 7A or (9.1)) and Stokes' Theorem (Theorems 14A and 18A). Though the definition and fundamental properties of the integral depend only on the affine character of the space E^n, the metric is a very useful tool; hence we take E^n to be Euclidean (App. I, 13).

As seen in the Introduction, the natural integrand for an r-dimensional integral in E^n is a differential r-form ω in E^n. (That ω is of necessity an r-form, with the simplest assumptions, will be seen in (V, Theorem 10A) and (IX, Theorem 5A).) The simplest domains of integration are convex polyhedral cells, and linear combinations of these, cellular r-chains. In the first sections of this chapter, we show how to integrate r-forms over cellular (or polyhedral (V, 1)) r-chains. In the case $r = n$, the definition reduces to the usual one for the Riemann integral. We use the fundamental properties of Grassmann algebra, in particular, the correspondence between simple r-vectors and oriented r-volumes (I, Theorem 9A). The concepts of mass and comass in Grassmann algebra (I, 13) are not needed here; in the few places they are used, for later purposes, we could clearly do with the norms of (I, 12).

The transformation formula, (7.1) or (9.1), is in its broadest form if we integrate over open sets; hence we first study improper integrals (§ 6). The basic reason for the transformation formula is best seen through the inequality (7.2). If the mapping is affine, this becomes an identity, as in (8.1); in the general case, since the mapping is locally almost affine, we have approximate equality. Our proof of the formula is carried out by considering this local approximation. The usual formulation (9.1), using Jacobians, follows at once from the more intrinsic formulation (7.1). It would not be hard to translate the proof of (7.1) into a form not using Grassman algebra, thus giving (9.1) directly.

Integration in a smooth manifold M may be defined with the help of coordinate systems in M (§ 10); that the result is independent of the coordinate system employed follows from the transformation formula. If M is an r-dimensional manifold in E^n, an r-form ω in E^n defines an r-form ω_1 in M, through the identity mapping of M into E^n; see §§ 11 and 12 of Chapter II. We may define $\int_M \omega = \int_M \omega_1$. This situation occurs, in particular, in Stokes' Theorem. (One could approximate to M by means of polyhedral chains A_i, and define $\int_M \omega = \lim \int_{A_i} \omega$, under very general circumstances; see Chapter X.)

The simplest case of Stokes' Theorem is taken up in § 11. Largely in order to show the intrinsic reason for the theorem, we give a direct proof. The extension given in § 12 is then sufficient for the proof of the much more general case studied later.*

The principal difficulty in the general case of Stokes' Theorem is concerned with the structure of the boundary B of the region or manifold considered; B is in general made up of pieces of various dimensions. Actually, the parts of B of lower dimension play no part; hence it is well to show how to eliminate them from the discussion. We do this with the help of the concept of "zero s-extent" of a set. Using also partitions of unity, we obtain a straightforward proof of the general theorem for bounded regions in § 14. The theorem for bounded manifolds (Theorem 18A) is easily reduced to this theorem. Greater generality is achieved by using "regular forms" (§§ 16, 17), introduced by E. Cartan and others.

In the final section, the iterated integral in E^n is considered in a geometric formulation; this could be easily generalized to the case of smooth manifolds.

1. The r-vector of an oriented r-simplex. Let σ be an oriented r-simplex in E^n, or, more generally, an oriented r-dimensional polyhedral subset of an r-plane. Taking the r-plane P of σ and the oriented r-volume of σ in P gives an oriented r-volume in E^n, and hence a simple r-vector in E^n (I, Theorem 9A). This is the *r-vector* $\{\sigma\}$ of σ. The definition does not use the Euclidean character of E^n. For $r = 0$, σ is a point p; set $\{p\} = 1$. We shall find a formula for $\{\sigma\}$ in terms of edge vectors of σ.

For any r-simplex $\sigma = p_0 \cdots p_r$ in E^n, the vectors $u_{ij} = p_j - p_i$ $(i \neq j)$ are the *edge vectors* of σ. Any set of r edge vectors of σ which are linearly independent we call a *defining set of edge vectors* of σ. If σ is oriented, we call the set an *orienting defining set* for σ if they are given in an order which defines the positive orientation of σ (App. II, 5).

It is easy to see that a set of r edge vectors of σ forms a defining set if and only if the corresponding set of edges of σ contains no closed path; also, if and only if any two vertices of σ are joined by a succession of these edges. With the orientation of $\sigma = p_0 \cdots p_r$ given by the ordered p_i (App. II, 5), two important orienting defining sets are

$$u_{01}, u_{02}, \cdots, u_{0r}; \qquad u_{01}, u_{12}, \cdots, u_{r-1,r}. \tag{1}$$

* For literature on the subject, see K. Krickeberg, Über den Gausschen und den Stokesschen Integralsatz. III. Math. Nachr. vol. 12, pp. 341–365 (1954).

THEOREM 1A. *Let $v_1, \cdots, v_r$ be an orienting defining set for σ. Then*

(2) $$\{\sigma\} = v_1 \vee \cdots \vee v_r / r!.$$

As a consequence, if $|\sigma| = |\sigma|_r$ denotes the r-volume of σ (using the Euclidean character of E^n),

(3) $$|\sigma| = |\{\sigma\}| = |v_1 \vee \cdots \vee v_r| / r!.$$

We shall prove (3); then (2) follows, since $v_1 \vee \cdots \vee v_r$ defines the positive orientation of σ.

Clearly (3) is true if $r = 1$. For the general case, we use induction on r. It is easy to see that there is a vertex of σ, say p_0, which is an end point of just one of the v_i, say v_1. Let σ' be the face of σ opposite p_0. Then $v_2, \cdots, v_r$ form a defining set for σ', and by induction,

$$|\sigma'| = |\beta| / (r-1)!, \qquad \beta = v_2 \vee \cdots \vee v_r.$$

The points $q_t = (1-t)p_0 + tp$ $(p \in \sigma')$, with t fixed, form a simplex σ'_t, for which clearly

$$|\sigma'_t| = t^{r-1} |\sigma'| = t^{r-1} |\beta| / (r-1)!.$$

We can find a vector $w = v_1 + c_2 v_2 + \cdots + c_r v_r$ such that $w \cdot v_i = 0$ for $i = 2, \cdots, r$; then $|w|$ is the altitude from p_0 to the plane of σ'. Using (I, 12.20), we find

$$|\sigma| = \int_0^1 |\sigma'_t| \, |w| \, dt = |\beta| \, |w| / r! = |w \vee v_2 \vee \cdots \vee v_r| / r!,$$

which gives (3).

LEMMA 1a. *For any simplex $\sigma = p_0 \cdots p_r$ in E^n and any point q of E^n, if $v_i = p_i - q$, then*

(4) $$\{\sigma\} = \frac{1}{r!} \sum_{i=0}^{r} (-1)^i v_0 \vee \cdots \hat{v}_i \cdots \vee v_r.$$

Since $u_{0i} = p_i - p_0 = v_i - v_0$, we have

$$\begin{aligned} r!\{\sigma\} &= u_{01} \vee \cdots \vee u_{0r} = (v_1 - v_0) \vee \cdots \vee (v_r - v_0) \\ &= v_1 \vee \cdots \vee v_r - \sum_{i=1}^{r} v_1 \vee \cdots \vee v_{i-1} \vee v_0 \vee v_{i+1} \vee \cdots \vee v_r \\ &= v_1 \vee \cdots \vee v_r - \sum_{i=1}^{r} (-1)^{i-1} v_0 \vee v_1 \vee \cdots \hat{v}_i \cdots \vee v_r, \end{aligned}$$

which gives (4).

2. The r-vector of an r-chain. By a *cellular r-chain* $\sum a_i \sigma_i^r$ we mean a set of oriented cells σ_i^r (App. II, 5), each with a real coefficient a_i; we set

$a(-\sigma) = (-a)\sigma$, and identify the oriented cell σ_i^r with the chain $1\sigma_i^r$. For the purposes of § 3, subdividing the σ_i^r will give a new chain; this is contrary to the situation for polyhedral chains used in Chapter V and later.

The *r-vector* of a cellular chain is defined by

$$\left\{\sum a_i\sigma_i^r\right\} = \sum a_i\{\sigma_i^r\}. \tag{1}$$

This gives a linear mapping of the linear space of cellular r-chains in E^n into $V_{[r]}(E^n)$. For $r = 0$,

$$\left\{\sum a_i p_i\right\} = \sum a_i. \tag{2}$$

THEOREM 2A. *The r-vector of a cellular chain is independent of subdivisions.*

This is clear for subdivisions of a cell; hence it holds for subdivisions of chains.

THEOREM 2B. *For any cellular chain A, with boundary ∂A,*

$$\{\partial A\} = 0. \tag{3}$$

REMARK. This clearly holds also for polyhedral chains defined in (V, 1).

It is sufficient to prove this for a simplex $\sigma = p_0 \cdots p_r$. Set $\sigma_i = p_0 \cdots \hat{p}_i \cdots p_r$; then (App. II, 7.1) $\partial\sigma = \sum(-1)^i\sigma_i$. Set $v_i = p_i - p_0$. By (1.2) and (1.4),

$$(r-1)!\{\sigma_i\} = v_1\vee\cdots\hat{v}_i\cdots\vee v_r \qquad (i > 0),$$

$$(r-1)!\{\sigma_0\} = \sum_{i=1}^{r}(-1)^{i-1}v_1\vee\cdots\hat{v}_i\cdots\vee v_r.$$

These show that $\{\partial\sigma\} = \sum(-1)^i\{\sigma_i\} = 0$.

For another proof (using later material), suppose $\{\partial\sigma\} = \alpha \neq 0$. Choose ω_1 so that $\omega_1\cdot\alpha \neq 0$ (I, Theorem 3A). Set $\omega(p) = \omega_1$ in E^n. Then $d\omega = 0$, and (4.1) and Stokes' Theorem give

$$\omega_1\cdot\alpha = \int_{\partial\sigma}\omega = \int_\sigma d\omega = 0,$$

a contradiction.

THEOREM 2C. *$\{A\}$ depends only on ∂A.*

For suppose $\partial A = \partial B$. Then $\partial(B - A) = 0$, and by (App. II, Lemma 10a), there is a cellular chain C with $B - A = \partial C$. Hence $\{B\} - \{A\} = \{\partial C\} = 0$.

3. Integration over cellular chains. First, supposing the r-form ω is defined at all points of the cellular r-chain $A \subset E^n$, we define an operation $\omega\circ A$ (which is not independent of subdivisions). With the "center" p_i^* of σ_i^r as in (App. II, 1.2) and the scalar product of (I, 2.2),

$$\omega\circ\sum_i a_i\sigma_i^r = \sum_i a_i\omega(p_i^*)\cdot\{\sigma_i^r\}. \tag{1}$$

(We shall use only simplexes σ_i^r in general.)

We next define $\int_\sigma \omega$ for any oriented r-cell σ, ω being defined and continuous in σ. Let $\mathfrak{S}_1\sigma, \mathfrak{S}_2\sigma, \cdots$ be any sequence of simplicial subdivisions of σ, whose meshes $\to 0$ (App. II, Lemma 3c); set

$$(2) \qquad \int_\sigma \omega = \int_\sigma \omega(p)\, dp = \lim_{k\to\infty} \omega\circ\mathfrak{S}_k\sigma.$$

That the limit exists and is independent of the sequence of subdivisions chosen follows at once from the following lemma and the uniform continuity of ω.

LEMMA 3a. *Suppose* $\epsilon > 0$, $\zeta > 0$ *are such that*

$$(3) \qquad |\omega(q) - \omega(p)|_0 \leqq \epsilon \quad \text{if} \quad p, q \in \sigma, \quad |q - p| \leqq \zeta.$$

Then for any two simplicial subdivisions $\sum\sigma'_i$, $\sum\sigma''_j$ *of* σ, *of mesh* $\leqq \zeta$,

$$(4) \qquad \left|\omega\circ\sum\sigma''_j - \omega\circ\sum\sigma'_i\right| \leqq 2\epsilon\,|\sigma|.$$

Let p'_i and p''_j be the centers of σ'_i and σ''_j respectively. Let $\sum\tau_k$ be a common simplicial refinement of the two subdivisions (App. II, Lemma 3*b*). For each σ'_i, let $\sum'_k\tau'_{ik}$ denote its subdivision, formed by the τ_l lying in it. Let q_{ik} be the center of τ_{ik}. Since

$$\{\sigma'_i\} = \sum_k{}'\{\tau'_{ik}\}, \qquad \sum_k{}'|\{\tau'_{ik}\}| = \sum_k{}'|\tau'_{ik}| = |\sigma'_i|,$$

and $\operatorname{diam}(\sigma'_i) \leqq \zeta$ (all i), using (I, 13.4) and (I, 13.9) (or (I, 12.10) if we use $|\ |$ in place of $|\ |_0$) gives

$$\begin{aligned}\left|\omega\circ\sum\tau_k - \omega\circ\sum\sigma'_i\right| &= \left|\sum_i\left[\sum_k{}'\omega\circ\tau'_{ik} - \omega\circ\sigma'_i\right]\right| \\ &= \left|\sum_i\sum_k{}'[\omega(q_{ik}) - \omega(p'_i)]\cdot\{\tau'_{ik}\}\right| \\ &\leqq \sum_i\sum_k{}'\epsilon\,|\{\tau'_{ik}\}| = \epsilon\sum_i|\sigma'_i| = \epsilon\,|\sigma|.\end{aligned}$$

Similarly $\left|\omega\circ\sum\tau_k - \omega\circ\sum\sigma''_j\right| \leqq \epsilon\,|\sigma|$, and (4) follows.

We now define integrals over cellular chains by

$$(5) \qquad \int_{\sum a_i\sigma_i} \omega = \sum a_i \int_{\sigma_i} \omega.$$

4. Some properties of integrals. From the proof of Lemma 3a it is clear that the integral is independent of subdivisions of the chain. (Hence it is defined for polyhedral chains; (V, 1).) The integral is bilinear:

$$\int_A (c_1\omega_1 + c_2\omega_2) = c_1\int_A \omega_1 + c_2\int_A \omega_2,$$

$$\int_{c_1A_1+c_2A_2} \omega = c_1\int_{A_1}\omega + c_2\int_{A_2}\omega.$$

In particular, $\int_{-\sigma}\omega = -\int_\sigma \omega$: reversing orientation reverses the sign of the integral.

The definition of the integral, applied to constant forms, gives

$$\text{(1)} \qquad \int_A \omega = \omega_1 \cdot \{A\} \quad \text{if} \quad \omega(p) = \omega_1, \quad \text{all } p.$$

The *mass* of a cellular chain is defined by

$$\text{(2)} \qquad \left| \sum a_i \sigma_i^r \right| = \sum |a_i| \, |\sigma_i^r| \qquad \text{if the } \sigma_i^r \text{ are non-overlapping;}$$

use a subdivision if the σ_i^r overlap. (See Chapter V for further details.) If $\sum \sigma_i$ is a subdivision of a cell σ, then, by (I, 13.4) and (I, 13.9),

$$\left| \omega \circ \sum \sigma_i \right| \leqq \sum | \omega(p_i) \cdot \{\sigma_i\} | \leqq \sum | \omega(p_i) |_0 \, | \sigma_i | \leqq | \omega |_0 \, | \sigma |;$$

the definition of the integral now gives

$$\text{(3)} \qquad \left| \int_A \omega \right| \leqq | \omega |_0 \, | A | \leqq | \omega | \, | A |.$$

Hence also

$$\text{(4)} \qquad \left| \int_A \omega - \int_A \xi \right| \leqq \epsilon \, | A | \quad \text{if} \quad | \omega(p) - \xi(p) |_0 \leqq \epsilon \quad \text{in } A.$$

We give a lemma on the degree of approximation to $\int_A \omega$ by sums like $\omega \circ A$:

LEMMA 4a. *Given the cellular chain $A = \sum a_i \sigma_i^r$, suppose*

$$\text{(5)} \qquad | \omega(q) - \omega(p) |_0 \leqq \epsilon, \qquad p \text{ and } q \text{ in the same } \sigma_i^r.$$

Then if p_i is a point of σ_i^r,

$$\text{(6)} \qquad \left| \int_A \omega - \sum a_i \omega(p_i) \cdot \{\sigma_i^r\} \right| \leqq \epsilon \, | A |.$$

For each i, (1) and (4) give

$$\left| \int_{\sigma_i^r} \omega - \omega(p_i) \cdot \{\sigma_i^r\} \right| \leqq \epsilon \, | \sigma_i^r |;$$

(6) follows at once from this.

5. Relation to the Riemann integral. In this section we consider the integration of an n-form ω over n-cells σ in E^n, *supposing E^n is Euclidean and oriented.* Then we may write $\omega(p) = \bar{\omega}(p)\omega_0$, as in (II, 3.1). Given any polyhedral region $\bar{\sigma}$ in E^n, let σ denote the corresponding oriented region, oriented like E^n; we define

$$\text{(1)} \qquad \int_{\bar{\sigma}} \bar{\omega} = \int_{\bar{\sigma}} \bar{\omega}(p) \, dp = \int_\sigma \omega.$$

From the definition of $\int_\sigma \omega$ it is clear that $\int_{\bar{\sigma}} \bar{\omega}$ is exactly the Riemann integral of the real function $\bar{\omega}$ over $\bar{\sigma}$.

Thus the integral of § 3 is a generalization of the Riemann integral, in the following manner. First, for r-dimensional integration in r-space, the real function $\bar{\omega}(p)$ is replaced by the r-form $\omega(p)$; then the integral is defined, independently of the metric character of E^r; note that we now integrate over oriented domains. Both integrals carry over at once to r-dimensional integration over r-dimensional polyhedral regions in n-space; in the Riemann integral, one integrates real functions over the domains, using the metric character of E^n, while in our integral, one integrates r-forms over oriented domains. Finally, our integral extends at once to an integral over cellular r-chains in E^n. The additivity of the Riemann integral over disjoint polyhedral regions is replaced by the linearity of the integral as a function of cellular chains.

In the definition of $\int_\sigma \omega$, there is no need for $\omega(p)$ to be an r-covector; we could use a function $\phi(p, \alpha)$, defined for points p and r-directions α, which is continuous in p for each α. However, if simple continuity properties are to hold, or if $d\phi$ is to be definable to be a bounded function (II, 8), then it turns out that ϕ must define an r-form (possibly not continuous); see (V, Theorem 10A) and (IX, Theorem 5A). The algebraic reason for this is given in (V, Theorem 9A).

In the future, when we are working with n-dimensional integration in n-space, we will use either integral in (1) interchangeably.

The following inequality is easily proved:

$$\left|\int_\sigma \omega\right| = \left|\int_\sigma \omega(p)\, dp\right| \leqq \int_{\bar{\sigma}} |\omega(p)|\, dp; \tag{2}$$

the last integral is a Riemann integral.

6. Integration over open sets. Let ω be a continuous n-form in the open set $R \subset E^n$; let E^n be oriented. Then for any polyhedral region $Q \subset R$, we can define $\int_Q \omega$, orienting Q like E^n. In defining $\int_R \omega$, questions of convergence arise. This section belongs to the theory of improper integrals.

We say ω is *summable* over R if for each $\epsilon > 0$ there is a compact set $P \subset R$ with the following property. For any polyhedral regions Q_1, Q_2,

$$\left|\int_{Q_2} \omega - \int_{Q_1} \omega\right| < \epsilon \quad \text{if} \quad P \subset Q_i \subset R, \qquad i = 1, 2. \tag{1}$$

We may clearly require P to be a polyhedron.

The following condition is clearly equivalent: For each $\epsilon > 0$ there is a compact set (or polyhedron) $P \subset R$ such that

$$\left|\int_Q \omega\right| < \epsilon \quad \text{if} \quad Q \subset R - P, \quad Q \text{ a polyhedron.} \tag{2}$$

Let ω be summable over R. Then clearly there is a unique number

$\int_R \omega$, the *integral* of ω over R, with the following property. For each $\epsilon > 0$ there is a compact set (in fact, a polyhedron) P such that

(3) $$\left| \int_Q \omega - \int_R \omega \right| < \epsilon \quad \text{if} \quad P \subset Q \subset R, \quad Q \text{ a polyhedron.}$$

Conversely, if a number $\int_R \omega$ exists with this property, then ω is summable over R, and $\int_R \omega$ is its integral.

We say the increasing sequence of sets $Q_1, Q_2, \cdots$ in R *converges* to R if their interiors cover R; we write $Q_1 \subset Q_2 \subset \cdots \to R$. It is easy to see that such a sequence exists, with the Q_i polyhedral.

LEMMA 6a. *If ω is summable over R, and $Q_1 \subset Q_2 \subset \cdots \to R$, the Q_i polyhedral, then*

(4) $$\lim_{i \to \infty} \int_{Q_i} \omega = \int_R \omega.$$

For given $\epsilon > 0$, choose P to satisfy (3). Now $P \subset Q_{i_0}$ for some i_0, and (3) holds with Q_i, if $i \geqq i_0$.

The usual definition of the Riemann integral $\int_R \phi$ of a real function ϕ is exactly like that of $\int_R \omega$.

LEMMA 6b. *ω is summable over R if and only if $| \omega(p) |$ is summable over R.*

Suppose that $| \omega(p) |$ is summable. Then using (2) and applying (5.2) shows that ω is. The converse is easily proved, if we consider separately the sets where $\bar{\omega}(p) > 0$ and where $\bar{\omega}(p) < 0$.

LEMMA 6c. *If ω is summable over R, and ϕ is a real bounded continuous function in R, then $\phi\omega$ is summable over R.*

This follows from the last lemma.

The volume (or Lebesgue measure, or mass) of the open set R may be defined by the relation

(5) $$| R | = \int_R dp = \sup \{| Q |: Q \subset R, Q \text{ a polyhedron}\},$$

if this is finite. With the Q_i as in Lemma 6a, we have $| R | = \lim | Q_i |$. With the n-direction ω_0 of E^n,

(6) $$| R | = \int_R \omega_0.$$

LEMMA 6d. *Let ω be continuous in the open set R, and let $| R |$ and $| \omega |$ be finite. Then ω is summable over R, and*

(7) $$\left| \int_R \omega \right| \leqq | \omega | \, | R |.$$

Let $Q_1 \subset Q_2 \subset \cdots \to R$, the Q_i polyhedral. Given $\epsilon > 0$, choose i so that $| R - Q_i | < \epsilon / | \omega |$. Then for any polyhedron $Q \subset R - Q_i$, $\left| \int_Q \omega \right| \leqq | \omega | \, | Q | < \epsilon$, and hence ω is summable over R. Since $\left| \int_{Q_i} \omega \right| \leqq | \omega | \, | R |$, (7) follows from (4).

For a polyhedron P, we can define both $\int_P \omega$ and $\int_{\text{int}(P)} \omega$; we wish to show that these are the same. We need

LEMMA 6e. *Any $(n-1)$-dimensional polyhedron Q in E^n is in the interior of a polyhedron Q' with $|Q'|$ arbitrarily small.*

This is clear for an $(n-1)$-simplex, and hence for Q.

LEMMA 6f. *If ω is continuous in the polyhedron P, then ω is summable in* int(P), *and*

$$\int_P \omega = \int_{\text{int}(P)} \omega. \tag{8}$$

Set $P^* = P - \text{int}(P)$. Given $\epsilon > 0$, choose Q' by the last lemma so that (assuming ω not identically 0)

$$P^* \subset \text{int}(Q'), \qquad |Q'| < \epsilon/|\omega|.$$

Now $P' = P - \text{int}(Q')$ is a polyhedron, and $P' \subset \text{int}(P)$. Take any polyhedron Q with $P' \subset Q \subset \text{int}(P)$. Then

$$\left|\int_P \omega - \int_Q \omega\right| \leqq |P - Q|\,|\omega| \leqq |Q'|\,|\omega| < \epsilon,$$

which gives the lemma.

The following lemma will be used in § 8 below:

LEMMA 6g. *Let ω be continuous in R, and let I be a number. Suppose that for each $\epsilon > 0$ there is a compact set $P \subset R$ with the following property. For any continuous function ϕ in R such that $0 \leqq \phi(p) \leqq 1$, $\phi(p) = 1$ in P, and* spt(ϕ) *is a compact set in R, we have $\left|I - \int_R \phi\omega\right| < \epsilon$. Then ω is summable over R, and $\int_R \omega = I$.*

Given $\epsilon > 0$, choose P as above, using $\epsilon/2$. It is sufficient to show that for any polyhedron Q,

$$\left|I - \int_Q \omega\right| < \epsilon \quad \text{if} \quad P \subset Q \subset R. \tag{9}$$

We may choose a number N and a polyhedron $Q' \subset R$ such that

$$Q \subset \text{int}(Q'), \qquad |Q' - Q| < \epsilon/2N, \qquad |\omega(p)| \leqq N \text{ in } Q'$$

(compare Lemma 6e). Choose ϕ so that spt $(\phi) \subset Q'$, $\phi = 1$ in Q (App. III, Lemma 1a). Now

$$\left|\int_R \phi\omega - \int_Q \omega\right| = \left|\int_{Q'-Q} \phi\omega\right| \leqq \int_{Q'-Q} |\omega| < \epsilon/2,$$

and (9) follows.

7. The transformation formula. We give here the fundamental formula for the transformation of the integral, and an indication of a direct proof. Since some details of this proof (on the approximation to a curvilinear triangulation by a true triangulation) are not too simple to present, we shall give the complete proof, by another method, in the next section. The usual formulation of the theorem, using Riemann integrals and Jacobians, will be found in § 9. Recall the definition of $f^*\omega$ in (II, 4) and of $\mathfrak{J}_f$ in (II, 6.3).

THEOREM 7A. *Let f be a one-one regular mapping of the open set $R \subset E$ onto the open set $R' \subset E'$ (E and E' oriented and of dimension n), with $\mathfrak{J}_f(p) > 0$ in R. Let ω be a continuous n-form in R', summable over R'. Then $f^*\omega$ is summable over R, and*

$$\int_R f^*\omega = \int_{R'} \omega. \tag{1}$$

REMARKS. If $\mathfrak{J}_f(p) < 0$ in R, then the formula holds with the minus sign, as we see by reversing the orientation of E'. If we think of R as an n-chain and write $f(R)$ in place of R', this takes care of both cases. In the general theory, (1) corresponds to (X, 8.1); but the whole of R cannot appear as an n-chain there.

A direct proof of (1) may be given along the following lines. Given $\epsilon > 0$, cut E into small cubes, form the regular subdivision (App. II, 3), and let Q be the polyhedron containing those simplexes σ_i in $\mathrm{int}_\lambda(R)$, for a certain $\lambda > 0$ (notation in App. II). The $\sigma_i' = f(\sigma_i)$ are curvilinear simplexes in R'. Let g be the affine approximation to f in Q (see (X, 1)), defined as follows: For each vertex p_i, $g(p_i) = f(p_i)$; let g be affine in each σ_k (App. I, 12). Then (if the cubes are small enough, λ remaining fixed), the $\tau_k = g(\sigma_k)$ form a triangulation of a polyhedron Q' filling most of R'.

We suppose that we have made each of

$$\left|\int_{R'} \omega - \int_{Q'} \omega\right|, \qquad \left|\int_R f^*\omega - \int_Q f^*\omega\right|,$$

$$\left|\int_{Q'} \omega - \omega\circ\sum \tau_k\right|, \qquad \left|\int_Q f^*\omega - f^*\omega\circ\sum \sigma_k\right|,$$

less than $\epsilon/5$; there remains to show that we can make

$$\sum_k |\,\omega\circ\tau_k - f^*\omega\circ\sigma_k\,| < \epsilon/5. \tag{2}$$

Take a typical $\sigma_k = p_0 \cdots p_n$, with center p_k^*. Say

$$q_i = f(p_i) = g(p_i), \qquad \tau_k = q_0 \cdots q_n,$$

$$u_i = p_i - p_{i-1}, \qquad v_i = q_i - q_{i-1}.$$

Since g is affine in σ_k, $v_i = \nabla g(p_k^*, u_i)$ (App. I, 12.4); hence, by (1.2) and (II, 4.1),

$$\{\tau_k\} = v_1 \vee \cdots \vee v_n/n! = \nabla g(p_k^*, u_1 \vee \cdots \vee u_n)/n! = \nabla g(p_k^*, \{\sigma_k\}). \tag{3}$$

Therefore, if q_k^* is the center of τ_k, using (II, 4.4) gives

$$|\,\omega\circ\tau_k - f^*\omega\circ\sigma_k\,| = |\,\omega(q_k^*)\cdot\{\tau_k\} - \omega(f(p_k^*))\cdot\nabla f(p_k^*, \{\sigma_k\})|$$

$$\leqq |\,\omega(q_k^*) - \omega(f(p_k^*))|\,|\,\tau_k\,| + |\,\omega(f(p_k^*))|\,|\,\nabla g(p_k^*, \{\sigma_k\}) - \nabla f(p_k^*, \{\sigma_k\})|,$$

which may be made an arbitrarily small multiple of $|\sigma_k|$ (compare (X, Lemma 3a)). Summing over k gives (2).

8. Proof of the transformation formula. First we consider the case that ω is compact; that is, if $\operatorname{spt}(\omega)$ is the closure of the set of points q where $\omega(q) \neq 0$, then $\operatorname{spt}(\omega)$ is in R and is compact. If f is affine (App. I, 12), the proof may be carried out at once, as follows. Take a polyhedron P such that $\operatorname{spt}(f^*\omega) \subset P \subset R$; then $\operatorname{spt}(\omega) \subset P' \subset R'$, where $P' = f(P)$. An arbitrarily fine subdivision $\sum \sigma_k$ of P gives an arbitrarily fine subdivision $\sum \tau_k$ of P', $\tau_k = f(\sigma_k)$. Now (see § 7) $f(p_k^*) = q_k^*$, and

$$(1) \qquad \{\tau_k\} = \nabla f(p_k^*, \{\sigma_k\}), \qquad \omega \circ \tau_k = f^*\omega \circ \sigma_k;$$

hence

$$\int_{R'} \omega = \lim \omega \circ \sum \tau_k = \lim f^*\omega \circ \sum \sigma_k = \int_R f^*\omega.$$

For the general case with ω compact, we shall express ω as a sum $\sum \omega_i$, with each $\operatorname{spt}(\omega_i)$ small; in a small region, f is nearly affine, and we shall obtain an approximation which will prove (7.1).

Let $C_1, C_2, \cdots$ be a subdivision of E' into cubes of diameter 1; let $C_1', C_2', \cdots$ be the concentric cubes of diameter 2. Let ψ_1' be a smooth function in E' such that (App. III, Lemma 1c)

$$0 \leqq \psi_1'(q) \leqq 1, \qquad \psi_1'(q) > 0 \quad \text{in } C_1, \qquad \psi_1'(q) = 0 \quad \text{in } E - C_1';$$

form ψ_i' by translating C_1 into C_i. Set

$$\psi_i(q) = \psi_i'(q) \Big/ \sum_j \psi_j'(q);$$

this is a "partition of unity" in E' (App. III, 2). Now for some L (see (II, 4.14)),

$$\sum \psi_i(q) = 1, \qquad \psi_i(q) = 0 \quad \text{in } E' - C_i', \qquad \mathfrak{L}_{\psi_i} = L \quad (\text{all } i).$$

Set

$$Q' = \operatorname{spt}(\omega), \qquad Q = f^{-1}(Q') = \operatorname{spt}(f^*\omega).$$

Since Q' and hence Q is compact, we may choose ρ_0 so that

$$R_1 = U_{\rho_0}(Q), \qquad R_1 \subset R.$$

We may choose ρ_0' etc. so that we have

$$R_1' = U_{\rho_0'}(Q') \subset R', \qquad L' = \mathfrak{L}_{f^{-1}|R_1'}, \qquad L'\rho_0' \leqq \rho_0.$$

Set (see (II, 6))

$$M = |R_1'|, \qquad J = \sup\{|J_f(p)|:\ p \in R_1\}.$$

Now take any $\epsilon > 0$. Set (we may assume $\omega \not\equiv 0$)

$$\epsilon_1 = \epsilon/2^n n^{n/2} M L'^n, \qquad \epsilon_2 = \epsilon_1/3LL'J\,|\omega|.$$

Choose ρ and $\rho' > 0$ so that (see (II, 2.3))

(2) $$\rho' = \rho/L' \leqq \rho_0'/3 \quad (\text{hence } \rho \leqq \rho_0/3),$$

(3) $$|f(p) - f(p_0) - \nabla f(p_0, p - p_0)| \leqq \epsilon_2 |p - p_0| \quad \text{if } |p - p_0| < \rho$$

(4) $$|J_f(p) - J_f(p_0)| \leqq \epsilon_1/3\, |\omega| \quad \text{and } p_0 \in R_1,$$

(5) $$|\omega(q) - \omega(q_0)| \leqq \epsilon_1/3J \qquad \text{if } q_0 \in R_1' \text{ and } |q - q_0| < \epsilon_2\rho.$$

By contracting E' by the factor ρ', we form (from the C_i and C_i') a cubical subdivision of E' with cubes $P_1, P_2, \cdots$ of diameter ρ' and concentric cubes $P_1', P_2', \cdots$ of diameter $2\rho'$; the ψ_i go into functions ϕ_i such that

(6) $$0 \leqq \phi_i \leqq 1, \qquad \phi_i = 0 \ \text{ in } E' - P_i', \qquad \sum \phi_i = 1, \qquad \mathfrak{L}_{\phi_i} = L/\rho'.$$

We use henceforth only those i, say $i = 1, \cdots, m$, such that P_i' touches Q'; by (2), these P_i' lie in R_1'. Let q_i be the center of P_i; set $p_i = f^{-1}(q_i)$.

Let F_i be the affine approximation to f at p_i; it is defined by

(7) $$F_i(p) = f(p_i) + \nabla f(p_i, p - p_i).$$

Since $\nabla f(p_i, v)$ is linear in v, (7) and (II, 1.1) give

(8) $$\nabla F_i(p, v) = \nabla f(p_i, v), \qquad \text{all } p, v.$$

Set

(9) $$\omega_i = \phi_i \omega.$$

We show that

(10) $$\operatorname{spt}(f^*\omega_i) \subset \bar{U}_\rho(p_i), \ \operatorname{spt}(F_i^*\omega_i) \subset \bar{U}_\rho(p_i).$$

Suppose $f^*\omega_i(p) \neq 0$. Then $\omega_i(f(p)) \neq 0$, hence $\phi_i(f(p)) \neq 0$, and $|f(p) - q_i| < \rho'$. By the choice of m, $f(p)$ and q_i are in R_i'; hence, by (2),

$$|p - p_i| \leqq \mathfrak{L}_{f^{-1}, R_1'} |f(p) - q_i| < \rho,$$

giving $\operatorname{spt}(f^*\omega_i) \subset \bar{U}_\rho(p_i)$. Also, by (8), (II, 4.16) for F_i^{-1} (defined in E'), and (II, 4.15) for f,

$$\mathfrak{L}_{F_i^{-1}, R_1'} = |\nabla F_i^{-1}| = |\nabla f^{-1}(q_i)| \leqq \mathfrak{L}_{f^{-1}|R_1'};$$

hence the same proof holds for $F_i^*\omega$.

We show next that if α_0 is the n-direction of E, then

(11) $$|[F_i^*\omega_i(p) - f^*\omega_i(p)] \cdot \alpha_0| < \epsilon_1 \quad \text{if} \quad |p - p_i| < \rho.$$

By (3),

$$|q' - q| \leqq \epsilon_2 |p - p_i| < \epsilon_2 \rho \quad \text{if} \quad q' = F_i(p), \quad q = f(p).$$

Hence, by (6), (2) and (5),

$$| \omega_i(q') - \omega_i(q) | \leqq | \phi_i(q') - \phi_i(q) | | \omega(q') | + | \phi_i(q) | | \omega(q') - \omega(q) |$$
$$\leqq \mathfrak{L}_{\phi_i} | q' - q | | \omega | + | \omega(q') - \omega(q) | < LL' | \omega | \epsilon_2 + \epsilon_1/3J.$$

By (II, 4.4), (8) and (II, 6.1),

$$F_i^* \omega_i(p) \cdot \alpha_0 = \omega_i(F_i(p)) \cdot \nabla F_i(p, \alpha_0) = \omega_i(q') \cdot J_f(p_i);$$

similarly $f^*\omega_i(p) \cdot \alpha_0 = \omega_i(q) \cdot J_f(p)$. Therefore, using (4),

$$| [F_i^* \omega_i(p) - f^* \omega_i(p)] \cdot \alpha_0 |$$
$$\leqq | \omega_i(q') - \omega_i(q) | | J_f(p_i) | + | \omega_i(q) | | J_f(p_i) - J_f(p) |$$
$$< (LL' | \omega | \epsilon_2 + \epsilon_1/3J) J + | \omega | \epsilon_1/3 | \omega |,$$

which gives (11).

Since $U_\rho(p_i)$ is contained in a cube of side 2ρ, (10) and (11) give

$$(12) \qquad \left| \int_R (F_i^* \omega_i - f^* \omega_i) \right| \leqq \int_{U_\rho(p_i)} | F_i^* \omega_i - f^* \omega_i | < 2^n \rho^n \epsilon_1.$$

Now F_i maps a neighborhood U_i of $\operatorname{spt}(F_i^* \omega_i)$ onto a neighborhood U_i' of $\operatorname{spt}(\omega_i)$; by (8), $J_{F_i}(p) > 0$. Since we have proved the theorem for affine F_i and compact ω_i, we have

$$\int_R F_i^* \omega_i = \int_{U_i} F_i^* \omega_i = \int_{U_i'} \omega_i = \int_{R'} \omega_i.$$

Therefore, since $\sum_{i=1}^m \phi_i(p) = 1$ in $\operatorname{spt}(\omega)$,

$$\sum_{i=1}^m \int_R F_i^* \omega_i = \sum_{i=1}^m \int_{R'} \omega_i = \int_{R'} \omega.$$

Also $\sum_{i=1}^m \int_R f^* \omega_i = \int_R f^* \omega$; therefore

$$\int_{R'} \omega - \int_R f^* \omega = \sum_{i=1}^m \int_R (F_i^* \omega_i - f^* \omega_i).$$

Since the P_i lie in R_1' and $| P_i | = (\rho'/n^{1/2})^n$, the number of values of i is

$$m \leqq | R_1' | / | P_1 | = M \cdot n^{n/2} / \rho'^n = n^{n/2} M L'^n / \rho^n.$$

Therefore

$$\left| \int_{R'} \omega - \int_R f^* \omega \right| \leqq m \cdot 2^n \rho^n \epsilon_1 \leqq \epsilon,$$

proving (7.1) for compact ω.

Consider now the general case; we shall apply Lemma 6g. Given $\epsilon > 0$, choose a compact set $P' \subset R'$ by Lemma 6b so that

$$\int_{R' - P'} | \omega | < \epsilon.$$

Set $P = f^{-1}(P')$. Now take any continuous function ϕ in R with $\mathrm{spt}(\phi) \subset R$ compact, such that $0 \leqq \phi \leqq 1$ and $\phi = 1$ in P. Set $\phi'(q) = \phi(f^{-1}(q))$ in R'. Then

$$\left| \int_{R'} \omega - \int_{R'} \phi' \omega \right| = \left| \int_{R'-P'} (1 - \phi')\omega \right| \leqq \int_{R'-P'} \left| \omega \right| < \epsilon.$$

Since $\phi'\omega$ is compact and $f^*(\phi'\omega) = \phi(f^*\omega)$ (which may be proved directly from (II, 4.4)), the proof above gives

$$\int_R \phi f^*\omega = \int_R f^*\phi'\omega = \int_{R'} \phi'\omega.$$

Hence

$$\left| \int_{R'} \omega - \int_R \phi f^*\omega \right| < \epsilon.$$

Therefore, by Lemma 6g, $f^*\omega$ is summable over R, and (7.1) holds. This completes the proof.

9. Transformation of the Riemann integral. We show how Theorem 7A gives the usual formula for the transformation of the Riemann integral under a change of coordinates. Let $\phi(p)$ be continuous and Riemann summable over the open set R' (see § 6). Let f be a one-one regular mapping of the open set R onto R'. With coordinate systems in the spaces of R and R', this is equivalent to choosing a new coordinate system in R' (see (II, 5)). We wish to express $\int_{R'} \phi(p)\, dp$ as a Riemann integral over R.

With the unit n-covectors ω_0 and ω_0' of the oriented spaces of R and R', define ω and ϕ^* by

$$\omega(q) = \phi(q)\omega_0', \qquad f^*\omega(p) = \phi^*(p)\omega_0.$$

Then, by (II, 6.8), $\phi^*(p) = \mathfrak{J}_f(p)\phi(f(p))$. Writing both sides of (7.1) as Riemann integrals (see § 5) gives

$$(1) \qquad \int_{R'} \phi(q)\, dq = \int_R \mathfrak{J}_f(p)\phi(f(p))\, dp \quad \text{if} \quad \mathfrak{J}_f(p) > 0 \quad \text{in } R;$$

the same formula holds with the minus sign if $\mathfrak{J}_f(p) < 0$ in R. Also, taking $\phi(p) = 1$ in R' gives

$$(2) \qquad |R'| = \int_{R'} dq = \int_R \mathfrak{J}_f(p)\, dp \quad \text{if} \quad \mathfrak{J}_f(p) > 0 \quad \text{in } R.$$

10. Integration in manifolds. In Euclidean space, the simplest domains of integration are oriented polyhedral cells. In smooth manifolds M (II, 10), the simplest domains are smooth images $f\sigma$ of oriented cells σ. If the continuous r-form ω is defined in a neighborhood of $f\sigma$, σ an oriented r-cell, we may define $f^*\omega$ as in (II, 12), and define (corresponding to (7.1))

$$(1) \qquad \int_{f\sigma} \omega = \int f^*\omega.$$

We suppose next that M is a compact oriented smooth manifold of dimension n, and ω is a continuous n-form in M; we shall define $\int_M \omega$.

Let spt (ω) denote the closure of the set of points $p \in M$ with $\omega(p) \neq 0$. Suppose first that this set lies in some coordinate system χ. Then we may define $\int_M \omega$ as in (1), using χ. Thus

$$\int_M \omega = \int_O \chi^* \omega \quad \text{if} \quad \text{spt}\,(\omega) \subset \chi(O). \tag{2}$$

Suppose we had spt $(\omega) \subset \chi'(O')$ also. Then there are neighborhoods R and R' of $\chi^{-1}(\text{spt}\,(\omega))$ and $\chi'^{-1}(\text{spt}\,(\omega))$ respectively, such that $\psi = \chi'^{-1}\chi$ is a one-one regular mapping of R onto R': since M is oriented, $J_\psi(x) > 0$ in R. By Theorem 7A,

$$\int_{O'} \chi'^* \omega = \int_{R'} \chi'^* \omega = \int_R \psi^* \chi'^* \omega = \int_O \chi^* \omega,$$

showing that the definition of $\int_M \omega$ is independent of the coordinate system employed.

To integrate any ω, one could take a triangulation of M (IV, Theorem 12A), and integrate separately over each cell. But it is simpler to use a partition $\sum \phi_i$ of unity (see § 8 and (App. III, 2)): Express ω as $\sum \omega_i$, $\omega_i = \phi_i \omega$, with each spt (ω_i) in a coordinate system, and sum the $\int_M \omega_i$. We show how to do this.

Let O and O' be the open balls about 0 in $\mathfrak{A}^n$, of radii 1 and 2 respectively. We may find a finite set $\chi_1, \cdots, \chi_m$ of coordinate systems, each defined in O', the $\chi_i(O)$ covering M. Define the smooth non-negative function $\Phi(x)$ in $\mathfrak{A}^n$, positive in O and zero in a neighborhood of $\mathfrak{A}^n - O'$, as in (App. III, 1). Set $\phi_i'(p) = \Phi(\chi_i^{-1}(p))$ in $\chi_i(O')$, and $= 0$ elsewhere in M; then ϕ_i' is smooth in M, and $\phi_i' > 0$ in $\chi_i(O)$. Hence we may set $\phi_i(p) = \phi_i'(p)/\sum \phi_j'(p)$ in M; now $\phi_i = 0$ outside $\chi_i(O')$, and $\sum \phi_i(p) = 1$ in M.

Set $\omega_i(p) = \phi_i(p)\omega(p)$ in M; then $\omega = \sum \omega_i$. We may define each $\int_M \omega_i$, and set $\int_M \omega = \sum \int_M \omega_i$. To show that the result is independent of the χ_i and ϕ_i employed, let χ_j' and ϕ_j'' $(j = 1, \cdots, m')$ be another such set. Using the invariance proof above, we find

$$\sum_i \int_{O'} \chi_i^*(\phi_i \omega) = \sum_i \int_{O'} \chi_i^* \Big(\sum_j \phi_j'' \phi_i \omega \Big) = \sum_{i,j} \int_{O'} \chi_i^*(\phi_j'' \phi_i \omega)$$

$$= \sum_{i,j} \int_{O'} \chi_j'^*(\phi_j'' \phi_i \omega) = \sum_j \int_{O'} \chi_j'^* \Big(\sum_i \phi_i \phi_j'' \omega \Big) = \sum_j \int_{O'} \chi_j'^*(\phi_j'' \omega),$$

which gives the result.

Now consider any open subset R of an oriented manifold M, with $\bar{R}$ compact. Let ω be a continuous n-form, defined in a neighborhood U of $\bar{R}$. We may choose coordinate systems $\chi_1, \cdots, \chi_m$ as above, so that the $\chi_i(O)$ cover $\bar{R}$. Define the ϕ_i' and the ϕ_i as before; for some neighborhood $U' \subset U$ of $\bar{R}$, the ϕ_i are defined in U', and $\sum \phi_i(p) = 1$ there. We may set

$$(3) \qquad \int_R \phi_i\omega = \int_{\chi_i^{-1}(R)} \chi_i^*\phi_i\omega, \qquad \int_R \omega = \sum_i \int_R \phi_i\omega;$$

for $|\chi_i^*\phi_i\omega|$ is finite in $U_i^* = \chi_i^{-1}(R)$, and $|U_i^*|$ is finite; now Lemma 6d shows that each term is defined. As before, the definition of $\int_R \omega$ is independent of the choice of the χ_i and ϕ_i.

Finally, take any open subset R of the oriented smooth n-manifold M, and suppose the n-form ω is continuous in R. We say ω is *summable* over R if for each $\epsilon > 0$ there is an open set R_0 with $\bar{R}_0$ compact, $\bar{R}_0 \subset R$, such that for any open set R' with $\bar{R}'$ compact,

$$(4) \qquad \left| \int_{R'} \omega - \int_{R_0} \omega \right| < \epsilon \quad \text{if} \quad R_0 \subset R', \quad \bar{R}' \subset R.$$

If this holds, then $\int_R \omega$ is uniquely definable, as in § 6, and the properties through Lemma 6a continue to hold, with polyhedra replaced by open sets with compact closures; we see easily that Lemma 6c holds also. The definition is equivalent to that in § 6 in case $M = E^n$, as follows from Lemma 6f.

We give a generalization of Theorem 7A to the present situation.

THEOREM 10A. *Let M and M' be oriented smooth manifolds of dimension n, and let f be a one-one regular orientation preserving mapping of the open subset R of M onto the open subset R' of M'. Let ω be a continuous n-form in R', summable over R'. Then $f^*\omega$ is summable over R, and* (7.1) *holds. If f reverses orientation,* (7.1) *holds, with the minus sign.*

With the help of coordinate systems, this follows at once from the discussion above and Theorem 7A.

Let us now consider the integration of r-forms ω in M. For a domain of integration, we take a subset R of an oriented r-dimensional submanifold M' of M, R being open in M'. We suppose ω is continuous in a neighborhood U of R in M (or simply, ω is an r-form in M, defined and continuous at all points of R). We may consider M' as the image fM', f being the identity. Now we may define $\int_R \omega$ to be $\int_R f^*\omega$ (which amounts to considering $\omega(p)\cdot\alpha$ just for α in the tangent space of M' at p).

11. Stokes' Theorem for a parallelepiped. We prove Stokes' Theorem for the simplest case; it then follows through approximations and partitions of unity for as general a case as desired. We say ω is *smooth in the cell* Q if $\nabla_v\omega(p)$ is continuous in Q for each v. It can be shown that this holds if and only if† ω can be extended to be smooth in a neighborhood of Q.

LEMMA 11a. *Let Q be an oriented n-dimensional parallelepiped in E^n, and let ω be a smooth $(n-1)$-form in Q. Then*

$$\int_Q d\omega = \int_{\partial Q} \omega. \tag{1}$$

We shall give a direct proof, not involving the theory of partial integration. For the classical treatment, see §§ 19–21 of the Introduction. The proof will be carried through for a more general case in (IX, Theorem 12B).

For some p_0 and independent vectors $v_1, \cdots, v_n$, Q consists of all points $p_0 + \sum a_i v_i$ $(0 \leqq a_i \leqq 1)$. For each i, there are faces

$$A_i^-: \text{ all points } p_0 + \sum_{j \neq i} a_j v_j, \qquad A_i^+: \text{ all points } p_0 + v_i + \sum_{j\neq i} a_j v_j.$$

We may suppose Q is oriented by the ordered set $(v_1, \cdots, v_n)$. Orient A_i^- and A_i^+ by the ordered set $(v_1, \cdots \hat{v}_i \cdots, v_n)$. By (App. II, 7),

$$\partial Q = \sum_{i=1}^{n} (-1)^{i-1}(A_i^+ - A_i^-). \tag{2}$$

Suppose $\epsilon > 0$ is given. Using (II, Lemma 2a), we may choose $\zeta > 0$ so that

$$|\omega(q) - \omega(p)|, \quad |\nabla_v\omega(q) - \nabla_v\omega(p)|, \quad |d\omega(q) - d\omega(p)| \text{ are } \leqq \epsilon$$
$$\text{if } \quad p, q \in Q, \quad |q - p| < \zeta, \quad |v| = 1.$$

For some m, if we subdivide Q into m^n equal parallelepipeds Q_k by means of $(n-1)$-planes parallel to the A_i^-, we will have diam $(Q_k) < \zeta$ for each k.

Take any Q_k; let B_{ki}^- and B_{ki}^+ be its faces, corresponding to A_i^- and A_i^+. Let p_k, p_{ki}^-, p_{ki}^+ be the centers of Q_k, B_{ki}^- and B_{ki}^+ respectively. Define v_i' by

$$p_{ki}^+ - p_{ki}^- = v_i' = v_i/m.$$

By § 1 and (I, Theorem 9A),

$$\{Q_k\} = v_1' \vee \cdots \vee v_n', \qquad \{B_{ki}^-\} = \{B_{ki}^+\} = \beta_i = v_1' \vee \cdots \hat{v}_i' \cdots \vee v_n'.$$

Applying the law of the mean to the function $\omega(p)\cdot\beta_i$ over the segment $p_{ki}^- p_{ki}^+$ (equivalently, using (II, 1.3) and the continuity of $\nabla_{v_i'}\omega$), we find a point p_{ki}^* on the segment such that

$$\omega(p_{ki}^+)\cdot\{B_{ki}^+\} - \omega(p_{ki}^-)\cdot\{B_{ki}^-\} = \nabla_{v_i'}\omega(p_{ki}^*)\cdot\beta_i.$$

† H. Whitney, Functions differentiable on the boundaries of regions, *Annals of Math.* *35* (1934) 482–485.

Hence, letting $\xi \circ Q_k$ denote $\xi(p_k)\cdot\{Q_k\}$ etc., using (II, 8.1) and (I, 12.16) gives

$$\left| d\omega\circ Q_k - \omega\circ\partial Q_k \right| = \left| \sum_i (-1)^{i-1}[\nabla_{v_i'}\omega(p_k) - \nabla_{v_i'}\omega(p^*_{ki})]\cdot\beta_i \right|$$
$$\leqq \sum_i \epsilon \,|\, v_i' \,|\,|\, \beta_i \,| \leqq n\epsilon \,|\, v_1 \,| \cdots |\, v_n \,|/m^n.$$

Now $\sum Q_k$ is a subdivision of Q, and letting B_j' denote those faces of the Q_k, properly oriented, which lie in ∂Q, $\sum B_j'$ is a subdivision of ∂Q. By the choice of ζ and Lemma 4a,

$$\left| \int_Q d\omega - d\omega\circ\sum Q_k \right| \leqq \epsilon \,|\, Q \,|,$$
$$\left| \int_{\partial Q} \omega - \omega\circ\sum B_j' \right| \leqq \epsilon \,|\, \partial Q \,|.$$

Also $\omega\circ\sum B_j' = \omega\circ\sum\partial Q_k$, and by the inequality proved above,

$$\left| d\omega\circ\sum Q_k - \omega\circ\sum B_j' \right| \leqq n\epsilon \,|\, v_1 \,| \cdots |\, v_n \,|.$$

These inequalities give

$$\left| \int_Q d\omega - \int_{\partial Q} \omega \right| \leqq \epsilon[\,|\, Q \,| + |\, \partial Q \,| + n \,|\, v_1 \,| \cdots |\, v_n \,|\,].$$

Since ϵ was arbitrary, (1) follows.

12. A special case of Stokes' Theorem. We shall extend Lemma 11a in two directions: We consider a region with one face curved instead of flat, and we do not require ω to be smooth on this face; however, we assume ω vanishes on the other faces. No other case of Stokes' Theorem is needed in the proofs of Theorem 14A and 18A below.

Let $h(x^2, \cdots, x^n)$ be a smooth function, defined for $-2 < x^i < 2$ $(i = 2, \cdots, n)$, such that $-1/2 < h(x^2, \cdots, x^n) < 1/2$. Let R be the region in $\mathfrak{A}^n$ defined by

$$h(x^2, \cdots, x^n) < x^1 < 1, \qquad -1 < x^i < 1 \quad (i = 2, \cdots, n).$$

Let A be the face where $x^1 = h(x^2, \cdots, x^n)$, and let B be the sum of the other faces, oriented so that $\partial R = A + B$, for some fixed orientation of R (App. II, 5).

LEMMA 12a. *Let R be as above. Let ω be a continuous $(n - 1)$-form in $\bar{R}$, such that ω is smooth in R, $d\omega$ is summable in R, and $\omega = 0$ in a neighborhood of the closed set B. Then*

(1) $$\int_A \omega = \int_R d\omega.$$

REMARK. In place of assuming that ω is smooth in R, we could simply assume that ω is regular there; see § 16 below. The proof would be slightly simplified, since we could then replace $h_k(x^2, \cdots)$ below by $h(x^2, \cdots) + 1/k$.

Let R' be the region defined by

$$0 < x^1 < 2, \qquad -1 < x^i < 1 \quad (i = 2, \cdots, n);$$

we shall approximate to R by regions R_k where ω is smooth, and compare each R_k with R'.

For each integer k, let $h_k(x^2, \cdots, x^n)$ be a 2-smooth function which approximates to $h(x^2, \cdots, x^n) + 1/k$, together with first partial derivatives, with an error $<1/k$, in the set where $-1 \leqq x^i \leqq 1$ $(i \geqq 2)$ (App. III, Lemma 4a). Let R_k be the region defined by

$$h_k(x^2, \cdots, x^n) < x^1 < h_k(x^2, \cdots, x^n) + 2, \qquad -1 < x^i < 1 \quad (\mathrm{i} = 2, \cdots, n).$$

Set $\omega = 0$ in $\bar{R}_k - R$; now ω is smooth in $\bar{R}_k$. Set

$$(2) \quad f_k(x^1, x^2, \cdots, x^n) = (x^1 + h_k(x^2, \cdots, x^n), x^2, \cdots, x^n) \quad \text{in } \bar{R}';$$

this is a one-one regular 2-smooth mapping of $\bar{R}'$ onto $\bar{R}_k$. Let A', A_k be the faces of R', R_k respectively corresponding to the face A of R. Since f_k is 2-smooth and ω is smooth in R, (II, 8.8) shows that $df_k^*\omega = f_k^* d\omega$ in R'; also $f_k^*\omega$ is smooth in $\bar{R}'$. Hence, by Lemma 11a and Theorem 7A,

$$\int_{A'} f_k^*\omega = \int_{\partial R'} f_k^*\omega = \int_{R'} df_k^*\omega = \int_{R'} f_k^* d\omega = \int_{R_k} d\omega.$$

Now let $k \to \infty$. Since $d\omega$ is summable over R, $\int_{R_k} d\omega \to \int_R d\omega$.

Define f like f_k in (2), with h_k replaced by h; we shall use this in A' only. Now $f_k \to f$ and $\nabla f_k \to \nabla f$, both uniformly, as $k \to \infty$. Hence

$$f_k^*\omega(p)\cdot\alpha = \omega(f_k(p))\cdot\nabla f_k(p, \alpha) \to \omega(f(p))\cdot\nabla f(p, \alpha) = f^*\omega(p)\cdot\alpha,$$

uniformly in A'. Therefore, by Theorem 10A,

$$\lim_{k\to\infty} \int_{A'} f_k^*\omega = \int_{A'} f^*\omega = \int_A \omega.$$

These relations give (1).

13. Sets of zero s-extent. We shall introduce a concept which, in a sense, expresses the smallness of the s-volume of a set. For an actual measure of s-dimensional volume, one should use Hausdorff measure (see Saks, p. 53; Halmos, p. 53); but the present notion is simpler to handle, and is sufficient for our purposes.

Let Q be a subset of E^n (or of any metric space). We say that Q has *zero s-extent* if the following is true. For each $\epsilon > 0$ there is a $\zeta_0 > 0$ such that for any $\zeta \leqq \zeta_0$ there are sets $Q_1, \cdots, Q_k$ (for some k) such that

$$(1) \quad Q = Q_1 \cup \cdots \cup Q_k, \qquad \operatorname{diam}(Q_i) \leqq \zeta \quad (\text{all } i), \qquad k\zeta^s < \epsilon.$$

Note that Q must be bounded. Clearly the union of a finite set of sets of zero s-extent is of zero s-extent.

LEMMA 13a. *A bounded subset Q of E^{s-1} is of zero s-extent; if f is a Lipschitz mapping* (II, 4) *of Q into E^n, then $f(Q)$ is of zero s-extent.*

To prove the second statement (from which the first follows), let Q' be a cube containing Q, of diameter δ. Given $\epsilon > 0$, let ζ_0 be the smaller of $\epsilon/2^s\mathfrak{L}_f^{s-1}\delta^{s-1}$, $\mathfrak{L}_f\delta$. Now take any $\zeta \leqq \zeta_0$. There is an integer $m \geqq 0$ such that

$$\mathfrak{L}_f\delta/2^m \leqq \zeta < \mathfrak{L}_f\delta/2^{m-1}.$$

Cut Q' into $2^{(s-1)m}$ equal cubes; these are of diameter $\delta/2^m \leqq \zeta/\mathfrak{L}_f$. This cuts Q into pieces $Q_1, \cdots, Q_k$, of diameter $\leqq \zeta/\mathfrak{L}_f$, with $k \leqq 2^{(s-1)m}$. Now

$$f(Q) = f(Q_1) \cup \cdots \cup f(Q_k), \qquad \text{diam}\,(f(Q_i)) \leqq \zeta,$$

$$k\zeta^s < 2^{(s-1)m}(\mathfrak{L}_f\delta/2^{m-1})^s \leqq 2^s\mathfrak{L}_f^{s-1}\delta^{s-1}\zeta \leqq \epsilon,$$

completing the proof.

The following lemma is of general interest, but will not be needed.

LEMMA 13b. *A bounded closed subset Q of E^n has zero n-extent if and only if it has zero Lebesgue measure.*

If Q has zero Lebesgue measure, it may be covered by a finite set of rectangular parallelepipeds, of arbitrarily small total volume; it follows easily that Q has zero n-extent. The converse is simple.

In the rest of this section we use a certain subdivision of a given open set $R \subset E^n$ into cubes, as follows. Take a cubical subdivision of E^n into cubes of side 1, hence of diameter $n^{1/2}$; let K_0' be this set of cubes, and let K_0 be the subset of these cubes in R whose distances from $E^n - R$ are at least $3n^{1/2}$. Having found $K_0', K_0, \cdots, K_m', K_m$, cut each cube of $K_m' - K_m$ into 2^n equal cubes, let K_{m+1}' be this set of cubes, and let K_{m+1} be the subset of these cubes whose distances from $E^n - R$ are at least $3n^{1/2}/2^{m+1}$. The cubes of $K_0, K_1, \cdots$ cover R.

Each cube C of K_m is of side $1/2^m$ and of diameter $n^{1/2}/2^m$; it lies in R, and

$$(2) \qquad 3n^{1/2}/2^m \leqq \text{dist}\,(C, E^n - R) < 7n^{1/2}/2^m \qquad (m > 0).$$

To prove the last inequality, suppose C was formed by subdividing C', of side $1/2^{m-1}$, in $K_{m-1}' - K_{m-1}$. Since C' is not in K_{m-1}, dist $(C', E^n - R) < 3n^{1/2}/2^{m-1}$. But $C' \subset \bar{U}_h(C)$, $h = n^{1/2}/2^m$; hence

$$\text{dist}\,(C, E^n - R) < 3n^{1/2}/2^{m-1} + n^{1/2}/2^m = 7n^{1/2}/2^m.$$

We prove

$$(3) \qquad \text{dist}\,(K_{m-1}, K_{m+1}) \geqq 2n^{1/2}/2^m.$$

For any C in K_{m+1}, dist $(C, E^n - R) < 7n^{1/2}/2^{m+1}$ and diam $(C) = n^{1/2}/2^{m+1}$; hence $K_{m+1} \subset U_h(E^n - R)$, $h = 4n^{1/2}/2^m$. Also dist $(C', E^n - R) \geqq 6n^{1/2}/2^m$ for cubes C' of K_{m-1}, and (3) follows.

LEMMA 13c. *Let R be an open set in E^n, subdivided as above. Let $Q \subset E^n - R$ be of zero s-extent. Then given any $\epsilon > 0$, there is an m_0 with the following property. Take any $m \geqq m_0$, and let N be the number of cubes of K_m whose distances from Q are at most $7n^{1/2}/2^m$. Then $N \leqq 2^{sm}\epsilon$.*

Set $a = 2 + 16n^{1/2}$, $\epsilon_1 = \epsilon/a^n$, and choose ζ_0 so that (1) holds with ϵ_1, if $\zeta \leqq \zeta_0$. Choose m_0 so that $1/2^{m_0} \leqq \zeta_0$. Now take any $m \geqq m_0$. Set $\zeta = 1/2^m$, and choose $Q_1, \cdots, Q_k$ to satisfy (1), with ϵ_1. For each i, let Q_i' be a cube with a point of Q_i as center, and of side $a/2^m$. Since diam $(Q_i) \leqq \zeta = 1/2^m$ and diam (C) $= n^{1/2}/2^m$ if C is in K_m, Q_i' contains all cubes C to be counted which are within $7n^{1/2}/2^m$ of Q_i. Since C is of side $1/2^m$, there are at most a^n such cubes. Hence

$$N \leqq a^n k < a^n \epsilon_1/\zeta^s = 2^{sm}\epsilon.$$

We shall define a partition of unity in R corresponding to the subdivision. Let $C_1, C_2, \cdots$ denote the cubes of all the K_m. For each i, let C_i' be the cube concentric with C_i and of twice the side length. Since $C_i' \subset U_h(C_i)$, $h = n^{1/2}/2^m$, if $C_i \in K_{m-1}$, and similarly for cubes of K_{m+1}, (3) shows that

$$C_i' \cap C_j' = 0 \quad \text{if} \quad C_i \in K_{m-1}, C_j \in K_{m+1}. \tag{4}$$

Let Φ be an ∞-smooth function in $\mathfrak{A}^n$ which is >0 within a cube C and is $= 0$ outside (App. III, Lemma 1b). Using an affine mapping of C into C_i' gives an ∞-smooth function ϕ_i' in E^n which is >0 within C_i' and $= 0$ outside. Clearly, for some N_0,

$$|\nabla \phi_i'| \leqq 2^m N_0 \quad \text{if} \quad C_i \in K_m.$$

Set $\phi_i(p) = \phi_i'(p)/\sum_k \phi_k'(p)$ in R; ϕ_i is ∞-smooth, >0 in C_i', and $= 0$ outside; $\sum \phi_i(p) = 1$ in R. Because of (4), there is a number c such that any point of R is in at most c cubes C_i'. Consequently there are but a finite number of combinations of shapes of cubes about any point of R. This, with the relation above, shows that there is a number N_1 such that

$$|\nabla \phi_i| \leqq 2^m N_1 \quad \text{in } R \qquad \text{if} \quad C_i \in K_m. \tag{5}$$

14. Stokes' Theorem for standard domains. In this section we state the Theorem of Stokes for domains in n-space which should be sufficiently general for all ordinary applications; the proof will be given in the next section.

By a *standard domain* in oriented n-space E^n, we mean a bounded connected open set R, with the following properties. Set $P^* = \bar{R} - R$. There is a closed set $Q \subset P^*$ of zero $(n - 1)$-extent. Set $P = P^* - Q$. For each point $p \in P$ there is a unit vector $v(p)$ such that if axes in E^n are chosen with $v(p)$ in the x^1-direction, then the set of points of P in some neighborhood of p is given by a smooth function $x^1 = h(x^2, \cdots, x^n)$, and the set of points of R in this neighborhood is given by the inequality $x^1 < h(x^2, \cdots, x^n)$. We may suppose $v(p)$ is the outward normal.

Since the topology of P is given by that of the surrounding space, P is separable; hence it consists of a finite or denumerable set of smooth manifolds. Thus a standard domain is a bounded connected open set, whose frontier is the union of a closed set of zero $(n-1)$-extent and a finite or denumerable set of smooth $(n-1)$-manifolds, each of which has the open set on just one side.

In the commonly used domains, Q consists of a finite set of manifolds, of dimensions $<n-1$; by Lemma 13a and the preceding remark, Q is of zero $(n-1)$-extent, and thus we have a standard domain. We could include, in Q, any closed subset of P of zero $(n-1)$-dimensional Lebesgue measure in an obvious sense; see Lemma 13b. (We could not include a larger set; compare Example 2 below.)

For each point $p \in P$, the orientation of R and the outward normal $v(p)$ define an orientation of P near p (App. II, 5). Thus P becomes a set of oriented manifolds.

THEOREM 14A*. *Let R be a standard domain in E^n, and let ω be an $(n-1)$-form such that*

(a) *ω is defined, continuous, and bounded in $\bar{R} - Q$, and is smooth in R,*

(b) *ω is summable over P,*

(c) *$d\omega$ is summable over R.*

Then

$$\int_P \omega = \int_R d\omega. \tag{1}$$

REMARKS. We could assume merely that ω is regular instead of smooth in R; see § 16 and Theorem 18A below. If $d\omega$ is bounded, then (c) holds automatically. In the usual applications, P will be of finite $(n-1)$-volume (defined through integration; we need not go into this here); then (*b*) holds automatically.

EXAMPLE 1. We cannot omit the assumption that ω is bounded, as we now show. Let R be the square $0 < x < 1$, $0 < y < 1$, in the plane; let Q be the set of four corners. For each point p of $\bar{R} - Q$, let $\theta(p)$ be its polar coordinate, $0 \leqq \theta(p) \leqq \pi/2$. Set $\omega(p) = \nabla\theta(p)$. Then all conditions are satisfied except that ω is not bounded. Clearly $d\omega = 0$ in R, $\int_P \omega = \pi/2$, so that (1) fails.

EXAMPLE 2. If we weaken the hypothesis about Q too much, the theorem may fail. To show this, take R as above. On the bottom side A, form a closed set Q_0 as follows. Remove the open interval of length $1/4$ from the center of A. Next remove the central interval of length $1/4^2$

* A very general theorem has been proved by H. Federer, The Gauss-Green Theorem, *Transactions of the Am. Math. Society*, *58* (1945) 44–76.

from each of the two remaining parts. Next remove the central interval of length $1/4^3$ from each of the four remaining parts, etc. Then Q_0 is the set of points remaining. Let Q be Q_0, together with the remaining corners of R. (Clearly we could let Q_0 be any closed set of positive Lebesgue measure; compare Lemma 13b.)

Let ϕ be a real function which is smooth in $\bar{R} - Q_0$, such that

$$\phi(x, y) = 1 \quad \text{if} \quad (x, 0) \in Q_0, \qquad \phi(p) = 0 \quad \text{in } A - Q_0,$$
$$0 \leqq \phi(p) \leqq 1, \qquad \partial\phi(p)/\partial y \geqq 0.$$

We may construct ϕ as follows. For each open interval $H_i = p_i p_i'$ of $A - Q_0$, let $q_i(t)$ be the point at a distance $t\,|\,p_i' - p_i\,|$ above the mid point of H_i. Let $C_i(t)$ be the part lying in R of the parabola through p_i, $q_i(t)$ and p_i' $(0 < t \leqq 1)$. Set $\kappa(t) = 2t - t^2$, and set $\phi(p) = \kappa(t)$ in $C_i(t)$ (all i), $\phi(p) = 0$ in $A - Q_0$, and $\phi(p) = 1$ elsewhere in $\bar{R} - Q$.

Define ω in $\bar{R} - Q$ by its components

$$\omega_1(p) = -\phi(p), \qquad \omega_2(p) = 0.$$

Then (a) and (b) of the theorem clearly hold. If R_h is the part of R with $y > h$, with top B and bottom A_h, then

$$\int_{R_h} d\omega = \int_{A_h} \omega - \int_B \omega, \qquad \left|\int_{A_h} \omega\right| = \left|\int_0^1 \phi(x, h)\, dx\right| \leqq 1,$$

and hence $\int_{R_h} d\omega$ is bounded. Also $(d\omega)_{12}(p) = -\partial\omega_1(p)/\partial y \geqq 0$; hence clearly (c) holds.

Take any $h > 0$, $\epsilon > 0$. Since $\phi(x, h) = 1$ if $(x, 0) \in Q_0$, $\phi(x, h) \geqq 1 - \epsilon$ except in a finite number of open intervals, of total length $< \sum_{i=0}^{\infty} 2^i/4^{i+1} = 1/2$; hence

$$\int_0^1 \phi(x, h)\, dh \geqq 1/2 \quad \text{if} \quad h > 0.$$

Since $\omega = 0$ in $A - Q_0$,

$$\int_R d\omega - \int_P \omega = \lim_{h\to 0} \int_{R_h} d\omega + \int_B \omega = \lim_{h \to 0} \int_{A_h} \omega \leqq -1/2,$$

and (1) fails.

15. Proof of the theorem. Take R, P, Q, ω as in Theorem 14A. We say ω is *Q-free* if $\omega = 0$ in some neighborhood of Q. We prove the theorem first in the case that ω is Q-free, then in the general case.

Suppose ω is Q-free; say $\omega = 0$ in $U \cap \bar{R}$, $Q \subset U$. Choose cubes about the points of $\tilde{R} = \bar{R} - U$, as follows. For each $p \in R \cap \tilde{R}$, let $U(p)$ and $U'(p)$ be concentric cubes about p, with $U(p) \subset U'(p) \subset R$. For each $p \in P \cap \tilde{R}$, let $v(p)$ be the outward normal at p; let $U(p) \subset U'(p)$ be concentric cubes with center p and one face perpendicular to $v(p)$;

if these are small enough, then we may choose rectangular coordinates in E^n so that the region $U'(p) \cap R$ is just the region R of § 12. A finite number of the $U(p)$ cover $\tilde{R}$; say these are $U_1, \cdots, U_k$, and the concentric cubes, $U_1', \cdots, U_k'$. Let $\phi_i'(p)$ be an ∞-smooth function $\geqq 0$ in E^n which is >0 in $\bar{U}_i$ and is 0 in a neighborhood of $E^n - U_i'$ (App. III, Lemma 1b). Set $\phi_i(p) = \phi_i'(p)/\sum_j \phi_j'(p)$ wherever defined. Then the ϕ_i are ∞-smooth in a neighborhood U^* of $\tilde{R}$, and $\sum \phi_i(p) = 1$ in $\tilde{R}$. Set

$$\omega_i(p) = \phi_i(p)\omega(p) \quad \text{in } \tilde{R}, \qquad \omega_i(p) = 0 \quad \text{in } E^n - \tilde{R}. \tag{1}$$

Consider any ω_i; first suppose U_i is about a point of P. Since $\nabla \phi_i$ and ω are bounded, $\nabla \phi_i \vee \omega$ is summable over $R \cap U_i'$; $\phi_i \, d\omega$ is also, by Lemma 6c. By (II, 8.6) and (II, 8.2),

$$d\omega_i = \nabla \phi_i \vee \omega + \phi_i \, d\omega; \tag{2}$$

this is summable over $R \cap U_i'$, and we may apply Lemma 12a, giving

$$\int_P \omega_i = \int_{P \cap U_i'} \omega_i = \int_{R \cap U_i'} d\omega_i = \int_R d\omega_i.$$

If $U_i' \subset R$, then by Lemma 11a,

$$\int_R d\omega_i = \int_{U_i'} d\omega_i = \int_{\partial U_i'} \omega_i = 0 = \int_P \omega_i.$$

Since $\sum \omega_i = \omega$, adding these relations gives

$$\int_P \omega = \sum_i \int_P \omega_i = \sum_i \int_R d\omega_i = \int_R d\omega.$$

We now consider the general case. Take the subdivision of the open set $E^n - Q$ (not the set R) into cubes, and the corresponding functions ϕ_i, as in § 13. For each integer m, set

$$\psi_m(p) = \sum_i^{(m)} \phi_i(p), \qquad \psi_m'(p) = 1 - \psi_m(p), \qquad \text{in} \quad E^n - Q, \tag{3}$$

the sum being over all i such that $C_i \in K_{m'}$ for some $m' \leqq m - 1$. Set

$$\omega_m = \psi_m \omega, \qquad \omega_m' = \psi_m' \omega = \omega - \omega_m, \qquad \text{in} \quad \bar{R} - Q. \tag{4}$$

Then ω_m is Q-free. Since ψ_m and ω are bounded, so is ω_m; since $\omega_m = 0$ outside a compact subset of P, ω_m is summable over P. Using (2) for $d\omega_m$ and the boundedness of $\nabla \psi_i$, we see that $d\omega_m$ is summable over R. Hence we may apply the theorem to ω_m, giving $\int_P \omega_m = \int_R d\omega_m$. We shall show that

$$\lim_{m \to \infty} \int_P \omega_m = \int_P \omega, \qquad \lim_{m \to \infty} \int_R d\omega_m = \int_R d\omega; \tag{5}$$

this will complete the proof.

If P has finite $(n-1)$-volume, the first relation in (5) is evident. In the general case, the method of proof of Lemma 6a applies (see also Lemma 6c and § 10).

We now prove the second relation in (5). Let H'_m be the union of all cubes C'_i such that $C_i \in K_{m'}$ for some $m' \geqq m$; set $H_m = H'_m \cap R$. Then by (4), (3) and (13.4),

$$(6) \qquad \omega'_m = 0 \quad \text{in} \quad R - H_m, \qquad \omega'_m = \omega \quad \text{in} \quad H_{m+1}.$$

Take any $\epsilon > 0$. Since $d\omega$ is summable over R, we may choose m_0 so that (see Lemma 6b)

$$(7) \qquad \int_{H_{m_0}} | d\omega | < \epsilon/2.$$

Each point of R is in at most b cubes C'_i, for some fixed b. Say $| \omega | \leqq N$ in R. With N_1 as in (13.5), set $\epsilon_1 = \epsilon/2^n b N_1 N$. By Lemma 13c, there is an $m_1 \geqq m_0$ such that if $m \geqq m_1$, then there are at most $2^{(n-1)m}\epsilon_1$ cubes in K_m. Since $H^*_m = H_m - H_{m+1}$ is covered by the cubes C'_i such that $C_i \in K_m$, and $| C'_i | = 1/2^{(m-1)n}$, we have

$$(8) \qquad | H^*_m | \leqq 2^{(n-1)m}\epsilon_1/2^{(m-1)n} = 2^{n-m}\epsilon_1 \quad \text{if} \quad m \geqq m_1.$$

For any $p \in R$, the sum in (3) contains at most b non-zero terms. Hence, by (13.5), $| \nabla \psi'_m | \leqq 2^{m-1} b N_1$, and if $m \geqq m_1$,

$$(9) \qquad \int_{H^*_m} | \nabla \psi'_m \vee \omega | \leqq 2^{m-1} b N_1 \cdot N \cdot 2^{n-m}\epsilon_1 = \epsilon/2.$$

Since $| \psi'_m | \leqq 1$, (7) gives

$$(10) \qquad \int_{H^*_m} | \psi'_m \, d\omega | + \int_{H_{m+1}} | d\omega | \leqq \int_{H_m} | d\omega | < \epsilon/2.$$

Therefore (see (2)), (6), (9) and (10) give

$$\int_R | d\omega'_m | = \int_{H^*_m} | \nabla \psi'_m \vee \omega + \psi'_m \, d\omega | + \int_{H_{m+1}} | d\omega | < \epsilon$$

for $m \geqq m_1$, completing the proof of (5) and hence of Theorem 14A.

16. Regular forms in Euclidean space.

Since the definition of an r-form ω $(r > 0)$ in a manifold M requires smoothness of M, and the definition of $d\omega$ requires 2-smoothness, the study of forms in manifolds in (II, 12) applies only to 2-smooth ones. To obtain a similar theory for smooth manifolds, we enlarge the definition of $d\omega$; this definition is due to E. Cartan.†

We shall need a lemma showing that a continuous form is determined by its integral over simplexes (we could equally well use parallelepipeds).

† See G. Papy, Formes différentielles exterieures · · · , Bull. Soc. Math. Belgique, 1953, pp. 62–69 (1954), and references given there. We follow the procedure given by H. Cartan, Notes of lectures delivered at Harvard University in 1948.

LEMMA 16a. *Let ω_1 and ω_2 be continuous r-forms in the open set R, and suppose* $\int_\sigma \omega_1 = \int_\sigma \omega_2$ *for all oriented r-simplexes σ in R. Then $\omega_1 = \omega_2$.*

Take any $p \in R$, and any r-direction α. There is a sequence of oriented simplexes $\sigma_1, \sigma_2, \cdots$ in R, with r-directions $\alpha = \{\sigma_i\}/|\sigma_i|$, such that $\sigma_i \in U_{\zeta_i}(p)$, $\zeta_i \to 0$. By (4.1) and (4.4), for any continuous r-form ω,

$$\frac{1}{|\sigma_i|}\int_{\sigma_i} \omega - \omega(p)\cdot\alpha = \frac{1}{|\sigma_i|}\int_{\sigma_i} [\omega(q) - \omega(p)]\, dq \to 0;$$

hence

$$\omega(p)\cdot\alpha = \lim_{i\to\infty} \frac{1}{|\sigma_i|}\int_{\sigma_i} \omega. \tag{1}$$

Applying this to ω_1 and to ω_2 shows that $\omega_1(p)\cdot\alpha = \omega_2(p)\cdot\alpha$ for all r-directions α; hence $\omega_1 = \omega_2$.

An r-form ω in the open set $R \subset E^n$ is *regular* if it is continuous there, and there is a continuous $(r + 1)$-form ω' in R such that

$$\int_{\partial\sigma} \omega = \int_\sigma \omega', \qquad \text{all } (r+1)\text{-simplexes} \quad \sigma \subset R \tag{2}$$

(we could use parallelepipeds). Then ω' is uniquely determined, by Lemma 16a; we call it the *derived* form $d\omega$ of ω. Note that it is sufficient to prove (2) in some neighborhood of each point of R. One could give a definition of regularity without using integrals; see Lemma 16d below.

If ω is regular, then so is $d\omega$, and $dd\omega = 0$; for if σ is an $(r + 2)$-simplex, then $\int_{\partial\sigma} d\omega = \int_{\partial\partial\sigma} \omega = 0$.

If ω is smooth, then (2) holds, with $\omega' = d\omega$ as previously defined; hence smooth forms are regular, and the present definition of $d\omega$ is an extension of the previous one.

A regular r-form ω need not be smooth, if $r > 0$. For example, with $n = 2$, $r = 1$, set $\omega = \omega_1(x^1)e^1 + \omega_2(x^2)e^2$, with real functions ω_1, ω_2 which are continuous but not differentiable. Then $d\omega = 0$. In general, if ω is smooth but not 2-smooth, then $d\omega$ will be regular but not smooth.

For $r = 0$, a regular form ω is always smooth. To show this, take any point p and vector $v \neq 0$. Set $p_t = p + tv$. Applying (2) with $\sigma = pp_t$ and using (1) gives

$$\nabla_v\omega(p) = \lim_{t\to 0+} \frac{\omega(p_t) - \omega(p)}{t} = \lim_{t\to 0+} \frac{1}{t}\int_{pp_t} \omega' = \omega'(p)\cdot v.$$

This is linear in v and continuous in p, showing that ω is smooth.

To prove various properties of regular forms ω, we shall smooth them by taking averages $A_i\omega$, as in (App. III, 3). Recall that for a set R, R_λ is the set of points p such that $\bar{U}_\lambda(p) \subset R$, and

$$A_i\omega(p) = \int_{E^n} \kappa_i'(q - p)\omega(q)\, dq = \int_{V^n} \kappa_i'(v)\omega(p + v)\, dv, \tag{3}$$

defined in $R_{1/i}$ if ω is defined in R. Also $A_i\omega$ is ∞-smooth. Since ω is continuous, (App. III, Lemma 3a) shows that

$$\lim_{i\to\infty} A_i\omega = \omega \qquad \text{u.c.s.,} \tag{4}$$

where "u.c.s." denotes "uniformly in compact sets".

LEMMA 16b. *If ω is regular in R, then*

$$A_i\, d\omega = dA_i\omega \quad \text{in } R_{1/i}. \tag{5}$$

For any $(r+1)$-simplex σ in $R_{1/i}$,

$$\begin{aligned}\int_\sigma A_i\, d\omega(p)\, dp &= \int_\sigma \int_V \kappa_i'(v)\, d\omega(p+v)\, dv\, dp\\ &= \int_V \kappa_i'(v) \int_\sigma d\omega(p+v)\, dp\, dv = \int_V \kappa_i'(v) \int_{\partial\sigma} \omega(p+v)\, dp\, dv\\ &= \int_{\partial\sigma}\int_V \kappa_i'(v)\omega(p+v)\, dv\, dp = \int_{\partial\sigma} A_i\omega(p)\, dp = \int_\sigma dA_i\omega(p)\, dp;\end{aligned}$$

(5) follows on using Lemma 16a.

LEMMA 16c. *If $\omega_1, \omega_2, \cdots$ are regular, and*

$$\lim \omega_i = \omega, \qquad \lim d\omega_i = \omega', \qquad \textit{both u.c.s.,} \tag{6}$$

then ω is regular, and $d\omega = \omega'$.

Because of the u.c.s. property, ω and ω' are continuous. To prove (2), we have

$$\int_\sigma \omega' = \lim \int_\sigma d\omega_i = \lim \int_{\partial\sigma} \omega_i = \int_{\partial\sigma} \omega.$$

LEMMA 16d. *ω is regular and $d\omega = \omega'$ if and only if there is a sequence $\omega_1, \omega_2, \cdots$ of smooth forms such that* $\lim \omega_i = \omega$ *and* $\lim d\omega_i = \omega'$ *u.c.s.*

This follows from (4), (5), and the last lemma.

THEOREM 16A. *If ω and ξ are regular in R, so is $\omega\vee\xi$, and $d(\omega\vee\xi)$ is given by* (II, 8.6). *In particular, $\phi\omega$ is regular if ϕ is smooth and ω is regular.*

For

$$\lim\, (A_i\omega\vee A_i\xi) = \lim A_i\omega \vee \lim A_i\xi = \omega\vee\xi,$$

also $\lim dA_i\omega = \lim A_i\, d\omega = d\omega$ etc., and hence

$$\begin{aligned}\lim d(A_i\omega\vee A_i\xi) &= \lim\, (dA_i\omega\vee A_i\xi \pm A_i\omega\vee dA_i\xi)\\ &= d\omega\vee\xi \pm \omega\vee d\xi.\end{aligned}$$

Each point p is in a neighborhood U such that for some i_0, $U \subset R_{1/i}$ for $i \geqq i_0$. The above limits are all u.c.s. in U; hence, by Lemma 16c, $\omega\vee\xi$ is regular as required in U, and hence in R.

In order to study smooth mappings f of open sets in E^n into E^m, we first smooth them further, by applying A_i. By (App. III, Lemma 3c), A_if is ∞-smooth, and

$$\lim A_i f(p) = f(p), \qquad \lim \nabla(A_if)(p, \alpha) = \nabla f(p, \alpha), \qquad \text{both u.c.s.} \tag{7}$$

LEMMA 16e. *Let f be as in the theorem below, and let ω be continuous in R'. Then*

$$\lim\,(A_i f)^*\omega = f^*\omega \qquad u.c.s.\ in\ R. \tag{8}$$

Take any $p \in R$ and r-vector α. For i large enough, $f_i = A_i f$ is defined near p, and (7) gives

$$\lim f_i^*\omega(p)\cdot\alpha = \lim\,[\omega(f_i(p))\cdot\nabla f_i(p, \alpha)] = \omega(f(p))\cdot\nabla f\,(p, \alpha) = f^*\omega(p)\cdot\alpha.$$

Clearly the limits are u.c.s., and (8) follows.

LEMMA 16f. *With f as in the theorem below, let $\omega_1, \omega_2, \cdots$ be continuous in R'. Then*

$$\lim f^*\omega_i = f^*\omega \quad u.c.s. \quad if \quad \lim \omega_i = \omega \quad u.c.s. \tag{9}$$

The proof is similar to that of the last lemma.

THEOREM 16B. *Let f be a smooth mapping of the open set $R \subset E^n$ into the open set $R' \subset E^m$, and let ω be regular in R'. Then $f^*\omega$ is regular in R, and*

$$df^*\omega = f^*\,d\omega. \tag{10}$$

First suppose that f is 2-smooth. Applying (10) to the smooth form $A_i\omega$ (II, 8.8) and using (5), (4), and the last lemma gives

$$\lim df^*A_i\omega = \lim f^*A_i\,d\omega = f^*\lim A_i\,d\omega = f^*\,d\omega;$$

also $\lim f^*A_i\omega = f^*\omega$; both limits are u.c.s. Now (10) follows from Lemma 16c.

Now take the general case. Using (10) with the 2-smooth mapping $A_i f$ and applying Lemma 16e gives

$$\lim d(A_i f)^*\omega = \lim\,(A_i f)^*\,d\omega = f^*\,d\omega;$$

also $\lim\,(A_i f)^*\omega = f^*\omega$; both are u.c.s. Again Lemma 16c applies.

17. Regular forms in smooth manifolds. Let ω be an r-form in the open set R in the smooth manifold M. We say ω is *regular* if ω is continuous, and there is a continuous $(r+1)$-form ω' in R with the following property. For any coordinate system χ, $\chi^*\omega$ is regular (where defined), and $d\chi^*\omega = \chi^*\omega'$. Set $d\omega = \omega'$; ω' is uniquely defined, and $d\chi^*\omega = \chi^*\,d\omega$.

If the property holds near p in one coordinate system χ, it holds near p in any other, χ_1. For if $\psi = \chi^{-1}\chi_1$, then $\chi_1 = \chi\psi$, $\chi_1^* = \psi^*\chi^*$, and using (16.10) gives

$$d\chi_1^*\omega = d\psi^*\chi^*\omega = \psi^*\,d\chi^*\omega = \psi^*\chi^*\omega' = \chi_1^*\omega',$$

as required.

A definition of regularity, using integrals, will be given in Lemma 17c below. If M is 2-smooth, then the condition of Lemma 16d could be used as a definition.

The elementary properties of regular forms continue to hold.

We prove the theorems of § 16 for smooth manifolds.

THEOREM 17A. *If ω and ξ are regular in the open set $R \subset M$, then so is $\omega \vee \xi$, and*

$$(1) \qquad d(\omega \vee \xi) = d\omega \vee \xi + (-1)^r \omega \vee d\xi, \qquad r = \deg(\omega).$$

For with any coordinate system χ, $\chi^*\omega$ and $\chi^*\xi$ are regular, hence so is their product, by Theorem 16A, and

$$d\chi^*(\omega \vee \xi) = d(\chi^*\omega \vee \chi^*\xi) = d\chi^*\omega \vee \chi^*\xi \pm \chi^*\omega \vee d\chi^*\xi$$
$$= \chi^*\, d\omega \vee \chi^*\xi \pm \chi^*\omega \vee \chi^*\, d\xi = \chi^*(d\omega \vee \xi \pm \omega \vee d\xi).$$

THEOREM 17B. *If f is a smooth mapping of the open subset R of M into M' (both manifolds smooth), and ω is regular in a neighborhood of $f(R)$, then $f^*\omega$ is regular in R, and*

$$(2) \qquad df^*\omega = f^*\, d\omega.$$

For given $p \in M$, choose coordinate systems χ about p and χ_1 about $f(p)$; then $g = \chi_1^{-1} f \chi$ is smooth. Now $\chi_1^*\omega$ is regular, and hence so is $g^*\chi_1^*\omega = \chi^* f^*\omega$ (Theorem 16B); since

$$d\chi^* f^*\omega = dg^*\chi_1^*\omega = g^*\chi_1^*\, d\omega = \chi^* f^*\, d\omega,$$

the theorem follows.

A *smooth simplex* $f\sigma$ in M is a one-one regular mapping f of σ into M. Define $\int_{f\sigma} \omega$ to be $\int_\sigma f^*\omega$. We may define $\int_{f\partial\sigma} \xi$ similarly.

LEMMA 17a. *If ω is a regular r-form in the open set $R \subset M$, and $f\sigma$ is a smooth $(r+1)$-simplex in R, then*

$$(3) \qquad \int_{\partial f\sigma} \omega = \int_{f\partial\sigma} \omega = \int_{f\sigma} d\omega.$$

For

$$\int_{f\partial\sigma} \omega = \int_{\partial\sigma} f^*\omega = \int_\sigma df^*\omega = \int_\sigma f^*\, d\omega = \int_{f\sigma} d\omega.$$

LEMMA 17b. *$\omega_1 = \omega_2$ in M if and only if $\int_{f\sigma} \omega_1 = \int_{f\sigma} \omega_2$ for all smooth r-simplexes.*

Suppose the condition holds. With a coordinate system χ, and simplexes σ in the domain of definition of χ,

$$\int_\sigma \chi^*\omega_1 = \int_{\chi\sigma} \omega_1 = \int_{\chi\sigma} \omega_2 = \int_\sigma \chi^*\omega_2;$$

hence $\chi^*\omega_1 = \chi^*\omega_2$, by Lemma 16a, and $\omega_1 = \omega_2$.

LEMMA 17c. *Let ω and ξ be continuous forms in the open set $R \subset M$, of degrees r and $r + 1$ respectively. Then ω is regular and $d\omega = \xi$ if and only if $\int_{f\partial\sigma} \omega = \int_{f\sigma} \xi$ for all smooth $(r + 1)$-simplexes σ in R.*

The necessity of the condition follows from Lemma 17a. To prove the sufficiency, note that the hypothesis gives $\int_{\partial\sigma} \chi^*\omega = \int_\sigma \chi^*\xi$ for simplexes σ in the domain of any coordinate system χ; hence $\chi^*\omega$ is regular and $d\chi^*\omega = \chi^*\xi$, giving the result.

18. Stokes' Theorem for standard manifolds. A "standard manifold" is locally like a standard domain. By a *partial standard domain* in E^n, we mean a set of sets R, P, Q, which have the properties of a standard domain, except that E^n is replaced by an open set O. Thus $P^* = \bar{R} \cap O - R$, and P^* and Q are closed in O, but in general are not closed in E^n. Again R is an oriented open set (which we need not assume connected), and P is a set of oriented smooth $(n - 1)$-manifolds. Clearly the intersection of any standard domain with any open set is a partial standard domain.

A *standard n-manifold M* is a system of the following sort. There is a connected compact topological space M, a closed subset ∂M of M, and a closed subset $\partial_0 M$ of ∂M. There is a finite set $(O'_i;\ R_i, P_i, Q_i)$ of partial standard domains, O'_i being an open ball. For each i, χ_i is a one-one continuous mapping of $\bar{R}_i \cap O'_i$ into M, such that

$$\chi_i(R_i) \subset M - \partial M, \qquad \chi_i(P_i) \subset \partial M - \partial_0 M, \qquad \chi_i(Q_i) \subset \partial_0 M.$$

There are interior concentric balls O_i such that the sets $\chi_i(\bar{R}_i \cap O_i)$ cover M. Set $\psi_{ij}(q) = \chi_j^{-1}(\chi_i(p))$, where defined. Then ψ_{ij} is smooth where defined in R_i, $\nabla \psi_{ij}$ having continuous boundary values in $R_i \cup P_i$, and (II, 4.14) $\mathfrak{L}_{\psi_{ij}}$ is finite.

Note that the coordinate systems χ_i make $M - \partial M$ into a smooth manifold; we say M is *oriented* if $M - \partial M$ is. By the hypothesis on $\nabla \psi_{ij}$, $\partial M - \partial_0 M$ is a set of smooth manifolds, all oriented if M is. We remark also that since Q_i is of zero $(n - 1)$-extent and $\mathfrak{L}_{\psi_{ij}}$ is finite, each $\psi_{ij}(Q_i)$ is of zero $(n - 1)$-extent, by Lemma 13a.

We may define r-forms ω in $M - \partial_0 M$. Say ω is *continuous* if each $\omega_i^* = \chi_i^* \omega$ is. Since

$$(1) \qquad \omega_j^*(p)\cdot\alpha = \psi_{ji}^*\omega_i^*(p)\cdot\alpha = \omega_i^*(\psi_{ji}(p))\cdot\nabla\psi_{ji}(p, \alpha),$$

and $\nabla \psi_{ji}$ is continuous where defined in $R_j \cup P_j$, the definition of continuity near a point of $\partial M - \partial_0 M$ is independent of the coordinate system chosen. Say ω is *bounded* if each ω_i^* is. Since $|\nabla \psi_{ji}| \leq \mathfrak{L}_{\psi_{ji}}$ is finite, we again have independence of the coordinate system.

THEOREM 18A. *Let M be an oriented standard n-manifold, and let ω be an $(n-1)$-form such that*

(a) *ω is defined, continuous, and bounded in $M - \partial_0 M$, and is regular in $M - \partial M$,*

(b) *ω is summable over $\partial M - \partial_0 M$,*

(c) *$d\omega$ is summable over $M - \partial M$.*

Then

$$\int_{\partial M - \partial_0 M} \omega = \int_{M - \partial M} d\omega. \tag{2}$$

To prove the theorem, we first reduce it to a local problem, as follows. For each i, let ϕ_i'' be a smooth function $\geqq 0$ in E^n, which is >0 in O_i and $=0$ in a neighborhood of $E^n - O_i'$. Set

$$\phi_i'(p) = \phi_i''(\chi_i^{-1}(p)) \quad \text{in } \chi_i(O_i'), \qquad = 0 \text{ elsewhere in } M.$$

Now ϕ_i' is continuous in M and smooth in $M - \partial_0 M$. The same is true of $\phi_i(p) = \phi_i'(p)/\sum_j \phi_j'(p)$, and $0 \leqq \phi_i(p) \leqq 1$, $\sum \phi_i(p) = 1$, in M. Set $\omega_i^* = \chi_i^* \omega$, and

$$\omega_i(p) = \phi_i(p)\omega(p) \quad \text{in} \quad M - \partial_0 M, \tag{3}$$

$$\omega_i'(q) = \chi_i^* \omega_i(q) = \phi_i(\chi_i(q))\omega_i^*(q) \quad \text{in } R_i. \tag{4}$$

We shall prove

$$\int_{P_i} \omega_i' = \int_{R_i} d\omega_i', \tag{5}$$

$$\int_{\partial M - \partial_0 M} \omega_i = \int_{P_i} \omega_i', \qquad \int_{M - \partial M} d\omega_i = \int_{R_i} d\omega_i', \tag{6}$$

everything being summable; this will prove (2) for ω_i, and summing over i gives (2) for ω.

First, since ω is summable over $\partial M - \partial_0 M$ and ϕ_i is bounded, $\phi_i \omega = \omega_i$ is summable over $\partial M - \partial_0 M$ (Lemma 6c, § 10). Also $\omega_i = 0$ outside $\chi_i(O_i')$. By Theorem 10A, $\chi_i^* \omega_i$ is summable over P_i, and the first part of (6) holds.

Next, set

$$\phi_i^*(q) = \chi_i^* \phi_i(q) = \phi_i(\chi_i(q)), \qquad \phi_{ij}(q) = \phi_j'(\chi_i(q)),$$

where defined in R_i. Then

$$\phi_{ij}(q) = \phi_j'(\chi_j(\psi_{ij}(q))) = \phi_j''(\psi_{ij}(q)),$$

and since $\nabla \phi_j''$ and $\nabla \psi_{ij}$ are bounded, so is $\nabla \phi_{ij}$ (outside Q_i). Also $\phi_i^* = \phi_{ii}/\sum_j \phi_{ij}$, showing that $\nabla \phi_i^*$ is bounded. By hypothesis, ω_i^* is bounded; therefore so is $\nabla \phi_i^* \vee \omega^*$. Since R_i is bounded, $\nabla \phi_i^* \vee \omega_i^* = \chi_i^*(\nabla \phi_i \vee \omega)$ is summable over R_i. Applying Theorem 10A to this form and the smooth mapping χ_i^{-1} gives

$$\int_{R_i} \nabla \phi_i^* \vee \omega_i^* = \int_{\chi_i(R_i)} \chi_i^{-1*} \chi_i^* (\nabla \phi_i \vee \omega) = \int_{M - \partial M} \nabla \phi_i \vee \omega.$$

Since $d\omega$ is summable over $M - \partial M$ and ϕ_i is bounded, $\phi_i\, d\omega$ is summable over $M - \partial M$. Applying Theorem 10A to the mapping χ_i gives

$$\int_{M-\partial M} \phi_i\, d\omega = \int_{R_i} \chi_i^*(\phi_i\, d\omega) = \int_{R_i} \phi_i^*\, d\omega_i^*.$$

Since $d\omega_i = \nabla\phi_i \vee \omega + \phi_i\, d\omega$, and similarly for $d\omega_i' = d(\phi_i^* \omega_i^*)$, adding these relations gives the rest of (6).

By the hypothesis of the theorem, (a) of Theorem 14A holds for the form ω_i' and the partial standard domain with R_i; also $\omega_i' = 0$ in a neighborhood U of $E^n - O_i'$. We have just proved (b) and (c) for ω_i'. We now follow the proof of Theorem 14A. Using only cubes U_j' so small that any U_j' touching $O_i' - U$ lies in O_i', we prove (5) in the case that ω_i' is Q_i-free. The last part of the proof of Theorem 14A goes through for the present case; hence (5) follows, and the present theorem is proved.

19. The iterated integral in Euclidean space. Let Q_1 and Q_2 be polyhedral regions in E^{n_1} and E^{n_2} respectively; their Cartesian product $Q = Q_1 \times Q_2$ is a polyhedral region in E^n, $n = n_1 + n_2$. If ω is a continuous n-form in Q, we wish to give meaning to, and prove,

$$(1) \qquad \int_Q \omega = \int_{Q_1 \times Q_2} \omega(p \times q)\, d(p \times q) = \int_{Q_2} \Big[\int_{Q_1 \times q} \omega(p \times q)\, dp \Big]\, dq.$$

The *partial integral*

$$(2) \qquad \Omega(q) = \int_{Q_1 \times q} \omega(p \times q)\, dp$$

must be defined in a generalized sense; it is not a number, but an n_2-covector in Q_2. Using subdivisions $\sum \sigma_i$ of Q_1 whose meshes $\to 0$, and taking $p_i \in \sigma_i$, we define

$$(3) \qquad \Omega(q) = \lim \sum_i \omega(p_i \times q) \wedge \{\sigma_i\},$$

using the interior product of (I, 7). Now Ω is an n_2-form, and the formula to be proved is

$$(4) \qquad \int_Q \omega = \int_{Q_2} \Omega(q)\, dq.$$

Given any n_2-vector β in E^{n_2}, we may write $\Omega(q)\cdot\beta$ as an integral:

$$(5) \qquad \Omega(q)\cdot\beta = (-1)^{n_1 n_2} \int_{Q_1 \times q} \omega(p \times q) \wedge \beta\, dp.$$

To show this, note that, by (I, 7.2) and (I, 6.2),

$$\Omega(q)\cdot\beta = \lim \sum_i [\omega(p_i \times q) \wedge \{\sigma_i\}]\cdot\beta$$

$$= (-1)^{n_1 n_2} \lim \sum_i [\omega(p_i \times q) \wedge \beta]\cdot\{\sigma_i\};$$

(5) now follows from Lemma 4a.

To prove (4), take any $\epsilon > 0$. Set $\eta = \epsilon/(|Q_2| + 2|Q|)$. From (5) we see that Ω is continuous; hence we may choose $\zeta > 0$ so that

$$|\omega(p'') - \omega(p')| < \eta \quad \text{if} \quad |p'' - p'| < 2^{1/2}\zeta,$$

$$|\Omega(q') - \Omega(q)| < \eta \quad \text{if} \quad |q' - q| < \zeta.$$

Now let $\sum \sigma_i$ and $\sum \tau_j$ be subdivisions of Q_1 and Q_2 respectively, of meshes $<\zeta$. Take $p_i \in \sigma_i$, $q_j \in \tau_j$; then $p_i \times q_j \in \sigma_i \times \tau_j$. By Lemma 4a,

$$\left| \int_{Q_2} \Omega - \sum_j \Omega(q_j)\cdot\{\tau_j\} \right| \leqq \eta\, |Q_2|.$$

If p and p' are in the same σ_i, then, by (I, 12.13) and the proof of (I, 14.4),

$$|\omega(p' \times q) \wedge \beta - \omega(p \times q) \wedge \beta| \leqq |\omega(p' \times q) - \omega(p \times q)|\, |\beta| \leqq \eta\, |\beta|;$$

hence, by (5) and Lemma 4a,

$$\left| \Omega(q)\cdot\beta - (-1)^{n_1 n_2} \sum_i [\omega(p_i \times q) \wedge \beta]\cdot\{\sigma_i\} \right| \leqq \eta\, |\beta|\, |Q_1|.$$

Therefore

$$\left| \sum_j \Omega(q_j)\cdot\{\tau_j\} - \sum_{i,j} \omega(p_i \times q_j)\cdot[\{\sigma_i\} \vee \{\tau_j\}] \right|$$
$$\leqq \sum_j \eta\, |\tau_j|\, |Q_1| = \eta\, |Q_1|\, |Q_2| = \eta\, |Q|.$$

Since $\sum_{i,j} \sigma_i \times \tau_j$ is a subdivision of Q of mesh $< 2^{1/2}\zeta$,

$$\left| \int_Q \omega - \sum_{i,j} \omega(p_i \times q_j)\cdot\{\sigma_i \times \tau_j\} \right| \leqq \eta\, |Q|.$$

Since $\{\sigma_i \times \tau_j\} = \{\sigma_i\} \vee \{\tau_j\}$, combining the above inequalities gives $\left| \int_Q \omega - \int_{Q_2} \Omega \right| \leqq \epsilon$, proving (4).

Suppose the spaces are oriented. Let α_0 be the unit n_1-vector of E^{n_1}. Then for any β, using (5) gives an expression for $\Omega(q)\cdot\beta$ as a Riemann integral:

(6) $$\Omega(q)\cdot\beta = (-1)^{n_1 n_2} \int_{Q_1 \times q} [\omega(p \times q) \wedge \beta]\cdot\alpha_0\, dp = \int_{Q_1 \times q} \omega(p \times q)\cdot(\alpha_0 \vee \beta)\, dp.$$

IV. Smooth Manifolds

This chapter is divided into three parts, each being devoted to a basic theorem in the theory of smooth (i.e. differentiable) manifolds. (Definitions and elementary properties of smooth manifolds are given in §§ 10–12 of Chapter II.)

The third part is the only one concerned directly with integration theory. Its purpose is to prove the Theorem of de Rham, which gives the cohomology structure of a smooth manifold M in terms of the differentiable forms in M. Integrating a closed form over cycles gives the periods of the form; through this, the cohomology spaces defined by differential forms become the linear functions on the homology spaces. Moreover, the products of forms correspond to the products in the algebraic cohomology spaces. For the use of very general forms in de Rham's Theorem, see the end of the introduction to Chapter IX.

It is habitual these days to treat algebraic topology from a very abstract point of view; using some basic properties of differential forms, de Rham's Theorem becomes a corollary of general theorems. However, the difficulty of grasping the large body of theory required for this, and the resulting lack of evident geometric meaning, makes a direct proof by elementary means desirable. The proof we give is closely related to de Rham's original proof. A proof making use of currents (and also a proof of the imbedding theorem) may be found in de Rham's book.

The initial topological study of a smooth manifold M by H. Poincaré was carried out with the help of a triangulation of M (cutting M into cells); that this can in fact always be done was first proved by S. S. Cairns. Perhaps due in part to the difficulty of the proof, using triangulations has gone out of fashion. However, both for the sake of geometric intuition and for various applications, triangulations are very useful. The definitions and proofs in the third part of this chapter are based on them. For this reason, the second part is devoted to the proof of existence of triangulations. The method of proof should bring the theorem into the position of being intuitively reasonable; the methods may clearly be used to prove related theorems.

In the triangulation theorem, we assume M is imbedded in Euclidean space E. That this is no restriction is proved in the first part. The relation of M to the surrounding space is also studied here. This imbedding theorem is of importance because it enables the simple analytic tools existing in E to be carried over to the subset M.

We end the chapter with a discussion of the Hopf invariant of a smooth mapping of a sphere S^{2n-1} into a sphere S^n.

A. Manifolds in Euclidean Space

1. The imbedding theorem. Let f be a smooth mapping of the smooth manifold M into Euclidean m-space E^m (II, 10). We say f is *regular* at p (compare (II, 5)) if independent vectors in M at p are carried into independent vectors in E^m (II, 11); f is *regular* if it is regular at all points of M. If M is compact, we say f is an *imbedding* if it is one-one and regular.

To consider the non-compact case, we introduce a further definition. The *limit set* L_f of the mapping f is the set of points $q \in E^m$ with the following property. There is a sequence of points $p_1, p_2, \cdots$ in M without limiting point in M, such that $f(p_i) \to q$. (If M is compact, then L_f is void.) For instance, if $M = \mathfrak{A}$, and f is the identity, then L_f is void; but if f is a one-one mapping onto the open interval $0 < t < 1$, then L_f contains $t = 0$ and $t = 1$.

The mapping f is *proper* if $L_f \cap f(M) = 0$. If, for instance, f maps $\mathfrak{A}$ into a figure 6 in E^2, then L_f contains a point in $f(M)$, and f is not proper. It is easy to see that a one-one mapping f is proper if and only if the inverse f^{-1} is continuous in $f(M)$, or, if and only if f^{-1} (with domain $f(M)$) carries compact sets into compact sets.

An *imbedding* is a one-one proper regular mapping.

Theorem 1A. *Let M be a μ-smooth manifold of dimension n, $\mu \geqq 1$ or $\mu = \infty$. Then there is a μ-smooth regular mapping f of M into E^m, without limit set, if $m \geqq 2n$, and there is a μ-smooth imbedding of M into E^m, without limit set, if $m \geqq 2n + 1$.*

2. The compact case. We give here a quick proof that if M is compact, it may be imbedded in some E^m.*

Let O_a be the open ball of radius a about 0 in $\mathfrak{A}^n$, and let $\Phi(x)$ be an ∞-smooth non-negative function in $\mathfrak{A}^n$ which equals 1 in $\bar{O}_1$, is <1 in $\mathfrak{A}^n - \bar{O}_1$, and is zero in $\mathfrak{A}^n - O_2$ (App. III, Lemma 1b). Let $\chi_1, \cdots, \chi_\nu$ be coordinate systems in M, defined in O_3, so that the $\chi_i(O_1)$ cover M. In each $U'_j = \chi_j(O_3)$, set

$$(1)\quad f_{0j}(p) = \Phi(x), \qquad f_{ij}(p) = x^i\Phi(x) \quad (i = 1, \cdots, n), \qquad p = \chi_j(x);$$

* It is then possible (if $\mu \geqq 2$) to project into some $E^{2n+1} \subset E^m$ to give an imbedding in E^{2n+1}; see H. Whitney, *Annals of Math. 38* (1937) 809–818, Appendix I, and Chapter I of de Rham. The proof of Theorem 1A we give below is a somewhat simplified version of the author's proof in *Annals of Math. 37* (1936) 645–680. See this paper for further theorems relating to imbedding and approximation. In particular, *the imbedded manifold may be made analytic.*

let the f_{ij} vanish in $M - U'_j$. This is a set of $m = \nu(n+1)$ μ-smooth real functions in M. Arranged in order, they are the components of a μ-smooth mapping F of M into E^m; this is the required imbedding.

To show that F is regular, take any $p \in M$; say $p = \chi_j(x)$, $x \in O_1$. Let F_j be the mapping of M into E^n whose components are $f_{1j}, \cdots, f_{nj}$. In O_1, the mapping $F'_j(x) = F_j(p)$ $(p = \chi_j(x))$ has components $f_{ij}(p) = x^i$; thus F'_j is an isometric mapping of O_1 into E^n, and hence is regular there. Therefore F_j and hence F is regular at p.

To show that F is one-one, let p and q be distinct points of M. If they are both in some $\bar{U}_j = \chi_j(\bar{O}_1)$, then the proof above shows that $F(p) \neq F(q)$. If not, say p is in $\bar{U}_j$ while q is not. Then $f_{0j}(p) = 1$ and $f_{0j}(q) < 1$, and again $F(p) \neq F(q)$.

3. Separation of subsets of E^m. We prove two lemmas, which form the essential parts of the imbedding theorem. Say a subset S of E^m is *nowhere dense* in E^m if int $(S) = 0$.

LEMMA 3a. *Let f be a Lipschitz mapping* (II, 4) *of a subset Q of E^s into E^m, $s < m$. Then $f(Q)$ is nowhere dense in E^m.*

Take a cube D in E^m of side length ϵ; we shall find a point $p \in D - f(Q)$. First suppose Q is bounded; let C be a cube containing Q, of diameter δ. Choose k so that $(2\mathfrak{L}_f\delta)^m/2^k < \epsilon^m$. Cut C into ν equal cubes $C_1, \cdots, C_\nu$, of diameter $\delta/2^k$; then $\nu = 2^{sk}$. Set $Q'_i = f(Q \cap C_i)$. Then diam $(Q'_i) \leqq \mathfrak{L}_f\delta/2^k$ if $Q'_i \neq 0$, and hence Q'_i lies in some cube D_i of side length $2\mathfrak{L}_f\delta/2^k$. Now the sum of the volumes of the D_i is

$$\sum |D_i| = 2^{sk}(2\mathfrak{L}_f\delta/2^k)^m \leqq (2\mathfrak{L}_f\delta)^m/2^k < \epsilon^m.$$

Hence there is a point (in fact, a cube) in D touching no D_i and hence not touching $f(Q)$.

Now consider the general case. Let $C_1, C_2, \cdots$ be concentric cubes in E^s such that diam $(C_i) \to \infty$. By the proof above, we may find a (closed) cube $D_1 \subset D$ not touching $f(Q \cap C_1)$, a closed cube $D_2 \subset D_1$ not touching $f(Q \cap C_2)$, etc. There is a point p in all the D_i; this is in $D - f(Q)$.

Given a set $S \subset E^m$ and a vector v, let $T_v(S)$ denote the set of points $p + v$, $p \in S$ (translation of S by v).

LEMMA 3b. *Let Q and Q' be subsets of E^n and $E^{n'}$ respectively, let f and f' be Lipschitz mappings of Q and Q' respectively into E^m, and suppose $n + n' = s < m$. Then there is an arbitrarily small vector v in E^m such that $T_v(f(Q))$ does not intersect $f'(Q')$.*

Set $F(q \times q') = f'(q') - f(q)$ for $q \in Q$, $q' \in Q'$; this s a Lipschitz mapping of the Cartesian product $Q \times Q' \subset E^s$ into $V^m = V(E^m)$. By the last lemma, there is an arbitrarily small vector v not in $F(Q \times Q')$; this is the required vector.

4. Regular approximations. We shall show how to find a regular mapping of $\bar{O}_1$ (see § 2) into E^m $(m \geqq 2n)$ approximating a given mapping.

Let ψ and ψ' be μ-smooth mappings of a neighborhood of the set $Q \subset \mathfrak{A}^n$ into E^m, and let $\eta(p)$ be a positive continuous function in Q. Using a fixed coordinate system in E^m, we say ψ' *approximates* (ψ, Q, μ, η) if the components of the function $\Psi(p) = \psi'(p) - \psi(p)$, together with all partial derivatives of order $\leqq \mu$, are at most $\eta(p)$ in absolute value at each point $p \in Q$.

LEMMA 4a. *Let ψ be a μ-smooth mapping of O_3 into E^m, $m \geqq 2n$. Then for any $\epsilon > 0$ there is a μ-smooth mapping ψ' of O_3 into E^m which approximates $(\psi, \bar{O}_2, \mu, \epsilon)$ and is regular in $\bar{O}_1$.*

If $\mu = 1$, let ψ_0 be a 2-smooth mapping of O_3 which approximates $(\psi, \bar{O}_2, 1, \epsilon')$ for an $\epsilon' > 0$ determined below (App. III, Lemma 4a); if $\mu > 1$, set $\psi_0 = \psi$. We shall find $\psi_1, \cdots, \psi_n$ in turn so that ψ_i approximates $(\psi_{i-1}, \bar{O}_2, \mu, \epsilon')$ and the vectors $\partial\psi_i/\partial x^1, \cdots, \partial\psi_i/\partial x^i$ are independent in $\bar{O}_1$; the ψ_i are μ'-smooth in O_3, $\mu' = \sup\{2, \mu\}$. If ϵ' is small enough, $\psi' = \psi_n$ is the required mapping.

Suppose ψ_{i-1} is found; we show how to find ψ_i. Set

$$v_j(x) = \partial\psi_{i-1}(x)/\partial x^j \qquad (j = 1, \cdots, i); \tag{1}$$

then $v_1, \cdots, v_{i-1}$ are independent in $\bar{O}_1$. Let $P(x)$ be the set of all vectors

$$\phi(\lambda^1, \cdots, \lambda^{i-1}; x) = \sum_{j=1}^{i-1} \lambda^j v_j(x) - v_i(x), \tag{2}$$

and let P be the union of the $P(x)$ for $x \in \bar{O}_1$. We shall find an arbitrarily small vector v not in P, and shall set

$$\psi_i(x) = \psi_{i-1}(x) + x^i v. \tag{3}$$

Then for v small enough, ψ_i approximates $(\psi_{i-1}, \bar{O}_2, \mu, \epsilon')$. Also

$$\partial\psi_i(x)/\partial x^j = v_j(x) \quad (j < i), \qquad \partial\psi_i(x)/\partial x^i = v_i(x) + v.$$

For each $x \in \bar{O}_1$, since v is not in $P(x)$, $v_i(x) + v$ is not a linear combination of $v_1(x), \cdots, v_{i-1}(x)$. Hence ψ_i has the required properties.

Since ψ_{i-1} is 2-smooth, the $v_j(x)$ are smooth; therefore ϕ is smooth in the Cartesian product $\mathfrak{A}^{i-1} \times O_3$. Since $i - 1 + n < m$, Lemma 3a shows that there is an arbitrarily small vector v not in $\phi(\mathfrak{A}^{i-1} \times O_1) = P$, completing the proof.

5. Proof of Theorem 1A, *M* compact. Let $\chi_1, \ldots, \chi_\nu$ be the coordinate systems of § 2. Choose $q_0 \in E^m$ and set $f_0(p) = q_0 (p \in M)$; f_0 is μ-smooth. Supposing $m \geqq 2n$, we shall define $f_1, \cdots, f_\nu = f'$, each being μ-smooth in M, so that f_i is regular in $Q_i = \bar{U}_1 \cup \cdots \cup \bar{U}_i$ $(U_i = \chi_i(O_1))$; then f' is regular in M.

Supposing we have f_{i-1}, we find f_i as follows. Set

$$(1) \qquad \psi(x) = f_{i-1}(\chi_i(x)), \qquad x \in O_3.$$

For a certain $\epsilon > 0$, we shall find a μ-smooth mapping ψ^* of O_3 into E^m such that:

(a) ψ^* approximates $(\psi, \bar{O}_2, 1, \epsilon)$,
(b) $\psi^* = \psi$ in $O_3 - O_2$,
(c) ψ^* is regular in $\bar{O}_1$.

Then if we set

$$(2) \qquad f_i(p) = \psi^*(\chi_i^{-1}(p)) \quad \text{in } U_i', \qquad f_i = f_{i-1} \quad \text{in} \quad M - U_i',$$

f_i will be μ-smooth in M and regular in $\bar{U}_i$. Also, since f_{i-1} is regular in the compact set Q_{i-1}, f_i will be also for small enough ϵ.

For a certain $\epsilon' > 0$, choose ψ' in O_3 by Lemma 4a. With Φ as in § 2, set

$$(3) \qquad \psi^*(x) = \psi(x) + \Phi(x)[\psi'(x) - \psi(x)], \qquad x \in O_3.$$

Then (a) holds if ϵ' is small enough, (b) holds, and since $\psi^* = \psi'$ in $\bar{O}_1$, (c) holds. Thus we find ψ^* and hence f_i, and finally, f'.

Supposing now that $m \geqq 2n + 1$, we find the imbedding f. Since f' is regular, it is one-one in a neighborhood of each point (II, Theorem 7B). Hence we may choose new coordinate systems $\chi_1, \cdots, \chi_\nu$ such that (with U_i and U_i' as before) the $\bar{U}_i$ cover M, and such that if $U_i' \cap U_j' \neq 0$, then f' is one-one in $U_i' \cup U_j'$. Let $\kappa_1, \cdots, \kappa_\beta$ denote the pairs (i, j), $i < j$, such that $U_i' \cap U_j' = 0$, arranged in some order. Set $f_0' = f'$. We shall find μ-smooth regular mappings $f_1', \cdots, f_\beta'$ so that each f_k' is one-one in each $U_i' \cup U_j'$ if $U_i' \cap U_j' \neq 0$, and such that for each $k' \leqq k$, if $\kappa_{k'} = (i', j')$, then $f_k'(\bar{U}_{i'}) \cap f_k'(\bar{U}_{j'}) = 0$. Then $f = f_\beta'$ is the required imbedding.

Having found f_{k-1}', we find f_k' as follows. Say $\kappa_k = (i, j)$. Set

$$(4) \qquad \psi(x) = f_{k-1}'(\chi_j(x)), \qquad x \in O_3.$$

For a certain $\epsilon > 0$, choose a vector v by Lemma 3b with $|v| < \epsilon$, such that

$$T_v(\psi(\bar{O}_1)) \cap f_{k-1}'(\bar{U}_i) = 0,$$

set $\psi'(x) = \psi(x) + v$, and define ψ^* by (3). (Thus $\psi^*(x) = \psi(x) + \Phi(x)v$.) As before, this gives f_k' in M; then $f_k' = f_{k-1}'$ in $M - U_j'$, in particular, in U_i'. Now $f_k'(\bar{U}_i) \cap f_k'(\bar{U}_j) = 0$. For ϵ small enough, the required conditions on f_{k-1}' continue to hold; thus f_k' is constructed. This completes the proof.

6. Admissible coordinate systems in M. We say a set of coordinate systems in M (each defined in O_3) is *admissible* if they are finite or denumerable in number, the U_i cover M, and any compact subset of M touches but a finite number of the U_i' (notations as in § 2).

LEMMA 6a. *M has an admissible set of coordinate systems.*

Since M may be covered by a denumerable set of coordinate systems, we may clearly find compact subsets $H_1, H_2, \cdots$ of M such that $H_i \subset$ int (H_{i+1}) and $H_1 \cup H_2 \cup \cdots = M$. For each i there is a finite set of coordinate systems such that the corresponding U_k cover $H_{i+1} - H_i$ (using $H_0 = 0$), and the U'_k do not touch H_{i-1} (if $i > 1$); the set of all of these forms an admissible set.

7. Proof of Theorem 1A, M not compact. Let $\chi_1, \chi_2, \cdots$ be an admissible set of coordinate systems in M. With the notations of § 2, set

$$(1) \qquad \rho_i(p) = \begin{cases} \Phi(\chi_i^{-1}(p)) & \text{in } U'_i, \\ 0 \quad \text{in} & M - U'_i, \end{cases} \qquad \rho(p) = \sum_{i=1}^{\infty} i\rho_i(p).$$

Choose $q_0 \in E^m$ and a vector $v_0 \neq 0$, and set

$$(2) \qquad f_0(p) = q_0 + \rho(p)v_0;$$

this is a μ-smooth mapping of M into E^m without limit set.

Define $f_1, f_2, \cdots$ as in § 5; f_i is regular in $\bar{U}_1 \cup \cdots \cup \bar{U}_i$. Since the set of coordinate systems used is admissible, $f' = \lim f_i$ exists and is regular in M, and has no limit set.

Now choose new coordinate systems $\chi_1, \chi_2, \cdots$ as in the second part of § 5; it is easy to see that we may take these denumerable in number, and with the $\bar{U}'_i$ compact. As in § 6, we may require these to be admissible. Define $\kappa_1, \kappa_2, \cdots$ as before; but we use only pairs (i, j) such that $U'_i \cap U'_j = 0$ and $f(\bar{U}_i) \cap f(\bar{U}_j) \neq 0$. Since the χ_i are admissible and $L_{f'}$ is void, each integer i occurs at most a finite number of times in pairs (i, j).

We now define $f'_1, f'_2 \cdots$ as before. As above we may set $f(p) = \lim f'_i(p)$; this is the required imbedding.

8. Local properties of M in E^m. Let M_0 be a μ-smooth manifold of dimension n, and let f be a μ-smooth imbedding of M_0 in E^m. We wish to show how the set of points $M = f(M_0)$ has a "differentiable structure" equivalent to that determined by M_0.

Take any $q_0 \in M_0$; set $p_0 = f(q_0)$. Since f is regular at q_0, the images $\nabla f(q_0, v)$ of all vectors v in M_0 at q_0 form a vector space V_{p_0}, of the same dimension n as M_0; these are the *tangent vectors* to M at p_0. The set of points $p_0 + w$ ($w \in V_{p_0}$) form the *tangent plane* P_{p_0} to M at p_0. Let π_{p_0} denote the orthogonal projection of E^m onto P_{p_0}, and set $F_{p_0}(q) = \pi_{p_0}(f(q))$, $q \in M_0$. Since $\nabla F_{p_0}(q_0, v) = \nabla f(q_0, v)$, F_{p_0} is regular at q_0; hence there is a neighborhood U_0 of q_0 in M_0 which maps under F_{p_0} in a one-one manner onto a neighborhood U of p_0 in P_{p_0} (II, Theorem 7A). Let F'_{p_0} be the inverse of F_{p_0}, mapping U onto U_0; set $\psi_{p_0}(p) = f(F'_{p_0}(p))$, mapping U onto $U^* \subset M$. Then

$$(1) \qquad \pi_{p_0}\psi_{p_0} = \pi_{p_0} f F'_{p_0} = F_{p_0} F'_{p_0} = \text{identity in } U;$$

thus π_{p_0} and ψ_{p_0} are inverses of each other. Both are μ-smooth.

The plane P_{p_0} and the projection π_{p_0} in M are defined by the point set M alone. We may consider the inverse ψ_{p_0}, in U, as a coordinate system in M. The corresponding mapping of U into M_0 is $f^{-1}\psi_{p_0} = F'_{p_0}$; since this is smooth, the coordinate system ψ_{p_0} in M is differentiably related to the coordinate systems in M_0. We may call M a *smooth manifold in* E^m.

The following lemmas will be used in the proof of the triangulation theorem. For convenience, we assume M is compact. In the contrary case, we would use a positive continuous function $\xi_0(p)$ in M in place of the number ξ_0, etc. Set

$$(2) \qquad P_{p,\xi} = P_p \cap U_\xi(p), \qquad M_{p,\xi} = \psi_p(P_{p,\xi}),$$

the latter being defined if ξ is small enough.

LEMMA 8a. *Let M in E^m be compact. Then there is a $\xi_0 > 0$ such that M_{p,ξ_0} is defined for all $p \in M$. Moreover,*

$$(3) \qquad \operatorname{dist}(p, M - M_{p,\xi}) \geqq \xi, \qquad \xi \leqq \xi_0.$$

It is clear that for each $p \in M$ there is an $\eta > 0$ such that $M_{p',\eta}$ is defined for p' in some neighborhood of p in M; since M is compact, M_{p,ξ'_0} exists for some ξ'_0 (all $p \in M$). Choose ξ_0 so that

$$\operatorname{dist}(p, M - M_{p,\xi'_0}) \geqq \xi_0, \qquad p \in M.$$

Since

$$\operatorname{dist}(p, M_{p,\xi'} - M_{p,\xi}) \geqq \xi, \qquad \xi < \xi'_0, \quad p \in M,$$

ξ_0 has the required property.

As in (I, 15), let $\pi_p v$ be the orthogonal projection into P_p of the vector v; this is $\nabla \pi_p(q, v)$ for any $q \in E^m$.

LEMMA 8b. *Let M in E^m be compact. Then for any $\lambda > 0$ there is a $\xi_1 > 0$ with the following property. For any $p \in M$ and any vector v tangent to M_{p,ξ_1},*

$$(4) \qquad |v - \pi_p v| \leqq \lambda |\pi_p v| \leqq \lambda |v|.$$

As a consequence, tangent planes at nearby points are nearly parallel. Since $\pi_p v = v$ for vectors v tangent to M at p, $\xi_1(p)$ exists for a given $p \in M$. Since M is compact, ξ_1 exists as required.

A *secant vector* to a point set Q in E^m is a vector of the form $a(q - p)$, p and q in Q, a real.

LEMMA 8c. *Let M, λ and ξ_1 be as in Lemma 8b. Then any secant vector v to M_{p,ξ_1} satisfies* (4). *Moreover,*

$$(5) \qquad |p' - \pi_p(p')| < \lambda\xi, \qquad p' \in M_{p,\xi}, \quad \xi \leqq \xi_1,$$

$$(6) \qquad M_{p,\xi} \subset U_{\lambda\xi}(P_{p,\xi}), \qquad P_{p,\xi} \subset U_{\lambda\xi}(M_{p,\xi}), \qquad \xi \leqq \xi_1.$$

Suppose $v = p_1 - p_0$, p_0 and p_1 in M_{p,ξ_1}. Set

$$q_0 = \pi_p(p_0), \qquad q_1 = \pi_p(p_1), \qquad w = q_1 - q_0 = \pi_p v,$$

$$q_t = (1 - t)q_0 + tq_1, \qquad v_t = \nabla_w \psi_p(q_t).$$

Since $\pi_p\psi_p$ is the identity in P_{p,ξ_1}, $\pi_p v_t = w = \pi_p v$. Applying (II, 1.3) to ψ_p shows that

$$v - \pi_p v = \int_0^1 (v_t - \pi_p v_t)\, dt.$$

Since v_t is a tangent vector to M_{p,ξ_1}, (4) holds for it; therefore it holds also for v, and hence for any secant vector av.

Now take any $p' \in M_{p,\xi}$, $\xi \leqq \xi_1$. Set $v = p' - p$. Then v is a secant vector, and hence

$$|p' - \pi_p p'| = |v - \pi_p v| \leqq \lambda |\pi_p v| < \lambda\xi,$$

proving (5). Relation (6) follows from this.

9. On n-directions in E^m. Take a fixed orthonormal coordinate system in E^m. Now an n-vector α in E^m is given by naming its components α^λ ($\lambda_1 < \cdots < \lambda_n$); there are $\nu = \binom{m}{n}$ of these (I, 3). Thus α corresponds to a point of $\mathfrak{A}^\nu$, and vice versa. By (I, 12.7), the metric in $V_{[n]}$ agrees with that in $\mathfrak{A}^\nu$.

A certain subset M_0 of $\mathfrak{A}^\nu$ corresponds to n-vectors which are n-directions (I, 12). We shall show that M_0 is an analytic manifold in $\mathfrak{A}^\nu$.

Take any n-direction α, and choose an orthonormal set $v_1, \cdots, v_m$ such that $\alpha = v_1 \vee \cdots \vee v_n$. Now take any n-direction α' such that $|\alpha' - \alpha| < 1/n$. Write $\alpha' = u_1 \vee \cdots \vee u_n$, the u_i orthonormal. Let π denote orthogonal projection into the subspace P of α, and set $u_i' = \pi u_i$. Then (I, 15.7) gives

$$|u_i' - u_i| \leqq |\alpha - \alpha'|\,|u_i| < 1/n,$$

and hence, by (I, 12.17),

$$|u_1' \vee \cdots \vee u_n' - u_1 \vee \cdots \vee u_n| < 1,$$

proving $u_1'\vee\cdots\vee u_n' \neq 0$. Therefore π maps the subspace P' of α' in a one-one manner onto P, and we may find vectors $v_1', \cdots, v_n'$ in P' such that $\pi v_i' = v_i$. Now $v_i' - v_i$ is orthogonal to P, and hence is a linear combination of $v_{n+1}, \cdots, v_m$. Also $\alpha' = c(v_1'\vee\cdots\vee v_n')$ for some c. Therefore

$$(1)\quad \alpha' = c\left(v_1 + \sum_{j=n+1}^{m} a_{1j}v_j\right)\vee\cdots\vee\left(v_n + \sum_{j=n+1}^{m} a_{nj}v_j\right) \quad \text{if} \quad |\alpha' - \alpha| < 1/n.$$

Now take any similar expression

$$\alpha' = c'(v_1''\vee\cdots\vee v_n''), \qquad v_i'' = v_i + \sum_{j=n+1}^{n} b_{ij}v_j.$$

Write $v_i'' = \sum_{j=1}^{n} A_{ij}v_j'$. Then

$$v_i = \pi v_i'' = \sum_{j=1}^{n} A_{ij}\pi v_j' = \sum_{j=1}^{n} A_{ij}v_j,$$

proving $A_{ij} = \delta_i^j$, and $v_i'' = v_i'$, $b_{ij} = a_{ij}$, $c' = c$. Thus the expression (1) for α' is unique. Any choice of a set of $\mu = n(m - n)$ numbers a_{ij} determines an n-direction by (1). Hence a neighborhood of α in M_0 is determined by a mapping of an open subset O of $\mathfrak{A}^\mu$ into M_0. We use this for a coordinate system in M_0.

Using (I, 2.5) to find α'^λ, we see that this is an analytic (in fact, algebraic) function of the a_{ij}. This gives an analytic mapping of O into $\mathfrak{A}^\nu$, which is easily seen to be regular. Hence also the relation between overlapping coordinate systems is analytic, and M_0 is analytic.

We shall find a "projection" π_0 of a neighborhood of M_0 in $\mathfrak{A}^\nu$ onto M_0; details of proof of properties of π_0 are as in the proof of Theorem 10A below. Since M_0 is compact, we may choose $\rho_0 > 0$ with the following property. For each $p \in M_0$, let P_p^* be the normal plane to M_0 at p, and set $Q_p^* = P_p^* \cap U_{\rho_0}(p)$. Set $\pi_0(q) = p$ for $q \in Q_p^*$. The Q_p^* fill up a neighborhood U_0 of M_0 in $\mathfrak{A}^\nu$ in a one-one manner, and π_0 is an analytic mapping of U_0 onto M_0.

Since the transformation $\phi(\alpha) = -\alpha$ sends n-directions into n-directions, it is clear that

$$(2)\qquad \pi_0(-\beta) = -\pi_0(\beta), \qquad \beta \text{ and } -\beta \text{ in } U_0.$$

10. The neighborhood of M in E^m. Let M be μ-smooth in E^m. For each $p \in M$, let P_p' be the normal plane to M at p; orient it, and let α_p' be its $(m - n)$-direction. With M_0 as in § 9 (but with n replaced by $m - n$), we consider α_p' as a point of M_0. Orienting the P_q' similarly in a neighborhood U of p, this is a mapping of U into M_0. Since α_p' depends on first derivatives in M, α_p' may be only $(\mu - 1)$-smooth in M. Phrasing this statement differently, P'_p may be only $(\mu - 1)$-smooth in M. We shall define a family P^*_p of approximately normal planes, which will be a μ-smooth function. (For $\mu = \infty$, we may take $P^*_p = P'_p$.) Recall the notations in § 8.

THEOREM 10A. *Let M be a μ-smooth n-manifold in E^m, and suppose $\lambda_0 > 0$. Then there is a μ-smooth family of $(m - n)$-planes P^*_p and a positive continuous function $\delta(p)$ $(p \in M)$ with the following properties. For each $p \in M$, P^*_p contains p, and*

$$(1) \qquad |\pi_p v| \leqq \lambda_0 |v| \qquad \textit{if } v \textit{ is in } P^*_p.$$

Set

$$(2) \qquad Q^*_p = P^*_p \cap U_{\delta(p)}(p).$$

The Q^*_p fill out a neighborhood U^* of M in a one-one way. Set

$$(3) \qquad \pi^*(q) = p \quad \text{if} \quad q \in Q^*_p.$$

This is a μ-smooth mapping of U^* onto M, and

$$(4) \qquad |\pi^*(q) - q| \leqq 2 \operatorname{dist}(q, M), \qquad q \in U^*.$$

We may choose $\lambda_1 \leqq \lambda_0/2$ so that, with M_0 etc. as in § 9,

$$(5) \qquad U_1 = U_{\lambda_1}(M_0) \subset U_0, \qquad |\pi_0\alpha - \alpha| < \lambda_0/2 \quad (\alpha \in U_1).$$

Since P'_p is continuous in M, we may choose an admissible set of coordinate systems $\chi_1, \chi_2, \cdots$ in M (§ 6) such that for a chosen point $p_i \in U'_i = \chi_i(O_3)$ and chosen orientations of the P'_p $(p \in U'_i)$,

$$(6) \qquad |\alpha'_p - \alpha'_{p_i}| < \lambda_1, \qquad p \in U'_i.$$

Let $\phi_1, \phi_2, \cdots$ be a corresponding μ-smooth partition of unity. (With ρ_i as in (7.1), set $\phi_i(p) = \rho_i(p)/\sum_j \rho_j(p)$; then $0 \leqq \phi_i \leqq 1$, $\phi_i > 0$ in $\bar{U}_i$, $\phi_i = 0$ in $M - U'_i$, and $\sum \phi_i = 1$ in M.)

Given $p \in M$, define P^*_p as follows. Let $U'_{\lambda_1}, \cdots, U'_{\lambda_s}$ be the U'_k containing p. Orient P'_p, and orient the $P'_{p_{\lambda_i}}$ similarly (this is determined by (6)). Using $\mathfrak{A}^\nu$ as a vector space, set

$$(7) \qquad \alpha''_p = \sum_i \phi_{\lambda_i}(p)\alpha'_{p_{\lambda_i}}, \qquad \alpha^*_p = \pi_0\alpha''_p;$$

the existence of α^*_p is proved below. Let P^*_p be the $(m - n)$-plane through p which, with proper orientation, has the $(m - n)$-direction α^*_p. If we had used the opposite orientation of P'_p, then, because of (9.2), the same P^*_p would be found.

Since the ϕ_i and π_0 are μ-smooth, P^*_p is μ-smooth in M. Because of

(6), and since $\phi_j(p) \geqq 0$, $\sum \phi_j(p) = 1$, we have $|\alpha''_p - \alpha'_p| < \lambda_1$; hence $\alpha''_p \in U_1$, and (5) gives $|\alpha^*_p - \alpha''_p| < \lambda_0/2$; therefore

(8) $$|\alpha^*_p - \alpha'_p| < \lambda_0.$$

Let π'_p denote orthogonal projection into P'_p. Then, for any v in P^*_p, (I, 15.7) gives

$$|\pi_p v| = |v - \pi'_p v| \leqq |\alpha^*_p - \alpha'_p|\,|v| \leqq \lambda_0 |v|,$$

which is (1).

Take any $p_0 \in M$. Define ψ_{p_0} as in § 8, and let π^{**}_p denote orthogonal projection into P^*_p. For any vector v in E^m, write

$$v = v' + v'', \qquad v' \in P_{p_0}, \quad v'' \in P^*_{p_0},$$

and define the mapping Ψ of a neighborhood of 0 in $V(E^m)$ into E^m by setting

(9) $$p = \psi_{p_0}(p_0 + v'), \qquad q = \Psi(v) = p + \pi^{**}_p(v'').$$

(We may suppose $\lambda_0 < 1$ in (1); then $P^*_{p_0}$ intersects P_{p_0} in a single point.) Since ψ_{p_0} and P^*_q are μ-smooth, so is π^{**}_p, and so is Ψ. Since

$$\nabla_{v'}\psi_{p_0}(p_0) = v', \qquad \nabla_{v'}\Psi(0) = v' \qquad (v' \in P_{p_0}),$$

$$\pi^{**}_{p_0}(v'') = v'', \qquad \nabla_{v''}\Psi(0) = v'' \qquad (v'' \in P^*_{p_0}),$$

$\nabla\Psi$ is the identity at $v = 0$; hence Ψ is regular at 0. Therefore (II, Theorem 7A) Ψ has a μ-smooth inverse near 0; we may write

(10) $$v = v' + v'' = \Gamma_1(q) + \Gamma_2(q), \qquad \Psi(\Gamma_1(q) + \Gamma_2(q)) = q.$$

Set

(11) $$\pi^*(q) = \Psi(\Gamma_1(q)) = \psi_{p_0}(p_0 + \Gamma_1(q))$$

near p_0; this is a μ-smooth mapping of a neighborhood of p_0 into M, such that $\pi^*(q) = p$ if $q \in P^*_p$.

This shows that, given $p_0 \in M$, $\delta_0 > 0$ exists such that the required properties of Q^*_p hold in $U_{\delta_0}(p_0)$. Hence clearly $\delta(p)$ exists as required. (This would not be so if M were the image of a manifold under an improper mapping.)

Clearly we may suppose λ_0 and $\delta(p)$ small enough so that (4) will hold. The proof may be given as follows. Take $\lambda_0 \leqq 1/4$. Using $\lambda = 1/8$, find $\xi(p)$ by Lemmas 8a and 8c (non-compact case). Take $\delta(p) \leqq \xi(p)/2$. Now suppose $p = \pi^*(q)$, $v = q - p$, $|v| = a$. Then

$$\operatorname{dist}(q, P_p) = |v - \pi_p v| \geqq |v| - \lambda_0 |v| \geqq 3a/4,$$

$$M_{p,2a} \subset U_{2a\lambda}(P_p), \qquad \operatorname{dist}(q, M_{p,2a}) \geqq 3a/4 - a/4 = a/2,$$

$$\operatorname{dist}(q, M - M_{p,2a}) \geqq \operatorname{dist}(p, M - M_{p,2a}) - a \geqq a,$$

and hence (4) holds.

LEMMA 10a. *Take* λ *and* $\xi_1 \leqq \xi_0$ *as in* § 8, *and suppose* $\lambda + \lambda_0 < 1$. *Take any* $p, p' \in M$ *such that* $|p' - p| < \xi_1$. *Then* $P^*_{p'}$ *intersects* P_p *in a unique point, and*

$$(12) \qquad |\pi_p v| \leqq (\lambda + \lambda_0)|v| \qquad \text{if } v \text{ is in } P^*_{p'}.$$

Write $v = \pi_{p'}v + w$. Take any unit vector u in P_p. Since w is orthogonal to $P_{p'}$, $\pi_{p'}u \cdot w = 0$. By (8.3), $p \in M_{p',\xi_0}$; hence (8.4) gives

$$|u \cdot w| = |(u - \pi_{p'}u) \cdot w| \leqq \lambda |w| \leqq \lambda |v|.$$

Therefore $|\pi_p w| \leqq \lambda |v|$. Also, by (1), $|\pi_p \pi_{p'} v| \leqq \lambda_0 |v|$. These relations give (12). If the statement about intersections were false, then there would be a unit vector u in both $P^*_{p'}$ and P_p. But then (12) would give $|u| = |\pi_p u| < |u|$, a contradiction.

EXAMPLE. Let M be the smooth 1-manifold in E^2 given by $y = |x|^{3/2}$. Then the normal line at the origin intersects neighboring normal lines in points arbitrarily near the origin.

11. Projection along a plane. Let P and P' be planes of dimensions n and $m - n$ respectively in E^m, with just one point in common. Then to each point $p \in E^m$ corresponds a unique point $q = \pi'(p) \in P$ such that $q - p$ is a vector in P'. We call π' the *projection into* P, *along* P'. Recall the definition of ind (P, P') in (App. II, 14).

LEMMA 11a. *Given* M *in* E^m, *let* λ *and* ξ_1 *be as in Lemma* 8c. *Take* $p \in M$, *and let* P' *be an* $(m - n)$*-plane such that*

$$(1) \qquad \text{ind}\,(P_p, P') \geqq \lambda' > \lambda.$$

Then π', *considered in* M_{p,ξ_1}, *is an imbedding in* P_p. *We have*

$$(2) \qquad |\pi'(q) - q| < \lambda\xi/\lambda' \quad \text{if} \quad q \in M_{p,\xi}, \quad \xi \leqq \xi_1,$$

$$(3) \qquad P_{p,c} \subset \pi'(M_{p,\xi}), \qquad c = (1 - \lambda/\lambda')\xi, \quad \xi \leqq \xi_1.$$

First, let $v \neq 0$ be any tangent or secant vector to M_{p,ξ_1}. Then (8.4) holds, and since $\lambda < \lambda'$, v is not in P'. Therefore π' is regular and one-one in M_{p,ξ_1}, and hence is an imbedding.

Next, take any $q \in M_{p,\xi}$, $\xi \leqq \xi_1$; set

$$q' = \pi'(q), \qquad q'' = \pi_p(q), \qquad v = q - q'.$$

Then v is in P', and hence

$$|q - q''| = |v - \pi_p v| \geqq \lambda' |v|.$$

Using (8.5) gives

$$|\pi'(q) - q| = |v| \leqq |q - q''|/\lambda' < \lambda\xi/\lambda'.$$

Suppose (3) were false; then take $p' \in P_{p,c}$, not in $\pi'(M_{p,\xi})$. The segment pp' contains a point p^* which is a limiting point of points in and points not in $\pi'(\bar{M}_{p,\xi})$. Since this set is compact, there is a $q \in \bar{M}_{p,\xi}$ with $\pi'(q) = p^*$. If $q \in M_{p,\xi}$, then since π' is regular at q, a neighborhood of p^* is covered; but this is not so. Hence $| \pi_p(q) - p | = \xi$. But

$$| p - \pi_p(q) | \leqq | p - p^* | + | \pi'(q) - q | < c + \lambda\xi/\lambda' = \xi,$$

again a contradiction; hence (3) holds.

B. TRIANGULATION OF MANIFOLDS

12. The triangulation theorem. Let M be a μ-smooth manifold. By a *μ-smooth triangulation* of M we mean the pair consisting of a simplicial complex K and a homeomorphism π^* of K onto M, with the following property. For each n-simplex σ of K there is a coordinate system χ in M (defined in some open set in $\mathfrak{A}^n$) such that χ^{-1} is defined in a neighborhood of $\pi^*(\sigma)$ in M, and $\chi^{-1}\pi^*$ is affine in σ (n is the dimension of M).

THEOREM 12A.† *Every μ-smooth manifold M has a μ-smooth triangulation.*

REMARKS. (*a*) In the proof below, M is a manifold in E^m. It should not be hard to extend the proof to show that there is a (curvilinear) triangulation of E^m, of which a subcomplex corresponds to M.

(*b*) Suppose M is a "manifold with boundary": It is composed of pieces of manifolds M_i^r of different dimensions, fitting together in a simple manner. One might first imbed this in E^m, and then triangulate it.

(*c*) One might ask if there is a triangulation of M with the following property. For each vertex p_i of K, there is a coordinate system χ containing $St(p_i)$, such that $\chi^{-1}\pi^*$ is affine in each simplex of the star. This cannot be done in general; it is impossible for any compact simply connected manifold, for instance, for the 2-sphere.‡

13. Outline of the proof. By Theorem 1A, we may suppose M is an n-manifold in E^m ($m = 2n + 1$), without limit set. We give the full details for the case that M is compact; this will save some minor complications (see below).

† S. S. Cairns, On the triangulation of regular loci, *Annals of Math. 35* (1934) 579–587; Triangulation of the manifold of class 1, *Bulletin Am. Math. Soc., 41* (1935) 549–552. (The affine property is not discussed here.) For an improvement in the proof, see J. H. C. Whitehead, On C^1-complexes, *Annals of Math. 41* (1940) 809–824.

‡ If M, of dimension n, is "locally affine" in this sense and is simply connected, it is easily seen that there is a mapping of M into E^n which is locally affine and locally one-one; hence M is not compact. (For this proof I am indebted to A. Nijenhuis.)

A fine subdivision L_0 of E^m into cubes is chosen; let L be the regular subdivision of L_0 (App. II, 3). We move the vertices of L slightly, forming a new triangulation L^* of E^m, which will be "in general position" with respect to M. In particular, M intersects no simplexes of L^* of dimension $<s = m - n$. Since the simplexes of L^* are quite small relative to the curvature of M, and M is at a certain positive distance from $\partial\sigma^s$ for each σ^s which M intersects, M cuts σ^s at a unique point, at an angle not too small. The intersections of M with the simplexes of L^* are approximately convex cells, and the desired complex K is like the regular subdivision of this set of cells. The homeomorphism of K onto M is given by the mapping π^* of Theorem 10A.

We make some remarks on the proof for the non-compact case. Take $p_0 \in E^m$. In each $R_i = U_{2^i}(p_0)$, we have a compact part of M, to which the above method of proof applies. We must choose the triangulation L of E^m so that the proof (which is local in character) applies throughout. Hence L_0 must contain cubes which are smaller and smaller as we go further from p_0. We may choose them so that the ratio of side lengths of adjacent cubes is either 1/2, 1, or 2. Then a fixed Θ_0 and N (see below) may be used. Other numbers will be correspondingly smaller. In Lemmas 8a and 8c, we use a continuous function $\xi(p)$ in place of the number ξ.

14. Fullness. Both in area theory and in the theory of integration in several variables, it is well known that working with r-dimensional sets such that the ratio of volume to rth power of the diameter is very small may lead to difficulties. For instance, given a smooth surface S in E^3, one may find, in an arbitrary neighborhood of a given point p of S where S is not flat, a triangle with vertices in S which is nearly perpendicular to S; but the above ratio must then be small. This is the basis of the example of Schwarz of a polyhedral surface inscribed in a cylinder, which has arbitrarily large area.† We give some properties of the above ratio here.

Given the r-simplex σ $(r > 0)$ in Euclidean space E^m (σ could be any set to which are attached a "dimension" r, a "volume" $|\sigma|$, and a "diameter" diam (σ)), define its *fullness* by

(1) $$\Theta(\sigma) = |\sigma|/\delta_\sigma^r, \qquad \delta_\sigma = \operatorname{diam}(\sigma).$$

If $v_1, \cdots, v_r$ form a defining set of vectors for σ, then (III, 1.3) and (I, 12.16) give

(2) $$|\sigma| = |v_1 \vee \cdots \vee v_r|/r! \leqq |v_1| \cdots |v_r|/r! \leqq \delta_\sigma^r/r!;$$

hence

(3) $$\Theta(\sigma) \leqq 1/r!, \qquad \dim(\sigma) = r.$$

† H. A. Schwarz, Sur une définition erronée de l'aire d'une surface courbe, *Gesammelte Math. Abhandlungen, I*, pp. 309–311. The example is described for instance in T. Rado, Length and area, *Am. Math. Soc. Colloquium Publications, 30* (1948) 6–7.

An *altitude* of an r-simplex σ in E^m is the distance from a vertex to the plane of the opposite $(r-1)$-face.

LEMMA 14a. *For any r-simplex σ and any altitude h of σ,*

$$h \geqq r!\Theta(\sigma)\delta_\sigma. \tag{4}$$

Say $\sigma = p_0 \cdots p_r$, $\sigma' = p_1 \cdots p_r$, and h is the altitude from p_0. Then

$$|\sigma| = h|\sigma'|/r, \qquad |\sigma'| \leqq \delta_\sigma^{r-1}/(r-1)!,$$

and hence

$$h = r|\sigma|/|\sigma'| = r\Theta(\sigma)\delta_\sigma^r/|\sigma'| \geqq r!\Theta(\sigma)\delta_\sigma.$$

An immediate consequence of this is

$$|p_j - p_i| \geqq r!\Theta(\sigma)\delta_\sigma \quad \text{if} \quad j \neq i, \quad \sigma = p_0 \cdots p_r. \tag{5}$$

Suppose σ^k is a face of $\sigma = \sigma^r$. Choose a defining set $v_1, \cdots, v_r$ of vectors for σ such that $v_1, \cdots, v_k$ is a defining set for σ^k. Using (I, 12.14) shows that

$$\Theta(\sigma) = |v_1 \vee \cdots \vee v_r|/r!\delta_\sigma^r \leqq |v_1 \vee \cdots \vee v_k|/r!\delta_\sigma^k,$$

and hence

$$r!\Theta(\sigma^r) \leqq k!\Theta(\sigma^k), \qquad \sigma^k \text{ a face of } \sigma^r. \tag{6}$$

LEMMA 14b. *For any r-simplex $\sigma = p_0 \cdots p_r$ and point $p = \mu_0 p_0 + \cdots + \mu_r p_r$ in σ,*

$$\operatorname{dist}(p, \partial\sigma) \geqq r!\Theta(\sigma)\delta_\sigma \inf\{\mu_0, \cdots, \mu_r\}. \tag{7}$$

Let q be a nearest point of $\partial\sigma$ to p; then pq is parallel to an altitude h of σ, say that from p_0. Now (4) gives

$$\operatorname{dist}(p, \partial\sigma) = |q - p| = \mu_0 h \geqq r!\Theta(\sigma)\delta_\sigma \inf\{\mu_i\}.$$

The function $\Theta(\sigma)$ is of course continuous with respect to the positions of the vertices of σ. We give a uniformity condition.

LEMMA 14c. *Given r, $\Theta_0 > 0$, and $\epsilon > 0$, there is a $\rho_0 > 0$ with the following property. Take any simplex $\sigma = p_0 \cdots p_r$ with $\Theta(\sigma) \geqq \Theta_0$, and take any points $q_0, \cdots, q_r$, with $|q_i - p_i| \leqq \rho_0\delta_\sigma$ (all i). Then $\sigma' = q_0 \cdots q_r$ is a simplex, and $\Theta(\sigma') \geqq \Theta_0 - \epsilon$.*

Choose $\rho_0 < 1/2$ so that

$$\frac{\Theta_0}{(1+2\rho_0)^r} - \frac{2\rho_0(1+2\rho_0)^{r-1}}{(r-1)!(1-2\rho_0)^r} > \Theta_0 - \epsilon. \tag{8}$$

Now take σ', and set $v_i = p_i - p_0$, $w_i = q_i - q_0$. Since $|w_i| \leqq (1+2\rho_0)\delta_\sigma$, (I, 12.17) gives

$$|w_1 \vee \cdots \vee w_r - v_1 \vee \cdots \vee v_r| \leqq r(1+2\rho_0)^{r-1}\delta_\sigma^{r-1} \cdot 2\rho_0\delta_\sigma;$$

hence, by (III, 1.3),

$$|\sigma'| \geqq |\sigma| - 2\rho_0(1+2\rho_0)^{r-1}\delta_\sigma^r/(r-1)!.$$

Since $(1-2\rho_0)\delta_\sigma \leqq \delta_{\sigma'} \leqq (1+2\rho_0)\delta_\sigma$, using (8) gives the result.

15. Linear combinations of edge vectors of simplexes. If $v_1, \cdots, v_r$ are orthonormal and $v = \sum a_i v_i$, then $|v| = (\sum a_i^2)^{1/2}$. If $v_1, \cdots, v_r$ are unit vectors but not orthogonal, $|v|$ may be quite different, in particular, much less (but not 0 if some a_k is $\neq 0$ and the v_i are independent). We wish some inequalities.

LEMMA 15a. *Given vectors $u_1, \cdots, u_r$ and numbers $a_1, \cdots, a_r$,*

(1) $\left|\sum a_i u_i\right| \geqq \sup\{|a_1|, \cdots, |a_r|\}\,|u_1 \vee \cdots \vee u_r|$ if each $|u_i| = 1$.

Suppose not. Say $a = |a_r|$ is the largest of the $|a_i|$. Set

$$w = b_1 u_1 + \cdots + b_{r-1} u_{r-1} + a_r u_r,$$

with the b_i chosen so as to minimize $|w|$. Then

$$\begin{aligned} |w| &\leqq \left|\sum a_i u_i\right| < a\,|u_1 \vee \cdots \vee u_r| = |u_1 \vee \cdots \vee u_{r-1} \vee w| \\ &\leqq |u_1| \cdots |u_{r-1}|\,|w| = |w|, \end{aligned}$$

a contradiction.

LEMMA 15b. *Let $u_1, \cdots, u_r$ be independent unit vectors parallel to edges of the r-simplex σ. Then*

$$\left|\sum a_i u_i\right| \geqq r!\,\sup\{|a_1|, \cdots, |a_r|\}\Theta(\sigma), \tag{2}$$

$$|a_i| \leqq \left|\sum a_j u_j\right|/r!\Theta(\sigma), \qquad i = 1, \cdots, r. \tag{3}$$

Let $v_1, \cdots, v_r$ be the corresponding defining set of edge vectors of σ. Then $u_i = v_i/|v_i|$, and the last lemma gives, if $a = \sup\{|a_i|\}$,

$$\begin{aligned} \left|\sum a_i u_i\right| &\geqq a\,|u_1 \vee \cdots \vee u_r| = a\,|v_1 \vee \cdots \vee v_r|/|v_1| \cdots |v_r| \\ &\geqq r!a\,|\sigma|/\delta_\sigma^r = r!a\Theta(\sigma), \end{aligned}$$

which is (2). The other relation follows from this.

We now show that if σ is near a plane and $\Theta(\sigma)$ is not too small, then σ is nearly parallel to the plane.

LEMMA 15c. *Let π denote orthogonal projection into a plane P. Let $\sigma = p_0 \cdots p_r$ be a simplex, and suppose*

$$\sigma \subset U_\zeta(P), \qquad |p_i - p_0| \geqq \delta > 0 \quad (i = 1, \cdots, r). \tag{4}$$

Then for any unit vector u in σ,

$$|u - \pi u| \leqq 2\zeta/(r-1)!\Theta(\sigma)\delta. \tag{5}$$

The vectors $v_i = p_i - p_0$ $(i = 1, \cdots, r)$ form a defining set for σ. Setting $u_i = v_i/|v_i|$, we have

$$|u_i - \pi u_i| = |v_i - \pi v_i|/|v_i| \leqq 2\zeta/\delta.$$

Say $u = \sum a_i u_i$. Then (3) gives

$$|u - \pi u| = \left|\sum a_i(u_i - \pi u_i)\right| \leqq r[1/r!\Theta(\sigma)][2\zeta/\delta],$$

which is (5).

16. The quantities used in the proof. We shall indicate how various quantities appearing in the proof enter. The cubes of L_0 are of side length h; the simplexes of L are of diameter δ. The vertices of L are pushed at most a distance $\rho_0\delta$ to give vertices of L^*; then the simplexes of L^* have fullness $\geqq \Theta_0$ (Lemma 14c). Each vertex of L^* is on at most N simplexes of L^*.

In pushing the vertex p_i of L into the vertex p_i^* of L^*, we must avoid at most N sets Q_j, whose total volume is less than that of a sphere of radius $\rho_0\delta$; hence p_i^* may be found. The volume of Q_j is determined by the number ρ_1. The distances from M to the vertices, the 1-simplexes, etc. of L^* are determined in succession, using (App. II, Lemma 14b) and induction; each bound on distances is $\rho = \rho_1\rho_0/4$ times the preceding. Finally, the distance from M to the $(s-1)$-simplexes of L^* is at least $4\alpha\delta = 2a$; the distance from the part of any tangent plane P_p to M at p which lies within $7\delta = \xi - \delta$ of p, to such simplexes, is at least a.

Each simplex σ of K has an altitude at least $b = \beta\delta$; its fullness is at least Θ_1. The center p_1 of any n-simplex σ_1 of K is at a distance $\geqq c = \gamma\delta$ from $\partial\sigma_1$; this is used in applying (App. II, Lemma 15a). The proof of the above properties, and making the simplexes of K nearly tangent to M, etc., give requirements on the rate of turn of tangent planes to M relative to the sizes of the simplexes of K. Thus a number λ is determined, giving ξ, and hence h and δ, by Lemmas 8a and 8c.

17. The complex L. If we take a cubical subdivision of E^m and the regular subdivision $\mathfrak{S}$ of this, all simplexes of $\mathfrak{S}$ have the same fullness, which we call $2\Theta_0$. (Actually, $2\Theta_0 = 1/m!m^{m/2}$.) Let N be the largest number of simplexes in any star of a vertex of $\mathfrak{S}$.

Choose $\rho_0 < 1/4m^{1/2}$ by Lemma 14c so that for any n-simplex $\sigma = p_0 \cdots p_n$, if $\Theta(\sigma) \geqq 2\Theta_0$, and $|q_i - p_i| \leqq \rho_0\delta_\sigma$, then $\tau = q_0 \cdots q_n$ is a simplex, with $\Theta(\tau) \geqq \Theta_0$; with ρ^* as in (App. II, Lemma 16a), we take $\rho_0 \leqq 2\rho^*/m^{1/2}$.

There is a number $\rho_1 > 0$ with the following property. Let Q be any ball in E^m, of any radius a, and let Q' be the part of Q between any two parallel $(m-1)$-planes whose distance apart is $\leqq 2\rho_1 a$. Then we have the inequality on volumes

(1) $$|Q'| < |Q|/N.$$

Set

(2) $$\rho = \rho_0\rho_1/4, \qquad \alpha_r = \rho^r\rho_0\rho_1/2, \quad \alpha = \alpha_{s-1}/4,$$

(3) $$\beta = \Theta_0\alpha/m^{1/2}N, \qquad \Theta_1 = \beta^n/2^n, \qquad \gamma = (n-1)!\Theta_1\beta/2.$$

Choose $\rho_0' \leqq 1/4$ by Lemma 14c, using n, Θ_1 and $\Theta_1/2$ in place of r, Θ_0 and ϵ. Set

(4) $$\lambda = \inf\{\alpha\gamma/128, \rho_0'\alpha\beta/8\}.$$

Say the projection π^* of Theorem 10A is defined in the neighborhood $U^* = U_{\delta_0}(M)$; we take $\lambda_0 \leqq 1/4$ in that theorem.

Choose ξ_0 by Lemma 8a, choose $\xi_1 \leqq \xi_0$ by Lemma 8b, and set

$$\xi = \inf\{\xi_1, \alpha\delta_0/3\lambda\}, \qquad \delta = \xi/8, \qquad h = 2\delta/m^{1/2}, \tag{5}$$

$$a = 2\alpha\delta, \qquad b = \beta\delta, \qquad c = \gamma\delta. \tag{6}$$

Let L_0 be a cubical subdivision of E^m, the cubes being of side length h, and let L be the regular subdivision of L_0. Then each 1-simplex of L is of length $\geqq h/2$, and the m-simplexes have diameter δ.

18. The complex L^*. Let the vertices of L be $p_1, p_2, \cdots$. We shall choose new points $p_1^*, p_2^*, \cdots$ with

$$|p_i^* - p_i| < \rho_0\delta, \qquad \text{all } i. \tag{1}$$

By the choice of ρ_0, this will define a new triangulation of E^m, and using $\rho_0\delta < h/8$ and (14.6) gives, for all simplexes τ of L^* of dimension $\geqq 1$,

$$h/4 < \operatorname{diam}(\tau) < 2\delta, \qquad \Theta(\tau) \geqq \Theta_0. \tag{2}$$

We shall obtain also

$$\operatorname{dist}(M, \tau^r) > \alpha_r\delta, \qquad \text{all } \tau^r \text{ in } L^*, \quad r \leqq s-1, \tag{3}$$

and hence, if L^{*s-1} denotes the $(s-1)$-dimensional part of L^*,

$$\operatorname{dist}(M, L^{*s-1}) > 2a. \tag{4}$$

Suppose $p_1^*, \cdots, p_{i-1}^*$ have been found, so that the complex L_{i-1}^* with these vertices satisfies (3); we shall find p_i^*, so that L_i^* satisfies (3).

Case I, $\operatorname{dist}(p_i, M) \geqq 3\delta$. Then we set $p_i^* = p_i$. Because of (2), (3) will hold for L_i^*.

Case II, there is a point $p \in M$, $|p - p_i| < 3\delta$. Let P_0 be the tangent plane P_p (§ 8). Let $\tau_1', \cdots, \tau_\nu'$ $(\nu \leqq N-1)$ be the simplexes of L_{i-1}^* of dimension $\leqq s-2$ such that $\tau_j = p_i^*\tau_j'$ will be a simplex of L_i^*. Let P_j be the plane spanned by τ_j' and P_p $(j \geqq 1)$; its dimension is at most $(s-2)+n+1 < m$. Set

$$Q_j = U_{\rho_0\delta}(p_i) \cap U_{\rho_1\rho_0\delta}(P_j), \qquad j = 0, 1, \cdots, \nu. \tag{5}$$

By the choice of ρ_1, $|Q_j| < |U_{\rho_0\delta}(p_i)|/N$; hence there is a point p_i^* satisfying (1), such that

$$\operatorname{dist}(p_i^*, P_j) > \rho_1\rho_0\delta, \qquad j = 0, 1, \cdots, \nu. \tag{6}$$

We show now that

$$\operatorname{dist}(\tau_j', P_p) > 2\alpha_{r-1}\delta/3 \quad \text{if} \quad \dim(\tau_j') = r-1. \tag{7}$$

Since τ_j' is in L^*_{i-1}, dist $(\tau_j', M) > \alpha_{r-1}\delta$. By (8.6), $P_{p,\xi} \subset U_{\lambda\xi}(M)$; since $\lambda < \alpha_{s-1}/24$, $\lambda\xi < \alpha_{r-1}\delta/3$, and (7) holds with $P_{p,\xi}$ in place of P_p. Since $|p - p_i| < 3\delta$ and diam $(p_i\tau_j') < 2\delta$, dist $(\tau_j', P_p - P_{p,\xi}) > 3\delta$, which gives (7).

Applying (App. II, Lemma 14b) gives

$$\text{dist}\,(\tau_j, P_p) \geqq \text{dist}\,(\tau_j', P_p)\,\text{dist}\,(p_i^*, P_j)/\text{diam}\,(\tau_j)$$
$$> (2\alpha_{r-1}\delta/3)(\rho_1\rho_0\delta)/2\delta = 4\alpha_{r-1}\rho\delta/3 = 4\alpha_r\delta/3.$$

Since $\lambda\xi < \alpha_r\delta/3$, using (8.6) and (8.3) and the same argument as above gives (3), for $\tau^r = \tau_j$, $j \geqq 1$. Using $j = 0$ in (6) and the same argument again gives (3) for $\tau^r = p_i^*$; hence (3) and (4) are proved.

19. The intersections of M with L^*. We prove some properties of the intersections of M and its tangent planes with the simplexes of L^*.

(*a*) For any $p \in M$ and r-simplex τ^r of L^*,

$$(1) \qquad \text{dist}\,(P_p, \tau^r) > a \quad \text{if} \quad \tau^r \subset U_{7\delta}(p), \qquad r \leqq s - 1.$$

For dist $(P_p - P_{p,\xi}, \tau^r) > \xi - 7\delta > a$, and $P_{p,\xi} \subset U_{\lambda\xi}(M)$, $\lambda\xi < a$; using (18.4) gives (1).

(*b*) If M intersects τ^r, $p \in M$, and $\tau^r \subset U_{7\delta}(p)$, then P_p intersects τ^r. For if $p' \in M \cap \tau^r$, then by (8.3), $p' \in M_{p,\xi}$. By (8.6), dist $(p', P_p) < \lambda\xi < a$. Let τ^t be a face of smallest dimension of τ^r with dist $(\tau^t, P_p) \leqq a$. By (1), $t \geqq s$, and by (App. II, Lemma 14a), P_p intersects τ^t.

(*c*) If $r = s$ in (*b*), and $P(\tau^s)$ is the plane of τ^s, then

$$(2) \qquad \text{ind}\,(P_p, P(\tau^s)) > \alpha.$$

This follows from the lemma quoted, (1), and (18.2).

(*d*) If $p \in M$, $\tau^r \subset U_{7\delta}(p)$, and P_p intersects τ^r, then $r \geqq s$, and $M_{p,\xi}$ intersects τ^r. Let τ^t be a smallest face of τ^r such that dist $(P_p, \tau^t) \leqq a$. By (1) and (App. II, Lemma 14a), $t = s$ (hence $r \geqq s$), P_p has a point p' in τ^s, and (2) holds. Let π' be the projection into P_p along planes parallel to τ^s (§ 11). By Lemma 11a, $\pi'(M_{p,\xi})$ covers $P_{p,\eta}$, with $\eta = (1 - \lambda/\alpha)\xi > 7\delta$. Since $|p' - p| < 7\delta$, there is a $p^* \in M_{p,\xi}$ with $\pi'(p^*) = p'$; hence $p^* \in P(\tau^s)$. By (11.2),

$$|p' - p^*| < \lambda\xi/\alpha \leqq \rho_0'\beta\delta < \beta\delta < a.$$

Since $p' \in \tau^s$, (18.4) shows that $p^* \in \tau^s$.

(*e*) M intersects any τ^s in at most one point. For suppose M had the distinct points p, p' in τ^s. Then by (8.3), $p' \in M_{p,\xi}$, and $M_{p,\xi}$ has a secant vector $v = p' - p$ in τ^s. By Lemma 8c, $|v - \pi_p v| \leqq \lambda |v|$. But (2) gives $|v - \pi_p v| > \alpha |v| > \lambda |v|$, a contradiction.

(*f*) If M intersects $\tau^r = q_0 \cdots q_r$, then for each k, M intersects some s-face of τ^r containing q_k. For take $p \in M \cap \tau^r$. Let τ^t be a face of smallest dimension of τ^r containing q_k which P_p intersects. Suppose $t > s$. Then if τ^{t-1} is the face of τ^t opposite q_k, P_p intersects some s-face of τ^{t-1}. Because of (c), P_p contains interior points of τ^t, and hence intersects $\partial\tau^t - \tau^{t-1}$, a contradiction. Hence $t = s$. By (d), M also intersects τ^s.

20. The complex K. In each simplex τ of L^* intersecting M we shall choose a point $\psi(\tau)$; these are the vertices of K. For each sequence $\tau_0 \subset \tau_1 \subset \cdots \subset \tau_r$ of distinct simplexes of L^* such that M intersects τ_0 (and hence all the τ_i),

(1) $$\sigma^r = \psi(\tau_0) \cdots \psi(\tau_r)$$

will be a simplex of K.

First, for each τ^s which M intersects, there is just one point of intersection, by (19(e)); let $\psi(\tau^s)$ be this point. Next, for any τ^r ($r > s$) which M intersects, let $\tau_1^s, \cdots, \tau_k^s$ be the s-faces of τ^r intersecting M (see (19(f))); set

(2) $$\psi(\tau^r) = (1/k)\psi(\tau_1^s) + \cdots + (1/k)\psi(\tau_k^s).$$

We show that for any $\tau^s = q_0 \cdots q_s$ of L^* intersecting M,

(3) $$\mu_k > 2\alpha \quad (k = 0, \cdots, s) \quad \text{if} \quad \psi(\tau^s) = \sum \mu_i q_i.$$

For let τ_k be the $(s - 1)$-face opposite q_k. Let A_k and A_k' be the altitudes from q_k and $\psi(\tau^s)$ respectively to $P(\tau_k)$. By (18.4) and (18.2),

$$\mu_k = A_k'/A_k > 2a/2\delta = 2\alpha.$$

Next, if M intersects $\tau^r = q_0 \cdots q_r$, then

(4) $$\mu_k > 2\alpha/N \quad (k = 0, \cdots, r) \quad \text{if} \quad \psi(\tau^r) = \sum \mu_i q_i.$$

Given k, let τ^s be an s-face of τ^r containing q_k, which intersects M (19(f)). By (3), the barycentric coordinate μ' of $\psi(\tau^s)$ corresponding to q_k is at least 2α. By (2), μ_k is the average of at most N barycentric coordinates, one of which is μ'; hence (4) holds.

The vertices of each simplex σ of K have a natural order; let alt (σ) be the altitude from the last vertex (vertex in the simplex of highest dimension of L^*). We prove

(5) $$\text{alt}\,(\sigma^r) \geqq rb.$$

For if σ^r is as in (1), the $(r - 1)$-face σ^{r-1} opposite $\psi(\tau_r)$ lies in τ_{r-1}. If $\dim\,(\tau_r) = t \geqq r$, Lemma 14b, (4) and (18.2) give

$$\begin{aligned} \text{alt}\,(\sigma^r) &\geqq \text{dist}\,(\psi(\tau_r), \partial\tau_r) \geqq t!\Theta(\tau_r)\delta_{\tau_r}(2\alpha/N) \\ &\geqq r\Theta_0(\delta/2m^{1/2})(2\alpha/N) = rb. \end{aligned}$$

Since $|\sigma^r| = \text{alt}(\sigma^r)|\sigma^{r-1}|/r$, we see at once, by induction, that $|\sigma^r| \geqq b^r$, and hence

$$(6) \qquad \Theta(\sigma^r) \geqq b^r/(2\delta)^r = \beta^r/2^r \geqq \Theta_1.$$

21. Imbedding of simplexes in M. We show that K is near M, and that any n-simplex near an n-simplex of K is imbedded in M by π^*.

We prove first, for any simplex σ of K,

$$(1) \qquad \text{if} \quad \sigma \subset U_{6\delta}(p), \quad p \in M, \qquad \text{then} \quad \sigma \subset U_{\lambda\xi}(P_{p,\xi}).$$

Say σ is in the simplex τ of L^*; then $\tau \subset U_{8\delta}(p)$. Each vertex p_i of σ is an average of points $\psi(\tau_j^s)$, which are in $M_{p,\xi}$ and hence in $U_{\lambda\xi}(P_{p,\xi})$; therefore (1) holds for the p_i and hence for σ. As a consequence of this and (10.4),

$$(2) \qquad K \subset U_{2\lambda\xi}(M), \qquad |\pi^*(q) - q| < 4\lambda\xi \quad (q \in K).$$

LEMMA 21a. *Let $\sigma = p_0 \cdots p_n$ be an n-simplex of K (vertices in increasing order), and let $p_0', \cdots, p_n'$ be any points such that*

$$(3) \qquad |p_i' - p_i| \leqq \lambda\xi/\alpha, \qquad i = 0, \cdots, n.$$

Then $\sigma' = p_0' \cdots p_n'$ is a simplex in U^*, and π^* imbeds σ' in M.

First, since $\lambda\xi/\alpha \leqq \rho_0'\beta\xi/8 = \rho_0' b$, $\Theta(\sigma) \geqq \Theta_1$, and $\text{diam}(\sigma) \geqq b$, by (20.5), the choice of ρ_0' gives $\Theta(\sigma') \geqq \Theta_1/2$.

Next, because of (2) and (17.5), $\sigma' \subset U_\eta(M)$, with

$$(4) \qquad \eta = \lambda\xi/\alpha + 2\lambda\xi < 3\lambda\xi/\alpha \leqq \delta_0;$$

hence $\sigma' \subset U^*$, and π^* is defined in σ'.

Now take any $q \in \sigma'$. Say $q \in Q_p^*$, $p \in M$; then $\pi^*(q) = p$. By (10.4) and part of (4),

$$|q - p| \leqq 2(3\lambda\xi/\alpha) < \delta;$$

hence $\sigma \subset U_{4\delta}(p)$, and (1) gives $\sigma \subset U_{\lambda\xi}(P_p)$. Therefore $\sigma' \subset U_{2\lambda\xi/\alpha}(P_p)$. Also, by (20.5), $|p_i - p_0| \geqq b$; hence

$$|p_i' - p_0'| \geqq b - 2\lambda\xi/\alpha \geqq b - 2\rho_0' b \geqq b/2.$$

Hence, if u is any unit vector in σ', Lemma 15c gives

$$|u - \pi_p u| \leqq 2(2\lambda\xi/\alpha)/(n-1)!(\Theta_1/2)(b/2) = 64\lambda/\alpha\gamma \leqq 1/2.$$

Since we took $\lambda_0 \leqq 1/4$, (10.1) shows that u is not in Q_p^*. Therefore π^* maps each non-zero vector in σ' at q into a non-zero vector, and π^* is regular at q. Also, if q' is another point of σ', then using $v = q' - q$ shows that q' is not in Q_p^*, and hence $\pi^*(q') \neq \pi^*(q)$. This completes the proof.

22. The complexes K_p. For each $p \in M$, let L_p^* be the subcomplex of L^* containing all simplexes which touch $\bar{U}_{4\delta}(p)$, together with their faces; then

$$(1) \qquad L_p^* \subset U_{6\delta}(p).$$

Let K_p'' be the complex in P_p formed by the intersections of P_p with the simplexes of L_p^*, and let K_p' be the regular subdivision of K_p''. By (*b*) and (*d*) of § 19, P_p intersects a simplex of L_p^* if and only if M does. Hence, if K_p is the subcomplex of K containing those simplexes with vertices $\psi(\tau)$, τ in L_p^*, there is a one-one correspondence ϕ_p of the vertices of K_p onto the vertices of K_p', and this defines a simplicial mapping ϕ_p which is an isomorphism of K_p onto K_p'.

We prove

$$(2) \qquad |\phi_p(q) - q| < \lambda\xi/\alpha, \qquad q \in K_p.$$

First suppose $q = \psi(\tau^s)$ for some τ^s in L_p^*. Then $v = q - \phi_p(q)$ is in τ^s, and using (8.6) and (19(*c*)) gives

$$\lambda\xi > |q - \pi_p(q)| = |v - \pi_p v| \geqq \alpha |v|,$$

giving (2). Next, if $q = \psi(\tau^r)$, $r > s$, then the definitions (20.2) and (App. II, 1.2) show that q and $\phi_p(q)$ are the same averages of sets of points, each corresponding pair satisfying (2); hence (2) holds for $q = \psi(\tau^r)$. Finally, for any simplex of K_p, (2) holds for its vertices and hence for all its points.

We must show that

$$(3) \qquad K \cap U_{2\delta}(p) \subset K_p.$$

For take any point q in a simplex $\sigma = \psi(\tau_0) \cdots \psi(\tau^r)$ of K, $|q - p| < 2\delta$. Then $\tau^r \subset U_{4\delta}(p)$, hence τ^r is in L_p^*, and σ and q are in K_p.

Choose an orientation of P_p, and orient all n-simplexes of K_p' accordingly. Now K_p' is an oriented n-pseudomanifold (App. II, 15), and (1) and the definition of L_p^* show that

$$(4) \qquad K_p' \subset U_{6\delta}(p), \qquad \partial K_p' \subset P_p - \bar{U}_{4\delta}(p).$$

Define the mapping π_p^* of K_p into P_p as follows. Each $q \in K_p$ is in a unique $Q_{p'}^*$; then $p' = \pi^*(q)$. By (1), $|q - p| \leqq 6\delta$, and by (21.2), $|p' - q| < 4\lambda\xi < \delta$; hence $|p' - p| < \xi$. By Lemma 10a, $P_{p'}^*$ intersects P_p in a unique point, which we call $\pi_p^*(q)$. We prove

$$(5) \qquad |\pi_p^*(q) - q| < 6\lambda\xi, \qquad q \in K_p.$$

Because of (21.2), we need merely prove $|v| < 2\lambda\xi$, where $v = p' - \pi_p^*(q)$. Since $\lambda + \lambda_0 < 1/2$, (10.12) gives $|\pi_p v| \leqq |v|/2$. By (8.5), $|v - \pi_p v| < \lambda\xi$. Therefore $|v| < |v|/2 + \lambda\xi$, and the statement follows.

23. Proof of the theorem. Given $p \in M$, choose an orientation of P_p, and orient the n-simplexes of K'_p and K_p correspondingly. Now K_p is an oriented pseudomanifold (App. II, 15). The proof of Theorem 12A rests on the following lemmas.

LEMMA 23a. *π^*_p is a simplexwise positive mapping of K_p into P_p.*

Take any n-simplex σ of K_p. Set

$$(1) \qquad \phi_{p,t}(q) = (1-t)q + t\phi_p(q) \quad \text{in } \sigma, \qquad \sigma_t = \phi_{p,t}(\sigma).$$

Since ϕ_p is affine in σ, so is $\phi_{p,t}$, and σ_t is a simplex. Say $\sigma = q_0 \cdots q_n$. For any t $(0 \leqq t \leqq 1)$, set $q_{it} = \phi_{p,t}(q_i)$; then $\sigma_t = q_{0t} \cdots q_{nt}$. By (22.2), $|q_{it} - q_i| < \lambda\xi/\alpha$. By Lemma 21a, π^* imbeds σ_t in M; hence (by the reasoning of that lemma) π^*_p imbeds σ_t in P_p. Since σ_1 is in P_p, π^*_p is the identity in σ_1, and hence $J_{\pi^*_p} > 0$ in σ_1. Since $J_{\pi^*_p} \neq 0$ in σ_t for all t, $J_{\pi^*_p} > 0$ in $\sigma_0 = \sigma$, as required.

For each $p \in M$, let R_p be the set of those points $q \in K_p$ such that $\pi^*_p(q) \in P_{p,3\delta}$.

LEMMA 23b. *For each $p \in M$, π^*_p, considered in R_p only, is one-one and onto $P_{p,3\delta}$.*

First, let σ_1 be an n-simplex of K'_p containing p; say $\sigma_1 = \phi_p(\sigma_0)$ (σ_0 in K_p), and let p_0 be the center of σ_0. By (14.7), (20.6), (20.5) and (17.3),

$$\operatorname{dist}(p_0, \partial\sigma_0) \geqq n!\Theta_1 b/(n+1) \geqq c.$$

Hence, by (22.2),

$$(2) \qquad \operatorname{dist}(\phi_p(p_0), \partial\sigma_1) > c - 2\lambda\xi/\alpha = c'.$$

Now take any $q \in K_p - \sigma_0$. Since ϕ_p is an isomorphism, (2) shows that $|\phi_p(q) - \phi_p(p_0)| > c'$. By (17.4), (17.5) and (17.6),

$$4\lambda\xi/\alpha + 12\lambda\xi < 16\lambda\xi/\alpha \leqq \gamma\delta = c.$$

Hence, by (22.2) and (22.5),

$$|\pi^*_p(q) - \pi^*_p(p_0)| > c' - 2(\lambda\xi/\alpha + 6\lambda\xi) > 0,$$

proving $\pi^*_p(q) \neq \pi^*_p(p_0)$. This shows that $p^* = \pi^*_p(p_0)$ is covered exactly once, under π^*_p, by simplexes of K_p. Also

$$|p^* - p| \leqq |\pi^*_p(p_0) - \phi_p(p_0)| + \operatorname{diam}(\sigma_1) < 3\delta,$$

and hence $p^* \in P_{p,3\delta}$.

By (22.4), (22.2) and (22.5), since $2(\lambda\xi/\alpha + 6\lambda\xi) < \delta$,

$$(3) \qquad \pi^*_p(\partial K_p) \subset P_p - \bar{U}_{3\delta}(p).$$

The lemma now follows from Lemma 23a and (App. II, Lemma 15a).

We now prove the theorem. First, given $p \in M$, the last lemma shows that $\pi^*_p(q) = p$ for some $q \in K_p$; hence $\pi^*(q) = p$, and π^* is onto. Next, suppose that $\pi^*(q') = p$ also, $q' \in K$. By (21.2), $|q' - p| < 4\lambda\xi < \delta$; hence, by (22.3), $q' \in K_p$, and therefore $q' \in R_p$. By Lemma 23b, $q' = q$. This proves that π^* is one-one.

Finally, take any n-simplex σ of K. By Lemma 21a, π^* is one-one and regular in σ, and hence in a neighborhood U of σ in the plane $P(\sigma)$ of σ. Let ϕ be an affine mapping of $\mathfrak{A}^n$ onto $P(\sigma)$; set $U_0 = \phi^{-1}(U)$. Then $\chi = \pi^*\phi$ is defined in U_0, and is a coordinate system in M, and $\chi^{-1}\pi^* = \phi^{-1}$ is affine in σ. This completes the proof.

C. Cohomology in Manifolds

24. μ-regular forms. The theorem of de Rham (Theorem 29A below) relates cohomology in a smooth manifold defined algebraically to cohomology defined by differential forms. We must use a class of forms which are invariant under the operation d of exterior differentiation. We choose "μ-regular forms". *We could clearly use ∞-smooth forms if M is ∞-smooth.*

For some integer $\mu \geq 0$, assume that M is $(\mu + 1)$-smooth. Then the r-form ω in an open subset R of M is *μ-regular* in R if the following is true. If $\mu = 0$, then ω is regular (III, 17); if $\mu > 0$, then both ω and $d\omega$ are μ-smooth. If this holds, then $d\omega$ is uniquely defined and is μ-regular. For $r = 0$, ω is μ-regular if and only if ω is $(\mu + 1)$-smooth, since the components of $d\omega$ in some coordinate system are the partial derivatives of ω.

We say ω is *closed* if $d\omega = 0$; ω is *derived* if $\omega = d\xi$ for some ξ; ω is *μ-derived* if $\omega = d\xi$, ξ being μ-regular.

Using (II, 8) (if $\mu > 0$) and (III, 17) (if $\mu = 0$), we have the following facts. If ω and ξ are μ-regular, so is $\omega \vee \xi$. If f is a $(\mu + 1)$-smooth mapping of the open subset R of the $(\mu + 1)$-smooth manifold M into the $(\mu + 1)$-smooth manifold M', and ω is μ-regular in a neighborhood of $f(R)$ in M', then $f^*\omega$ is μ-regular in R. The usual formulas for $d(\omega \vee \xi)$ and $df^*\omega$ hold.

We say a given property of forms holds *near* a given closed set Q if it holds in some neighborhood of Q.

Lemma 24a. *Let ω be a μ-regular form near the closed set Q in M. Then there is a μ-regular form ω_1 in M such that $\omega_1 = \omega$ near Q. We may make $\omega_1 = 0$ outside an arbitrary neighborhood U of Q.*

Choose a neighborhood $U_1 \subset U$ of Q such that ω is μ-regular in a neighborhood of $\bar{U}_1$. Let ϕ be a $(\mu + 1)$-smooth real function in M such that $\phi(p) = 1$ in Q and $\phi(p) = 0$ in $M - U_1$. (For instance, let f imbed M in E^m, by Theorem 1A, define ϕ' in E^m so that $\phi'(q) = 1$ in $f(Q)$ and $\phi'(q) = 0$ in $f(M - U_1)$, by (App. III, Lemma 1c), and set $\phi(p) = \phi'(f(p))$ in M.) Set $\omega_1(p) = \phi(p)\omega(p)$ in U_1 and $\omega_1(p) = 0$ in $M - U_1$; then ω_1 has the required properties.

25. Closed forms in star shaped sets. (Compare § 22 of the Introduction.) We say the set $R \subset E^n$ is *star shaped* if there is a point $p_0 \in R$ such that for each $p \in R$, the segment $p_0 p$ is in R.

LEMMA 25a. *Let R be a star shaped open set in E^n, and let ω be a μ-regular closed r-form in R, $r > 0$. Then ω is μ-derived in R.*

We shall define a μ-regular $(r - 1)$-form ω_1 in R such that

$$\int_{\partial\sigma} \omega_1 = \int_\sigma \omega, \qquad \text{all } r\text{-simplexes } \sigma \text{ in } R; \tag{1}$$

Then $d\omega_1 = \omega$, by definition if $\mu = 0$, and by Stokes' Theorem and (III, Lemma 16a) if $\mu > 0$.

Say R is star shaped from p_0. Define the affine mapping g of the Cartesian product $\mathfrak{A} \times E^n$ into E^n by

$$g(t \times p) = (1 - t)p_0 + tp. \tag{2}$$

For each $p \in R$ there is a number $\eta_p > 0$ such that $g(t \times p) \in R$ for $-\eta_p < t < 1 + \eta_p$. Hence there is a neighborhood U of $I \times R$ in $\mathfrak{A} \times E^n$ ($I =$ interval $(0, 1)$) such that g maps U into R. Then $g^*\omega$ is regular in U.

Define $\omega_1(p)$ as the partial integral (III, 19)

$$\omega_1(p) = \int_{I \times p} g^*\omega(t \times p)\,dt, \qquad p \in R; \tag{3}$$

then ω_1 is μ-regular in R. (For $\mu = 0$, μ-regularity will follow from (1); for $\mu > 0$, it is clear from (3) or (6) that ω_1 is μ-smooth; μ-regularity will then follow from $d\omega_1 = \omega$.)

Take any oriented $(r - 1)$-simplex τ in R. First suppose its plane $P(\tau)$ does not contain p_0; let $p_0\tau = J(p_0, \tau)$ be the join of p_0 with τ (App. II, 10). With the obvious choice of orientations, the mapping g, considered in int $(I \times \tau)$, is an orientation preserving smooth homeomorphism onto int $(p_0\tau)$. Hence, by (III, Theorem 7A), (III, Lemma 6f) and (III, 19.1),

$$\int_{p_0\tau} \omega = \int_{I \times \tau} g^*\omega = \int_\tau \omega_1. \tag{4}$$

If $p_0 \in P(\tau)$, then $p_0\tau = 0$ (App. II, 10), and $\int_{p_0\tau} \omega$ is defined to be 0; also, considering g in $I \times \tau$ only, $J_g = 0$, hence (II, 6.6) $g^*\omega = 0$, and $\int_\tau \omega_1 = 0$.

Now take any r-simplex σ in R. By (4), (App. II, 10.3) and Stokes' Theorem,

$$\int_\sigma \omega - \int_{\partial\sigma} \omega_1 = \int_{\sigma - J(p_0, \partial\sigma)} \omega = \int_{\partial J(p_0, \sigma)} \omega = \int_{J(p_0, \sigma)} d\omega = 0,$$

proving (1).

We shall give an explicit formula for $\omega_1(p)\cdot\beta$ for any $(r-1)$-direction β. Say $\beta = e_1\vee\cdots\vee e_{r-1}$. Let e_0 be the unit vector in I. Now

$$\nabla g(t\times p, e_0) = \frac{\partial}{\partial t} g(t\times p) = p - p_0,$$

$$\nabla g(t\times p, e_i) = \lim_{h\to 0} \frac{1}{h} [g(t\times(p + he_i)) - g(t\times p)] = te_i$$

for $i \geqq 1$, and hence

(5) $$\nabla g(t\times p, e_0\vee\beta) = t^{r-1}(p - p_0)\vee\beta.$$

We may define ω_1 by

(6) $$\omega_1(p)\cdot\beta = \int_0^1 t^{r-1}\omega(g(t\times p))\cdot[(p - p_0)\vee\beta]dt.$$

To show that this is the same ω_1 as before, we prove (4). Setting $\beta = \{\tau\}/|\tau|$ and using the Riemann integral (III, 5) and (II, 4.4),

$$\int_{p_0\tau} \omega = \int_{I\times\tau} g^*\omega(t\times p)\cdot(e_0\vee\beta) = \int_{I\times\tau} \omega(g(t\times p))\cdot\nabla g(t\times p, e_0\vee\beta)$$
$$= \int_{I\times\tau} t^{r-1}\omega(g(t\times p))\cdot[(p - p_0)\vee\beta]dtdp = \int_\tau \omega_1(p)dp,$$

at least if $p_0\tau$ is non-degenerate. In the contrary case, $\int_{p_0\tau}\omega = 0$, and since $(p - p_0)\vee\beta = 0$, $\int_\tau \omega_1\cdot\beta = 0$.

26. Extensions of forms. We prove two lemmas on the extension through the neighborhood of a simplex of forms defined near the boundary of the simplex.

LEMMA 26a. *Let ω be a closed μ-regular r-form near $\partial\sigma$ ($\sigma = \sigma^s \subset E^n$), with $r \geqq 0$, $s \geqq 1$. Suppose that*

(1) $$\int_{\partial\sigma} \omega = 0 \qquad \text{if } s = r + 1.$$

Then there is a closed μ-regular form ω' near σ which equals ω near $\partial\sigma$.

LEMMA 26b. *Let ω be a closed μ-regular r-form near $\sigma = \sigma^s \subset E^n$, with $r \geqq 1$, $s \geqq 1$, and let ξ be a μ-regular $(r-1)$-form near $\partial\sigma$ such that $d\xi = \omega$ near $\partial\sigma$. Suppose that*

(2) $$\int_{\partial\sigma} \xi = \int_\sigma \omega \qquad \text{if } s = r.$$

Then there is a μ-regular form ξ' near σ such that $\xi' = \xi$ near $\partial\sigma$ and $d\xi' = \omega$ near σ.

Let (a_r) and (b_r) denote the lemmas, using r-forms (all s). We shall prove (a_0); then (b_r), assuming (a_{r-1}); then (a_r), assuming (b_r). This will prove the lemmas.

Proof of (a_0): Since $d\omega = 0$, ω is constant near any connected part of $\partial\sigma$. If $s > 1$, then ω is constant near $\partial\sigma$, and we may let ω' be the same constant near σ. If $s = 1$, and $\sigma = p_0p_1$, then

$$\omega(p_1) - \omega(p_0) = \int_{\partial\sigma} \omega = 0,$$

and again ω' exists.

Proof of (b_r): By Lemma 25a there is a μ-regular $(r - 1)$-form ξ_1 near σ such that $d\xi_1 = \omega$ near σ. Set $\eta = \xi - \xi_1$ near $\partial\sigma$; then η is closed near $\partial\sigma$. If $s = r$, then

$$\int_{\partial\sigma} \eta = \int_{\partial\sigma} \xi - \int_{\partial\sigma} \xi_1 = \int_\sigma \omega - \int_\sigma d\xi_1 = 0.$$

Hence, by (a_{r-1}), there is a closed μ-regular form η' near σ which equals η near $\partial\sigma$. Set $\xi' = \xi_1 + \eta'$ near σ; the required properties hold.

Proof of (a_r), $r > 0$: Say $\sigma = p_0 \cdots p_s$; set $\sigma' = p_1 \cdots p_s$. Let Q be the union of all proper faces of σ with p_0 as a vertex. (Thus, for $s = 2$, $Q = p_0p_1 \cup p_0p_2$.) For some $\epsilon > 0$, ω is defined and closed in $U = U_\epsilon(Q)$. Clearly U is star shaped (from p_0); hence, by Lemma 25a, there is a μ-regular ξ_0 such that $d\xi_0 = \omega$ in U. This holds, in particular, near $\partial\sigma'$.

If $s - 1 = r$, then letting A be the chain $\partial\sigma - \sigma'$, we have $\partial\sigma' = -\partial A$, and hence

$$\int_{\sigma'} \omega - \int_{\partial\sigma'} \xi_0 = \int_{\sigma'} \omega + \int_A d\xi_0 = \int_{\partial\sigma} \omega = 0;$$

hence we may apply (b_r), using σ', giving a μ-regular form ξ_1 near σ' such that $\xi_1 = \xi_0$ near $\partial\sigma'$ and $d\xi_1 = \omega$ near σ'. There is a neighborhood U' of $\partial\sigma'$ in which ξ_1 and ξ_0 are both defined and are equal; let ξ' be their common value here, and set $\xi' = \xi_0$ near $Q - U'$ and $\xi' = \xi_1$ near $\sigma' - U'$. Then ξ' is μ-regular and $d\xi' = \omega$ near $\partial\sigma$. By Lemma 24a there is a μ-regular form ξ near σ which equals ξ' near $\partial\sigma$. Set $\omega' = d\xi$ near σ; the required properties hold.

27. Elementary forms. By Theorem 12A, there is a triangulation of M, as follows. K is a simplicial complex, and π is a homeomorphism of K onto M. For each simplex τ of K, $\sigma = \pi(\tau)$ is a smooth simplex in M (III, 17), and there is a coordinate system χ in M containing σ, such that $\chi^{-1}\pi$ is affine in τ.

The smooth simplexes $\sigma = \pi(\tau)$ form a "curvilinear complex" L; we may define algebraic chains $A = \sum a_i\sigma_i^r$ in L, and integrate r-forms ω over A; Stokes' Theorem $\int_{\partial\sigma} \omega = \int_\sigma d\omega$ holds (III, 17.3).

Given the r-form ω in M, the function $\int_A \omega$ of r-chains of L is linear, and hence defines an r-cochain $\psi\omega$ of L (App. II, 6):

$$\psi\omega \cdot A = \int_A \omega. \tag{1}$$

Since, for $(r + 1)$-chains B,

$$\psi d\omega \cdot B = \int_B d\omega = \int_{\partial B} \omega = \psi\omega \cdot \partial B = d\psi\omega \cdot B,$$

we have

$$\psi d\omega = d\psi\omega. \tag{2}$$

Conversely, we wish to define forms ϕX corresponding to cochains X of L, with the following properties:

$$\phi\sigma = 0 \quad \text{near} \quad M - \mathrm{St}\,(\sigma) \qquad (\sigma \text{ in } L), \tag{3}$$

$$\phi dX = d\phi X, \tag{4}$$

$$\psi\phi X = X, \tag{5}$$

$$\phi I^0 = \mathbf{1}, \tag{6}$$

I^0 being the unit 0-cochain of L (App. II, 6), and $\mathbf{1}$ being the real function equal to 1 in M. The "elementary forms" $\phi\sigma$ will be sufficient to generate the differential cohomology spaces (§ 29).

We construct a partition of unity (compare (App. III, 2), (III, 10)) in M as follows. Let $p_1, p_2, \cdots$ be the vertices of K; the $q_i = \pi(p_i)$ are the vertices of L. For each $q = \pi(p)$ in M, write $p = \sum \mu_i(p)p_i$, as in (App. II, 2.1); write

$$\nu_i(q) = \mu_i(p), \qquad q = \sum \nu_i(q)q_i. \tag{7}$$

Thus we have introduced "barycentric coordinates" in L. For each i, let Q_i and Q'_i be the subsets of M such that

$$Q_i\colon\ \nu_i(q) \geqq 1/(n + 1), \qquad Q'_i\colon\ \nu_i(q) \leqq 1/(n + 2). \tag{8}$$

Then

$$Q_i \subset \mathrm{St}\,(q_i), \qquad \mathrm{int}\,(Q'_i) \supset M - \mathrm{St}\,(q_i). \tag{9}$$

Let $\phi'_i(p)$ be a $(\mu + 1)$-smooth non-negative real function in M which is positive in Q_i and is zero in Q'_i (see the proof of Lemma 24a); set

$$\phi_i(p) = \phi'_i(p) \Big/ \sum_j \phi'_j(p). \tag{10}$$

Take any $p \in M$; since p has at most $n + 1$ non-zero barycentric coordinates, at least one of these, say $\nu_j(p)$, is $\geqq 1/(n + 1)$; hence $p \in Q_j$, $\phi'_j(p) > 0$, and $\phi_i(p)$ is defined (all i). Now ϕ_i is $(\mu + 1)$-smooth, and

$$\sum \phi_i(p) = 1, \qquad \sum d\phi_i(p) = 0 \qquad \text{in } M. \tag{11}$$

For each oriented simplex $\sigma = q_{\lambda_0} \cdots q_{\lambda_r}$ of L, set*

$$\phi(q_{\lambda_0} \cdots q_{\lambda_r}) = r! \sum_{i=0}^{r} (-1)^i \phi_{\lambda_i} d\phi_{\lambda_0} \vee \cdots \widehat{d\phi_{\lambda_i}} \cdots \vee d\phi_{\lambda_r}. \tag{12}$$

* Compare A. Weil, Sur les théorèmes de de Rham, *Commentarii Math. Helvetici*, *26* (1952), formula on p. 127.

For instance,

$$(13)\qquad \phi(q_i) = \phi_i, \qquad \phi(q_iq_j) = \phi_i d\phi_j - \phi_j d\phi_i.$$

By (I, 1.13) and (App. II, Lemma 5a), a permutation of the vertices either leaves unchanged or changes the sign of both $q_{\lambda_0} \cdots q_{\lambda_r}$ and the right hand side of (12); hence $\phi\sigma$ is well defined, and we may set $\phi\sum a_i\sigma_i = \sum a_i\phi\sigma_i$.

To prove (3), take any $\sigma = q_{\lambda_0} \cdots q_{\lambda_r}$, and any p such that $\nu_{\lambda_j}(p) < 1/(n+2)$ for some j. Then $p \in Q'_{\lambda_j}$, $\phi_{\lambda_j} = 0$ near p, and $(\phi\sigma)(p) = 0$.

Because of (13), (6) is simply the first part of (11).

To prove (4), we need consider only $X = \sigma = q_{\lambda_0} \cdots q_{\lambda_r}$. Note first that applying d to (12) gives

$$(14)\qquad d\phi(q_{\lambda_0} \cdots q_{\lambda_r}) = (r+1)!\, d\phi_{\lambda_0} \vee \cdots \vee d\phi_{\lambda_r}.$$

Using (App. II, 7.5), both parts of (11), and the facts that $\phi_j = 0$ outside of St (p_j) and $d\eta \vee d\eta = 0$, we find

$$\phi d(q_{\lambda_0} \cdots q_{\lambda_r})/(r+1)! = \sum{}_k^* \phi(q_k q_{\lambda_0} \cdots q_{\lambda_r})/(r+1)!$$

$$= \sum{}^* [\phi_k d\phi_{\lambda_0} \vee \cdots \vee d\phi_{\lambda_r} - \sum_{i=0}^{r} (-1)^i \phi_{\lambda_i} d\phi_k \vee d\phi_{\lambda_0} \vee \cdots \hat{\imath} \cdots \vee d\phi_{\lambda_r}]$$

$$= \sum_{k \neq \text{all } \lambda_i} \phi_k d\phi_{\lambda_0} \vee \cdots \vee d\phi_{\lambda_r} + \sum_{i=0}^{r} (-1)^i \phi_{\lambda_i} \sum_{j=0}^{r} d\phi_{\lambda_j} \vee d\phi_{\lambda_0} \vee \cdots \hat{\imath} \cdots \vee d\phi_{\lambda_r}]$$

$$= \sum_{k \neq \text{all } \lambda_i} \phi_k d\phi_{\lambda_0} \vee \cdots \vee d\phi_{\lambda_r} + \sum_{i=0}^{r} \phi_{\lambda_i} d\phi_{\lambda_0} \vee \cdots \vee d\phi_{\lambda_r}$$

$$= d\phi_{\lambda_0} \vee \cdots \vee d\phi_{\lambda_r} = d\phi(q_{\lambda_0} \cdots q_{\lambda_r})/(r+1)!,$$

as required.

To prove (5), suppose first that $r = 0$. Taking $X = q_i$ and using (1) with $A = q_j$ and (13) shows that we must prove the last part of

$$\psi\phi(q_i)\cdot q_j = [\phi(q_i)](q_j) = \phi_i(q_j) = \delta_i^j.$$

Since $\phi_i = 0$ outside St (q_i), this follows from (11).

We now use induction on r. We need merely prove

$$(15)\qquad \psi\phi\sigma_i^r \cdot \sigma_j^r = \int_{\sigma_j^r} \phi\sigma_i^r = \delta_i^j.$$

If $j \neq i$, this follows from (3); suppose $j = i$. Say $\sigma = \sigma_i^r$, $\partial\sigma = \sigma' + \cdots$. Using (4), and (3) in both dimensions r and $r-1$, we find

$$\int_\sigma \phi\sigma = \int_\sigma \phi d\sigma' = \int_\sigma d\phi\sigma' = \int_{\partial\sigma} \phi\sigma' = \int_{\sigma'} \phi\sigma' = 1.$$

28. Certain closed forms are derived. We prove

LEMMA 28a. *Let ω be a closed μ-regular r-form in M $(r > 0)$, and let Y be an algebraic $(r-1)$-cochain of L, such that $\psi\omega = dY$. Then there is a μ-regular $(r-1)$-form ξ in M such that $d\xi = \omega$ and $\psi\xi = Y$.*

We define forms $\xi_0, \xi_1, \cdots, \xi_n$ in turn such that:

(a) ξ_s is defined and μ-regular near the s-dimensional part L^s of L.

(b) $d\xi_s = \omega$ near L^s, and $\xi_s = \xi_{s-1}$ near L^{s-1} $(s > 0)$.

(c) $\psi\xi_{r-1} = Y$.

Then $\xi = \xi_n$ is the required form.

To begin with, choose ξ'_0 near each vertex q_i of L so that $d\xi'_0 = \omega$ there, by Lemma 25a (use χ as in the proof below). If $r = 1$, choose numbers a_i and set $\xi_0 = \xi'_0 + a_i$ near q_i, so that $\psi\xi_0 = Y$; if $r > 1$, set $\xi_0 = \xi'_0$.

Now suppose ξ_{s-1} has been constructed; we construct ξ_s. It is sufficient to define $\xi_s = \xi_{s,i}$ near each s-cell σ_i^s of L; since these agree near L^{s-1}, the single form ξ_s near L^s may be built up from them.

In the case that $s = r$, note that (c) gives, for any $\sigma = \sigma_i^r$,

$$\int_{\partial\sigma} \xi_{r-1} = \psi\xi_{r-1}\cdot\partial\sigma = Y\cdot\partial\sigma = \psi\omega\cdot\sigma = \int_\sigma \omega. \tag{1}$$

Let χ be a coordinate system containing $\sigma = \sigma_i^s$ such that if $\theta = \chi^{-1}$, then $\theta\pi$ is affine in $\pi^{-1}(\sigma)$; then $\tau = \theta(\sigma)$ is an s-simplex in $\mathfrak{A}^n$. Set

$$\omega^* = \chi^*\omega, \qquad \xi^*_{s-1} = \chi^*\xi_{s-1} \qquad \text{near } \partial\tau.$$

Then $d\xi^*_{s-1} = \chi^* d\xi_{s-1} = \omega^*$ near $\partial\tau$, and if $s = r$, (1) and (III, Theorem 10A) (applied to int (σ) and to int (σ_j^{r-1}) for each face of σ) give

$$\int_{\partial\tau} \xi^*_{r-1} = \int_{\partial\sigma} \xi_{r-1} = \int_\sigma \omega = \int_\tau \omega^*.$$

Hence, by Lemma 26b, there is a μ-regular form $\xi^*_{s,i}$ near τ such that $\xi^*_{s,i} = \xi^*_{s-1}$ near $\partial\tau$ and $d\xi^*_{s,i} = \omega^*$ near τ. Set

$$\xi'_{s,i} = \theta^*\xi^*_{s,i} \qquad \text{near } \sigma;$$

then $d\xi'_{s,i} = \theta^*\omega^* = \omega$ near σ, and we may set $\xi_s = \xi'_{s,i}$ near σ if $s \neq r-1$.

Suppose $s = r - 1$. Form ξ'_{r-1} from the $\xi'_{r-1,i}$. We may define $\int_{\sigma_j^{r-1}} \xi'_{r-1}$, and hence define $\psi\xi'_{r-1}$. Set

$$Z = Y - \psi\xi'_{r-1}, \qquad \xi_{r-1} = \xi'_{r-1} + \phi Z \qquad \text{near } L^{r-1}.$$

By (27.3), $\xi_{r-1} = \xi'_{r-1}$ and hence $\xi_s = \xi_{s-1}$ near $L^{r-2} = L^{s-1}$. Also, using (27.4), and the fact that $\phi dZ = 0$ near L^s (by (27.3)),

$$d\xi_{r-1} = d\xi'_{r-1} + \phi dZ = \omega \qquad \text{near } L^s.$$

Also, by (27.5),

$$\psi\xi_{r-1} = \psi\xi'_{r-1} + Z = Y,$$

completing the requirements on $\xi_{r-1} = \xi_s$. The lemma is now proved.

29. Isomorphism of cohomology rings. Given the $(\mu + 1)$-smooth manifold M, triangulated as in § 27, we have the complex L, and hence "algebraic" cohomology spaces $\mathbf{H}^r(L)$, as in (App. II, 8); these form a ring of operators on the homology spaces $\mathbf{H}_r(L)$, as in (App. II, 9).

In a similar fashion, we define the *μ-regular differential cohomology ring* $\mathbf{H}^*_\mu$ of M, as follows. For each r, the closed μ-regular r-forms in M form a linear space, and the derived μ-regular forms form a subspace; the quotient space (App. I, 5) is $\mathbf{H}^r_\mu$; $\mathbf{H}^*_\mu$ is the direct sum of these spaces. Then each closed μ-regular form ω defines an element of $\mathbf{H}^r_\mu$, and ω and ω' define the same element if and only if $\omega' - \omega$ is μ-derived.

Suppose $\mathbf{h} \in \mathbf{H}^r_\mu$ and $\mathbf{h}' \in \mathbf{H}^s_\mu$ are defined by ω and ω' respectively. Then $\omega \vee \omega'$ is a closed μ-regular $(r + s)$-form, defining an element $\mathbf{h}''$ of $\mathbf{H}^{r+s}_\mu$. Using (II, 8.6) shows that this element depends on $\mathbf{h}$ and $\mathbf{h}'$ only; we call it the *product* $\mathbf{h} \smile \mathbf{h}'$ of $\mathbf{h}$ and $\mathbf{h}'$. With this product, $\mathbf{H}^*_\mu$ becomes a ring.

A 0-form is closed if and only if it is constant (assuming M is connected); no 0-form $\neq 0$ is derived. Hence (pretending the 0-form identically zero is derived) $\mathbf{H}^0_\mu$ is isomorphic to the reals $\mathfrak{A}$; the element corresponding to the function $\mathbf{1}$ identically 1 in M is the unit in $\mathbf{H}^*_\mu$.

Because of (27.2), the linear transformation ψ defines a linear transformation Ψ of $\mathbf{H}^r_\mu$ into $\mathbf{H}^r$ for each r; if $\mathbf{h}$ is defined by the closed μ-regular form ω, then $\Psi\mathbf{h}$ is the cohomology class of $\psi\omega$.

The theorem of de Rham† is

THEOREM 29A. *Let M be a $(\mu + 1)$-smooth manifold ($\mu \geqq 0$), and let $L = \pi(K)$ be a triangulation of M as in Theorem* 12A. *Then the above Ψ is a ring isomorphism of $\mathbf{H}^*_\mu$ onto $\mathbf{H}^*$.*

This proves, incidentally, that any two such triangulations give isomorphic algebraic cohomology rings, and also that the $\mathbf{H}^r_\mu$ are independent of μ.

Proof that Ψ is onto: Given $\mathbf{h} \in \mathbf{H}^r$, defined by the cocycle X, set $\omega = \phi X$. Then $d\omega = \phi dX = 0$, and ω defines an element $\mathbf{h}'$ of $\mathbf{H}^r_\mu$. By (27.5), $\psi\omega = X$; hence $\Psi\mathbf{h}' = \mathbf{h}$.

Proof that Ψ is one-one: Suppose $\Psi\mathbf{h} = 0$, $\mathbf{h} \in \mathbf{H}^r_\mu$. Let ω define $\mathbf{h}$; set $X = \psi\omega$. Then X defines $\Psi\mathbf{h}$. First suppose $r = 0$. Then $X = 0$, and hence, for each vertex q_i of L, $\omega(q_i) = X \cdot q_i = 0$. Since $d\omega = 0$, ω is constant, hence $\omega = 0$, and $\mathbf{h} = 0$. Now suppose $r > 0$. Then $X = dY$ for some $(r - 1)$-cochain Y. By Lemma 28a, ω is μ-derived, and $\mathbf{h} = 0$.

† G. de Rham, Sur l'analysis situs des variétés a n dimensions, *Journal de Math. Pures et Appliqués* (*9*) *10* (1931) 115–200. Compare Chapter IV of de Rham's book.

Note that the proof shows that ϕ defines an isomorphism Φ of $\mathbf{H}^*$ onto $\mathbf{H}^*_\mu$, inverse to Ψ'.

Proof that Ψ' preserves products: Define a product operation in $\mathbf{H}^*$ by setting

$$X \smile Y = \psi(\phi X \vee \phi Y). \tag{1}$$

By (27.3), $\phi\sigma^r \vee \phi\sigma^s \neq 0$ only in $\mathrm{St}\,(\sigma^r) \cap \mathrm{St}\,(\sigma^s)$, and property (a) of (App. II, 9) holds. Property (b) follows from (II, 8.6), using (27.2) and (27.4). Property (c) follows from (27.6) and (27.5). Hence this product defines the product operation in $\mathbf{H}^*$. Now (1) (with the fact that Φ and Ψ' are inverses) shows that Ψ' preserves this operation: $\Psi'(\mathbf{h}_1 \smile \mathbf{h}_2) = \Psi'\mathbf{h}_1 \smile \Psi'\mathbf{h}_2$. This completes the proof.

Remark. If M is not compact, then L is infinite; the cochains may have an infinite number of non-zero coefficients. We may, however, restrict ourselves to finite cochains on the one hand, and compact forms (forms vanishing outside of some compact set) on the other. The proof shows that the theorem, with these restrictions, continues to hold.

30. Periods of forms. If ω is a closed r-form, then the integrals $\int_A \omega$ for r-cycles A in M (for instance, cycles of L) are the "periods" of ω. It is immediate that if $A - B = \partial C$, then $\int_A \omega = \int_B \omega$, and if $\omega - \xi = d\eta$, then $\int_A \omega = \int_A \xi$; hence the periods depend only on the cohomology class of ω and the homology class of A. Since $\int_A \omega = \psi\omega \cdot A$, this operation corresponds to that which shows $\mathbf{H}^*$ to be isomorphic to the conjugate space of the homology space $\mathbf{H}$; see (App. II, 8.3). Thus, the various linear functions in $\mathbf{H}$ are given by the various cohomology classes of the closed μ-regular forms in M. This is the original formulation of the theorem (without mentioning products).

31. The Hopf Invariant. As an example of the use of forms in smooth manifolds, we shall find a number γ_f defined by a smooth mapping f of an oriented $(2n - 1)$-sphere S' into an oriented n-sphere S $(n \geqq 2)$, which is invariant† under homotopies of f; γ_f is an integer (see § 33). We follow the procedure given by J. H. C. Whitehead.‡

The theory is capable of great generalization;†† also, we could clearly use "flat forms" and the technique of Chapter X.

All forms we use will be regular; see § 24 and (III, 17). For examples of mappings f with $\gamma_f \neq 0$, see Hopf and Whitehead, loc. cit.

† H. Hopf, Über die Abbildungen der dreidimensionalen Sphäre auf die Kugelfläche, *Math. Annalen, 104* (1931) 639–665; Über die Abbildungen von Sphären auf Sphären niedriger Dimension, *Fundamenta Math. 25* (1935) 427–440.

‡ J. H. C. Whitehead, An expression of Hopf's invariant as an integral, *Proc. Nat. Ac. of Sci. 33* (1947) 117–123.

†† See N. Steenrod, Cohomology invariants of mappings, *Annals of Math. 50* (1949) 954–988.

Given f, we define γ_f as follows. Let ω be an n-form in S such that $\int_S \omega = 1$. Now $f^*\omega$ is an n-form in S' (III, 17), and $df^*\omega = f^*d\omega = 0$. Since the cohomology groups of S' vanish in dimensions between 0 and $2n - 1$ (this is elementary in algebraic topology), de Rham's Theorem shows that there is an $(n - 1)$-form ξ in S' such that $d\xi = f^*\omega$. Set

$$\gamma_f = \int_{S'} \xi \vee f^*\omega. \tag{1}$$

We show first that this is independent of the choice of ξ. Suppose $d\xi' = f^*\omega$ also. Then $d(\xi' - \xi) = 0$, and hence there is an $(n - 2)$-form η in S' such that $d\eta = \xi' - \xi$. Considering S' as a chain, we have $\partial S' = 0$; since $df^*\omega = 0$, using the formula for $d(\eta \vee f^*\omega)$ gives

$$\int_{S'} \xi' \vee f^*\omega - \int_{S'} \xi \vee f^*\omega = \int_{S'} d\eta \vee f^*\omega$$
$$= \int_{S'} [d(\eta \vee f^*\omega \pm \eta \vee df^*\omega] = \int_{\partial S'} \eta \vee f^*\omega = 0.$$

Now suppose we had started with ω' instead of ω. Then $\int_S (\omega' - \omega) = 0$. If X is an n-cochain corresponding to $\omega' - \omega$, then $X \cdot S = 0$; hence X is cohomologous to 0, and $\omega' - \omega = d\zeta$ for some $(n - 1)$-form ζ in S. Say $d\xi = f^*\omega$. Since $\zeta \vee \omega'$ and $\omega \vee \zeta$ are forms of degree $2n - 1 > n$ in S, they vanish; hence also $d\xi \vee f^*\zeta = f^*(\omega \vee \zeta) = 0$,

$$(\xi + f^*\zeta) \vee f^*\omega' - \xi \vee f^*\omega = \xi \vee f^*(\omega' - \omega) + f^*(\zeta \vee \omega')$$
$$= \xi \vee f^*d\zeta = \pm d(\xi \vee f^*\zeta) \pm d\xi \vee f^*\zeta = \pm d(\xi \vee f^*\zeta),$$

and

$$\int_{S'} (\xi + f^*\zeta) \vee f^*\omega' = \int_{S'} \xi \vee f^*\omega.$$

Also

$$d(\xi + f^*\zeta) = d\xi + f^*(\omega' - \omega) = f^*\omega';$$

hence the two integrals above define γ_f with the help of ω' and ω respectively, and these definitions of γ_f are the same.

We must show still that

$$\gamma_{f_0} = \gamma_{f_1} \qquad \text{if } f_0 \text{ is homotopic to } f_1. \tag{2}$$

Let I be the unit interval $(0, 1)$. There is a mapping F of the Cartesian product $I \times S'$ (which is a manifold with boundary) into S such that $F(0, p) = f_0(p)$ and $F(1, p) = f_1(p)$. We first suppose that F is smooth.

Choose ω in S with $\int_S \omega = 1$. Since $dF^*\omega = 0$, we may find ξ in $I \times S'$ such that $d\xi = F^*\omega$. (We may define F in $I' \times S'$, where I' is an open interval containing I; then $I' \times S'$ is a smooth manifold, to which we may apply de Rham's Theorem to give ξ.) Since

$$\partial(I \times S') = 1 \times S' - 0 \times S', \qquad dF^*\omega = 0, \qquad \omega \vee \omega = 0,$$

we have

$$\int_{1 \times S'} \xi \vee F^*\omega - \int_{0 \times S'} \xi \vee F^*\omega = \int_{I \times S'} d(\xi \vee F^*\omega)$$
$$= \int_{I \times S'} d\xi \vee F^*\omega = \int_{I \times S'} F^*(\omega \vee \omega) = 0.$$

The part of $F^*\omega$ in $0\times S'$ is the same as the form $f_0^*\omega$ in S', and the part of ξ in $0\times S'$ gives a form ξ_0 in S' such that $d\xi_0 = f_0^*\omega$; similarly we have $d\xi_1 = f_1^*\omega$. Hence

$$\gamma_{f_0} = \int_{S'} \xi_0 \vee f_0^*\omega = \int_{0\times S'} \xi\vee F^*\omega = \int_{1\times S'} \xi\vee F^*\omega = \gamma_{f_1}.$$

The case that F is merely continuous is taken care of by the following lemma.

LEMMA 31a. *If the smooth mappings f_0, f_1 of the smooth manifold M' into the smooth manifold M are homotopic, then they are smoothly homotopic.*

That is, there is a smooth mapping F of $I\times M'$ into M such that $F(i, p) = f_i(p)$ $(i = 0, 1)$.

We may suppose that M lies in a Euclidean space E (Theorem 1A), and that there is a smooth mapping π of a neighborhood U of M onto M (Theorem 10A); similarly with M', E', U', π'. Let G be the given mapping of $I\times M'$. We may clearly suppose G defined in $I'\times M'$ ($I\subset \operatorname{int}(I')$), and such that

$$G(t, p) = f_0(p) \ (t \leqq 1/3), \qquad G(t, p) = f_1(p) \ (t \geqq 2/3).$$

Set

$$G'(t, q) = G(t, \pi'(q)) \qquad (q \in U');$$

this is a mapping of an open subset $I'\times U'$ of a Euclidean space $\mathfrak{A}\times E'$ into M. By the Weierstrass Approximation Theorem (App. III, Lemma 4a), we may find a smooth mapping G^* of $I'\times U'$ into E approximating arbitrarily closely to G' in $I\times M'$.

Let $\lambda(t)$ be a real smooth function which equals 0 if $t\leqq 1/6$ or $t \geqq 5/6$, and equals 1 if $1/4 \leqq t \leqq 3/4$. Set

$$F'(t, p) = G(t, p) + \lambda(t)[G^*(t, p) - G(t, p)]$$

in $I'\times M'$. For t in a neighborhood of $(1/3, 2/3)$, $F' = G^*$ is smooth. For $t < 1/3$ or $t > 2/3$, $G(t, p) = f_0(p)$ or $f_1(p)$, and hence G and G^* and therefore F' are smooth. Thus F' is smooth in $I'\times M'$. We may suppose the approximation good enough so that $F'(I\times M')$ lies in U. Then we may set

$$F(t, p) = \pi(F'(t, p)),$$

which is the required mapping.

32. On smooth mappings of manifolds. This section will be used in § 33. Let f be a smooth mapping of the smooth manifold M' of dimension m into the smooth manifold M of dimension n, with $m \geqq n$. For each $p\in M'$, $\nabla f(p)$ is a linear transformation of the tangent space $V(M', p)$ into $V(M, f(p))$; we say f is *normal at* p if this transformation is onto. For any $q \in M$, we say f is *normal above* q if f is normal at each $p \in f^{-1}(q)$.

LEMMA 32a. *Given f and q_0 as above, there is a smooth mapping g homotopic to f, arbitrarily close to f, which is normal above q_0.*

Using coordinate systems about q_0 and about any point of M', we see from the considerations of § 5 that the deformation can be found if we prove the following. Let E' and E be Euclidean spaces of dimensions m and n respectively, and let f be a smooth mapping of a neighborhood U' of the m-cell $\sigma' \subset E'$ into E. Take $q_0 \in E$. Then we may find g arbitrarily near f which is normal in σ' above q_0.

There are two ways in which this may be proved:

(a) Let S be the set of points of E above which f is normal; then S is open. As a consequence of f being m-smooth, S is dense; this follows at once from a theorem of A. P. Morse and A. Sard.† Hence a small translation of f gives the required g.

(b) A small translation plus rotation will give the required g, provided merely that f is smooth.‡

LEMMA 32b. *Let f be a smooth mapping of M' into M ($m \geqq n$), and suppose f is normal above $q_0 \in M$. Then*

$$M'_{q_0} = f^{-1}(q_0) \tag{8}$$

is a smooth manifold (not necessarily connected) in M, of dimension $m - n$.

Suppose $f(p_0) = q_0$. Using coordinate systems about p_0 in M' and q_0 in M, we may replace neighborhoods of these points by open sets R', R in E', E respectively; let p_1, q_1 be the corresponding points, and g, the corresponding mapping of R' into R. Choose a base $e'_1, \cdots, e'_m$ in E' such that

$$\nabla g(p_1, e'_i) = 0 \qquad (i = n + 1, \cdots, m).$$

With a base $e_1, \cdots, e_n$ in E, we may write g in terms of components:

$$y^j = g^j(x^1, \cdots, x^n, x^{n+1}, \cdots, x^m) \qquad (j = 1, \cdots, n);$$

Since the $\nabla g(p_1, e'_i)$ are independent for $i = 1, \cdots, n$,

$$\begin{vmatrix} \partial g^1/\partial x^1 \cdots \partial g^1/\partial x^n \\ \cdots\cdots\cdots \\ \partial g^n/\partial x^1 \cdots \partial g^n/\partial x^n \end{vmatrix} = \nabla g(p, e'_1 \vee \cdots \vee e'_n)$$

is $\neq 0$ at p_1. Hence, by the implicit function theorem (see (II, 7)), the set of all solutions of $g(p) = q_1$ near p_1 is given by

$$x^j = \phi^j(x^{n+1}, \cdots, x^m) \qquad (j = 1, \cdots, n),$$

the ϕ^j being smooth. Set also $\phi^j(x^{n+1}, \cdots, x^m) = x^j$ for $x > n$; the set of all ϕ^j gives a smooth mapping of $\mathfrak{A}^{m-n}$ onto a neighborhood of p_1 in

† See R. Thom, Quelques propriétés globales des variétés différentiables. *Commentarii Math. Helvetici 28* (1954), 17–86, Théorème I. 3 (p. 20).

‡ See H. Whitney, Differentiable manifolds, *Annals of Math. 37* (1936), 645–680, property (D) on p. 655. It is clearly not necessary that f be locally one-one for this.

M_{q_0}, which is a coordinate system in M_{q_0} about q_0. Overlapping coordinate systems are clearly smoothly related.

33. Other expressions for the Hopf Invariant. We give further formulas for the γ_f of § 31 (see Whitehead, loc. cit.), relating it to the original definition of Hopf.

Take S', S, f as in § 31. Choose $q \in S$, and suppose that f is normal above q. By Lemma 32b, $M_q = f^{-1}(q)$ is a smooth manifold (not necessarily connected) of dimension $n - 1$. We orient M_q as follows. Choose a base $e'_1, \cdots, e'_m$ in $V(S', p)$, where $f(p) = q$, so as to determine the positive orientation of S', so that $\nabla f(p, e'_i) = 0$ for $i \leqq n - 1$, and so that

$$e_1 = \nabla f(p, e'_n), \cdots, e_n = \nabla f(p, e'_{2n-1})$$

determine the positive orientation of S; then $e'_1, \cdots, e'_{n-1}$ determine the positive orientation of M_q.

We shall show that if ω is an n-form in S such that $\int_S \omega = 1$, $d\xi = f^*\omega$ in S', and f is normal above q, then

$$\gamma_f = \int_{S'} \xi \vee f^*\omega = \int_{M_q} \xi. \tag{1}$$

The definition of normality shows that there is a neighborhood U of q such that f is normal above each q' in U; then each $M_{q'} (q' \in U)$ is a smooth manifold, and these manifolds form a "fibering" of $f^{-1}(U)$. Take U to be connected. We suppose first that $\omega = 0$ outside U. Given $q' \in U$, let A be a smooth arc in U joining q to q'. The set $f^{-1}(A)$ is clearly a smooth bounded n-manifold $M_A \subset S'$, with $\partial M_A = M_{q'} - M_q$. Letting f_A denote f, considered in M_A alone, f_A maps M_A into A; since $n > 1$, $J_{f_A} = 0$ at all points of M_A, and hence $f_A^*\omega = 0$ in M_A. Therefore

$$\int_{M_{q'}} \xi - \int_{M_q} \xi = \int_{\partial M_A} \xi = \int_{M_A} d\xi = \int_{M_A} f_A^*\omega = 0. \tag{2}$$

Given $p \in M_q$, choose $e'_1, \cdots, e'_m$ and $e_1, \cdots, e_n$ as above. Since $\nabla f(p, e'_i) = 0$ for $i < n$,

$$f^*\omega(p)\cdot e'_{n,\cdots,2n-1} = \omega(f(p))\cdot e_{1\cdots n}$$

is the only non-zero component of $f^*\omega(p)$. Hence, by (I, 6.6),

$$[\xi(p) \vee f^*\omega(p)]\cdot e'_{1,\cdots,2n-1} = [\xi(p)\cdot e'_{1,\cdots,n-1}][\omega(f(p))\cdot e_{1\cdots n}].$$

Hence, using an expression of $\int_{f^{-1}(U)} \xi \vee f^*\omega$ as an iterated integral and remembering (2),

$$\int_{S'} \xi \vee f^*\omega = \int_{f^{-1}(U)} \xi \vee f^*\omega = \int_U \omega(q') \int_{M_{q'}} \xi(p) dp dq'$$
$$= \int_{M_q} \xi \int_U \omega = \int_{M_q} \xi,$$

proving (1).

Now take any ω and corresponding ξ. Choose ω' like the above ω, and choose ζ as in § 31. If f_q is f, considered in M_q alone, then $f_q^*\zeta = 0$; hence

$$\int_{M_q} \xi - \int_{M_q} (\xi + f^*\zeta) = -\int_{M_q} f_q^*\zeta = 0.$$

Since $d(\xi + f^*\zeta) = f^*\omega'$ (see § 31), we have

$$\int_{M_q} \xi = \int_{M_q} (\xi + f^*\zeta) = \gamma_f.$$

Finally, we interpret γ_f as the degree of a mapping, which shows that γ_f is an integer. We suppose f is normal above q (see Lemma 32a), and suppose A is a smooth chain bounded by M_q. (That is, there is a simplicial complex K, an algebraic n-chain A_0 in K, and a simplexwise smooth mapping ϕ of K into S', such that $A = \phi A_0$ and $M_q = \phi\partial A_0$. This exists, since S' is a sphere.) Now

$$\gamma_f = \int_{M_q} \xi = \int_{\partial A} \xi = \int_A f^*\omega = \int_{fA} \omega. \tag{3}$$

Since f maps the boundary $\partial A = M_p$ into a point, fA is an n-cycle in S and hence equals (algebraically) DS, D being the "degree" of the mapping f in A. By (3),

$$\gamma_f = \int_{DS} \omega = D\int_S \omega = D. \tag{4}$$

PART II

GENERAL THEORY

V. Abstract Integration Theory

We now introduce r-dimensional integration in n-space E^n from a postulation point of view. As seen in the Introduction, §§ 1, 2, domains of integration must include polyhedral r-chains, and a given integrand, which we call a "cochain" X, must be a linear function of chains. According to what further conditions are put on X, we obtain "flat" or "sharp" cochains. The basic facts, though derived with the help of the Euclidean character of E^n (App. I, 13), clearly hold irrespective of a choice of metric, and hold therefore in an affine space; compare (VII, 10). In this chapter, we assume a knowledge of the major portion of Chapter I, part of Chapter II, the beginning of Chapter III, and parts of the appendices.

There is a natural definition of the "mass" $|A|$ of the polyhedral chain A. In terms of this, we give explicit definitions of the "flat norm" $|A|^\flat$ and the "sharp norm" $|A|^\sharp$ of A. Completing the space of polyhedral r-chains in these norms gives Banach spaces (App. I, 14) $\mathbf{C}_r^\flat$ and $\mathbf{C}_r^\sharp$ respectively; the elements of these spaces are called "flat chains" and "sharp chains" respectively. The requirement that the function $X \cdot A$ of polyhedral chains A be bounded in one of these norms is the condition that X be an element of the conjugate space of one of these spaces; X is a "flat" or "sharp" cochain correspondingly. The sharp norm has simple analytical properties (see § 10), but is less important from the topological point of view, since the boundary ∂A of a sharp chain A is not defined in general.

It is elementary that in the dimension 0, the notions of "flat" and "sharp" coincide. A flat (or sharp) 0-cochain X corresponds to a real function $\phi(p)$ which is "sharp" (we could call it "flat"), i.e. ϕ is bounded and satisfies a Lipschitz condition. In the dimension n, the flat norm and mass of a chain are the same.

The definition of the "comass" $|X|$ of X is dual to that of mass. Formulas for the flat and sharp norms of a cochain X are found in terms of $|X|$, $|dX|$, and the "Lipschitz constant" $\mathfrak{L}_X$ of X; see (4.8), (7.2) and (7.8). A characterization of the norms by simple inequalities is given in § 8.

An algebraic fact is proved in § 9 which is basic in § 10. In the latter section, we show that there is a one-one correspondence between sharp cochains and differential forms which are bounded and satisfy a Lipschitz condition. A similar theorem in the flat norm (Wolfe's Theorem) requires Lebesgue methods; it will be given in Chapter IX.

A fundamental problem is to obtain results of this sort with weakened conditions on the cochains. We give one example. Say the linear function X of polyhedral r-chains A is *semi-sharp* if the following hold:

(*a*) $|X|$ is locally finite; that is, for each bounded set R there is an N_R such that

$$|X \cdot A| \leqq N_R |A| \text{ for any polyhedral chain } A \subset R.$$

(*b*) For each point p and $\epsilon > 0$ there is a $\zeta > 0$ such that for any $(r + 1)$-simplex σ,

$$|X \cdot \partial\sigma| \leqq \epsilon |\partial\sigma| \quad \text{if} \quad \sigma \subset U_\zeta(p).$$

(*c*) We may choose ζ in (*b*) such that for any r-simplex σ and vector v,

$$|X \cdot (T_v\sigma - \sigma)| \leqq \epsilon |\sigma| \quad \text{if} \quad \sigma \subset U_\zeta(p), \qquad |v| \leqq \zeta.$$

The proof in § 10 goes through with these assumptions, without any essential modifications; we find that *there is a one-one correspondence between semi-sharp r-cochains and continuous r-forms.*

The object of § 13 is to show that any flat cochain is the "weak limit" of a sequence of sharp cochains. This has various important applications; for example, any flat chain may be considered as a sharp chain.

The definitions of the norms treat $|D|$ and $|\partial D|$ as quantities of the same sort (numbers), though D and ∂D are of different dimensions. In the "ρ-norms" of § 15, thinking of ρ as having the dimension of distance brings back dimensional consistency.

The "mass" of a general (flat or sharp) chain may be defined in any of three ways; we show in § 16 that all definitions agree. (In general, mass will be infinite.) This gives a generalization of the notion of area of a "rectifiable" surface; compare (X, 4).

We end the chapter by showing that the spaces of chains are separable, while the spaces of cochains are not; hence these spaces are not reflexive.

1. Polyhedral chains. First we consider polyhedral r-chains in r-space E^r. Let $\sigma_1, \cdots, \sigma_m$ be bounded non-overlapping oriented polyhedral regions in E^r (App. II, 2). To each σ_i we assign a coefficient a_i; then we call the expression $\sum a_i\sigma_i$ a *polyhedral r-chain A* in E^r. We may determine A by orienting E^r, and defining a function $A(p)$ in E^r, which equals a_i or $-a_i$ in int (σ_i) according as σ_i is oriented like or opposite to E^r, and equals 0 elsewhere in E^r. We let $A = B$ if the corresponding functions are equal except in a finite set of polyhedral cells of dimension $<r$. Note that if $\sum_j \tau_{ij}$ is a subdivision of σ_i, then the polyhedral chains $\sum_i a_i\sigma_i$, $\sum_{i,j} a_i\tau_{ij}$ are the same.

Define aA by the function $aA(p)$, and $A + B$ by the function $A(p) + B(p)$; the chains obviously form a linear space. Given two chains

$A = \sum a_i \sigma_i$, $B = \sum b_i \tau_i$, we may find a common sub-division, with cells σ_i', of the σ_i and τ_i (App. II, Lemma 3b); then if $A = \sum a_i' \sigma_i'$, $B = \sum b_i' \sigma_i'$, we have $A + B = \sum (a_i' + b_i') \sigma_i'$.

A *polyhedral r-chain* A in E^n consists of a finite set of distinct r-planes, together with a polyhedral r-chain in each. We may drop out an r-plane if the part of A in it is 0. We may define A by orienting each r-plane, and defining $A(p)$ in each.

The definition of aA is obvious. To define $A + B$, take all r-planes occurring in A and B, and add $A(p)$ and $B(p)$ in each. Again the chains form a linear space, which we call $\mathbf{C}_r^{\text{pol}}(E^n)$.

As before, any expression $\sum a_i \sigma_i^r$ determines a polyhedral r-chain; subdividing the σ_i^r does not alter the chain, and we may drop out or add the term $a_i \sigma_i^r$ if $a_i = 0$. The σ_i^r may overlap; of course an expression may be chosen with the σ_i^r non-overlapping. We could allow "degenerate" σ_i^r (App. II, 1), this representing a set with no interior points in the corresponding E^r; we count $a_i \sigma_i^r$ as 0 in this case. (This arises for instance in the proof of Theorem 3A below.)

For $r = 0$, we have 0-chains $\sum a_i \sigma_i^0$; the σ_i^0 are points, which need not be "oriented". For $r = n$, the definition reduces to that given above.

For any representation $\sum a_i \sigma_i^r$ of A such that the σ_i^r are non-overlapping cells and the a_i are $\neq 0$, the set of points on the (closed) cells σ_i^r is the same. This set of points is the *support* spt (A) of A. We say *A lies in R*, and write $A \subset R$, if spt $(A) \subset R$.

The *boundary* ∂A of $A = \sum a_i \sigma_i^r$ is defined as $\sum a_i \partial \sigma_i^r$. That this is independent of the expression of A is easily seen by computing $(\partial A)(p)$ from $A(p)$. For instance, if $A = a\sigma + b\tau$, σ and τ in E^r and oriented like E^r, and $\partial \sigma = \sigma^{r-1} + \cdots$, $\partial \tau = -\sigma^{r-1} + \cdots$, then in the oriented plane E^{r-1} of σ^{r-1}, $(\partial A)(p) = a - b$ in σ^{r-1}. Of course $\partial(aA) = a\partial A$, $\partial(A + B) = \partial A + \partial B$, $\partial\partial A = 0$.

2. Mass of polyhedral chains. Recall that for an r-cell σ, $|\sigma| = |\sigma|_r$ denotes the r-dimensional volume of σ. For $r = 0$, $|\sigma| = 1$. The *mass* of a polyhedral chain $\sum a_i \sigma_i^r$ is defined by

$$(1) \qquad \left|\sum a_i \sigma_i^r\right| = \sum |a_i| \, |\sigma_i^r| \qquad \text{if the } \sigma_i^r \text{ are non-overlapping.}$$

If A is defined by the functions $A_1(p), \cdots$ on the distinct r-planes $P_1, \cdots$, an obviously equivalent expression is

$$(2) \qquad |A| = \sum_i \int_{P_i} |A_i(p)| \, dp.$$

For $r = 0$, $|\sum a_i \sigma_i^0| = \sum |a_i|$ if the σ_i^0 are distinct. Clearly

$$(3) \qquad |aA| = |a| \, |A|, \qquad |A + B| \leq |A| + |B|,$$

and $|A| \neq 0$ if $A \neq 0$; hence the mass is a norm (App. I, 8).

LEMMA 2a. *Let π be the orthogonal projection of E^n onto an s-plane P^s. Then*

(4) $$|\pi A| \leqq |A|.$$

By definition, $\pi \sum a_i \sigma_i = \sum a_i \pi \sigma_i$. Either $\pi\sigma_i$ is degenerate (App. II, 1), in which case it drops out of πA (§ 1), or $|\pi\sigma| \leqq |\sigma|$; thus (4) holds.

LEMMA 2b. *Let Q be a convex polyhedral cell in E^n, and let π be the projection of E^n onto Q defined by*

(5) $$\pi(p) = \text{point of } Q \text{ nearest to } p.$$

Then (4) *holds.*

It is not hard to see that E^n breaks up into polyhedral regions, in each of which π is affine; write $A = \sum a_i\sigma_i$, each σ_i lying in one of these regions. Consider any σ_i. We see easily that π increases no distances; using coordinates in σ_i, as in the end of (I, 12), we see that $|\pi\sigma_i| \leqq |\sigma_i|$. Thus (4) holds. (Compare the proof of (VIII, 1.29).)

3. The flat norm. The *flat norm* $|A|^\flat$ of the polyhedral r-chain A in E^n is defined by

(1) $$|A|^\flat = \inf\{|A - \partial D| + |D|\},$$

using all polyhedral $(r+1)$-chains D. See also § 8. Clearly

(2) $$|aA|^\flat = |a|\,|A|^\flat, \qquad |A + B|^\flat \leqq |A|^\flat + |B|^\flat;$$

we prove later (§ 12) that

(3) $$|A|^\flat = 0 \quad \text{if and only if} \quad A = 0.$$

It follows that $|\quad|^\flat$ is a norm. With this norm, the space $\mathbf{C}_r^{\text{pol}}(E^n)$, when completed, becomes a Banach space $\mathbf{C}_r^\flat(E^n)$ (App. I, 14). We call elements of this space *flat chains*. We define mass in this space in § 16.

From Lemma 2b it follows that in (1), we may require D to lie in the smallest convex set containing spt (A). See also (VII, Lemma 5b) and (VIII, 1.1).

The logical structure of the next sections is as follows. *Read through Section* 12, *using polyhedral chains only, with the semi-norm* $|A|^\flat$; *this being proved a norm, all facts in these sections now follow. Similarly for* $|A|^\sharp$.

We prove two elementary inequalities, at first for polyhedral chains:

(4) $$|\partial A|^\flat \leqq |A|^\flat \leqq |A|.$$

The second follows at once on taking $D = 0$ in (1). To prove the first, given $\epsilon > 0$, choose D so that $|A - \partial D| + |D| < |A|^\flat + \epsilon$. Now setting $D' = A - \partial D$, we have

$$|\partial A - \partial D'| + |D'| = |A - \partial D| < |A|^\flat + \epsilon;$$

hence $|\partial A|^\flat < |A|^\flat + \epsilon$, and (4) follows.

We now define ∂A for any flat chain A, and prove the first part of (4). Say $A = \lim^\flat A_i$, the A_i polyhedral. Then $|A_j - A_i|^\flat \to 0$, and hence $|\partial A_j - \partial A_i|^\flat \to 0$, by (4). Therefore $\partial A_1, \partial A_2, \cdots$ is a Cauchy sequence, and defines an element of $\mathbf{C}^\flat_{r-1}$, which we call ∂A; the element is independent of the sequence chosen. Since $|\partial A_i|^\flat \leqq |A_i|^\flat$, letting $i \to \infty$ proves the first part of (4) for A. Clearly $\partial\partial A = 0$. Because of (4), ∂ is a continuous operation.

Given an oriented r-cell σ in E^n, and a vector v, let $T_v\sigma$ denote the similarly oriented cell containing all points $p + v$, $p \in \sigma$. Set

$$T_v \sum a_i\sigma_i^r = \sum a_i T_v\sigma_i, \qquad T_v \lim{}^\flat A_i = \lim{}^\flat T_v A_i.$$

Let $\mathscr{D}_v\sigma$ (called $\mathscr{D}_v(I \times \sigma)$ in (App. II, 13)) be the oriented $(r + 1)$-cell containing all points $p + tv$, $p \in \sigma$, $0 \leqq t \leqq 1$; define $\mathscr{D}_v A$ for polyhedral A by linearity. By (App. 11, 13.5) (for polyhedral A),

$$\partial\mathscr{D}_v A = T_v A - A - \mathscr{D}_v\partial A.$$

Clearly

(5) $$|\mathscr{D}_v A| \leqq |v|\,|A|.$$

These relations give, using (4),

(6) $$|T_v A - A|^\flat \leqq |v|\,(|A| + |\partial A|).$$

(This holds for any flat A; see (VIII, 3.7).)

THEOREM 3A. *If $\sigma_1, \cdots, \sigma_s$ are parallel and similarly oriented, and the a_i are $\geqq 0$, then*

(7) $$\left|\sum a_i\sigma_i\right|^\flat = \left|\sum a_i\sigma_i\right| = \sum a_i\,|\sigma_i|\,;$$

in particular,

(8) $$|\sigma|^\flat = |\sigma|.$$

Set $a = |\sum a_i\sigma_i|$; this equals $\sum a_i\,|\sigma_i|$, even if the σ_i overlap. Suppose that for some D,

$$\left|\sum a_i\sigma_i - \partial D\right| + \left|D\right| < a.$$

Choose an r-plane P parallel to the σ_i, and let π be the projection into P. Then, by Lemma 2a,

$$\left|\sum a_i\pi\sigma_i - \partial\pi D\right| + \left|\pi D\right| < a.$$

Since D is $(r + 1)$-dimensional, $\pi D = 0$; hence also $\partial\pi D = 0$. Also $|\sum a_i\pi\sigma_i| = \sum a_i\,|\sigma_i| = a$, a contradiction. Therefore (7) holds. For another method of proof, see Theorem 6B below.

For small simplexes σ, $|\partial\sigma|^\flat = |\sigma|$; we shall not need this property. Note that, for 1-simplexes σ^1,

$$(9) \qquad |\partial\sigma^1|^\flat = \inf\{|\partial\sigma^1|, |\sigma^1|\} = \inf\{2, |\sigma^1|\}.$$

THEOREM 3B. *If π is as in either Lemma 2a or Lemma 2b, then*

$$(10) \qquad |\pi A|^\flat \leqq |A|^\flat .$$

For, for any D, the lemmas quoted give

$$|\pi A - \partial\pi D| + |\pi D| \leqq |A - \partial D| + |D| .$$

See, in this connection, Theorem 16C.

We give one more property of the norm:

$$(11) \qquad |A|^\flat = \inf\{|A - \partial D|^\flat + |D|^\flat\} .$$

Setting $D = 0$ proves $\geqq$. The reverse inequality follows from (2) and (4): For any D,

$$|A|^\flat \leqq |A - \partial D|^\flat + |\partial D|^\flat \leqq |A - \partial D|^\flat + |D|^\flat.$$

4. Flat cochains. A *flat r-cochain* X in E^n is a linear function, which we write as $X \cdot A$, of the flat r-chains A in E^n, such that for some N, $|X \cdot A| \leqq N|A|^\flat$. Thus it is an element of the conjugate space (App. I, 8)

$$(1) \qquad \overline{\mathbf{C}}_r^\flat = \mathbf{C}^{\flat r} = \mathbf{C}^{\flat r}(E^n)$$

of the Banach space of flat r-chains in E^n. (*Until we have read* § 12, we must consider $\mathbf{C}^{\flat r}$ as the semi-conjugate space of $\mathbf{C}_r^\flat$; see (App. I, 15).) For chains in open sets $R \subset E^n$, see (VIII, 1). The *flat norm* $|X|^\flat$ is, by definition (App. I, 8),

$$(2) \qquad |X|^\flat = \sup_{|A|^\flat \neq 0} |X \cdot A|/|A|^\flat = \sup_{|A|^\flat = 1} |X \cdot A| .$$

We may require A in (2) to be polyhedral; see (App. I, Lemma 14a). From general theory (App. I, Lemma 8c), for any flat r-chain A,

$$(3) \qquad |A|^\flat = \sup_{|X|^\flat \neq 0} |X \cdot A|/|X|^\flat = \sup_{|X|^\flat = 1} |X \cdot A|;$$

in fact (App. I, Lemma 8b) there is a flat r-cochain X such that $|X|^\flat = 1$ and $X \cdot A = |A|^\flat$.

We define the *comass* $|X|$ of X by

$$(4) \qquad |X| = \sup_{A \neq 0} |X \cdot A|/|A| = \sup_{|A| = 1} |X \cdot A| ,$$

with A polyhedral; because of (3.4), this is always finite, and $|X| \leqq |X|^\flat$. For given X, $|X|^\flat$ and $|X|$ are the smallest numbers such that

$$(5) \qquad |X \cdot A| \leqq |X|^\flat |A|^\flat, \qquad |X \cdot A| \leqq |X||A| ,$$

for all polyhedral A, respectively. The first of course holds for all A; the second will be proved for all A in § 16.

The *coboundary* dX of X is the linear function of $(r+1)$-chains defined by

$$dX \cdot A = X \cdot \partial A. \tag{6}$$

As in (App. II, 7), $ddX = 0$. By (5) and (3.4),

$$|dX \cdot A| = |X \cdot \partial A| \leqq |X|^\flat |\partial A|^\flat \leqq |X|^\flat |A|^\flat$$

for all $(r+1)$-chains A; hence dX is a flat cochain, and $|dX|^\flat \leqq |X|^\flat$

Now

$$|X| \leqq |X|^\flat, \qquad |dX| \leqq |dX|^\flat \leqq |X|^\flat. \tag{7}$$

We now prove a converse of these inequalities; in fact,

$$|X|^\flat = \sup\{|X|, |dX|\}. \tag{8}$$

We must prove $\leqq$. Take any polyhedral A, and any $\epsilon > 0$. Choose D so that $|A - \partial D| + |D| < |A|^\flat + \epsilon$. Now if M is the right hand side of (8),

$$\begin{aligned}|X \cdot A| &\leqq |X \cdot (A - \partial D)| + |X \cdot \partial D| \leqq |X||A - \partial D| + |dX||D| \\ &\leqq M(|A - \partial D| + |D|) \leqq M(|A|^\flat + \epsilon),\end{aligned}$$

hence $|X \cdot A| \leqq M|A|^\flat$, and the inequality follows.

THEOREM 4A. *If X is any linear function of polyhedral r-chains such that $|X|$ and $|dX|$ are finite, it defines a unique flat r-cochain, and* (8) *holds.*

For the proof above shows that considering X over polyhedral chains only, $|X|^\flat$ is finite and satisfies (8); therefore X is uniquely extendable over $\mathbf{C}_r^\flat$ (App. I, Lemma 14a), and (8) holds.

We show that $|X|$ is determined by using simplexes in (4):

$$|X| = \sup\{|X \cdot \sigma|/|\sigma|: \quad \text{simplexes } \sigma\}. \tag{9}$$

Since $\geqq$ holds, it is sufficient to show that for each $\epsilon > 0$ there is a simplex σ such that $|X \cdot \sigma|/|\sigma| > |X| - \epsilon$. We may choose a polyhedral chain A, expressed as $A = \sum a_i \sigma_i$, with non-overlapping simplexes σ_i, such that

$$|A| = \sum |a_i||\sigma_i| = 1, \qquad |X \cdot A| > |X| - \epsilon.$$

Then the required property holds for some σ_i; for otherwise,

$$|X \cdot A| \leqq \sum |a_i||X \cdot \sigma_i| \leqq \sum |a_i|(|X| - \epsilon)|\sigma_i| = |X| - \epsilon,$$

a contradiction.

Using dX in place of X above shows that

$$(10) \qquad |dX| = \sup\{|X\cdot\partial\sigma|/|\sigma|: \text{ simplexes } \sigma\}.$$

Consider the case $r = 0$. Since each point p of E^n is a 0-cell, each flat 0-cochain X determines a real valued function D_X: $D_X(p) = X\cdot p$.

We say a function ϕ with values in a Banach space is *sharp* if $\mathfrak{L}_\phi$ (II, 4) and $|\phi|$ are both finite; compare (7.9) and § 10.

THEOREM 4B. *The flat 0-cochains X correspond to the real valued sharp functions D_X; in fact,*

$$(11) \qquad |X| = |D_X|, \qquad\qquad |dX| = \mathfrak{L}(D_X).$$

The first relation follows at once from (9), since $|\sigma^0| = 1$. For any 1-simplex $\sigma^1 = pq$,

$$X\cdot\partial\sigma^1 = X\cdot q - X\cdot p = D_X(q) - D_X(p);$$

since $|\sigma^1| = |q - p|$ (length of the vector $q - p$), (10) gives the second part of (11).

For $r = n$, each bounded continuous function ϕ determines an n-cochain X, with the definition $X\cdot\sigma = \int_\sigma \phi$. More generally, the n-cochains correspond to the bounded measurable functions; see (IX, 1). Since every polyhedral $(n + 1)$-chain in E^n is 0, $dX = 0$ for n-cochains X, and $|dX| = 0$. Hence, by (3.1) and (8),

$$(12) \qquad |A|^\flat = |A|, \qquad |X|^\flat = |X|, \qquad \text{for } r = n.$$

5. Examples. (*a*) Let A be a 1-chain, formed of the two oriented segments p_1p_2, p_2p_3, with an angle at p_2. Insert a small triangle $D = p_1'p_2p_3'$, p_i' in p_ip_2, so that $A - \partial D = p_1p_1' + p_1'p_3' + p_3'p_3$; if D is small enough, then $|A - \partial D| + |D| < |A|$, showing that $|A|^\flat < |A|$. We may now insert smaller triangles in the corners at p_1' and p_3', etc. We know no formula for $|A|^\flat$.

(*b*) Let σ be a 1-simplex, $|\sigma| = a < 2$, let $|v| = b < a/2$, and let $A = T_v\sigma - \sigma$. Now $D = \mathscr{D}_v\sigma$ is a parallelogram, and (3.6) gives

$$|A|^\flat \leqq b(a + 2) < 2a = |A|.$$

We may replace two sides by polygonal lines which cut somewhat into the above D, showing that $|A|^\flat < b(a + 2)$.

(*c*) In E^1, let σ_i and τ_i be the points $-1/2^i$ and $1/2^i$ respectively. With these points as 0-cells, define the 0-chains

$$A_i' = \tau_i - \sigma_i, \qquad A_k = A_0' + A_1' + \cdots + A_k'.$$

Let B_i be the segment from σ_i to τ_i. Then $|A_i'|^\flat = |\partial B_i|^\flat = |B_i| = 2/2^i$, and hence $A = \lim^\flat A_k$ exists (limit in the flat norm). Clearly we cannot assign A a finite mass.

(*d*) In Example (*c*), let A_{mk} contain the points $\sigma_0, \cdots, \sigma_k$ with the coefficient -1, and the points $\tau_0, \cdots, \tau_{m+k}$ with the coefficient 1. Then $A^*_m = \lim^\flat_{k\to\infty} A_{mk}$ exists and "contains" the same points as A. (The supports are the same; see (VII, 3).) However, $A^*_m \neq A$ for $m \neq 0$. For instance, if X is the 0-cochain defined by $D_X(p) = 1$ (all p), then $X\cdot A = 0$, $X\cdot A^*_m = m$. See in this connection (VI, 1).

(*e*) We can construct similar 1-chains in E^n with pairs of segments.

(*f*) One can easily take a sequence $A_1, A_2, \cdots$ of oriented broken lines in the plane so that the limit $\lim^\flat A_i$ exists, and is represented by an arc which is non-rectifiable between any two of its points. We may use the graph of a non-differentiable function.

(*g*) In E^1, let the 1-cochain X be defined by

$$D_X(t) = 1/(1 + t^2), \qquad X\cdot\sigma = \int_\sigma D_X.$$

Then $|X| = 1$, $dX = 0$, and hence $|X|^\flat = 1$. For any 1-simplex σ, $X\cdot\sigma < |\sigma|$. In fact, $X\cdot A < |A| = |A|^\flat$ for all A. For clearly A would have to be "concentrated" at $t = 0$, which is impossible (see (VII, 9)).

6. The sharp norm. In example (*b*) above, if we push one segment σ (of length a) into the other, $T_v\sigma$, (through the distance b), the chain A is reduced to zero. This suggests defining a new norm for A, in amount $\leqq ab < |A|^\flat$. We carry out this idea as follows.

Given the polyhedral r-chain A, consider all expressions $\sum a_i\sigma_i$ of A; we translate each σ_i as we like, replacing A by $\sum a_i T_{v_i}\sigma_i$, and find the flat norm of the result; adding the "amount of push," $\sum |a_i||\sigma_i||v_i|$, except for a factor, gives a quantity we reduce to the minimum. Thus the definition of the *sharp norm* of A is (see also § 8)

$$(1) \qquad |A|^\sharp = \inf\left\{\frac{\sum |a_i||\sigma_i||v_i|}{r+1} + \left|\sum a_i T_{v_i}\sigma_i\right|^\flat : \quad A = \sum a_i\sigma_i\right\}.$$

The reason for the factor $r + 1$ is so that we may prove (7.8); see Example (*e*) of § 11. Note that cells σ_i may be used which are not in spt (A); see for instance the proof of (7.7). Because of (VIII, 1.29), we may require all the σ_i and $T_{v_i}\sigma_i$ to lie in any given convex open set containing A.

Taking the $v_i = 0$ shows that

$$(2) \qquad |A|^\sharp \leqq |A|^\flat.$$

Clearly $|aA|^\sharp = |a||A|^\sharp$, $|A + B|^\sharp \leqq |A|^\sharp + |B|^\sharp$; we prove $|A|^\sharp \neq 0$ if $A \neq 0$ in § 12. Complete the space of polyhedral chains with this norm to obtain $\mathbf{C}^\sharp_r(E^n)$, the space of *sharp r-chains* in E^n.

We do not define ∂A for sharp chains A; see § 11, Examples (*c*) and (*d*).

THEOREM 6A. *For any polyhedral r-chain A and any vector v,*

$$(3) \qquad |T_v A - A|^\sharp \leqq |A||v|/(r + 1).$$

First suppose $A = \sigma$. Set

$$\sigma_1 = \sigma, \quad \sigma_2 = T_v\sigma, \quad v_1 = v, \quad v_2 = 0, \quad a_1 = -1, \quad a_2 = 1.$$

Then $a_1 T_{v_1}\sigma_1 + a_2 T_{v_2}\sigma_2 = 0$, and (1) gives $| T_v\sigma - \sigma |^\sharp \leqq | \sigma | \, | v |/(r+1)$, as required. The relation now follows for any polyhedral chain A, expressed as $\sum a_i\sigma_i$, the σ_i non-overlapping.

THEOREM 6B. *With the conditions of Theorem* 3A,

$$\left| \sum a_i\sigma_i \right|^\sharp = \left| \sum a_i\sigma_i \right| = \sum a_i \, | \sigma_i | . \tag{4}$$

We prove this with the help of later theorems. Define P (oriented like the σ_i) and π as in the proof of Theorem 3A. Define X_0 in P by $X_0 \cdot \sigma = | \sigma |$ (σ oriented like P). Define X in E^n by $X \cdot A = X_0 \cdot \pi A$. Then $dX = 0$, $\mathfrak{L}(X) = 0$, and hence, by (7.8), $| X |^\sharp = | X | = 1$. Clearly $X \cdot \sum a_i\sigma_i = \sum a_i \, | \sigma_i |$. Hence $\geqq$ holds in (4); the opposite inequality is known.

Consider the case $r = 0$. We show that

$$| A |^\sharp = | A |^\flat \quad \text{if} \quad r = 0. \tag{5}$$

Because of (2), we need merely prove $\geqq$. Take any expression $\sum a_i\sigma_i^0$ for A, and any vectors v_i. Let σ_i' be the 1-simplex from σ_i^0 to $T_{v_i}\sigma_i^0$; then

$$\partial\sigma_i' = T_{v_i}\sigma_i^0 - \sigma_i^0, \qquad | \sigma_i' | = | v_i | .$$

Now setting $D = -\sum a_i\sigma_i'$ and using (3.11) gives

$$\begin{aligned} \sum | a_i | \, | \sigma_i^0 | \, | v_i | + \left| \sum a_i T_{v_i}\sigma_i^0 \right|^\flat &= \sum | a_i | \, | \sigma_i' | + \left| \sum a_i\sigma_i^0 + \sum a_i \partial\sigma_i' \right|^\flat \\ &\geqq | D | + | A - \partial D |^\flat \geqq | A |^\flat, \end{aligned}$$

as required. For another proof, we may use (7.9) below.

For $r = n$, we note merely that we may have $| A |^\sharp < | A |^\flat = | A |$. For instance, if σ and $T_v\sigma$ are disjoint and $| v | < 2$, then

$$| T_v\sigma - \sigma |^\flat = | T_v\sigma - \sigma | = 2 \, | \sigma | , \qquad | T_v\sigma - \sigma |^\sharp \leqq | v | \, | \sigma | < 2 \, | \sigma | .$$

WARNING. Theorem 3B is not true in the sharp norm. For take $r = n = 1$. Let Q be the interval (0, 1). Let σ and σ' be the oriented intervals $(1 - \epsilon, 1)$ and $(1, 1 + \epsilon)$ respectively. Set $A = \sigma' - \sigma$. Then, by (3), $| A |^\sharp \leqq \epsilon^2/2$. But $| \pi A |^\sharp = | -\sigma |^\sharp = \epsilon > | A |^\sharp$ for $\epsilon < 1$.

7. Sharp cochains. A *sharp r-cochain* is an element of the conjugate space $\mathbf{C}^{\sharp r}(E^n)$ of $\mathbf{C}_r^\sharp(E^n)$. Define $| X |$ as in § 4. Relations (4.2), (4.3) and (4.5) hold with $^\sharp$ in place of $^\flat$. Because of (6.2),

$$| X | \leqq | X |^\flat \leqq | X |^\sharp. \tag{1}$$

Hence *every sharp cochain is a flat cochain.*

Define the *Lipschitz constant* of a cochain by

$$(2)\qquad \mathfrak{L}(X) = \mathfrak{L}_X = \sup\left\{\frac{|X\cdot(T_vA - A)|}{|A||v|} : v \neq 0,\ A \neq 0 \text{ polyhedral}\right\}.$$

By (6.3), this is finite for sharp cochains, and

$$(3)\qquad (r+1)\mathfrak{L}(X) \leqq |X|^\sharp.$$

We show that

$$(4)\qquad \mathfrak{L}(X) = \sup\left\{\frac{|X\cdot(T_v\sigma - \sigma)|}{|\sigma||v|} : \text{ simplexes } \sigma\right\}.$$

Given $\epsilon > 0$, choose $A \neq 0$ and $v \neq 0$ so that

$$|X\cdot(T_vA - A)| \geqq [\mathfrak{L}(X) - \epsilon]|A||v|.$$

Write $A = \sum a_i\sigma_i$, the σ_i non-overlapping and the $a_i \neq 0$. It is sufficient to show that for some i,

$$|X\cdot(T_v\sigma_i - \sigma_i)| \geqq [\mathfrak{L}(X) - \epsilon]|\sigma_i||v|.$$

If this were false for all i, we would find

$$\begin{aligned}|X\cdot(T_vA - A)| &= \left|\sum a_iX\cdot(T_v\sigma_i - \sigma_i)\right|\\ &< \sum |a_i|[\mathfrak{L}(X) - \epsilon]|\sigma_i||v| = [\mathfrak{L}(X) - \epsilon]|A||v|,\end{aligned}$$

a contradiction.

The coboundary dX of a sharp cochain need not be sharp; see § 11, Example (*a*). We have

$$(5)\qquad |dX| \leqq |X|^\flat \leqq |X|^\sharp.$$

Corresponding to (4.8), we now prove

$$(6)\qquad |X|^\sharp = \sup\{|X|^\flat, (r+1)\mathfrak{L}(X)\};$$

see also (8) below. Because of (1) and (3), we need merely prove $\leqq$. Take any $\epsilon > 0$, and any polyhedral chain A. Write $A = \sum a_i\sigma_i$, and choose vectors v_i, so that

$$\frac{\sum|a_i||\sigma_i||v_i|}{r+1} + \left|\sum a_iT_{v_i}\sigma_i\right|^\flat \leqq |A|^\sharp + \epsilon.$$

Now

$$\begin{aligned}|X\cdot A| &= \left|X\cdot\left[-\sum a_i(T_{v_i}\sigma_i - \sigma_i) + \sum a_iT_{v_i}\sigma_i\right]\right|\\ &\leqq \sum|a_i||X\cdot(T_{v_i}\sigma_i - \sigma_i)| + \left|X\cdot\sum a_iT_{v_i}\sigma_i\right|\\ &\leqq \mathfrak{L}(X)\sum|a_i||\sigma_i||v_i| + |X|^\flat\left|\sum a_iT_{v_i}\sigma_i\right|^\flat\\ &\leqq \sup\{(r+1)\mathfrak{L}(X), |X|^\flat\}(|A|^\sharp + \epsilon),\end{aligned}$$

which gives the required inequality.

THEOREM 7A. *Any flat cochain with finite* $\mathfrak{L}(X)$ *is sharp.*

This follows at once from the inequalities above.

REMARK. A linear function of polyhedral chains with $|X|$ and $\mathfrak{L}(X)$ finite defines a sharp cochain, provided some local finiteness of $|dX|$ is assumed; compare Theorem 8B below. We cannot omit such a condition; see § 11, Example (*g*).

We now prove that

$$|dX| \leqq (r+1)\mathfrak{L}(X) \qquad \text{if } X \text{ is sharp.} \tag{7}$$

Because of (4.10), we need merely show that for any $(r+1)$-simplex σ,

$$|X\cdot\partial\sigma| \leqq (r+1)\mathfrak{L}(X)|\sigma|.$$

Given $\epsilon > 0$, we may cut the plane P of σ into equal cubes, so that if $\sigma_1, \cdots, \sigma_m$ are the cubes lying in σ, and $\sigma_{m+1}, \cdots, \sigma_l$ are the parts of cubes filling up the rest of σ, then $\sum_{i=m+1}^{l}|\sigma_i| < \epsilon/|dX|$. Each cube σ_i $(i \leqq m)$ has $2(r+1)$ faces, which can be expressed as σ_{ik} and $T_{v_k}\sigma_{ik}$ $(k = 1, \cdots, r+1)$. Also, with suitable orientations, $\partial\sigma_i = \sum_k(T_{v_k}\sigma_{ik} - \sigma_{ik})$, as in (III, 11.2). Now for $i \leqq m$,

$$\begin{aligned}|X\cdot\partial\sigma_i| &\leqq \sum_k |X\cdot(T_{v_k}\sigma_{ik} - \sigma_{ik})| \\ &\leqq \sum_k \mathfrak{L}(X)|\sigma_{ik}||v_k| = (r+1)\mathfrak{L}(X)|\sigma_i|.\end{aligned}$$

Consequently

$$\begin{aligned}|X\cdot\partial\sigma| = \left|\sum_{i=1}^{l} X\cdot\partial\sigma_i\right| &\leqq \sum_{i=1}^{m}(r+1)\mathfrak{L}(X)|\sigma_i| + \sum_{i=m+1}^{l}|dX||\sigma_i| \\ &\leqq (r+1)\mathfrak{L}(X)|\sigma| + \epsilon,\end{aligned}$$

giving the required inequality.

Relations (6), (4.8) and (7) give

$$|X|^\sharp = \sup\{|X|, (r+1)\mathfrak{L}(X)\} \qquad \text{if } X \text{ is sharp.} \tag{8}$$

Because of (6.5), we have

$$|X|^\sharp = |X|^\flat \quad \text{if} \quad r = 0. \tag{9}$$

A direct proof may be given as follows. If the 0-cochain X corresponds to the real function D_X: $X\cdot p = D_X(p)$, (4) shows that $\mathfrak{L}(X) = \mathfrak{L}(D_X)$. Now (8), (4.11) and (4.8) give the result.

For $r = n$, we show near the end of § 10 below that X corresponds to a real sharp function $\bar{D}_X$; hence:

THEOREM 7B. *In Euclidean* E^n, *both sharp* 0*-cochains and sharp* n*-cochains correspond exactly to real sharp functions.*

For n-cochains, the correspondence between X and $\bar{D}_X$ depends on the metric.

8. Characterization of the norms. In the linear space of polyhedral r-chains in E^n, there are various norms; we shall consider semi-norms $| \, |'$ satisfying one or more of the following properties (σ^s denoting an s-simplex):

(1) $$|\sigma^r|' \leqq |\sigma^r|,$$

(2) $$|\partial\sigma^{r+1}|' \leqq |\sigma^{r+1}|,$$

(3) $$|T_v\sigma^r - \sigma^r|' \leqq |\sigma^r| \, |v|/(r+1).$$

Recall that the supremum of a set of semi-norms, if finite, is a semi-norm (App. I, Lemma 15b).

THEOREM 8A. *For each dimension r, the mass, the flat norm, and the sharp norm are the largest semi-norms in the space of polyhedral chains satisfying* (1), (1) *and* (2), *and* (1) *and* (2) *and* (3) *respectively.*

Since the norms satisfy the stated conditions, we need merely show that any semi-norm satisfying the stated conditions is less than or equal to the stated norm.

For the mass, the proof is immediate: Given any polyhedral r-chain A, write $A = \sum a_i\sigma_i$, the σ_i being non-overlapping simplexes. Then since $| \, |'$ is a semi-norm,

$$|A|' \leqq \sum |a_i| \, |\sigma_i|' \leqq \sum |a_i| \, |\sigma_i| = |A|.$$

For the flat norm, we note first that if $| \quad |'$ satisfies (1) and (2), then for r-chains C and $(r+1)$-chains D,

$$|C|' \leqq |C|, \qquad |\partial D|' \leqq |D|.$$

The first relation was proved above. For the second, write $D = \sum d_i\sigma_i$, the σ_i being non-overlapping; then

$$|\partial D|' \leqq \sum |d_i| \, |\partial\sigma_i|' \leqq \sum |d_i| \, |\sigma_i| = |D|.$$

Now take any polyhedral A. Given $\epsilon > 0$, choose D so that $|A - \partial D| + |D| < |A|^\flat + \epsilon$. Then

$$|A|' \leqq |A - \partial D|' + |\partial D|' \leqq |A - \partial D| + |D| < |A|^\flat + \epsilon,$$

proving $|A|' \leqq |A|^\flat$, as required.

For the sharp norm, take any A, and $\epsilon > 0$. Write $A = \sum a_i\sigma_i$, and choose vectors v_i, so that

$$\sum |a_i| \, |\sigma_i| \, |v_i| / (r+1) + \left|\sum a_i T_{v_i}\sigma_i\right|^\flat < |A|^\sharp + \epsilon.$$

Assuming (1), (2) and (3) hold, and hence $|C|' \leqq |C|^\flat$, we have

$$|A|' \leqq \sum |a_i| \, |T_{v_i}\sigma_i - \sigma_i|' + \left|\sum a_i T_{v_i}\sigma_i\right|',$$

which is bounded by the quantity above; hence $|A|' \leqq |A|^\sharp$.

LEMMA 8a. *For each s-simplex σ there is a number N with the following property. Let K be any cubical subdivision of the plane P of σ, the cubes being of diameter $\delta \leqq \operatorname{diam}(\sigma)$. Let $\tau_1, \cdots, \tau_\mu$ be the parts contained in σ of those cubes which touch $\partial\sigma$. Let $\tau_1', \cdots, \tau_\nu'$ be the s-simplexes of the regular subdivision* (*App.* II, 3) *of the τ_j. Then*

$$\nu\delta^{s-1} \leqq N. \tag{4}$$

Note that each $(s-1)$-face of each τ_i' is of diameter $\leqq \delta$ and hence of mass $\leqq \delta^{s-1}$; hence $|\partial\tau_i'| \leqq (s+1)\delta^{s-1}$, and setting $N' = (s+1)N/|\partial\sigma|$, we have ($N'$ depending on σ)

$$\sum |\partial\tau_i'| \leqq N' |\partial\sigma|. \tag{5}$$

Set $r = s - 1$. Any r-face of any τ_j is either part of a face of a cube of K or part of a face of $\partial\sigma$; hence τ_j has at most $M_1 = 3s + 1$ r-faces. Each face of τ_j of lower dimension is an intersection of r-faces; hence there are at most M_2 faces of all dimensions, for some fixed M_2. Each simplex τ_i' in τ_j is defined by a sequence of incident faces of τ_j; hence there are at most M_3 simplexes τ_i' in any τ_j, for some fixed M_3.

Set $\delta_0 = \operatorname{diam}(\sigma)$. Say $\partial\sigma = \sum\sigma_k^r$. Let P_k be the plane of σ_k^r; set

$$R_k = P_k \cap U_{\delta_0}(\sigma_k^r), \qquad N = 2s^{s/2}M_3 \sum |R_k|.$$

Now take the τ_j and τ_i' as above. Let R_k' be the set of those points of the plane P of σ which are distant at most δ from P_k, and whose orthogonal projections into P_k lie in R_k; then $|R_k'| = 2\delta |R_k|$. Let τ_j^* be the cube containing τ_j. Each τ_j^* lies in some R_k'. Since $|\tau_j^*| = (\delta/s^{1/2})^s$, the number ν_k of τ_j^* in R_k' is at most $s^{s/2}|R_k'|/\delta^s$. Hence

$$\nu \leqq M_3 \sum \nu_k \leqq M_3 \sum 2s^{s/2}\delta |R_k|/\delta^s = N/\delta^{s-1}.$$

LEMMA 8b. *For any s-simplex σ, if $\delta = \operatorname{diam}(\sigma)$,*

$$|\sigma| \leqq \delta |\partial\sigma|/s(s+1) \leqq \delta^s/s!. \tag{6}$$

Write $\sigma = p_0\sigma_0$, σ_0 being the smallest r-face of σ $(r = s - 1)$. Let $v_2, \cdots, v_s$ be vectors along the sides of σ_0, and v_1, a vector from a vertex of σ_0 to p_0. Then by (III, 1.3) and (I, 12.14),

$$|\partial\sigma| \geqq (s+1)|\sigma_0| = (s+1)|v_2\vee\cdots\vee v_s|/r!,$$

$$|\sigma| = |v_1\vee\cdots\vee v_s|/s! \leqq \delta|v_2\vee\cdots\vee v_s|/s!,$$

and (6) follows.

LEMMA 8c. *Let* $|\ \ |'$ *be a semi-norm in the space of polyhedral r-chains which satisfies* (1) *and* (3), *and also the following*:

(H) *For each point p and $\epsilon > 0$ there is a $\zeta > 0$ such that*

$$|\partial\sigma|' \leqq \epsilon[\operatorname{diam}(\sigma)]^r \qquad \text{for } (r+1)\text{-simplexes} \quad \sigma \subset U_\zeta(p). \tag{7}$$

then the semi-norm satisfies (2) *also*: $|\partial\sigma|' \leqq |\sigma|$.

REMARK. Since $|\partial\sigma| \leqq (r+2)[\text{diam}(\sigma)]^r$, we could replace (7) by

(8) $$|\partial\sigma|' \leqq \epsilon|\partial\sigma| \quad \text{if} \quad \sigma \subset U_\zeta(p).$$

Given the s-simplex σ_0 ($s = r + 1$), define N by Lemma 8a. Given $\epsilon > 0$, set $\epsilon_1 = \epsilon/N$. Using ϵ_1 in place of ϵ, let $\zeta(p)$ satisfy (7), for each p. Choose a cubical subdivision K of the plane P of σ_0 so that each cube (of diameter δ) which touches $\partial\sigma_0$ lies in some $U_{\zeta(p)}(p)$. With the notations in the proof of (7.7), we have

$$|\partial\sigma_i|' \leqq \sum_{k=1}^{r+1} |T_{v_k}\sigma_{ik} - \sigma_{ik}|' \leqq \sum_{k=1}^{r+1} |\sigma_{ik}|\,|v_k|/s = |\sigma_i|,$$

for $i = 1, \cdots, m$. Let $\tau'_1, \cdots, \tau'_\nu$ be the simplexes of the regular subdivision of the σ_j for $j > m$ (the τ_j of Lemma 8a); then $\nu\delta^r \leqq N$. Also, by the choice of K, $|\partial\tau'_i|' \leqq \epsilon_1\delta^r$. Hence

$$|\partial\sigma_0|' \leqq \sum_{i=1}^{m} |\partial\sigma_i|' + \sum_{i=1}^{\nu} |\partial\tau'_i|'$$

$$\leqq \sum_{i=1}^{m} |\sigma_i| + \nu\epsilon_1\delta^r \leqq |\sigma_0| + \epsilon,$$

giving $|\partial\sigma_0|' \leqq |\sigma_0|$, as required.

THEOREM 8B. *The sharp norm in the space of polyhedral r-chains is the largest semi-norm satisfying* (1), (3) *and* (H).

First, the sharp norm satisfies (H), because of (6). The theorem now follows from the last lemma and Theorem 8A.

REMARKS. We cannot omit the condition (H); see Example (f) in § 11. In place of (7), we could use (8); or (see (6)) we could assume that for each p there are numbers N and $\zeta > 0$ such that

(9) $$|\partial\sigma|' \leqq N|\sigma| \quad \text{if} \quad \sigma \subset U_\zeta(p).$$

9. An algebraic criterion for a multi-covector. In the next section we shall find, at each point p, a function $D_X(p, \alpha)$ of simple r-vectors α, with certain properties. We wish to show that this defines an r-covector. Recall (III, 2), in particular, (III, Theorem 2B).

THEOREM 9A. *Let $\phi(\alpha)$ be a real function of simple r-vectors such that*

(1) $$\phi(a\alpha) = a\phi(\alpha),$$

(2) $$\sum_{i=0}^{r+1}(-1)^i\phi(\{p_0 \cdots \hat{p}_i \cdots p_{r+1}\}) = 0 \qquad \textit{for any simplex} \quad p_0 \cdots p_{r+1}.$$

Then there is a unique r-covector ξ such that $\xi \cdot \alpha = \phi(\alpha)$ if α is simple.

The uniqueness is clear; we prove existence. First note that

(3) $$\phi\{\sigma\} = \sum \phi\{\sigma_i\} \qquad \text{if } \sum \sigma_i \text{ is a subdivision of } \sigma.$$

For $\{\sigma_i\} = a_i\{\sigma\}$ for some a_i, and since $\{\sigma\} = \sum\{\sigma_i\} = \sum a_i\{\sigma\}$, $\sum a_i = 1$. Therefore

$$\sum \phi\{\sigma_i\} = \sum \phi(a_i\{\sigma\}) = \sum a_i \phi\{\sigma\} = \phi\{\sigma\}.$$

We now define a function of polyhedral r-chains A in V by setting

$$\Phi\Big(\sum a_i \sigma_i\Big) = \sum a_i \phi\{\sigma_i\}. \tag{4}$$

This is independent of the representation of A. For if $A = \sum a_j' \sigma_j'$ also, then, using (App. II, Lemma 3b), we may write $A = \sum b_k \tau_k$, where each σ_i and σ_j' has a simplicial subdivision with simplexes τ_k. The statement is now an easy consequence of (3) and (1). Now Φ is a linear function of polyhedral r-chains.

We now prove

$$\Phi(\partial B) = 0 \qquad \text{for polyhedral } (r+1)\text{-chains } B. \tag{5}$$

For if B is an $(r+1)$-simplex $\sigma = p_0 \cdots p_{r+1}$, this is an immediate consequence of (2) and the linearity of Φ. Hence it holds for all B.

Define a function F of ordered sets of r vectors by

$$F(v_1, \cdots, v_r) = \phi(v_1 \vee \cdots \vee v_r); \tag{6}$$

because of (1) and (I, 1.13), F is alternating. We shall show that it is linear in v_1; it will follow that it is linear in each v_i.

Consider the following points and simplex in V:

$$p_0 = 0, \qquad p_1 = v_1, \qquad p_2 = v_1 + v_1', \qquad \sigma^2 = p_0 p_1 p_2.$$

If $r = 1$, then

$$F(v_1) = \Phi(p_0 p_1), \qquad F(v_1') = \Phi(p_1 p_2), \qquad F(v_1 + v_1') = \Phi(p_0 p_2),$$

and (5) gives $F(v_1) + F(v_1') - F(v_1 + v_1') = 0$. Supposing $r > 1$, let τ^{r-1} be the oriented parallelopiped formed by the vectors $v_2, \cdots, v_r$, and let σ^{r+1} be the Cartesian product

$$\sigma^{r+1} = \sigma^2 \times \tau^{r-1}.$$

Now σ^{r+1} consists of all vectors $u_1 + u_2$, $u_1 \in \sigma^2$, $u_2 \in \tau^{r-1}$; thus $u_1 = a_1 v_1 + a_1' v_1'$, $u_2 = a_2 v_2 + \cdots + a_r v_r$, the coefficients ranging between 0 and 1, with $a_1' \leq a_1$. The cells may be degenerate (App. II, 1). Now $\tau = \tau^{r-1}$ has $2(r-1)$-faces τ_i^- and τ_i^+ $(i = 2, \cdots, r)$, and as in (III, 11.2),

$$\partial \tau^{r-1} = \sum (-1)^i (\tau_i^+ - \tau_i^-).$$

Hence (App. II, 12.1)

$$\partial \sigma^{r+1} = p_0 p_1 \times \tau + p_1 p_2 \times \tau - p_0 p_2 \times \tau + \sum_i (-1)^i (\sigma^2 \times \tau_i^+ - \sigma^2 \times \tau_i^-).$$

Since τ_i^+ and τ_i^- are parallel, $\{\tau_i^+\} = \{\tau_i^-\}$, and hence $\{\sigma^2 \times \tau_i^+\} = \{\sigma^2 \times \tau_i^-\}$. Therefore $\Phi(\partial\sigma^{r+1}) = 0$ gives

$$\Phi(p_0p_1 \times \tau) + \Phi(p_1p_2 \times \tau) = \Phi(p_0p_2 \times \tau).$$

This is equivalent to

$$F(v_1, v_2, \cdots) + F(v_1', v_2, \cdots) = F(v_1 + v_1', v_2, \cdots),$$

which is the required linearity.

By (I, Theorem 4A), there is an r-covector ξ such that

$$\xi\cdot(v_1\vee\cdots\vee v_r) = F(v_1, \cdots, v_r) = \phi(v_1\vee\cdots\vee v_r);$$

this completes the proof.

Example. Taking $r = 1$, let $\phi(ae_1) = a$, and $\phi(v) = 0$ for all other v. Then (1) holds, but ϕ corresponds to no covector. Compare § 11, Example (g) below.

10. Sharp r-forms. We shall show how sharp r-cochains in E^n correspond exactly to certain differential forms. For a corresponding theorem in open subsets of E^n, see (VIII, 1). For the case of flat cochains, see Chapter IX. Recall the definition of the comass $|\omega|_0$ of an r-form in (II, 3.2). The *Lipschitz comass constant* $\mathfrak{L}_0(\omega)$ we define by

(1) $$\mathfrak{L}_0(\omega) = \sup \frac{|\omega(q) - \omega(p)|_0}{|q - p|}.$$

We say ω is *sharp* if both $|\omega|_0$ and $\mathfrak{L}_0(\omega)$ are finite; the *sharp norm* of ω is

(2) $$|\omega|^\# = \sup\{|\omega|_0, (r + 1)\mathfrak{L}_0(\omega)\}.$$

For the existence of $d\omega$, see (IX, Theorem 12B).

The normed linear space of sharp r-forms is easily seen to be complete (this follows also from Theorem 10A).

Theorem 10A. *To each sharp r-cochain in X in E^n there corresponds a unique sharp r-form D_X such that*

(3) $$X\cdot\sigma = \int_\sigma D_X \qquad \textit{for all oriented r-simplexes } \sigma.$$

This correspondence is one-one onto, and

(4) $$|D_X|_0 = |X|, \qquad \mathfrak{L}_0(D_X) = \mathfrak{L}(X), \qquad |D_X|^\# = |X|^\#.$$

Uniqueness follows from (III, Lemma 16a). Given X, we find a corresponding form $\omega = D_X$ as follows. First, take any p and any r-direction α (I, 12); let $\sigma_1, \sigma_2, \cdots$ be a sequence of r-simplexes containing p whose r-directions are $\alpha = \{\sigma_i\}/|\sigma_i|$ (III, 1), and whose diameters $\to 0$, and set

(5) $$D_X(p, \alpha) = \lim_{i\to\infty} X\cdot\sigma_i/|\sigma_i|.$$

The existence and uniqueness of the limit follow at once from the following relation:

$$\left|\frac{X\cdot\sigma}{|\sigma|}-\frac{X\cdot\sigma'}{|\sigma'|}\right|\leqq 4\mathfrak{L}(X)\zeta \quad \text{if} \quad \sigma, \sigma' \subset U_\zeta(p), \tag{6}$$

σ and σ' having the r-direction α.

To prove (6), set

$$\epsilon = 4\mathfrak{L}(X)\zeta, \quad \epsilon_1 = \inf \quad \left\{\frac{\epsilon|\sigma|}{8}, \frac{\epsilon|\sigma'|}{8}\right\}, \epsilon_2 = \frac{\epsilon_1}{|X|},$$

and choose an r-cube τ through p with r-direction α so that the following holds. There are vectors $v_1, \cdots, v_s$ such that $T_{v_1}\tau, \cdots, T_{v_s}\tau$ are in σ and are non-overlapping, and

$$|\sigma - T_{v_1}\tau \cup \cdots \cup T_{v_s}\tau| \leqq \epsilon_2;$$

also a similar inequality holds for σ'. Now

$$\left|X\cdot\sigma - \sum X\cdot T_{v_k}\tau\right| \leqq |X|\,\epsilon_2 = \epsilon_1,$$

$$|X\cdot T_{v_k}\tau - X\cdot\tau| \leqq \mathfrak{L}(X)|\tau|\,|v_k|;$$

and since $|v_k| < \zeta$,

$$|X\cdot\sigma - sX\cdot\tau| \leqq \epsilon_1 + s\mathfrak{L}(X)|\tau|\zeta \leqq \epsilon_1 + \mathfrak{L}(X)|\sigma|\zeta.$$

Therefore

$$\begin{aligned} &|(X\cdot\sigma)|\tau| - (X\cdot\tau)|\sigma|| \\ &\qquad \leqq |X\cdot\sigma - s(X\cdot\tau)|\,|\tau| + |X\cdot\tau|\,|s|\tau| - |\sigma|| \\ &\qquad \leqq [\epsilon_1 + \mathfrak{L}(X)|\sigma|\zeta]\,|\tau| + |X|\,|\tau|\,\epsilon_2, \end{aligned}$$

$$\left|\frac{X\cdot\sigma}{|\sigma|} - \frac{X\cdot\tau}{|\tau|}\right| \leqq \frac{\epsilon_1}{|\sigma|} + \mathfrak{L}(X)\zeta + \frac{\epsilon_1}{|\sigma|} \leqq \frac{\epsilon}{2}.$$

A similar inequality holds for σ'; these give (6).

Using the definition of $D_X(p, \alpha)$ for r-directions α alone, we can define $\mathfrak{L}_0(D_X)$; see (I, 13.2). Since

$$\left|\frac{X\cdot T_v\sigma}{|T_v\sigma|} - \frac{X\cdot\sigma}{|\sigma|}\right| \leqq \mathfrak{L}(X)|v|,$$

we find

$$\mathfrak{L}_0(D_X) \leqq \mathfrak{L}(X), \tag{7}$$

incidentally proving that $D_X(p, \alpha)$ is continuous in p.

We shall show below that if σ has the r-direction α, and $\delta = \text{diam}\,(\sigma)$,

$$|D_X(p, \alpha) - X\cdot\sigma/|\sigma|| \leqq \mathfrak{L}(X)\,\delta \quad \text{if} \quad p \in \sigma. \tag{8}$$

For the moment, it follows at once from (6) if we put in a factor 4.

We now show that for any σ, with r-direction α,

$$X\cdot\sigma = \int_\sigma D_X(p, \alpha)\, dp, \tag{9}$$

using the Riemann integral (III, 5). Given $\epsilon > 0$, cut σ into simplexes $\sigma_1, \cdots, \sigma_s$ of diameter $< \zeta = \epsilon/5\mathfrak{L}(X)|\sigma|$. For each σ_i, choosing $p_i \in \sigma_i$ and using (8) (with the factor 4) and (7) gives

$$\begin{aligned}\left| X\cdot\sigma_i - \int_{\sigma_i} D_X(p, \alpha)\, dp \right| \\ &\leqq \left| X\cdot\sigma_i - D_X(p_i, \alpha)|\sigma_i| \right| + \left| \int_{\sigma_i} [D_X(p_i, \alpha) - D_X(p, \alpha)]\, dp \right| \\ &\leqq 5\mathfrak{L}(X)\zeta\, |\sigma_i|\,;\end{aligned}$$

hence $\left| X\cdot\sigma - \int_\sigma D_X(p, \alpha)\, dp \right| \leqq \epsilon$, giving (9). Now (8) follows at once on using (7) and

$$D_X(p, \alpha)|\sigma| - X\cdot\sigma = \int_\sigma [D_X(p, \alpha) - D_X(q, \alpha)]\, dq.$$

For any simple r-vector $\alpha \neq 0$, set

$$D_X(p, \alpha) = |\alpha|\, D_X(p, \alpha/|\alpha|); \tag{10}$$

set $D_X(p, 0) = 0$. It is evident from (5) that $D_X(p, -\alpha) = -D_X(p, \alpha)$ for r-directions α; hence $D_X(p, a\alpha) = aD_X(p, \alpha)$ for all simple α and real a.

We now show that for any p and $(r + 1)$-simplex σ, if $\partial\sigma = \sum\sigma_i$, then

$$\sum D_X(p, \{\sigma_i\}) = 0. \tag{11}$$

(See § 8 of the Introduction for a slightly different proof.) We may suppose $p \in \sigma$. Let σ_λ be σ contracted towards p by the factor λ; say $\partial\sigma_\lambda = \sum\sigma_{\lambda i}$. Now

$$|\sigma_\lambda| = \lambda^{r+1}|\sigma|, \qquad |\sigma_{\lambda i}| = \lambda^r |\sigma_i|.$$

If $\delta = \text{diam}\,(\sigma)$ and $\alpha_i = r$-direction of σ_i, using (7) gives

$$\begin{aligned}\left|\sum_i D_X(p, \{\sigma_{\lambda i}\}) - \int_{\partial\sigma_\lambda} D_X \right| &= \left| \sum_i \int_{\sigma_{\lambda i}} [D_X(p, \alpha_i) - D_X(q, \alpha_i)]\, dq \right| \\ &\leqq \sum \mathfrak{L}(X)\lambda\delta\, |\sigma_{\lambda i}| = \lambda^{r+1}\mathfrak{L}(X)\delta\, |\partial\sigma|\,.\end{aligned}$$

Also, using (9),

$$\begin{aligned}\left| \int_{\partial\sigma_\lambda} D_X \right| &= \left| \sum_i \int_{\sigma_{\lambda i}} D_X(q, \alpha_i)\, dq \right| = \left| \sum_i X\cdot\sigma_{\lambda i} \right| \\ &= |X\cdot\partial\sigma_\lambda| = |dX\cdot\alpha_\lambda| \leqq \lambda^{r+1}\,|dX|\,|\sigma|\,.\end{aligned}$$

Combining these inequalities gives, on dividing by λ^r,

$$\left| \sum_i D_X(p, \{\sigma_i\}) \right| \leqq \lambda[\mathfrak{L}(X)\delta \mid \partial\sigma \mid + \mid dX \mid \mid \sigma \mid].$$

Since λ is arbitrary, (11) follows.

Because of Theorem 9A, there is a unique r-covector $D_X(p)$ for each p such that $D_X(p)\cdot\alpha = D_X(p, \alpha)$ for simple α. Now (3) follows from (9).

Since $\mid X\cdot\sigma_i \mid \leqq \mid X \mid \mid \sigma_i \mid$, (5) gives $\mid D_X \mid_0 \leqq \mid X \mid$. Conversely, given $\epsilon > 0$, (4.9) shows that we may choose σ so that $\mid X\cdot \sigma \mid \geqq (\mid X \mid - \epsilon) \mid \sigma \mid$; using (9) shows that $\mid D_X(p)\cdot\alpha \mid \geqq \mid X \mid - \epsilon$ for some $p \in \sigma$, proving $\mid D_X \mid_0 \geqq \mid X \mid - \epsilon$; thus $\mid D_X \mid_0 = \mid X \mid$. Next, given $\epsilon > 0$, we may choose σ and v by (7.4) so that

$$\mid X\cdot T_v\sigma - X\cdot\sigma \mid \geqq [\,\mathfrak{L}(X) - \epsilon] \mid v \mid \mid \sigma \mid;$$

using (9) shows that, if α is the r-direction of σ,

$$\mid D_X(p + v, \alpha) - D_X(p, \alpha) \mid \geqq [\mathfrak{L}(X) - \epsilon] \mid v \mid$$

for some $p \in \sigma$, proving $\mathfrak{L}_0(D_X) \geqq \mathfrak{L}(X) - \epsilon$; hence, using (7), we have $\mathfrak{L}_0(D_X) = \mathfrak{L}(X)$. Now (7.8) and (2) give $\mid D_X \mid^\# = \mid X \mid^\#$, proving (4).

Now take any sharp form ω. Set $X\cdot\sigma = \int_\sigma \omega$; this determines X uniquely. If we show that setting $X\cdot\sum a_i\sigma_i = \sum a_i X\cdot\sigma_i$ defines a sharp cochain X, the proof will be complete. Clearly $\mid X \mid \leqq \mid \omega \mid_0$, $\mathfrak{L}(X) \leqq \mathfrak{L}_0(\omega)$. Because of Theorems 7A and 4A, it is sufficient to show that $\mid dX \mid$ is bounded. We may suppose $\omega \neq 0$. Because of (4.10), given an $(r + 1)$-simplex σ, it is sufficient to prove

$$\mid X\cdot\partial\sigma \mid = \left| \int_{\partial\sigma} \omega \right| \leqq \mid \omega \mid^\# \mid \sigma \mid. \tag{12}$$

Define a semi-norm in the space of polyhedral r-chains by setting

$$\mid A \mid' = \frac{\mid X\cdot A \mid}{\mid \omega \mid^\#} = \frac{1}{\mid \omega \mid^\#} \left| \int_A \omega \right|. \tag{13}$$

The conditions for a semi-norm are clearly fulfilled. We prove the conditions of Lemma 8c. First,

$$\mid \sigma \mid' = \frac{1}{\mid \omega \mid^\#} \left| \int_\sigma \omega \right| \leqq \frac{\mid \omega \mid_0}{\mid \omega \mid^\#} \mid \sigma \mid \leqq \mid \sigma \mid,$$

$$\mid T_v\sigma - \sigma \mid' = \frac{1}{\mid \omega \mid^\#} \left| \int_\sigma [\omega(p + v) - \omega(p)]\, dp \right| \leqq \frac{\mathfrak{L}_0(\omega) \mid v \mid \mid \sigma \mid}{\mid \omega \mid^\#} \leqq \frac{\mid v \mid \mid \sigma \mid}{r + 1}.$$

To prove (H), take any p and $\epsilon > 0$. Choose $\zeta > 0$ so that

$$\mid \omega(q) - \omega(p) \mid < \epsilon_1 = \mid \omega \mid^\# \epsilon \quad \text{if} \quad \mid q - p \mid < \zeta.$$

Take any $(r+1)$-simplex $\sigma \subset U_\zeta(p)$. Say $\partial\sigma = \sum \sigma_i$, $\alpha_i = \{\sigma_i\}/|\sigma_i|$. Using (III, 4.1) and (III, 2.3), we find

$$\left|\int_{\partial\sigma} \omega\right| = \left|\sum_i \int_{\sigma_i} [\omega(q) - \omega(p)]\cdot\alpha_i \, dq\right| \leqq \epsilon_1 |\partial\sigma|.$$

Hence

$$|\partial\sigma|' = \frac{1}{|\omega|^\#}\left|\int_{\partial\sigma} \omega\right| \leqq \epsilon |\partial\sigma|,$$

giving (8.8), and hence (8.7). By Lemma 8c, we have, for any $(r+1)$-simplex σ,

$$|X\cdot\partial\sigma| = |\omega|^\# |\partial\sigma|' \leqq |\omega|^\# |\sigma|,$$

completing the proof of the theorem.

Let ω_0 be the n-direction of oriented E^n (I, 12). Then (II, 3.1) there is a one-one correspondence between n-forms $\omega(p)$ and real functions $\bar{\omega}(p)$, given by

$$\omega(p) = \bar{\omega}(p)\omega_0; \tag{14}$$

clearly

$$\mathfrak{L}(\bar{\omega}) = \mathfrak{L}(\omega) = \mathfrak{L}_0(\omega), \qquad \deg(\omega) = n. \tag{15}$$

Theorem 10A now gives the last part of Theorem 7B.

We say X is *smooth* if the corresponding differential form D_X is smooth. We may then define the exterior differential dD_X as in (II, 8). More generally, we may do this if D_X is regular, as in (III, 16). See also (IX, Theorem 12B).

THEOREM 10B. *Let X be smooth, or, let it be sharp and let D_X be regular* (III, 16). *Then D_{dX}, defined as in* (5), *exists, and*

$$D_{dX} = dD_X. \tag{16}$$

Given the point p and $(r+1)$-direction α, let $\sigma_1, \sigma_2, \cdots$ be a corresponding sequence of $(r+1)$-simplexes as in (5). Using the Theorem of Stokes (III, 14) if X is smooth, or (III, 16.2) if D_X is regular, we have

$$dX\cdot\sigma_i = X\cdot\partial\sigma_i = \int_{\partial\sigma_i} D_X = \int_{\sigma_i} dD_X.$$

Since dD_X is continuous, we find

$$D_{dX}(p, \alpha) = \lim_{i\to\infty} \frac{dX\cdot\sigma_i}{|\sigma_i|} = \lim_{i\to\infty} \frac{1}{|\sigma_i|}\int_{\sigma_i} dD_X(q, \alpha)\, dq$$
$$= (dD_X)(p, \alpha),$$

proving (16).

11. Examples. (*a*) Taking $n = 1$, $r = 0$, set

$$\phi(x) = \inf(|x|, 1), \qquad X\cdot x = \phi(x). \tag{1}$$

Then $|X| = 1$, $|dX| = 1$, $\mathfrak{L}(X) = 1$, and X is a sharp 0-cochain with $|X|^\sharp = |X|^\flat = 1$. Within the interval $(0, 1)$, dX is sharp, and (see (10.14)) (10.16) and (II, 8.2) show that $\bar{D}_{dX}(x) = d\phi(x)/dx = 1$; in the interval $(-1, 0)$, $\bar{D}_{dX}(x) = -1$. Hence $\mathfrak{L}(\bar{D}_{dX})$ is not finite, and dX is not sharp. In fact, if σ is the oriented segment $(-\epsilon, 0)$, and v is the number ϵ, $0 < \epsilon < 1$,

$$dX\cdot\sigma = X\cdot\partial\sigma = -\epsilon, \qquad dX\cdot T_v\sigma = \epsilon,$$

$$\frac{dX\cdot(T_v\sigma - \sigma)}{|v||\sigma|} = \frac{2\epsilon}{\epsilon^2} = \frac{2}{\epsilon}.$$

(*b*) Define σ, v, X as in (*a*). Set $A = T_v\sigma - \sigma$. By (4.12), $|A|^\flat = |A|$. By (6.3), $|A|^\sharp \leqq |\sigma||v|/2 = \epsilon^2/2$. Define the sharp 1-cochain Y by $\bar{D}_Y(x) = x/2$ if $|x| \leqq 2$, and $\bar{D}_Y(x) = 1$ $(x \geqq 2)$ and $= -1$ $(x \leqq -2)$. Then $\mathfrak{L}(\bar{D}_Y) = 1/2$, and (10.4) and (10.2) show that $|Y|^\sharp = 1$. Since

$$Y\cdot T_v\sigma = \int_{T_v\sigma} D_Y = \int_0^\epsilon \frac{x}{2}\,dx = \frac{\epsilon^2}{4}, \qquad Y\cdot\sigma = -\frac{\epsilon^2}{4},$$

we have $Y\cdot A = \epsilon^2/2$, and hence $|A|^\sharp \geqq \epsilon^2/2$. Thus

$$|A|^\flat = |A| = 2\epsilon, \qquad |A|^\sharp = \epsilon^2/2.$$

Since $X\cdot\partial A = 2\epsilon$, we have $|\partial A|^\sharp \geqq 2\epsilon$. Also $|\partial A|^\sharp = |\partial A|^\flat \leqq |A|^\flat = 2\epsilon$; hence $|\partial A|^\sharp = 2\epsilon$. Note that

$$|\partial A|^\sharp = (4/\epsilon)|A|^\sharp, \qquad |T_v\partial\sigma - \partial\sigma|^\sharp = (2/\epsilon)|\partial\sigma|^\flat|v|,$$

showing that (3.4) fails for the sharp norm, and that $|A|$ cannot be replaced by $|A|^\flat$ in the right hand side of (6.3).

(*c*) The impossibility of defining ∂A properly for sharp A may be seen as follows. Let σ_i and σ_i' be the segments $(0, 1/2^i)$ and $(-1/2^i, 0)$ respectively. Set $A_i = 2^i\sigma_i$, $A_i' = 2^i\sigma_i'$. Now, by (b), $|A_i' - A_i|^\sharp = 1/2^{i+1}$. Also, cutting A_i into 2^k pieces and sliding each into A_{i+k} (a distance $<1/2^i$), we see that $|A_{i+k} - A_i|^\sharp < 1/2^{i+1}$; similarly for the A_j'. Therefore $A_1, A_1', A_2, A_2', \cdots$ is a Cauchy sequence, with a limit A in the sharp norm. Now with X as above,

$$X\cdot\partial A_i = 1, \qquad X\cdot\partial A_i' = -1;$$

hence we cannot reasonably define $X\cdot\partial A$. In this connection, compare (VI, 1) and (VII, 7).

(*d*) Let $S_1, S_2, \cdots$ be a sequence of squares in E^2, of side lengths $b_1, b_2, \cdots$ with $b_i \to 0$, such that the concentric squares of three times the side lengths $3b_1, 3b_2, \cdots$ are disjoint. (We may keep them in a bounded part of E^2.) Choose $a_1, a_2, \cdots \geqq 0$ so that $\sum a_i b_i^2$ is convergent but $\sum a_i b_i$ is divergent (say $a_i = 1$, $b_i = 1/i$). Let A_i be the part of ∂S_i

formed of two opposite sides, oppositely oriented with the coefficient a_i. Then $|A_i|^\sharp \leqq a_i b_i^2/2$, and hence the sum in the sharp norm $A = \sum^\sharp A_i$ exists. But $\sum^\sharp \partial A_i$ does not exist. For, choose a corner p_i of S_i ($i = 1, 2, \cdots$ such that $\partial A_i = a_i p_i + \cdots$, let U be the union of the $U_{b_i}(p_i)$, and define the sharp 0-cochain X by $D_X(p) = \text{dist}(p, E^2 - U)$. Then $D_X(p_i) = b_i$, and we should have $X \cdot \partial A = \sum X \cdot \partial A_i = \sum a_i b_i$.

(*e*) We show why the factor $r + 1$ is necessary in (7.7). We shall define a sharp r-cochain X in the unit sphere in E^n (we could define it in E^n) so that $\mathfrak{L}(X) = 1$, $|dX| = r + 1$. Let $D_X = \omega$. With an orthonormal coordinate system, set

$$\omega_{1\cdots\hat{i}\cdots r+1}(x) = (-1)^{i-1}x^i, \qquad \text{other} \quad \omega_\lambda(x) = 0.$$

By (II, 8.4), we find $d\omega_{1\cdots r+1}(x) = r + 1$; the other $(d\omega)_\lambda$ are 0. Hence $|dX| = r + 1$. It is easy to see that $\mathfrak{L}(X) = \mathfrak{L}_0(\omega) = 1$.

We can define $X \cdot \sigma$ geometrically with $n = r + 1$, as follows. Let $\phi(\sigma)$ be the distance from the origin to the oriented plane P of σ, with the $+$ or $-$ sign according as the half plane (oriented like E^{r+1}) bounded by P does or does not contain the origin; set $X \cdot \sigma = \phi(\sigma)|\sigma|$. For a cube C with faces parallel to the coordinate planes, it is easy to see directly that $dX \cdot C = (r + 1)|C|$. That $\mathfrak{L}(X) = 1$ is evident.

(*f*) We show the need of (H) in Theorem 8B. Take $n = 2$, $r = 1$, and let L be an oriented line in E^2. Any polyhedral 1-chain A can be written as $\sum a_i \sigma_i + \sum b_i \sigma_i'$, the σ_i parallel to and oriented like L and the σ_i' not parallel to L; set $|A|' = |\sum a_i|$. Then $|A|' \leqq |A|$ and $|T_v A - A|' = 0$. But we may take a 2-simplex τ with one side σ lying on L; then $|\partial\tau|' = |\sigma|$, and we may make $|\tau|$ arbitrarily small, keeping σ and hence $|\partial\tau|'$ fixed. Hence (8.2) fails.

(*g*) The need for boundedness of $|dX|$ in the remark following Theorem 7A may be seen as follows. Given any polyhedral 1-chain A in E^2, express it as in (*f*), and set $X \cdot A = \sum a_i$. Then $|X| = 1$, $\mathfrak{L}(X) = 0$; but considering τ as above shows that $|dX|$ is not finite.

(*h*) See the polyhedral chain A_0 in (XI, 17) for further information on the sharp norm.

12. The semi-norms $|A|^\flat$, $|A|^\sharp$ are norms. We prove this with the help of sharp forms.

LEMMA 12a. *For any r-simplex $\sigma \subset E^n$ and any $\zeta > 0$ there is a sharp r-cochain X such that*

$$(1) \qquad |X| = 1, \qquad \mathfrak{L}(X) = 1/\zeta, \qquad X \cdot \sigma = |\sigma|,$$

$$(2) \qquad D_X(p) = 0 \qquad \text{in} \quad E^n - U_\zeta(\sigma).$$

Choose an orthonormal base $e_1, \cdots, e_n$ in E^n such that $e_1, \cdots, e_r$ determines the oriented r-plane of σ. Set $\omega_0 = e^{1\cdots r}$. Then

$$|\omega_0|_0 = 1, \qquad \omega_0 \cdot \{\sigma\} = |\sigma|.$$

Define the real valued function

$$\phi(p) = \phi_{\sigma,\zeta}(p) = \sup \{1 - \text{dist}\,(p, \sigma)/\zeta, 0\}; \tag{3}$$

then $\phi(p) = 1$ on σ and $\phi(p)$ goes to 0 as p moves away from σ. The triangle inequality shows that

$$\mathfrak{L}(\phi) = 1/\zeta. \tag{4}$$

Set $\omega(p) = \phi(p)\omega_0$. Then $|\,\omega\,|_0 = 1$, $\mathfrak{L}_0(\omega) = 1/\zeta$, $\omega(p) = 0$ outside $U_\zeta(p)$, and hence, by Theorem 10A, the corresponding X has the required properties. That $\int_\sigma \omega = \omega_0 \cdot \{\sigma\} = |\,\sigma\,|$ is clear; see (III, 4.1).

LEMMA 12b. *Let $\sigma_1, \cdots, \sigma_m$ be r-simplexes in E^n such that*

$$\text{dist}\,(\sigma_i, \sigma_j) \geqq 2\zeta > 0, \qquad i \neq j. \tag{5}$$

Then there is a sharp r-cochain X such that

$$|\,X\,| = 1, \qquad \mathfrak{L}(X) = 1/\zeta, \qquad X \cdot \sigma_i = |\,\sigma_i\,| \quad (i = 1, \cdots, m). \tag{6}$$

Define X_i for σ_i by the last lemma, and set $X = \sum X_i$.

LEMMA 12c. *For any polyhedral r-chain $A \subset E^n$ and any $\epsilon > 0$ there is a sharp r cochain X such that*

$$|\,X\,| = 1, \qquad X \cdot A \geqq |\,A\,| - \epsilon. \tag{7}$$

REMARK. We will prove this for sharp chains in § 16 below.

We suppose $r > 0$; the proof is simpler if $r = 0$. We can write $A = \sum a_i \sigma_i$ with the σ_i non-overlapping and the $a_i > 0$. Set $\epsilon' = \epsilon/2|\,A\,|$. For each i, let τ_i be a simplex interior to σ_i such that

$$|\,\sigma_i - \tau_i\,| \leqq \epsilon' \,|\,\sigma_i\,|.$$

Then the τ_i are disjoint, and hence, for some $\zeta > 0$, (5) holds for the τ_i. Now if $B = \sum a_i \tau_i$ and X is as in Lemma 12b, then

$$|\,X \cdot A - X \cdot B\,| = \left|\sum a_i X \cdot (\sigma_i - \tau_i)\right| \leqq \sum a_i |\,X\,|\,|\,\sigma_i - \tau_i\,|$$

$$\leqq \sum a_i \epsilon' \,|\,\sigma_i\,| = \epsilon/2,$$

$$|\,A\,| - X \cdot B = \sum a_i (|\,\sigma_i\,| - X \cdot \tau_i) \leqq \sum a_i \epsilon' |\,\sigma_i\,| = \epsilon/2,$$

which gives (7).

THEOREM 12A. *If $A \neq 0$ is polyhedral, then $|\,A\,|^\flat \geqq |\,A\,|^\sharp > 0$.*

Since $A \neq 0$, $|\,A\,| > 0$. Set $\epsilon = |\,A\,|/2$, and choose X by the last lemma. Then $X \cdot A > 0$. Since $|\,X \cdot A\,| \leqq |\,X\,|^\sharp\,|\,A\,|^\sharp$, $|\,A\,|^\sharp > 0$.

13. Weak convergence. Corresponding to the two norms, we have two kinds of limits:

$$(1)\qquad \lim_{i\to\infty}{}^{\flat} A_i = A \quad \text{if} \quad |A - A_i|^{\flat} \to 0,$$

$$(2)\qquad \lim_{i\to\infty}{}^{\sharp} A_i = A \quad \text{if} \quad |A - A_i|^{\sharp} \to 0,$$

with similar definitions of $\lim^{\flat} X_i$, $\lim^{\sharp} X_i$. We also write $A_i \xrightarrow{\flat} A$, etc. We shall use also weak limits: The function X of flat r-chains is the *weak flat limit* of the sequence $X_1, X_2, \cdots$ of flat r-cochains provided

$$(3)\qquad \lim_{i\to\infty} (X_i\cdot A) = X\cdot A, \qquad \text{all flat } A.$$

We write

$$(4)\qquad X = \operatorname{wkl}_{i\to\infty}^{\flat} X_i.$$

(We use, generally, the criterion in the lemma below.) Then (Banach, p. 123) X is bounded and linear, and hence is a flat cochain. Clearly

$$(5)\quad |X| \leqq \liminf |X_i|, |dX| \leqq \liminf |dX_i|, \qquad |X|^{\flat} \leqq \liminf |X_i|^{\flat}.$$

Lemma 13a. *Let $X_1, X_2, \cdots$ be a sequence of flat cochains, and let X be a function of simplexes σ, such that for some N,*

$$(6)\qquad |X_i|^{\flat} \leqq N \quad (\textit{all } i), \qquad \lim (X_i\cdot\sigma) = X\cdot\sigma \quad (\textit{all } \sigma).$$

Then X is uniquely extendable to be a flat cochain, and (4) *holds. Conversely,* (3) *implies* (6).

Clearly X is extendable to be linear over polyhedral chains A, and $|X\cdot A| \leqq N|A|^{\flat}$; hence X is a flat cochain (§ 4). We must prove (3) for any flat A. Given $\epsilon > 0$, choose a polyhedral B such that $|B - A|^{\flat} < \epsilon/4N$, and choose i_0 so that $|(X_i - X)\cdot B| < \epsilon/2$ if $i \geqq i_0$. Now for $i \geqq i_0$,

$$(7)\quad |(X_i - X)\cdot A| \leqq (|X_i|^{\flat} + |X|^{\flat})|B - A|^{\flat} + |(X_i - X)\cdot B| < \epsilon,$$

as required. For the converse, see Banach, p. 123.

Remark. In place of simplexes, we may use any set of chains whose linear combinations are dense in $\mathbf{C}_r^{\flat}$.

Example (*a*). In E^1, the real functions

$$\phi_i(t) = (\sin it)/i, \qquad \phi_i'(t) = d\phi_i(t)/dt = \cos it,$$

define the 0-cochain X_i and 1-cochain Y_i respectively, with $Y_i = dX_i$. Clearly these converge weakly to 0.

Define the *weak flat limit* (12) with the continuous variable η by the criterion of the lemma; then (3) holds, with $\eta \to 0$. Define *weak sharp limits* similarly.

In the dimension 0, the conditions in Lemma 13a for weak convergence, stated in terms of the corresponding real functions D_{X_i}, D_X, become:

$$|D_{X_i}(p)| \leqq N, \qquad \mathfrak{L}(D_{X_i}) \leqq N, \qquad \lim_{i\to\infty} D_{X_i}(p) = D_X(p) \qquad \text{(all } p).$$

We can strengthen the last condition:

LEMMA 13b. *If* wkl$^\flat$ $X_i = X$, *in the dimension* 0, *then* $\lim D_{X_i}(p) = D_X(p)$ *uniformly in compact sets.*

For let Q be compact, and let $\epsilon > 0$ be given. Choose $\delta > 0$ so that $\delta N \leqq \epsilon/3$, and choose $p_1, \cdots, p_m$ in Q so that the sets $U_\delta(p_k)$ cover Q. Choose i_0 so that

$$|D_{X_i}(p_k) - D_X(p_k)| < \epsilon/3, \qquad k = 1, \cdots, m, \qquad \text{if} \quad i \geqq i_0.$$

Now take any $p \in Q$, and any $i \geqq i_0$. Choose k so that $|p - p_k| < \delta$. Then

$$|D_{X_i}(p) - D_X(p)| < \mathfrak{L}(D_{X_i})|p_k - p| + \epsilon/3 + \mathfrak{L}(D_X)|p_k - p| \leqq \epsilon.$$

Given the flat (or sharp) r-cochain X in E^n, we smooth it by an averaging process: With $\kappa_\eta(v)$ as in (App. III, 3), set

$$(8) \qquad X_\eta \cdot A = \int_V \kappa_\eta(v)(X \cdot T_v A)\, dv, \qquad A \text{ polyhedral.}$$

By (3.6), $X \cdot T_v A$ is continuous in v (it is, in fact, for all flat A, by (X, Theorem 7B)); hence this is well defined.

THEOREM 13A. *For any flat r-cochain X in* E^n, X_η $(\eta > 0)$ *is a sharp r-cochain,* D_{X_η} *is* ∞*-smooth, and*

$$(9) \qquad dX_\eta = (dX)_\eta,$$

$$(10) \qquad |X_\eta| \leqq |X|, \qquad |dX_\eta| \leqq |dX|, \qquad |X_\eta|^\flat \leqq |X|^\flat,$$

$$(11) \qquad \mathfrak{L}(X_\eta) \leqq a_\eta \mathfrak{L}(\kappa_\eta)|X|,$$

$$(12) \qquad \operatorname*{wkl}^\flat_{\eta\to 0} X_\eta = X,$$

a_η *being the volume of a ball of radius* η.

REMARK. If X is defined in the open set R only, then X_η is defined in $\operatorname{int}_\eta(R)$; see (VIII, 1(g)).

To prove (9), we have (for A polyhedral)

$$dX_\eta \cdot A = X_\eta \cdot \partial A = \int_V \kappa_\eta(v)(X \cdot T_v \partial A)\, dv$$
$$= \int_V \kappa_\eta(v)(dX \cdot T_v A)\, dv = (dX)_\eta \cdot A.$$

Since $\int_V \kappa_\eta(v)\, dv = 1$, the first part of (10) is immediate, and the rest follows; hence, by Theorem 4A, X_η is a flat cochain. Because of (IX, 5.6) and (App. III, Lemma 3c), D_{X_η} is ∞-smooth (we do not need this fact here).

To prove (11) (which shows that X_η is sharp), take any vector u. We have

$$X_\eta \cdot T_u \sigma = \int_V \kappa_\eta(v)(X \cdot T_{v+u}\sigma)\, dv = \int_V \kappa_\eta(v - u)(X \cdot T_v \sigma)\, dv,$$

$$|X_\eta \cdot (T_u \sigma - \sigma)| \leqq \int_V |\kappa_\eta(v - u) - \kappa_\eta(v)|\, |X|\, |\sigma|\, dv$$

$$\leqq \mathfrak{L}(\kappa_\eta) |u|\, |X|\, |\sigma| a_{\eta+|u|},$$

since the integrand vanishes outside $U_\zeta(0)$, $\zeta = \eta + |u|$. Using (7.4), (11) follows easily (apply this to $\sigma, T_w\sigma, \cdots, T_{(m-1)w}\sigma$, with $w = u/m$, m large).

Take any r-simplex σ. Set $M = |\sigma| + |\partial\sigma|$. By (3.6),

$$|T_v \sigma - \sigma|^\flat \leqq M\zeta \quad \text{if} \quad |v| \leqq \zeta.$$

Using (App. III, 3.2) gives

$$|(X_\eta - X)\cdot\sigma| = \left| \int_V \kappa_\eta(v) X \cdot (T_v \sigma - \sigma)\, dv \right| \leqq |X|^\flat M\zeta \quad \text{if} \quad \eta \leqq \zeta.$$

Hence $\lim_{\eta \to 0} X_\eta \cdot \sigma = X \cdot \sigma$, and (12) follows.

EXAMPLE (b). Take $r = n = 1$. Set $\phi(t) = 0$ if $t < 0$ and $\phi(t) = 1$ if $t > 0$, and set $X \cdot \sigma = \int_\sigma \phi$. Then $|X_\eta - X| \geqq 1/2$ for $\eta > 0$; $\lim^\flat X_\eta = X$ is false, though $\text{wkl}^\flat X_\eta = X$ holds. Note that $\mathfrak{L}(X_\eta) \to \infty$ as $\eta \to 0$.

The following relation is an immediate consequence of the theorem and (4.3):

$$(13) \qquad |A|^\flat = \sup \{|X \cdot A| : \; X \text{ sharp}, \; |X|^\flat = 1\}.$$

Hence (see also (VI, Lemma 8d))

THEOREM 13B. *The flat chains A and B are equal if and only if $X \cdot A = X \cdot B$ for all sharp X.*

14. Some relations between the spaces of chains and cochains.

THEOREM 14A. *$\mathbf{C}^{\sharp r}$ is weakly dense in $\mathbf{C}^{\flat r}$.*

First, every sharp r-cochain is a flat r-cochain; hence $\mathbf{C}^{\sharp r} \subset \mathbf{C}^{\flat r}$. The theorem now follows from Theorem 13A.

REMARK. $\mathbf{C}^{\sharp r}$ is not in general dense in $\mathbf{C}^{\flat r}$; the cochain X of (13, Example (b)) is clearly not the flat limit of any sequence of sharp cochains. The subspace of $\mathbf{C}^{\sharp r}$ consisting of those X with D_X smooth is complete and separable; $\mathbf{C}^{\sharp r}$ is non-separable (§ 18).

THEOREM 14B. *For any flat chain $A = \lim^\flat A_i$ (the A_i polyhedral), set $\Psi A = \lim^\sharp A_i$. Then Ψ is a linear one-one mapping of $\mathbf{C}_r^\flat$ into $\mathbf{C}_r^\sharp$. Hence we may consider any flat chain as being also a sharp chain. $\Psi \mathbf{C}_r^\flat$ is dense in $\mathbf{C}_r^\sharp$, in the sharp norm.*

Since $|A_j - A_i|^\sharp \leqq |A_j - A_i|^\flat$, Ψ exists and is unique; it is clearly linear. Now take any flat $A \neq 0$. Then $|A|^\flat \neq 0$, and by (13.13), we may choose the sharp cochain X so that $X \cdot A > 0$. The definition of Ψ gives $X \cdot \Psi A = X \cdot A$; hence

$$| X |^{\#} | \Psi A |^{\#} \geqq | X \cdot \Psi A | = | X \cdot A | > 0,$$

hence $| \Psi A |^{\#} > 0$, and $\Psi A \neq 0$; therefore Ψ is one-one. Since the set of polyhedral chains is dense in $\mathbf{C}_r^{\#}$, so is $\Psi \mathbf{C}_r^{\flat}$.

15. The ρ-norms. For each $\rho > 0$, define the *flat ρ-norm* and the *sharp ρ-norm* of a polyhedral r-chain A by

$$(1) \qquad | A |_\rho^\flat = \inf \{| A - \partial D | + | D |/\rho\},$$

$$(2) \qquad | A |_\rho^{\#} = \inf \left\{ \frac{\sum | a_i | | \sigma_i | | v_i |}{(r+1)\rho} + \left| \sum a_i T_{v_i} \sigma_i \right|_\rho^\flat ; \quad A = \sum a_i \sigma_i \right\}.$$

One may think of the mass and norms of an r-chain as having the dimension of distance to the rth power, and of ρ as having the dimension of distance. As in §§ 3, 6, we find (at present for polyhedral chains)

$$(3) \qquad | \partial A |_\rho^\flat \leqq | A |_\rho^\flat / \rho,$$

$$(4) \qquad | T_v A - A |_\rho^{\#} \leqq | A | | v | / (r+1)\rho.$$

The corresponding norms of cochains are clearly

$$(5) \qquad | X |_\rho^\flat = \sup \{| X |, \rho | dX |\},$$

$$(6) \qquad | X |_\rho^{\#} = \sup \{| X |, (r+1)\rho \mathfrak{L}(X)\}.$$

Letting $| \ |^{\odot}$ denote either $| \ |^{\flat}$ or $| \ |^{\#}$, the following relations are evident:

$$(7) \qquad | A |_{\rho_1}^{\odot} \leqq | A |_{\rho_2}^{\odot} \leqq (\rho_1/\rho_2) | A |_{\rho_1}^{\odot} \qquad (\rho_2 \leqq \rho_1),$$

$$(8) \qquad (\rho_2/\rho_1) | X |_{\rho_1}^{\odot} \leqq | X |_{\rho_2}^{\odot} \leqq | X |_{\rho_1}^{\odot} \qquad (\rho_2 \leqq \rho_1).$$

For each $\rho > 0$, the space of polyhedral r-chains in the flat ρ-norm has a completion $\mathbf{C}_{\rho,r}^\flat$, etc. The inequalities above lead at once to the following Theorem (compare Theorem 14B):

THEOREM 15A. *The elements of $\mathbf{C}_{\rho,r}^\flat$ are independent of ρ, only the norms differing. For each ρ, the function $\phi_\rho(A) = | A |_\rho^\flat$ of flat r-chains is continuous in each space $\mathbf{C}_{\rho',r}^\flat$. The same is true for $\mathbf{C}_{\rho,r}^{\#}$, $\mathbf{C}_\rho^{\flat r}$, $\mathbf{C}_\rho^{\# r}$.*

Relations (5) and (6) give

THEOREM 15B. *For each flat cochain X there is a $\rho_0 > 0$ such that*

$$(9) \qquad | X |_\rho^\flat = | X | \quad \text{if} \quad \rho \leqq \rho_0.$$

The same is true for the sharp ρ-norm.

We prove a slightly weaker theorem about polyhedral chains. (For general chains, see the next section.)

THEOREM 15C. *For any polyhedral chain A,*

(10) $$\lim_{\rho\to 0} | A |_\rho^\flat = \lim_{\rho\to 0} | A |_\rho^\sharp = | A |.$$

Given $\epsilon > 0$, choose X by Lemma 12c; choose ρ_0 by Theorem 15B, using the sharp ρ_0-norm. Then for $\rho \leqq \rho_0$,

$$| A |_\rho^\flat \geqq | A |_\rho^\sharp = | X |_\rho^\sharp | A |_\rho^\sharp \geqq | X\cdot A | \geqq | A | - \epsilon.$$

Since $| A |_\rho^\flat \leqq | A |$, (10) follows.

16. The mass of chains. As noted in Example (*c*) of § 5, we may not be able to assign a finite mass to a flat or sharp chain. There are various ways of assigning a mass, finite or infinite, as follows:

(1) $$| A |_\flat = \inf \{\liminf_{i\to\infty} | A_i |: \quad A_i \text{ polyhedral}, A_i \xrightarrow{\flat} A\},$$

(2) $$| A |_{(\flat)} = \lim_{\rho\to 0} | A |_\rho^\flat,$$

(3) $$| A |_{[\flat]} = \sup \{| X\cdot A | : X \text{ flat}, | X | \leqq 1\}.$$

There are corresponding definitions of $| A |_\sharp$, $| A |_{(\sharp)}$, $| A |_{[\sharp]}$. All six have meaning for flat A (see Theorem 14B).

THEOREM 16A. *For a polyhedral chain, these all equal $| A |$. For any flat chain, all six definitions are the same. For any sharp chain, the three corresponding definitions are the same.*

We shall let $| A |$ denote any of the masses that are defined.

First, for any flat A and X with $| X | = 1$, there is a $\rho > 0$ such that $| X |_\rho^\flat = | X | = 1$ (Theorem 15B). Now using (15.7) gives

$$| X\cdot A | \leqq | X |_\rho^\flat | A |_\rho^\flat = | A |_\rho^\flat \leqq | A |_{(\flat)};$$

hence $| A |_{[\flat]} \leqq | A |_{(\flat)}$. Next, given $\epsilon > 0$, choose $\rho > 0$ so that $| A |_\rho^\flat > | A |_{(\flat)} - \epsilon/2$ (take $| A |_\rho^\flat > 1/\epsilon$ if $| A |_{(\flat)}$ is infinite). By (4.3) for the ρ-norm, there is a flat cochain X (compare Theorem 15A) such that

$$| X |_\rho^\flat = 1, \qquad X\cdot A > | A |_\rho^\flat - \epsilon/2.$$

Hence $X\cdot A > | A |_{(\flat)} - \epsilon$, and $| A |_{[\flat]} \geqq | A |_{(\flat)}$, proving that these two masses are equal. Similarly $| A |_{[\sharp]} = | A |_{(\sharp)}$ for sharp A.

Clearly $| A |_{[\sharp]} \leqq | A |_{[\flat]}$ for flat chains. To prove the reverse inequality, given the flat chain A and $\epsilon > 0$, choose a flat cochain X such that $| X | = 1$ and $X\cdot A > | A |_{[\flat]} - \epsilon/2$ (take $X\cdot A > 1/\epsilon$ if $| A |_{[\flat]}$ is infinite). By Theorem 13A, (13.10) and (13.12), we may find a sharp cochain Y such that

$$| Y | \leqq | X | = 1, \qquad | Y\cdot A - X\cdot A | < \epsilon/2.$$

Hence $Y\cdot A > | A |_{[\flat]} - \epsilon$, proving $| A |_{[\sharp]} \geqq | A |_{[\flat]}$; thus these are equal.

Suppose $\lim^\flat A_i = A$, the A_i being polyhedral. Given $\epsilon > 0$, choose ρ so that $|A|^\flat_\rho > |A|_{(\flat)} - \epsilon$ (take $|A|^\flat_\rho > 1/\epsilon$ if $|A|_{(\rho)}$ is infinite). Since $|A_i| \geqq |A_i|^\flat_\rho$ and $\lim_{i\to\infty} |A_i|^\flat_\rho = |A|^\flat_\rho$ (Theorem 15A), $\liminf |A_i| > |A|_{(\flat)} - \epsilon$, proving $|A|_{(\flat)} \leqq |A|_\flat$. Similarly $|A|_{(\sharp)} \leqq |A|_\sharp$ for sharp A.

To prove $|A|_\flat \leqq |A|_{(\flat)}$ for flat A, we may suppose $|A|_{(\flat)}$ finite; taking any $\epsilon > 0$, we need merely find a polyhedral chain B such that

$$(4) \qquad |B - A|^\flat < \epsilon, \qquad |B| < |A|_{(\flat)} + \epsilon.$$

Choose $\rho > 0$ such that $\rho(|A|_{(\flat)} + \epsilon/2) < \epsilon/2$. Choose a polyhedral chain A' so that

$$|A' - A|^\flat_\rho < \epsilon/2;$$

then $|A'|^\flat_\rho < |A|^\flat_\rho + \epsilon/2$. Hence there is a polyhedral chain D such that

$$|A' - \partial D| + |D|/\rho < |A|^\flat_\rho + \epsilon/2.$$

Set $B = A' - \partial D$; then the second relation in (4) holds. Also

$$|B - A|^\flat \leqq |A' - A|^\flat + |\partial D|^\flat \leqq |A' - A|^\flat_\rho + |D|$$
$$< \epsilon/2 + \rho(|A|^\flat_\rho + \epsilon/2) < \epsilon,$$

as required.

To prove $|A|_\sharp \leqq |A|_{(\sharp)}$ for sharp A (assuming $|A|_{(\sharp)}$ finite), we prove (4) with $^\sharp$ in place of $^\flat$. Choose ρ so that $2\rho(|A|_{(\sharp)} + \epsilon/2) < \epsilon/2$. Choose a polyhedral A' so that

$$|A' - A|^\sharp_\rho < \epsilon/2;$$

then $|A'|^\sharp_\rho < |A|^\sharp_\rho + \epsilon/2$. Hence we may write

$$A' = \sum a_i\sigma_i, \qquad \frac{\sum |a_i||\sigma_i||v_i|}{(r+1)\rho} + \left|\sum a_i T_{v_i}\sigma_i\right|^\flat_\rho < |A|^\sharp_\rho + \epsilon/2.$$

Choose a polyhedral D so that

$$\left|\sum a_i T_{v_i}\sigma_i - \partial D\right| + |D|/\rho < |A|^\sharp_\rho + \epsilon/2,$$

and set

$$B = \sum a_i T_{v_i}\sigma_i - \partial D.$$

Then $|B| < |A|_{(\sharp)} + \epsilon$, and using (6.3) gives

$$|B - A|^\sharp \leqq |A - A'|^\sharp + \left|\sum a_i(T_{v_i}\sigma_i - \sigma_i)\right|^\sharp + |\partial D|^\sharp$$
$$< \frac{\epsilon}{2} + \frac{\sum |a_i||\sigma_i||v_i|}{r+1} + |D| < \frac{\epsilon}{2} + 2\rho\left(|A|^\sharp_\rho + \frac{\epsilon}{2}\right) < \epsilon,$$

as required.

Finally, for polyhedral A, Theorem 15C shows that the above masses equal $|A|$.

Note that (3) gives

$$(5) \qquad |X \cdot A| \leqq |X||A| \qquad (X \text{ and } A \text{ flat, or } X \text{ and } A \text{ sharp}).$$

THEOREM 16B. *In each space $\mathbf{C}^{\sharp}_{\rho,r}$ or $\mathbf{C}^{\flat}_{\rho,r}$, mass is a lower semi-continuous function.*

By Theorem 15A, $\phi_\eta(A) = |A|^{\sharp}_{\eta}$ is continuous in each space $\mathbf{C}^{\sharp}_{\rho,r}$. Since $\phi_\eta(A)$ increases and $\to |A|$ as $\eta \to 0$, $|A|$ is lower semi-continuous. Since the imbedding of $\mathbf{C}^{\flat}_{\rho,r}$ in $\mathbf{C}^{\sharp}_{\rho,r}$ (Theorem 14B) is continuous, the conclusion is true in $\mathbf{C}^{\flat}_{\rho,r}$ also.

THEOREM 16C. *Let E' be a subspace of E, and let A be a flat chain in the space E'. Then A may be considered as a flat chain in E, and $|A|^{\flat}_{\rho,E'} = |A|^{\flat}_{\rho,E}$, $|A|_{E'} = |A|_E$. The same is true in the sharp norm.*

A polyhedral chain B in E' is also in E, and clearly $|B|^{\flat}_{\rho,E} \leqq |B|^{\flat}_{\rho,E'}$. The opposite inequality follows from (2.4) (compare Theorem 3B). If $\lim^{\flat}_{\rho,E'} A_i = A$ (polyhedral A_i in E'), then $|A_j - A_i|^{\flat}_{\rho,E} \to 0$, and A may be considered as a chain of E; also $|A|^{\flat}_{\rho,E} = \lim |A_i|^{\flat}_{\rho,E} = \lim |A_i|^{\flat}_{\rho,E'} = |A|^{\flat}_{\rho,E'}$. Using (16.2) shows that $|A|_{E'} = |A|_E$. The same proof holds in the sharp norm, using the relation $\pi T_{v_i} \sigma_i = T_{\pi v_i} \pi \sigma_i$.

17. Separability of spaces of chains.

We prove

THEOREM 17A. *The spaces $\mathbf{C}^{\flat}_r$, $\mathbf{C}^{\sharp}_r$ are separable.*

Let $p_1, p_2, \cdots$ be a sequence of points dense in E^n. The set of polyhedral chains $A = \sum a_i \sigma_i$, where the a_i are rational and the vertices of the σ_i are among the p_j, is denumerable. We shall show that it is dense in $\mathbf{C}^{\flat}_r$; hence it is dense also in $\mathbf{C}^{\sharp}_r$.

It is sufficient to show that this set is dense in the set of polyhedral chains; this will follow if we show that for each simplex $\sigma = q_0 \cdots q_r$ and $\epsilon > 0$ there is a simplex $\sigma' = p_{\lambda_0} \cdots p_{\lambda_r}$ such that $|\sigma' - \sigma|^{\flat} < \epsilon$. Set

$$c = \text{diam}(\sigma) + 2, \qquad N = (r+1)c^{r-1}/(r-1)! + c^r/r!.$$

We shall show in fact that

$$(1) \qquad |\sigma' - \sigma|^{\flat} \leqq N\zeta \quad \text{if} \quad |p_{\lambda_i} - q_i| \leqq \zeta \leqq 1 \qquad (\text{all } i).$$

Set

$$\tau_i = q_0 \cdots q_i p_{\lambda_i} \cdots p_{\lambda_r}, \qquad A = \sum (-1)^i \tau_i;$$

then (App. II, 12.4), $\partial A = \sigma' - \sigma - \sum B_i$, where B_i is formed from $\sigma_i = q_0 \cdots \hat{q}_i \cdots q_r$ the same way A is formed from σ. By (III, 1.3),

$$|\tau_i| \leqq |p_{\lambda_i} - q_i| [\text{diam}(\tau_i)]^r/(r+1)! \leqq \zeta c^r/(r+1)!,$$

and hence $|A| \leqq \zeta c^r/r!$. Similarly $|B_i| \leqq \zeta c^{r-1}/(r-1)!$. Therefore

$$|(\sigma' - \sigma) - \partial A| + |A| \leqq \zeta[(r+1)c^{r-1}/(r-1)! + c^r/r!] = N\zeta,$$

which gives (1), and completes the proof of separability.

18. Non-separability of spaces of cochains. We shall show that though $\mathbf{C}_r^\flat$ and $\mathbf{C}_r^\sharp$ are separable, their conjugate spaces are not; hence these spaces are not reflexive (App. I, 14). Note that for $r = n = 1$, Theorems 18D and 10A show that the space of functions ϕ with $|\phi|_0$ and $\mathfrak{L}_\phi$ finite, which vanish outside an interval I, with the norm $|\phi|^\sharp$, is non-separable. (Make use of (VIII, 1(f)), with an interval R, $\bar{R} \subset I$.) However, the subspace of functions which are smooth is easily seen to be separable.

We prove first two general theorems. In the first theorem, we shall consider a normed linear space of elements Z, each of which has a "support" $[Z]$, which is in E^n, such that the following holds:

$$(1) \qquad [aZ] \subset [Z], \qquad [Z + Z'] \subset [Z] \cup [Z'].$$

THEOREM 18A. *Let n be $\geqq 1$, and let $\mathbf{C}$ be a normed linear space, with supports satisfying* (1), *and with the following properties*:

(P_1) *To each open set $R \subset E^n$ corresponds some $Z \neq 0$ with $[Z] \subset R$.*

(P_2) *If $[Z_1]$ and $[Z_2]$ lie in disjoint cubes, then $|Z_1 + Z_2| \geqq |Z_1|$.*

(P_3) *If $[Z_1], [Z_2], \cdots$ lie in disjoint cubes, the union of these cubes is bounded, and there is an N such that $|Z_i| \leqq N$ for all i, then $\sum_{i=1}^{\infty} Z_i$ exists in $\mathbf{C}$.*

Then $\mathbf{C}$ is non-separable.

Let $Q_1, Q_2, \cdots$ be a sequence of disjoint cubes with centers on a line in E^n, and lying in a bounded set. By (P_1), we may choose Z_i so that $|Z_i| = 1$ and $[Z_i] \subset Q_i$. For each sequence of numbers $\mathbf{a} = (a_1, a_2, \cdots)$, set

$$Z_{\mathbf{a}}^k = a_1 Z_1 + \cdots + a_k Z_k, \qquad Z_{\mathbf{a}} = \lim_{k \to \infty} Z_{\mathbf{a}}^k;$$

by (P_3), $Z_{\mathbf{a}}$ exists if the a_i are bounded. The $Z_{\mathbf{a}}$ with each $a_i = \pm 1$ are non-denumerable in number; if we show that $|Z_{\mathbf{b}} - Z_{\mathbf{a}}| \geqq 2$ if $\mathbf{b} \neq \mathbf{a}$ (using only ± 1), the conclusion of the theorem will follow.

Set $c_i = b_i - a_i$; since $Z_{\mathbf{c}}^k = Z_{\mathbf{b}}^k - Z_{\mathbf{a}}^k$, and all three limits exist as $k \to \infty$, we have $Z_{\mathbf{c}} = Z_{\mathbf{b}} - Z_{\mathbf{a}}$. Let j be the first integer such that $c_j \neq 0$; then $|c_j| = 2$. We may enclose $Q_{j+1} \cup Q_{j+2} \cup \cdots$ in a cube Q_j' disjoint from Q_j. Since $[c_{j+1}Z_{j+1} + \cdots + c_k Z_k] \subset Q_j'$, ($P_2$) gives

$$|Z_{\mathbf{c}}^k| = |c_j Z_j + (c_{j+1}Z_{j+1} + \cdots + c_k Z_k)| \geqq |c_j Z_j| = 2.$$

Letting $k \to \infty$ proves the statement.

THEOREM 18B. *Let* $\mathbf{C}$ *be the linear space of polyhedral r-chains in* E^n $(n \geqq 1,\ 0 \leqq r \leqq n)$, *with a norm* $|A|'$. *Let* $|Z|$ *denote the norm in the conjugate space* $\bar{\mathbf{C}}$ *of* $\mathbf{C}$. *Suppose*:

(P_4) *For each* ∞-*smooth r-form* ω *in* E^n *with bounded support, the linear function* $\phi(A) = \int_A \omega$ *of polyhedral chains is bounded in the above norm, and hence defines an element* Z_ω *of* $\bar{\mathbf{C}}$.

(P_5) *For each* ω *as above*, $|Z_\omega| \geqq |\omega|_0$.

(P_6) *For* ω_1 *and* ω_2 *as above, if their supports lie in disjoint cubes, then*

$$|Z_{\omega_1} + Z_{\omega_2}| = \sup\{|Z_{\omega_1}|, |Z_{\omega_2}|\}. \tag{2}$$

Then $\bar{\mathbf{C}}$ *is non-separable.*

Let $\bar{\mathbf{C}}'$ be the subspace of $\bar{\mathbf{C}}$ consisting of all Z_ω, ω as above. Set $[Z_\omega] = \text{spt}(\omega)$; then (1) holds. We shall prove (P_1) and (P_2) in $\bar{\mathbf{C}}'$, and shall show that (P_3) holds if $Z_i = Z_{\omega_i}$ and the cubes are as in the proof of Theorem 18A. Then that proof gives the present theorem.

Given the open set R, we may choose an ∞-smooth $\omega \neq 0$ with $\text{spt}(\omega) \subset R$. Then $[Z_\omega] \subset R$. Also, for some σ, $\int_\sigma \omega \neq 0$; hence $Z_\omega \cdot \sigma \neq 0$, and $Z_\omega \neq 0$. Property (P_2) follows from (P_6).

Now take cubes $Q_1, Q_2, \cdots$ as described, and suppose

$$\omega_i \text{ is } \infty\text{-smooth}, \qquad \text{spt}(\omega_i) \subset Q_i, \qquad |Z_{\omega_i}| \leqq N.$$

Set

$$Z'_k = Z_{\omega_1} + \cdots + Z_{\omega_k};$$

we must show that $\lim Z'_k$ exists in $\bar{\mathbf{C}}$.

First we show that $\psi(\sigma) = \lim (Z'_k \cdot \sigma)$ exists for each r-simplex σ. Let τ_i be the part of σ in Q_i. Using (P_5), we find

$$\begin{aligned} |(Z_{\omega_{j+1}} + \cdots + Z_{\omega_k}) \cdot \sigma| &= \left| \int_{\tau_{j+1}} \omega_{j+1} + \cdots + \int_{\tau_k} \omega_k \right| \\ &\leqq \sum_{i=j+1}^{k} |\omega_i|_0 |\tau_i| \leqq \sum_{i=j+1}^{k} |Z_{\omega_i}| |\tau_i| \leqq N \sum_{i=j+1}^{k} |\tau_i|; \end{aligned}$$

given $\epsilon > 0$, we can choose j_0 so that the last sum is $< \epsilon/N$ if $j_0 \leqq j < k$; this shows that $\psi(\sigma)$ exists.

Now $\psi(\sum a_i \sigma_i) = \sum a_i \psi(\sigma_i)$ defines the linear function $\psi(A)$ of polyhedral chains uniquely, as is easily seen, and $\psi(A) = \lim (Z'_k \cdot A)$. To complete the proof, we need merely show that this function is bounded. We show in fact that $|\psi(A)| \leqq N |A|'$. Take any k. Then repeated use of (P_6) gives (using the cubes Q'_j as in the proof of Theorem 18A)

$$\begin{aligned} |Z'_k \cdot A| &= |(Z_{\omega_1} + \cdots + Z_{\omega_k}) \cdot A| \leqq |Z_{\omega_1} + \cdots + Z_{\omega_k}| |A|' \\ &= \sup\{|Z_{\omega_1}|, \cdots, |Z_{\omega_k}|\} |A|' \leqq N |A|'. \end{aligned}$$

Letting $k \to \infty$ gives the desired inequality.

The following properties are of interest:

(P_7) $|A|' \leqq |A|$ for all polyhedral A. (This follows at once from: $|\sigma|' \leqq |\sigma|$ for all σ.)

(P_8) If spt $(\omega) \subset Q$ (Q a cube), then for each $\epsilon > 0$ there is a polyhedral chain $A \subset Q$ such that $|A|' = 1$, $\int_A \omega > |Z_\omega| - \epsilon$.

(P_9) If Q is an n-cube, and π is a projection of E^n onto Q, defined for instance as in Lemma 2b, then $|\pi A|' \leqq |A|'$ for all polyhedral A.

THEOREM 18C. *In the last theorem, we can replace* (P_5) *by* (P_7). *We can replace* (P_6) *by the weakened hypothesis* (P_6'), *defined by using* $\leqq$ *in* (2), *and either of* (P_8), (P_9).

First, assuming (P_7), we prove (P_5). Take any $\epsilon > 0$. Since ω is continuous, the definition of $|\omega|_0$ shows that we may choose a simplex σ such that $\int_\sigma \omega \geqq (|\omega|_0 - \epsilon)|\sigma|$. Using ($P_7$) gives

$$|Z_\omega \cdot \sigma| = \left|\int_\sigma \omega\right| \geqq (|\omega|_0 - \epsilon)|\sigma|',$$

proving $|Z_\omega| \geqq |\omega|_0 - \epsilon$ and hence (P_5).

Next, using (P_8), we prove $\geqq$ in (2). We need merely prove $|Z_{\omega_1} + Z_{\omega_2}| \geqq |Z_{\omega_1}|$. Say spt $(\omega_i) \subset Q_i$. Given $\epsilon > 0$, choose A as in (P_8), using ω_1, Q_1. Then

$$(Z_{\omega_1} + Z_{\omega_2}) \cdot A = \int_A \omega_1 + \int_A \omega_2 = \int_A \omega_1 > |Z_{\omega_1}| - \epsilon,$$

giving $|Z_{\omega_1} + Z_{\omega_2}| \geqq |Z_{\omega_1}| - \epsilon$ and hence $\geqq |Z_{\omega_1}|$.

Finally, (P_8) follows from (P_9). For the definition of $|Z_\omega|$ shows that for some polyhedral B, $|B|' = 1$ and $\int_B \omega > |Z_\omega| - \epsilon$. Set $C = \pi B$. Then $\int_C \omega = \int_B \omega$. We may suppose $|Z_\omega| - \epsilon > 0$; then $C \neq 0$, and we may set $A = C/|C|'$. Now $|A|' = 1$, and since $|C|' \leqq |B|' = 1$, we have $Z_\omega \cdot A \geq Z_\omega \cdot C > |Z_\omega| - \epsilon$, as required.

It seems not easy to find conditions on the norm $|A|'$ from which $\leqq$ in (2) may be proved. In this connection, the following example is of interest.

EXAMPLE. Take $r = n = 1$; let $p_0 = 0$. Let $|A|'$ be the least norm satisfying (8.1) and (8.2), and also (8.3) in case σ and $T_v\sigma$ lie on opposite sides of p_0. We leave this for the reader to study.

THEOREM 18D. *Suppose* $0 \leqq r \leqq n$, $n \geqq 1$. *Then* $\mathbf{C}^{\flat r}$ *and* $\mathbf{C}^{\sharp r}$ *are non-separable. The same holds for* $\mathbf{C}^{\flat r}(R)$ *and* $\mathbf{C}^{\sharp r}(R)$, *for any open set* R (see (VIII, 1)).

This follows from the theorems above, for instance as follows. Property (P_4) is clear in both cases, since ω is sharp (use Theorem 10A and (7.1)). Also (P_7) holds. (P_6) for $\sharp$ is clear from Theorem 10A and

(10.2). (It holds for $\flat$ also, by Chapter IX.) Since (P_9) holds for $\flat$, by Theorem 3B, there remains to prove $\leqq$ in (2), for $\flat$.

Given spt $(\omega_i) \subset Q_i$ $(i = 1, 2)$, $Z_{\omega_i} = Z_i$, a polyhedral chain A, and $\epsilon > 0$, it suffices to prove

$$|(Z_1 + Z_2)\cdot A| \leqq \sup\{|Z_1|^\flat, |Z_2|^\flat\}(|A|^\flat + \epsilon).$$

We may choose C and D so that

$$C = A - \partial D, \qquad |C| + |D| < |A|^\flat + \epsilon.$$

Let C_1 and D_1 be the parts of C and D respectively in Q_1, and set $C_2 = C - C_1$, $D_2 = D - D_1$. Then

$$|C| = |C_1| + |C_2|, \qquad |D| = |D_1| + |D_2|.$$

Now

$$|(Z_1 + Z_2)\cdot A| \leqq |Z_1\cdot C_1| + |dZ_1\cdot D_1| + |Z_2\cdot C_2| + |dZ_2\cdot D_2|$$
$$\leqq \sup\{|Z_1|, |dZ_1|, |Z_2|, |dZ_2|\}(|C_1| + |D_1| + |C_2| + |D_2|),$$

which, with (4.8), gives the required inequality.

VI. Some Relations between Chains and Functions

In the last chapter, norms were introduced into the linear space of polyhedral r-chains in E^n; completing the space gave a Banach space. The new chains thus formed (flat or sharp) have in general no obvious geometric or analytic representation. Chains of finite mass, however, may be represented by additive set functions; see Chapter XI. In this chapter we study some special representations of certain chains. Section 7 is the only section used directly in the rest of the book.

The first six sections are devoted to the study of chains on the real line; such chains are of dimension 0 or 1. In § 1, we show how a real function ϕ may correspond to an r-chain A, both for $r = 0$ and for $r = 1$. Suppose ϕ is continuous, and vanishes outside the interval (a, b). Then an approximating polyhedral chain A' to the chain A may be determined as follows. Take a fine subdivision of (a, b), by the points $t_0 = a, t_1, \cdots, t_m = b$; let I_k denote the interval (t_{k-1}, t_k). Then, letting $\bar{t}$ denote the point t, considered as a 0-simplex,

$$A' = \sum_{k=1}^{n} a_k \bar{t}_{k-1}, \qquad a_k = \int_{t_{k-1}}^{t_k} \phi(x)\, dx, \quad \text{if} \quad r = 0,$$

$$A' = \sum_{k=1}^{n} b_k I_k, \qquad b_k = \phi(t_{k-1}), \quad \text{if} \quad r = 1.$$

In §§ 2–5, we discuss a different correspondence between 0-chains and real functions. We show that there is a one-one correspondence between 0-chains of finite mass and normalized functions of bounded variation. If we replace the function of bounded variation γ by the corresponding additive set function β, defined by $\beta(I_{ab}) = \gamma(b) - \gamma(a)$ (I_{ab} = interval (a, b)), we have the special case $n = 1$, $r = 0$ of the more general theory in Chapter XI. We give further relations with standard theorems of analysis in § 6.

In § 7 we show how a cellwise continuous summable function $\alpha(p)$ in E^n whose values are r-vectors defines a flat r-chain A. Chains A of this sort are dense in the space of chains (sharp or flat); hence norms of cochains X are determined by their values $X \cdot A$ on these "continuous chains". Using these in place of polyhedral chains gives an analytic rather than a geometric background to the theory; this is of importance in various ways. More general chains will be used in (IX, 15) and (XI, 14).

After some lemmas in § 8, we show in § 9 that if α is smooth, then the boundary of the corresponding chain A is given in terms of the partial derivatives of α; this is the "dual" of the operation of exterior differentiation of forms. A brief discussion of integration in smooth manifolds (where polyhedral chains are not defined) is presented in § 10. This relates the exposition to methods such as are used in the theory of harmonic integrals (see de Rham).

1. Continuous chains on the real line. Recall (V, Theorem 7B) that in E^1, both sharp 0-cochains and sharp 1-cochains correspond to real-valued sharp functions in a one-one manner. We shall show that a cellwise continuous summable function ϕ (App. III) corresponds both to a sharp 0-chain and to a sharp 1-chain. The converse is not true; see the end of this section for a summary of facts. We use here a particular E^1, namely $\mathfrak{A}$ itself.

For both $r = 0$ and $r = 1$, we say the cellwise continuous summable function ϕ *corresponds* to the sharp or flat r-chain A if

$$\int_{E^1} \bar{D}_X(t)\phi(t)\, dt = X \cdot A \tag{1}$$

for all sharp cochains X of the same dimension. (Define $D_X(t)$ as in (V, 10).) This corresponds to a real-valued function $\bar{D}_X(t)$, as in (V, 10.14), for $r = 1$; for $r = 0$, $\bar{D}_X = D_X$.) Clearly A is unique; we denote it by $\tilde{\phi}$, and call it a *continuous chain*.

For $r = 1$, given ϕ, we define approximating polyhedral chains A_1, A_2, $\cdots$ as follows. Given the integer k, choose an interval Q_k in E^1 so that

$$\int_{E^1 - Q_k} \langle \phi \rangle < 1/2^{k+1}.$$

Express Q_k as the union of a set of non-overlapping subintervals $Q_{k1}, \cdots, Q_{km_k}$, so that

$$|\phi(t') - \phi(t)| < 1/2^{k+1}|Q_k|, \qquad t, t' \text{ in the same int } (Q_{ki}).$$

We may suppose $Q_{k-1} \subset Q_k$ $(k > 1)$, and that each interval of Q_{k-1} is the union of intervals of Q_k. Say $Q_{ki} = (t_{k,i-1}, t_{ki})$. Let a_{ki} be a number nearest 0 which is a limit of values $\phi(t)$ with $t \in \text{int}\,(Q_{ki})$; set

$$A_k = \sum_i a_{ki} Q_{ki}. \tag{2}$$

With the Q_{ki} oriented like E^1, this is a polyhedral 1-chain.

We now prove that $A = \lim^\flat A_k$ exists, and hence is a flat (hence also a sharp) chain. Take any k, and any $l > k$. For some set of indices j, the interval Q_{ki} is cut into intervals Q_{lj}. We have the inequality on mass

$$\left|\sum_j a_{lj}Q_{lj} - a_{ki}Q_{ki}\right| = \left|\sum_j (a_{lj} - a_{ki})Q_{lj}\right| < |Q_{ki}|/2^{k+1}|Q_k|.$$

Let $\sum' a_{lj}Q_{lj}$ denote the part of A_l outside of Q_k. By the choice of Q_k and the a_{ki},

$$\left|\sum{}' a_{lj}Q_{lj}\right| = \sum{}' |a_{lj}||Q_{lj}| \leqq \int_{Q_l - Q_k} \langle\phi\rangle < 1/2^{k+1}.$$

Combining these inequalities gives

$$|A_l - A_k|^\flat = |A_l - A_k| < 1/2^k, \qquad l > k, \tag{3}$$

proving the existence of $A = \lim^\flat A_k$.

Take any 1-cochain X. Since

$$\left|\int_{Q_{ki}} \bar{D}_X\phi - X\cdot a_{ki}Q_{ki}\right| = \left|\int_{Q_{ki}} \bar{D}_X(t)[\phi(t) - a_{ki}]\,dt\right|$$
$$\leqq |X||Q_{ki}|/2^{k+1}|Q_k|,$$

we find at once

$$\left|\int_{E^1} \bar{D}_X\phi - X\cdot A_k\right| \leqq |X|/2^k;$$

letting $k \to \infty$ gives (1). Hence $A = \tilde{\phi}$.

THEOREM 1A. *The mapping $\phi \to \tilde{\phi}$ is a one-one linear mapping of cellwise continuous summable functions into flat (and hence sharp) 1-chains. We have*

$$|\tilde{\phi}| = \int \langle\phi\rangle, \tag{4}$$

and for the approximating polyhedral chains A_k above,

$$|\tilde{\phi} - A_k| \leqq 1/2^k, \qquad |A_k| \leqq |\tilde{\phi}|. \tag{5}$$

The 1-chains $\tilde{\phi}$ with ϕ continuous are dense in $\mathbf{C}_1^\flat$ and hence in $\mathbf{C}_1^\sharp$.

Clearly the mapping is one-one and linear. Since

$$|A_k| = \sum_i |a_{ki}||Q_{ki}| \leqq \int \langle\phi\rangle,$$

we have $|\tilde{\phi}| \leqq \int \langle\phi\rangle$ (V, Theorem 16B). (We could prove this also by using a sharp X such that $|X| \leqq 1$ and $X\cdot\tilde{\phi} > |\tilde{\phi}| - \epsilon$.) To prove the reverse inequality, take any sharp X such that $|X| \leqq 1$. Then

$$|X\cdot\tilde{\phi}| = \left|\int_{E^1} \bar{D}_X\phi\right| \leqq \int_{E^1} \langle\phi\rangle,$$

which gives the result. The inequalities (5) are clear.

To show that the $\tilde{\phi}$ are dense, we need merely find one arbitrarily near a 1-simplex $\sigma = (t_1, t_2)$. Let $\phi(t) = 0$ for $t \leqq t_1 - \epsilon$ and $t \geqq t_2 + \epsilon$, let $\phi(t) = 1$ for $t_1 \leqq t \leqq t_2$, and let ϕ be linear in the two remaining intervals. Comparing $\int \bar{D}_X\phi$ with $X\cdot\sigma$ for any sharp X shows that $\tilde{\phi}$ is as close to σ as desired (see (V, 10.3)).

We now show that the boundary of a smooth 1-chain is found simply by differentiating:

THEOREM 1B. *If ϕ is smooth and summable, and $\psi = d\phi/dt$ is summable, then*

$$\partial\tilde{\phi} = -\tilde{\psi}. \tag{6}$$

First we prove

$$\lim_{t\to\pm\infty} \phi(t) = 0. \tag{7}$$

Given $\epsilon > 0$, choose b so that $\int_b^\infty \langle\psi\rangle < \epsilon$. Then for $b' > b$,

$$|\phi(b') - \phi(b)| = \left|\int_b^{b'} \psi(t)\, dt\right| \leqq \int_b^{b'} \langle\psi\rangle < \epsilon;$$

hence $\lim_{t\to\infty} \phi(t)$ exists. Since ϕ is summable, the limit is 0; similarly at $-\infty$.

Now let X be any 0-cochain with smooth $D_X = w$. Then, using (V, 10.16),

$$\int_{-\infty}^{\infty} D_X\psi = \lim_{-a,b\to\infty} \int_a^b w\frac{d\phi}{dt}$$

$$= \lim_{-a,b\to\infty} \left[w(b)\phi(b) - w(a)\phi(a) - \int_a^b \frac{dw}{dt}\phi\right]$$

$$= -\lim_{-a,b\to\infty} \int_a^b \bar{D}_{dX}\phi = -dX\cdot\tilde{\phi} = X\cdot(-\partial\tilde{\phi}),$$

which proves (6).

A flat 0-chain need not correspond to a summable function if its mass is not finite. There may be a function simply related to the chain, but the chain need not be uniquely determined by the function. For example, consider first the function

$$\phi_a(t) = \begin{cases} \log t + a, & 0 < t < 1, \\ \log(-t), & -1 < t < 0, \end{cases}$$

with $\phi_a(t) = 0$ elsewhere. This is summable; even though it is not cellwise continuous, it clearly determines uniquely a flat 1-chain $A_a = \tilde{\phi}_a$. Since

$$d\phi_a/dt = 1/t, \qquad 0 < |t| < 1,$$

we may think of the function $1/t$ $(0 < |t| < 1)$ as corresponding to any of the 0-chains $-\partial\tilde{\phi}_a$. We may define $B_a = -\partial\tilde{\phi}_a$ as the limit of a sequence of 0-chains $B_{ai} = \tilde{\psi}_{ai}$, each corresponding to a function ψ_{ai} which equals

$1/t$ in $(-1, 1)$, except in an interval Q_{ai} about 0, where $\psi_{ai} = 0$; but the intervals depend on a. For $a = 0$, we may take the intervals with center at 0. Compare (V, 5(d)).

There is an extension of Theorem 1B to the case that ψ is Denjoy integrable, convergent from $-\infty$ to ∞; compare Saks, Chapter VIII, in particular, the theorem on p. 246.

We shall show in (IX, Theorem 15B) that flat 1-chains in E^1 (in fact, flat n-chains in E^n) correspond in a one-one manner to measurable summable functions. Recall that the flat norm equals the mass in this case.

We show in the sections below that 0-chains (for which the flat and sharp norms are the same) of finite mass in E^1 correspond in a one-one manner to normalized functions of bounded variation (equivalently, to additive set functions; see Chapter XI).

In Chapter XI we show that sharp r-chains of finite mass in E^n correspond in a one-one manner to additive set functions whose values are r-vectors; flat r-chains of finite mass correspond to a subset of these.

WARNING: For $E^n \subset E^m$ $(m > n)$, there can be sharp r-chains $\neq 0$ in E^n $(r > n)$; for instance sharp r-chains "at a point"; see (VII, 7) below. This cannot happen in the flat norm; see (VII, 9).

2. 0-chains in E^1 defined by functions of bounded variation. Our object in the next few sections is to show that there is a one-one correspondence between 0-chains (for which the flat and sharp norms are the same) and normalized functions of bounded variation on the real line $E^1 = \mathfrak{A}$; see Theorem 5A below.

We recall some facts about functions of bounded variation. Let $\gamma(t)$ be defined for all real t. Given a set of points $t_0 < t_1 < \cdots < t_m$, consider the expression

$$\sum_{i=1}^{m} | \gamma(t_i) - \gamma(t_{i-1}) | ;$$

the *variation* $|\gamma|$ of γ is the least upper bound of all such sums. The function γ is of *bounded variation* if $|\gamma|$ is finite. In this case, γ is continuous except at an at most denumerable set of points; the right and left hand limits exist at these points, and the sum of the jumps is $\leqq |\gamma|$.

We shall say γ is *normalized* if

$$(1) \qquad \gamma(-\infty) = \lim_{t\to-\infty} \gamma(t) = 0, \qquad \lim_{\lambda\to 0+} \gamma(t-\lambda) = \gamma(t) \qquad (\text{all } t).$$

If ϕ is bounded and continuous, then there is a uniquely defined number $\int_{ba} \phi \, d\gamma$, the integral of ϕ with respect to γ, with the following property.

For every $\epsilon > 0$ there is a $\zeta > 0$ such that, if $a = t_0 < t_1 < \cdots < t_m = b$, and $t_i - t_{i-1} < \zeta$ and $t_{i-1} \leqq t'_i \leqq t_i$ for each i, then

$$\left| \int_a^b \phi \, d\gamma - \sum_{i=1}^{m} \phi(t'_i)[\gamma(t_i) - \gamma(t_{i-1})] \right| < \epsilon.$$

For $\int_{-\infty}^b \phi \, d\gamma$, we may either take $\lim_{a\to-\infty} \int_a^b \phi \, d\gamma$, or use the above definition with any $t_0 \leqq a'$ for $-a'$ large enough; similarly for $\int_a^\infty$ and $\int_{-\infty}^\infty$. A different normalization of γ would not affect the integral.

We say the function γ of bounded variation *corresponds* to the 0-chain A if (in contrast to the definition in § 1)

$$\int_{-\infty}^{\infty} D_X(t) \, d\gamma(t) = X \cdot A \tag{2}$$

for all 0-cochains X. If $\phi(t) = d\gamma(t)/dt$ exists and is continuous, then ϕ corresponds to A, as in (1.1). Clearly A is unique.

To prove the existence of A, let $\mathfrak{S}_k$ be the subdivision of the interval $(-2^k, 2^k)$ into equal intervals of length $1/2^k$. For any number t, let $\bar{t}$ denote the point t, considered as a 0-chain. (Thus $t_2 - t_1$ is a number, $\bar{t}_2 - \bar{t}_1$ is a 0-chain.) Set

$$\left\{ \begin{array}{ll} A_k = \sum_{i=1}^{s_k} a_{ki}\bar{t}_{ki}, & t_{ki} = -2^k + i/2^k, \\ a_{ki} = \gamma(t_{ki}) - \gamma(t_{k,i-1}), & s_k = 2^{2k+1}. \end{array} \right. \tag{3}$$

If $\phi(t)$ exists, then (in contrast to (1.2), the case $r = 1$)

$$A_k = \sum_i a_{ki}\bar{t}_{ki}, \qquad a_{ki} = \int_{t_{k,i-1}}^{t_{ki}} d\gamma = \int_{t_{k,i-1}}^{t_{ki}} \phi(t) \, dt. \tag{4}$$

Let us denote by $\bar{\gamma}(a, b)$ the variation of γ from a to b, i.e. the least upper bound of the sums defining $|\gamma|$, except that we require $a \leqq t_0$, $t_m \leqq b$. We prove

$$|A_l - A_k|^\flat \leqq \bar{\gamma}(-\infty, -2^k) + |\gamma|/2^k + \bar{\gamma}(2^k, \infty), \qquad l > k. \tag{5}$$

Write $A_l = A_l^1 + A_l^2 + A_l^3$, corresponding to the parts of A_l to the left of $Q = (-2^k, 2^k)$, the parts in Q, and the parts to the right of Q, respectively. Clearly

$$|A_l^1| \leqq \bar{\gamma}(-\infty, -2^k), \qquad |A_l^3| \leqq \bar{\gamma}(2^k, \infty).$$

Take a typical interval $(t_{k,i-1}, t_{k,i})$ of $\mathfrak{S}_k$, divided into intervals $(t_{l,j-1}, t_{l,j})$ of $\mathfrak{S}_l$. Letting $\sum_j^{(i)}$ denote the sum over such values of j, define the 1-chains

$$D_{kli} = \sum_j{}^{(i)} [\gamma(t_{l,j}) - \gamma(t_{l,j-1})]\bar{t}_{lj}\bar{t}_{ki}, \qquad D_{kl} = \sum_i D_{kli}.$$

Then

$$\partial D_{kli} = [\gamma(t_{k,i}) - \gamma(t_{k,i-1})]\bar{t}_{ki} - \sum_j{}^{(i)}[\gamma(t_{l,j}) - \gamma(t_{l,j-1})]\bar{t}_{lj},$$

$$\partial D_{kl} = A_k - A_l^2.$$

Since

$$| D_{kl} | \leqq \sum_i \sum_j{}^{(i)} | \gamma(t_{l,j}) - \gamma(t_{l,j-1}) | /2^k \leqq | \gamma | /2^k,$$

we find that $| A_k - A_l - \partial D_{kl} | + | D_{kl} |$ is bounded by the right hand side of (5); hence (5) follows.

Because of (5), $A = \lim^\flat A_k$ exists. For any X,

(6) $$X\cdot A_k = \sum_i a_{ki}(X\cdot \bar{t}_{ki}) = \sum_i D_X(t_{ki})a_{ki};$$

by definition of the integral,

$$X\cdot A = \lim_{k\to\infty} X\cdot A_k = \int_{-\infty}^{\infty} D_X \, d\gamma;$$

hence γ corresponds to A.

Let $\gamma^\odot$ denote the 0-chain corresponding to γ. We prove

(7) $$| \gamma^\odot | = | \gamma | .$$

If $\phi = d\gamma/dt$ exists and is continuous, and $\tilde{\phi} = \gamma^\odot$ denotes the corresponding 0 chain, (7) gives (as in (1.4) for 1-chains)

(8) $$| \tilde{\phi} | = \int_{-\infty}^{\infty} | \phi(t) | \, dt.$$

Since, by (6),

$$| X\cdot A_k | \leqq | X | \sum_i | a_{ki} | \leqq | X | | \gamma | ,$$

we have $| \gamma^\odot | \leqq | \gamma |$; see (V, 16.1).

To prove the reverse inequality, we note first that because of the normalization (1),

$$\lim_{\lambda\to 0+} \int_{a-\lambda}^{b-\lambda} d\gamma = \int_a^b d\gamma = \gamma(b) - \gamma(a).$$

Suppose $\xi_\lambda(t) = 1$ for $a \leqq t \leqq b - 2\lambda$, $\xi_\lambda(t) = 0$ for $t \leqq a - \lambda$ and for $t \geqq b - \lambda$, and ξ_λ is linear in the two remaining intervals. Then it is clear from the above that

$$\lim_{\lambda\to 0+} \int_{-\infty}^{\infty} \xi_\lambda(t) \, d\gamma(t) = \gamma(b) - \gamma(a).$$

Now given $\epsilon > 0$, choose points $t_0 < t_1 < \cdots < t_m$ such that

$$\sum_{i=1}^{m} | \gamma(t_i) - \gamma(t_{i-1}) | > | \gamma | - \epsilon.$$

For small enough λ, define ξ_λ as above in each interval $(t_{i-1} - \lambda, t_i - \lambda)$ if $\gamma(t_i) - \gamma(t_{i-1}) \geqq 0$, but with the minus sign otherwise; set $\xi_\lambda = 0$ elsewhere. This defines a 0-cochain X_λ, with $|X_\lambda| = 1$. Then

$$\lim_{\lambda\to 0+} X_\lambda \cdot \gamma^{\odot} = \sum_{i=1}^{m} |\gamma(t_i) - \gamma(t_{i-1})|,$$

and it follows that $|\gamma^{\odot}| > |\gamma| - \epsilon$, and hence $\geqq |\gamma|$. This completes the proof of (7).

3. Sharp functions times 0-chains. We shall define a chain ϕA for any sharp function ϕ (V, 4) and any 0-chain A in E^n; for the case of r-chains, see (VII, 1). The following inequalities will hold:

$$|\phi A| \leqq |\phi|\,|A|, \tag{1}$$

$$|\phi A|^\flat \leqq (|\phi| + \mathfrak{L}_\phi)|A|^\flat. \tag{2}$$

The operation will be bilinear.

First, for any polyhedral chain $\sum a_i p_i$, set

$$\phi \sum a_i p_i = \sum \phi(p_i) a_i p_i; \tag{3}$$

clearly (1) holds. To prove (2) for $A = \sum a_i p_i$, given $\epsilon > 0$, choose D so that

$$|A - \partial D| + |D| < |A|^\flat + \epsilon.$$

Say $D = \sum d_j(p_j' p_j'')$, the $p_j' p_j''$ non-overlapping. Set

$$D' = \sum \phi(p_j'')\, d_j(p_j' p_j'').$$

Now

$$A - \partial D = \sum a_i p_i - \sum d_j p_j'' + \sum d_j p_j',$$

$$\begin{aligned}\phi A - \partial D' = \sum \phi(p_i) a_i p_i - \sum \phi(p_j'')\, d_j p_j'' + \sum \phi(p_j')\, d_j p_j' \\ + \sum [\phi(p_j'') - \phi(p_j')]\, d_j p_j'.\end{aligned}$$

Part of the last expression is simply $\phi(A - \partial D)$. Hence, using (1),

$$\begin{aligned}|\phi A - \partial D'| + |D'| &\leqq |\phi|\,|A - \partial D| + \mathfrak{L}_\phi |D| + |\phi|\,|D| \\ &\leqq (|\phi| + \mathfrak{L}_\phi)(|A|^\flat + \epsilon),\end{aligned}$$

which gives (2).

Now let A be any 0-chain. Say $A = \lim^\flat A_k$, the A_k polyhedral. Then set $\phi A = \lim^\flat \phi A_k$. Because of (2), the limit exists and is independent of the sequence chosen. Now (2) holds in the general case. We may suppose $\limsup |A_k| \leqq |A|$; because of (V, 16.1), applying (1) to each ϕA_k gives (1) for ϕA.

As a further consequence of (2), we see that if $\lim^\flat A_k = A$ (the A_k not necessarily polyhedral), then $\lim^\flat \phi A_k$ exists, and

$$\lim^\flat \phi A_k = \phi \lim^\flat A_k. \tag{4}$$

We give a condition under which the mass of the sum of two 0-chains equals the sum of the masses; see (VII, 3.22) for a generalization. The *support* spt (ψ) of ψ is the closure of the set of points where ψ is not zero.

LEMMA 3a. *For any sharp* ψ_1, ψ_2 *and* 0-*chain* A *in* E^n,

$$|(\psi_1 + \psi_2)A| = |\psi_1 A| + |\psi_2 A| \quad \textit{if} \quad \text{dist}(\text{spt}(\psi_1), \text{spt}(\psi_2)) > 0. \tag{5}$$

We suppose that $|\psi_1 A|$ and $|\psi_2 A|$ are finite; an easy modification of the proof takes care of the contrary case.

Given $\epsilon > 0$, choose X_1, X_2 so that

$$|X_i| \leqq 1, \qquad X_i\cdot\psi_i A \geqq |\psi_i A| - \epsilon, \qquad i = 1, 2.$$

The definition of $\psi_i A$ shows that if we alter D_{X_i} outside $Q_i = \text{spt}(\psi_i)$, $X_i\cdot\psi_i A$ will be unchanged. Hence we may suppose $D_{X_i} = 0$ outside a set Q_i', with $Q_1' \cap Q_2' = 0$. Now D_{X_1} and D_{X_2} together form a sharp function, and hence a 0-cochain X, with

$$|X| \leqq 1, \qquad X\cdot\psi_i A = X_i\cdot\psi_i A \qquad (i = 1, 2).$$

Hence

$$X\cdot(\psi_1 + \psi_2)A \geqq |\psi_1 A| + |\psi_2 A| - 2\epsilon,$$

proving $\geqq$ in (5). The reverse inequality is clear, and the lemma is proved.

4. The part $<T$ of a chain of finite mass. Given a 0-chain A of finite mass in E^1, and a number T, we shall define the "part A_T of A which is $<T$." We shall define the part of an r-chain of finite mass (defined in E^n) which lies in a given Borel set in (XI, 13).

EXAMPLE (a). Let $A = \bar{T}$ (notation in § 2). Then naturally the part of A which is $<T$ is 0, while the part $\leqq T$ is $A \neq 0$. If A is continuous (see § 1), it is easy to see with the definition below that the part $<T$ equals the part $\leqq T$.

EXAMPLE (b). Let A be the 0-chain in Example (c) of (V, 5). We cannot define A_0, for it would have to consist of $-\sum_{i=1}^{\infty}\sigma_i$,which is divergent.

EXAMPLE (c). With A_i' as in (V, 5(c)), set $A' = \lim^\flat_{i\to\infty} A_i'$; then $A' = 0$, so that $(A')_0 = 0$. We cannot define $(A')_0$ to be $\lim^\flat_{i\to\infty}(A_i')_0$, for $(A_i')_0 = \sigma_i$, and $\lim^\flat \sigma_i = \bar{0}$.

With T fixed, define continuous functions $\phi_0, \phi_1, \cdots$ as follows:

$$\phi_i(t) = \begin{cases} 1, & t \leqq T - 2/4^i, \\ 0, & t \geqq T - 1/4^i, \end{cases} \tag{1}$$

and let $\phi_i(t)$ be linear in the remaining interval. Set

$$(2) \qquad A_T = \lim_{i \to \infty} \phi_i A.$$

To prove the existence and properties of A_T, define the functions

$$(3) \qquad \phi'_0(t) = \phi_0(t), \qquad \phi'_i(t) = \phi_i(t) - \phi_{i-1}(t) \qquad (i = 1, 2, \cdots).$$

Since spt (ϕ'_0), spt (ϕ'_2), spt (ϕ'_4), $\cdots$ are disjoint, as are spt (ϕ'_1), spt (ϕ'_3), $\cdots$, repeated use of Lemma 3a gives

$$|(\phi'_0 + \phi'_2 + \cdots + \phi'_{2j})A| = |\phi'_0 A| + |\phi'_2 A| + \cdots + |\phi'_{2j} A|,$$

$$|(\phi'_1 + \phi'_3 + \cdots + \phi'_{2j+1})A| = |\phi'_1 A| + |\phi'_3 A| + \cdots + |\phi'_{2j+1} A|.$$

Also, since $\sum \phi'_i(t) \leqq 1$, (3.1) shows that the two left hand sides are each $\leqq |A|$. Hence

$$|\phi'_0 A| + |\phi'_1 A| + \cdots + |\phi'_i A| \leqq 2|A| \qquad (i = 1, 2, \cdots).$$

Therefore the series of masses converges, and hence

$$(4) \qquad |\phi_j A - \phi_i A| \leqq |\phi'_{i+1} A| + \cdots + |\phi'_j A| \to 0,$$

proving the existence of A_T; *in fact*, $\lim |A_T - \phi_i A| = 0$.

We prove some properties of A_T:

$$(5) \qquad |A_T| + |A - A_T| = |A|,$$

$$(6) \qquad \lim_{\lambda \to 0+} A_{T-\lambda} = A_T,$$

$$(7) \qquad \lim_{T \to \infty} |A'_T| = 0, \quad \text{where} \quad A'_T = A_{-T} + (A - A_T).$$

For any i, spt (ϕ_i) and spt $(1 - \phi_{i+1})$ are disjoint. Also $\phi_i + (1 - \phi_{i+1}) \leqq 1$. Hence Lemma 3a and (3.1) give

$$|\phi_i A| + |(1 - \phi_{i+1})A| = |(\phi_i + 1 - \phi_{i+1})A| \leqq |A|.$$

Letting $i \to \infty$ gives $\leqq$ in (5); the reverse inequality is clear.

Given $\epsilon > 0$, choose i such that $|\phi_j A - \phi_i A| < \epsilon/2$ for all $j > i$. Now take any $T' = T - \lambda$, $0 < \lambda < 1/4^i$. Choose j such that $T' < T - 2/4^j$. Now for any k such that $T' - 2/4^k \geqq T - 1/4^i$, we have (if ϕ'_k is defined for T' like ϕ_i for T)

$$\phi'_k \phi_i = \phi_i, \qquad \phi'_k \phi_j = \phi'_k;$$

hence, using (V, Theorem 16B),

$$|(\phi'_k - \phi_i)A| = |\phi'_k[(\phi_j - \phi_i)A]| \leqq |(\phi_j - \phi_i)A| < \epsilon/2,$$

$$|A_{T'} - \phi_i A| \leqq \liminf_{k \to \infty} |(\phi'_k - \phi_i)A| \leqq \epsilon/2.$$

Also $|\phi_i A - A| \leqq \epsilon/2$, hence $|A_{T'} - A| \leqq \epsilon$, and (6) holds.

Choose ζ so that $|A - A'|^\flat \leqq \zeta$ implies $|A'| \geqq |A| - \epsilon$ (V, Theorem 16B). Choose a polyhedral B, $|A - B|^\flat \leqq \zeta/3$. Choose ϕ with spt (ϕ) bounded, and $|\phi| = 1$, $\mathfrak{L}_\phi \leqq 1$, $\phi = 1$ on B; then

$$|\phi A - B|^\flat = |\phi(A - B)|^\flat \leqq (|\phi| + \mathfrak{L}_\phi)|A - B|^\flat \leqq 2\zeta/3.$$

Hence

$$|\phi A - A|^\flat \leqq \zeta, \qquad |\phi A| \geqq |A| - \epsilon.$$

Now take any T such that spt $(\phi) \subset (-T, T - 2)$. If $A_{-T} = \lim \phi_i' A$, $A_T = \lim \phi_i A$, as before, then

$$A_T' = \lim \psi_i A, \qquad \psi_i = \phi_i' + (1 - \phi_i).$$

Now by Lemma 3a,

$$|\psi_i A| + |\phi A| = |(\psi_i + \phi)A| \leqq |A|,$$

and hence $|\psi_i A| \leqq \epsilon$, all i. Therefore $|A_T'| \leqq \epsilon$, proving (7).

We note a generalization of (5): Let

(8) $$t_0 < \cdots < t_m, \qquad A_i^* = A_{t_i} - A_{t_{i-1}} \quad (i = 1, \cdots, m);$$

then

(9) $$|A| = |A_{t_0}| + |A_1^*| + \cdots + |A_m^*| + |A - A_{t_m}|.$$

For $(A_{t_i})_{t_{i-1}} = A_{t_{i-1}}$ clearly, and hence (5) gives

$$|A_{t_i}| = |A_{t_{i-1}}| + |A_{t_i} - A_{t_{i-1}}|;$$

combining these with (5) for $T = t_m$ gives (9).

5. Functions of bounded variation in E^1 defined by 0-chains. Given the 0-chain A of finite mass in E^1, we define a corresponding function γ_A by

(1) $$\gamma_A(t) = I^0 \cdot A_t,$$

I^0 being the unit 0-cochain: $I^0 \cdot \bar{t} = 1$. We show first that γ_A is of bounded variation. Take any points $t_0 < t_1 < \cdots < t_m$; then (4.9) gives

$$\sum_{i=1}^m |\gamma_A(t_i) - \gamma_A(t_{i-1})| = \sum_{i=1}^m |I^0 \cdot A_i^*| \leqq \sum_{i=1}^m |A_i^*| \leqq |A|;$$

hence $|\gamma_A| \leqq |A|$. We shall prove that A corresponds to γ_A (§ 2); then $|\gamma_A| = |A|$.

We need a lemma which compares a chain near a point with the chain "squashed" into the point; compare (VII, 7.2) below.

LEMMA 5a. *Let $C = \lim^\flat C_k$, the C_k being polyhedral 0-chains in (t', t''), and let $t' \leqq t \leqq t''$. Then*

(2) $$|C - (I^0 \cdot C)\bar{t}|^\flat \leqq N(t'' - t'), \qquad N = \limsup_{k \to \infty} |C_k|.$$

Say $C_k = \sum_i c_{ki}\bar{t}_{ki}$, $\sum_i |c_{ki}| \leqq N + \epsilon_k$, $\epsilon_k \to 0$. Set $\sigma_{ki} = \bar{t}_{ki}\bar{t}$. Then since $I^0 \cdot C_k = \sum_i c_{ki}$,

$$C_k - (I^0 \cdot C)\bar{t} = -\partial \sum_i c_{ki}\sigma_{ki} + [I^0 \cdot (C_k - C)]\bar{t},$$

$$|C_k - (I^0 \cdot C)\bar{t}|^\flat \leqq \sum_i |c_{ki}|\,|\sigma_{ki}| + |I^0 \cdot (C_k - C)|$$

$$\leqq (N + \epsilon_k)(t'' - t') + |C_k - C|^\flat;$$

letting $k \to \infty$ gives (2). (Taking a subsequence, we see that lim sup can be replaced by lim inf.)

Using (4.8), we now prove

$$\text{(3)} \qquad |A_i^* - (I^0 \cdot A_i^*)\bar{t}_i|^\flat \leqq |A_i^*|\,(t_i - t_{i-1}).$$

Say $A_{t_k} = \lim_{j\to\infty} \phi_{kj} A$ (the ϕ_{kj} like the former ϕ_i); then

$$A_i^* = \lim_{j\to\infty} \phi_{ij}^* A, \qquad \phi_{ij}^* = \phi_{ij} - \phi_{i-1,j}.$$

Let $\psi_\lambda(t) = 1$ for $t_{i-1} - \lambda \leqq t \leqq t_i$, let it $= 0$ for $t \leqq t_{i-1} - 2\lambda$ and $t \geqq t_i + \lambda$, and let it be linear in the remaining intervals. Then for j large enough, $\psi_\lambda \phi_{ij}^* = \phi_{ij}^*$; hence, using (3.4),

$$A_i^* = \lim_{j\to\infty} \phi_{ij}^* A = \lim_{j\to\infty} \psi_\lambda \phi_{ij}^* A = \psi_\lambda A_i^*.$$

Choose polyhedral chains A_{ij}^* such that

$$A_i^* = \lim_{k\to\infty} A_{ik}^*, \qquad |A_{ik}^*| < |A_i^*| + \epsilon_k, \qquad \lim_{k\to\infty} \epsilon_k = 0;$$

then, using (3.4) again and (3.1),

$$A_i^* = \lim_{k\to\infty} \psi_\lambda A_{ik}^*, \qquad |\psi_\lambda A_{ik}^*| < |A_i^*| + \epsilon_k.$$

Since $\psi_\lambda A_{ik}^*$ lies in $(t_{i-1} - 2\lambda, t_i + \lambda)$, Lemma 5a gives

$$|A_i^* - (I^0 \cdot A_i^*)\bar{t}_i|^\flat \leqq |A_i^*|\,(t_i - t_{i-1} + 3\lambda).$$

Since λ is arbitrary, (3) holds.

Let B be the 0-chain corresponding to γ_A. To prove $B = A$, take any $\epsilon > 0$. By the results of § 2, and by (4.7), we may choose k so that the following is true. Let $t_0 = -2^k$, $t_1 = -2^k + 1/2^k, \cdots, t_m = 2^k$, as in § 2, and let B' be the corresponding polyhedral 0-chain. Then, with A'_{2^k} as in (4.7),

$$|B' - B|^\flat < \epsilon/3, \qquad |A'_{2^k}| < \epsilon/3, \qquad |A|/2^k < \epsilon/3.$$

By definition of B',

$$B' = \sum_{i=1}^m b_i \bar{t}_i, \qquad b_i = \gamma_A(t_i) - \gamma_A(t_{i-1}) = I^0 \cdot A_i^*.$$

Hence

$$A - B' = A_{t_0} + \sum_{i=1}^{m}(A_i^* - b_i\bar{t}_i) + (A - A_{t_m}),$$

and because of (3) and (4.9),

$$|A - B'|^\flat \leqq |A_{t_0}| + \sum_{i=1}^{m} |A_i^*| (t_i - t_{i-1}) + |A - A_{t_m}| < 2\epsilon/3.$$

Hence $|A - B|^\flat < \epsilon$; therefore $A = B$.

THEOREM 5A. *There is a one-one linear correspondence between* 0*-chains* A *of finite mass in* E^1 *and normalized functions* γ_A *of bounded variation, defined by* (2.2); *we have*

(4) $$|\gamma_A| = |A|.$$

Given A, γ_A was determined above; γ_A is normalized because of (4.6) and (4.7). The correspondence is clearly one-one and linear. Relation (4) follows from (2.7).

6. Some related analytical theorems. We first give an analytical formulation of some preceding theorems. Let **M** be the linear space of normalized functions of bounded variation in E^1; the variation may be used as norm. Another norm in **M** is the *sharp norm*, defined by

(1) $$|\gamma|^\# = \sup\left\{\int_{-\infty}^{\infty} D_X d\gamma\colon\ X \text{ sharp}, |X|^\# \leqq 1\right\}.$$

(The "flat norm" is the same, since $r = 0$.) Because of Theorem 5A and (2.2),

(2) $$|\gamma_A|^\# = |A|^\# = |A|^\flat.$$

Now (V, 4.1), with Theorem 5A and (V, Theorem 4B), gives

THEOREM 6A. *The space* **C** *of real sharp functions in* E^1 *is the conjugate space of* **M**, *with the sharp norm.*

The following theorem states that, conversely, **M** is the space of linear functions in **C**, which satisfy a certain continuity condition.

THEOREM 6B. *Let* Λ *be any real valued linear function in* **C**, *with the following property. Let* $\phi_1, \phi_2, \cdots$ *be any sequence of functions in* **C** *such that*

(a) $|\phi_i| \leqq N$ *(all* i*) for some* N,

(b) $\phi_i(p) \to 0$ *uniformly on all compact subsets of* E^1.

Then there is a unique $\gamma \in$ **M** *such that* $\Lambda(\phi) = \int_{-\infty}^{\infty} \phi\, d\gamma$ *for all* $\phi \in$ **C**. *We have*

(3) $$|\gamma| = \sup\{\Lambda(\phi)\colon\ |\phi| \leqq 1\}.$$

This theorem is the case $n = 1$ of (XI, Lemma 7b). See also (XI, Theorem 8B).

Take any 0-chain A of finite mass, and set $\Lambda_A(D_X) = X \cdot A$. Using (4.7), we see easily that Λ_A satisfies the condition of the theorem; hence γ_A exists such that $X \cdot A = \int_{-\infty}^{\infty} D_X \, d\gamma_A$. We thus have an alternative proof of Theorem 5A. Compare (XI, Theorem 11A).

Consider the space **M**$'$ of normalized functions of bounded variation in (0, 1), with variation as norm, and the space **C**$'$ of continuous functions ϕ in (0, 1), with $|\phi|$ as norm. Then, in contrast to the situation in Theorem 6B, one can show that **M**$'$ is the conjugate space of **C**$'$. This is a famous theorem of F. Riesz (see Banach, p. 61).

7. Continuous r-chains in E^n. We generalize the definition in § 1 as follows. Let α be a cellwise continuous function in E^n whose values are r-vectors. We say α is *summable* if $\int \langle\alpha\rangle_0$ is finite; equivalently, if $\int \langle\alpha\rangle$ is finite, or, if $\int \langle\alpha^\lambda\rangle$ is finite for each λ $(\lambda_1 < \cdots < \lambda_r)$. We say the cellwise continuous summable function α *corresponds* to the flat r-chain A if (using the Riemann integral)

$$(1) \qquad \int_{E^n} D_X(p) \cdot \alpha(p) \, dp = X \cdot A \qquad \text{for all sharp } X.$$

We prove below that A exists; it is unique, by (V, Theorem 13B). Let us denote it by $\tilde{\alpha}$; we call it a *continuous chain*. We prove furthermore that

$$(2) \qquad |\tilde{\alpha}| = \int_{E^n} \langle\alpha\rangle_0.$$

For more general theorems, see (IX, Theorem 15A) and (XI, Theorem 14A).

We use sharp X only in (1) because the proof of existence of D_X (V, 10) is relatively simple in this case, and this suffices for the definition of $\tilde{\alpha}$.

For $r = 0$, the definition becomes that in § 1 (for $n = 1$). For $r = n$, we may write $\alpha(p) = \phi(p)\alpha_0$, α_0 being the unit n-vector; then $\int D_X \cdot \alpha = \int \phi \bar{D}_X$, and (1) again reduces to (1.1) (if $n = 1$).

With $r = n$, if we allow all measurable summable α, we obtain all flat n-chains (IX, Theorem 15B). For $r < n$, we do not obtain all r-chains; for instance, we obtain no polyhedral r-chains, and no r-chains of infinite mass. But we do obtain a set of chains which is dense (Theorem 7A). If, in place of functions α, we consider additive set functions whose values are r-vectors, we obtain in a similar manner all r-chains of finite mass (flat or sharp); see (XI, Theorem 11A). This generalizes Theorem 5A.

To define $\tilde{\alpha}$ in terms of α, we consider first the following special case. Let Q be an n-cube, with an oriented r-face Q'' and a complementary $(n - r)$-face Q'. Let α be the function equal to the r-direction of Q'' in Q, and $=0$ outside. For each $q \in Q'$, let $P(q)$ be the oriented r-cube through

q in Q, formed by translating Q''. Then for any sharp r-cochain X, (V, 10.3) shows that

$$\int_Q D_X(p)\cdot\alpha\, dp = \int_{Q'}\int_{P(q)} D_X(p)\cdot\alpha\, dp\, dq = \int_{Q'} X\cdot P(q)\, dq. \tag{3}$$

Let $\mathfrak{S}$ be a subdivision of Q' into cells $Q'_1, \cdots, Q'_m$, of mesh $\leqq \eta$ (App. II, 3). Choose $q'_i \in Q'_i$, and set

$$A(\mathfrak{S}) = \sum_i |\, Q'_i \,|\, P(q'_i); \tag{4}$$

this is an approximation to $\tilde{\alpha}$. We shall show that for any sharp X,

$$\left| \int_Q D_X(p)\cdot\alpha\, dp - X\cdot A(\mathfrak{S}) \right| \leqq |\, X\,|^\flat (|\, Q''\,| + |\, \partial Q''\,|)\, |\, Q'\,|\, \eta. \tag{5}$$

For any $q \in Q'_i$, (V, 3.6) gives

$$|\, P(q) - P(q'_i)\,|^\flat \leqq b\eta, \qquad b = |\, Q''\,| + |\, \partial Q''\,|.$$

Hence

$$\left| \int_Q D_X(p)\cdot\alpha\, dp - X\cdot A(\mathfrak{S}) \right| = \left| \sum_i \int_{Q'_i} X\cdot[P(q) - P(q'_i)]\, dq \right|$$

$$\leqq \sum |\, X\,|^\flat b\eta\, |\, Q'_i\,| = |\, X\,|^\flat b |\, Q'\,|\, \eta.$$

THEOREM 7A. *There is a one-one linear mapping of cellwise continuous summable r-vector functions α into corresponding flat r-chains $\tilde{\alpha}$;* (2) *holds. The images $\tilde{\alpha}$ with α continuous are dense in $\mathbf{C}_r^\flat$, and hence* (V, Theorem 14B) *also in $\mathbf{C}_r^\sharp$.*

Given α, which we suppose continuous at present, we choose a Cauchy sequence $A_1, A_2, \cdots$ of polyhedral r-chains which will define $\tilde{\alpha}$, as follows. Take any k and $\lambda = (\lambda_0, \cdots, \lambda_r)$. We may choose a cube Q with edges parallel to a chosen set of axes, so that

$$\int_{E^n - Q} \langle \alpha^\lambda \rangle < \epsilon'_k = 1/3\cdot 2^k \binom{n}{r}, \qquad \text{all } \lambda.$$

Since α^λ is uniformly continuous in Q, we may cut Q into cubes $Q_1, \cdots, Q_m$, and define the function β_k^λ, constant within each Q_i and 0 outside Q, so that

$$\int_{E^n} \langle \beta_k^\lambda - \alpha^\lambda \rangle < 2\epsilon'_k, \qquad \text{all } \lambda.$$

For each Q_i, (5) shows that we may define a polyhedral r-chain B_{ki}^λ in Q_i so that for any sharp X,

$$\left| \int_{Q_i} D_X\cdot\beta_k^\lambda e_\lambda - X\cdot B_{ki}^\lambda \right| \leqq |\, X\,|^\flat\, \epsilon'_k/m.$$

Therefore, setting $B_k^\lambda = \sum_i B_{ki}^\lambda$, $B_k = \sum_{(\lambda)} B_k^\lambda$, we find

$$\left| \int_{E^n} D_X \cdot \alpha^\lambda e_\lambda - X \cdot B_k^\lambda \right| \leqq \int_{E^n} \langle D_X \rangle_0 \langle \alpha^\lambda - \beta_k^\lambda \rangle + \sum_i \left| \int_{Q_i} D_X \cdot \beta_k^\lambda e_\lambda - X \cdot B_{ki}^\lambda \right|$$

$$\leqq 2 \,|\, X \,|\, \epsilon_k' + \sum_i |\, X \,|^\flat \epsilon_k'/m \leqq 3 \,|\, X \,|^\flat \epsilon_k',$$

(6) $$\left| \int_{E^n} D_X \cdot \alpha - X \cdot B_k \right| \leqq |\, X \,|^\flat / 2^k.$$

Take any integers k and l, $k < l$. Then (6) gives

$$|\, X \cdot (B_l - B_k) \,| \leqq 2 \,|\, X \,|^\flat / 2^k,$$

and (V, 13.13) shows that

(7) $$|\, B_l - B_k \,|^\flat \leqq 2/2^k, \qquad l > k.$$

Hence $A = \lim^\flat B_k$ exists. Because of (6), (1) holds; hence $A = \tilde{\alpha}$ corresponds to α. If α is not continuous, we need merely choose α_k continuous so that $\int \langle \alpha_k - \alpha \rangle < 1/2^k$, find B_k corresponding to α_k as above, etc.

REMARK. In the terminology of (VII, 7), we can form an approximation to $\tilde{\alpha}$ as follows. For a sufficiently large cube Q, subdivided into sufficiently small cubes Q_i, choose $q_i \in Q_i$, let (β, q) denote the sharp chain at q corresponding to the r-vector β, and set

(8) $$B = \sum |\, Q_i \,| \,(\alpha(q_i), q_i).$$

A sequence of such chains will converge to $\tilde{\alpha}$ in the sharp norm.

The mapping $\alpha \to \tilde{\alpha}$ is clearly one-one and linear. To show that the $\tilde{\alpha}$ with α continuous are dense, it is sufficient to show that for any oriented r-simplex σ and any $\epsilon > 0$ there is an α such that

$$\left| \int D_X \cdot \alpha - X \cdot \sigma \right| \leqq |\, X \,|^\flat \epsilon$$

for any sharp X; for this gives $|\, \tilde{\alpha} - \sigma \,|^\flat \leqq \epsilon$. Choose an $(n - r)$-cell Q' through $q_0 \in \sigma$ orthogonal to σ. The simplexes $\sigma(q) = T_{q - q_0} \sigma$ with $q \in Q'$ form a cell Q. Take Q' so small that

$$|\, \sigma(q) - \sigma \,|^\flat < \epsilon/2, \qquad q \in Q'.$$

Then (compare the proof above) if $\beta = \{\sigma\}/|\, Q \,|$ in Q and $\beta = 0$ in $E^n - Q$,

$$\left| \int D_X \cdot \beta - X \cdot \sigma \right| = \left| \int_{Q'} \frac{1}{|\, Q' \,|} X \cdot [\sigma(q) - \sigma] \right| \leqq |\, X \,|^\flat \epsilon/2.$$

We may choose a continuous function α in E^n so that $\int \langle \beta - \alpha \rangle_0 < \epsilon/2$; then the required inequality follows.

To prove (2) (see (XI, 14.7) for the general case), we note first that for any sharp X,

$$|X\cdot\tilde{\alpha}| = \left|\int D_X\cdot\alpha\right| \leqq |X|\int\langle\alpha\rangle_0;$$

hence (V, Theorem 16A) $|\tilde{\alpha}| \leqq \int\langle\alpha\rangle_0$.

To prove the reverse inequality, suppose first that α is a constant α_1 in a cube Q, and vanishes outside. Choose an r-covector ω_1 so that $|\omega_1|_0 = 1$ and $\omega_1\cdot\alpha_1 = |\alpha_1|_0$ (I, Theorem 13A), (App. I, Lemma 8b), and set $D_X(p) = \omega_1$ in E^n. Then

$$|X| = 1, \qquad \int_Q D_X\cdot\alpha_1 = |\alpha_1|_0\,|Q| = \int_{E^n}\langle\alpha\rangle_0.$$

Now given any cellwise continuous summable α, choose a cube Q, subdivided into cubes Q_i, and choose $\beta(p)$ (compare a proof above), constant within each Q_i and zero outside Q, so that $\int_{E^n}\langle\beta - \alpha\rangle_0 < \epsilon/3$. Choose X_i as above so that

$$|X_i| = 1, \qquad \int_{Q_i} D_{X_i}\cdot\beta = \int_{Q_i}\langle\beta\rangle_0.$$

By altering D_{X_i} near the faces of Q_i and outside Q_i, we may make it vanish outside, and thus define a cochain X_i' with $|X_i'| = 1$ and $\int_{E^n} D_{X_i'}\cdot\beta$ arbitrarily near $\int_{Q_i} D_{X_i}\cdot\beta$. Thus, setting $X = \sum X_i'$, we obtain $|X| = 1$ and

$$\int_{E^n} D_X\cdot\alpha \geqq \int_{E^n} D_X\cdot\beta - \frac{\epsilon}{3} \geqq \sum\int_{Q_i} D_{X_i}\cdot\beta - \frac{2\epsilon}{3} \geqq \int_{E^n}\langle\alpha\rangle_0 - \epsilon;$$

hence $|\tilde{\alpha}| \geqq \int\langle\alpha\rangle_0$, completing the proof.

8. On compact cochains. The first lemma below is used in the proof of Theorem 9B; the last lemma is also necessary if the chain A is not compact (see § 9). We say ω is *compact* if it vanishes outside some compact set. Say X is *smooth* if D_X is; X is *compact* if spt (D_X) is.

LEMMA 8a. *For any compact smooth $(n - 1)$-form ω in E^n,*

$$\int_{E^n} d\omega = 0. \tag{1}$$

For let Q be a cube containing spt (ω). Then by the Theorem of Stokes,

$$\int_{E^n} d\omega = \int_Q d\omega = \int_{\partial Q}\omega = 0.$$

The next lemma will be considered in a more general setting in (VII, 1). Recall the definition of $|\omega|^\sharp$ in (V, 10.2).

LEMMA 8b. *For any sharp function ϕ and sharp r-form ω,*

(2) $$|\phi\omega|^\# \leqq |\phi||\omega|^\# + (r+1)\mathfrak{L}(\phi)|\omega|_0 \leqq (r+2)|\phi|^\#|\omega|^\#.$$

For an easy calculation gives

(3) $$\mathfrak{L}_0(\phi\omega) \leqq \mathfrak{L}(\phi)|\omega|_0 + |\phi|\mathfrak{L}_0(\omega);$$

setting $s = r + 1$, we find

$$\begin{aligned}|\phi\omega|^\# &= \sup\{|\phi\omega|_0, s\mathfrak{L}_0(\phi\omega)\}\\ &\leqq \sup\{|\phi||\omega|_0, s|\phi|\mathfrak{L}_0(\omega) + s\mathfrak{L}(\phi)|\omega|_0\}\\ &\leqq |\phi|\sup\{|\omega|_0, s\mathfrak{L}_0(\omega)\} + s\mathfrak{L}(\phi)|\omega|_0,\end{aligned}$$

which gives the result.

Given the sharp function ϕ and the sharp r-cochain X, define the sharp r-cochain ϕX by setting

(4) $$D_{\phi X} = \phi D_X;$$

compare (VII, 2) and (IX, 7.6). Then (2) and (V, 10.4) give

(5) $$|\phi X|^\# \leqq |\phi||X|^\# + (r+1)\mathfrak{L}(\phi)|X| \leqq (r+2)|\phi|^\#|X|^\#.$$

In (V, Theorem 13A) we showed how a flat r-cochain may be expressed as the weak limit of smooth cochains. Here we show that any sharp r-cochain is the weak limit of compact sharp cochains.

LEMMA 8c. *Let $\phi_1, \phi_2, \cdots$ be sharp functions such that, for some N and open sets R_i,*

(6) $$|\phi_i|^\# \leqq N, \qquad \phi_i(p) = 1 \text{ in } R_i, \quad R_i \subset R_{i+1}, \qquad \bigcup R_i = E^n.$$

Then for any sharp r-cochain X,

(7) $$\text{wkl}^\# \phi_i X = X.$$

We may choose the R_i and ϕ_i so that ϕ_i and hence $\phi_i X$ is compact.

Suppose (6) holds. Then, by (5),

$$|\phi_i X|^\# \leqq (r+2)N|X|^\#.$$

For any r-simplex σ, $\phi_i(p) = 1$ in σ if $\sigma \subset R_i$; hence $\lim_{i\to\infty}(\phi_i X\cdot\sigma) = X\cdot\sigma$. Now (7) is a consequence of (V, Lemma 13a). The last part of the lemma is simple.

REMARK. The lemma holds also for flat X, using $\text{wkl}^\flat$ in (7). We need merely use (VII, 2.2) below in place of (5).

We show that $|A|^\#$ may be found with the help of compact smooth cochains (see also (VII, 4.6)):

(8) $$|A|^\# = \sup\{|X\cdot A| : X \text{ compact and smooth}, |X|^\# \leqq 1\}.$$

Let R_0 and R_0' be concentric spheres; we may define a smooth function ϕ_0 which is 1 in R_0 and is 0 in R_0', and such that $|\phi_0| = 1$; by means of an expansion of E^n, we may define R_i, R'_i and ϕ_i so that $|\phi_i| = 1$ and $\lim \mathfrak{L}(\phi_i) = 0$. Then (5) gives

$$\limsup_{i\to\infty} |\phi_i Y|^\sharp \leqq |Y|^\sharp, \qquad \text{all sharp } Y.$$

Given $\epsilon > 0$, choose X so that $|X|^\sharp \leqq 1$, $X\cdot A > |A|^\sharp - \epsilon$. By (V, Theorem 13A), we may choose η so that $X_\eta \cdot A > |A|^\sharp - 2\epsilon$. Using (7), we may choose i so that

$$\phi_i X_\eta \cdot A > |A|^\sharp - 3\epsilon, \qquad |\phi_i X_\eta|^\sharp \leqq |X|^\sharp + \epsilon \leqq 1 + \epsilon.$$

Since ϵ is arbitrary, (8) follows.

LEMMA 8d. *Let the sharp or flat r-chain A be such that $X\cdot A = 0$ for all compact smooth X. Then $A = 0$.*

This is an immediate consequence of (8); recall that any flat chain is also sharp.

9. The boundary of a smooth chain. We say the continuous r-chain $A = \tilde{\alpha}$ is *smooth* if α is smooth. We shall show how to find ∂A by differentiating α. Define the continuous $(r-1)$-chain $\tilde{\beta}$, $\beta = d^*\alpha$ having the components, in an orthonormal coordinate system,

$$(1) \qquad (d^*\alpha)^{\lambda_1\cdots\lambda_{r-1}} = \sum_{k\neq\lambda_1\cdots\lambda_{r-1}} \frac{\partial}{\partial x^k}\alpha^{\lambda_1\cdots\lambda_{r-1}k}.$$

We do not require $\lambda_1 < \cdots < \lambda_{r-1}$.

The following theorem shows that the result is independent of the coordinate system used.

THEOREM 9A. *With the dual operations of* (I, 11), *applied at each point,*

$$(2) \qquad d^*\alpha = \mathscr{D}' d\mathscr{D}\alpha.$$

Take any $\mu_1, \cdots, \mu_{n-r+1}$ in the natural order. Let $\mu'_1, \cdots, \mu'_{r-1}$ be the complementary set, in natural order. For any $i = 1, \cdots, n-r+1$, let $\mu^i_1, \cdots, \mu^i_r$ be complementary to $\mu_1, \cdots, \hat{\mu}_i, \cdots, \mu_{n-r+1}$, in natural order. By (II, 8.4) and (I, 11.3),

$$(d\mathscr{D}\alpha)^{\mu_1\cdots\mu_{n-r+1}} = \sum_{i=1}^{n-r+1} (-1)^{i-1} \frac{\partial}{\partial x^{\mu_i}} (\mathscr{D}\alpha)_{\mu_1\cdots\hat{\mu}_i\cdots\mu_{n-r+1}}$$

$$= \sum_{i=1}^{n-r+1} (-1)^{i-1} \epsilon_{\mu^i_1\cdots\mu^i_r\mu_1\cdots\hat{\mu}_i\cdots\mu_{n-r+1}} \frac{\partial}{\partial x^{\mu_i}} \alpha^{\mu^i_1\cdots\mu^i_r}$$

The index μ_i appears in $\mu^i_1\cdots\mu^i_r$. Moving it to the right in both ϵ and α gives

$$(d\mathscr{D}\alpha)_{\mu_1\cdots\mu_{n-r+1}} = \sum_{i=1}^{n-r+1} \epsilon_{\mu'_1\cdots\mu'_{r-1}\mu_1\cdots\mu_{n-r+1}} \frac{\partial}{\partial x^{\mu_i}} \alpha^{\mu'_1\cdots\mu'_{r-1}\mu_i}.$$

Using (I, 11.4) gives

$$(\mathscr{D}'d\mathscr{D}\alpha)^{\mu'} = \epsilon_{\mu'\mu}(d\mathscr{D}\alpha)_\mu = \sum_{i=1}^{n-r+1} \frac{\partial}{\partial x^{\mu_i}} \alpha^{\mu'\mu_i} = (d^*\alpha)^{\mu'}.$$

Let α_0 denote the unit n-vector in E^n. Recall from (I, 11.9) that for any r-form ω and any r-vector function α, $\omega\cdot\alpha = (\omega\vee\mathscr{D}\alpha)\cdot\alpha_0$. Hence (compare (III, 5.1))

$$\int \omega\cdot\alpha = \int (\omega\vee\mathscr{D}\alpha)\cdot\alpha_0 = \int \omega\vee\mathscr{D}\alpha. \tag{3}$$

We shall use the notation

$$\Phi\beta = \tilde{\beta}. \tag{4}$$

Theorem 9B. *For any smooth summable r-vector function* α,

$$\partial\Phi\alpha = (-1)^r\Phi\, d^*\alpha. \tag{5}$$

By Lemma 8d (or an easier argument if A is compact), it is sufficient to show that for any compact smooth X,

$$X\cdot[\partial\Phi\alpha - (-1)^r\Phi d^*\alpha] = 0.$$

Set $\xi = D_X$, $\omega = \mathscr{D}\alpha$; these are smooth forms. By (7.1), (V, 10.16) and (3),

$$X\cdot\partial\Phi\alpha = dX\cdot\Phi\alpha = \int D_{dX}\cdot\alpha = \int d\xi\vee\omega.$$

Also, since $\mathscr{D}\mathscr{D}'\eta = \eta$ (I, 11.2),

$$X\cdot\Phi\, d^*\alpha = \int D_X\cdot\mathscr{D}'d\mathscr{D}\alpha = \int \xi\vee\mathscr{D}\mathscr{D}'d\omega = \int \xi\vee d\omega.$$

Now $\xi\vee\omega$ is smooth and compact. Hence, by Lemma 8a and (II, 8.6),

$$X\cdot[\partial\Phi\alpha - (-1)^r\Phi d^*\alpha] = \int [d\xi\vee\omega + (-1)^{r-1}\xi\vee d\omega]$$
$$= \int d(\xi\vee\omega) = 0,$$

completing the proof.

10. Continuous chains in smooth manifolds. In a smooth n-manifold M, polyhedral chains are not defined; hence we cannot set up a theory of chains and cochains in M as in Chapter V. In place of r-cochains X, we may use r-forms ω, as in (V, 10). We shall show how, in place of r-chains A, one may use "continuous chains" α, define an operation $\omega\circ\alpha$, and replace $X\cdot A$ by $\int_M \omega\circ\alpha$. (We shall not discuss here requirements of continuity and integrability.)

CASE I. M is Riemannian and oriented. We may let α be an r-vector valued function in M, set $(\omega\circ\alpha)(p) = \omega(p)\cdot\alpha(p)$, and use the Riemann integral $\int_M \omega\cdot\alpha$, as in (7.1).

CASE II. M is oriented, but not Riemannian. Then the Riemann integral $\int_M \phi$ of a real function ϕ is not defined. We may let α be an $(n - r)$-form, and use $\int_M \omega\vee\alpha$; the definition in (III, 10) applies.

The artificiality of α being an $(n - r)$-form is due to the artificiality of integrating over M. To make the integral look more like the scalar product, use a polyhedral region Q in E^n in place of M; then, as in (III, 3.2), using subdivisions $\Sigma\sigma_i$ of Q,

$$(1)\qquad \int_Q \omega\vee\alpha = \lim \sum [\omega(p_i)\vee\alpha(p_i)]\cdot\{\sigma_i\} = \lim \sum \omega(p_i)\cdot\tilde{\alpha}_i,$$

where, by (I, 7.1),

$$(2)\qquad \tilde{\alpha}_i = \alpha(p_i)\wedge\{\sigma_i\}, \qquad \text{an } r\text{-vector.}$$

If α is continuous and the σ_i are small, $\tilde{\alpha}_i$ is approximately the "part of α" in σ_i. (See (XI, 13) for a discussion of this concept.)

CASE III. M is not assumed to be orientable; then there is no unique expression of M as an n-chain. Let $\xi(p)$ be a function of the following sort. Given $p \in M$, and an orientation of a neighborhood of p, $\xi(p)$ is an n-covector; it is replaced by its negative if the opposite orientation is chosen. Now for any orientable part R of M, $\int_R \xi$ may be defined as follows: Choose an orientation ϵ of R, forming an n-chain A, let $\xi_\epsilon(p)$ be the n-form which uses the orientation ϵ near each $p \in R$, and form $\int_A \xi_\epsilon$. If the opposite orientation had been chosen, we would obtain

$$\int_{-A} \xi_{-\epsilon} = \int_{-A} (-\xi_\epsilon) = \int_A \xi_\epsilon.$$

Thus $\int_R \xi = \int_A \xi_\epsilon$ is independent of the orientation chosen.

If ξ is defined in M, we may define $\int_M \xi$ in the following manner. Let $\phi_1, \phi_2, \cdots$ be a partition of unity in M (III, 10), each $\operatorname{spt}(\phi_i)$ lying in an orientable part U_i of M (for instance, in a coordinate system). Then set $\int_M \xi = \sum \int_{U_i} \phi_i \xi$.

Suppose that $\alpha(p)$ is like the $\xi(p)$ above, except that it is an $(n - r)$-covector depending on orientation. Then for any r-form ω in M, $\omega(p)\vee\alpha(p)$ is like the $\xi(p)$ above, and we may define $\int_M \omega\vee\alpha$. Following de Rham, p. 21, we may call α an "$(n - r)$-form of odd kind." For $r = n$, α is a "scalar of odd kind"; $\alpha(p)$ is a number depending on a chosen orientation about p, changing sign if the orientation is reversed.

VII. *General Properties of Chains and Cochains*

We give a variety of techniques and theorems in this chapter that are important for further study. First we consider the multiplication of chains and cochains by real sharp functions; also supports of chains and cochains, and some approximation theorems. Next we consider "sharp r-chains at a point," and show that there is no such thing in the flat case for $r > 0$. Finally we study cohomology in complexes.

Up to the present, we have considered the space of flat r-cochains in E^n as a Banach space, without further operations. We now introduce the real sharp functions as a ring of operators on this space. Let $A = \sum a_i \sigma_i^r$ be polyhedral, and let ϕ be sharp. For each i, $\psi(p) = \phi(p)a_i$ is a continuous function in σ_i, and defines a "continuous chain" B_i in σ_i and hence in E^n, by (VI, 7); set $\phi A = \Sigma B_i$. We show that the definition of ϕA is extendable to all sharp or flat A. By requiring $\phi X \cdot A$ to equal $X \cdot \phi A$, the real sharp functions become operators also on the spaces of flat or sharp cochains. (A more general theory of products will be given in Chapter IX.)

The "support" of a chain or cochain is, roughly speaking, the smallest closed set outside of which it has no effect on cochains or chains respectively. With the help of sharp functions as operators, the basic properties of supports are derived. We next show how to approximate to a chain A by a chain A' which is compact (i.e. has compact support); we may make $|A'|$ close to $|A|$ if the latter is finite, and similarly for $|\partial A'|$. A theorem of the same nature is given, requiring A' to be polyhedral.

The definition of the r-vector $\{A\}$ of a polyhedral chain A given in (III, 2) extends at once to any sharp or flat chain. We show that sharp r-chains $A \neq 0$ exist whose support contains only a single point p, even for $r > 0$; then A is determined by the pair p, $\{A\}$. One may approximate to any sharp chain A of finite mass by splitting it up, writing $A = \sum A_i$, $A_i = \phi_i A$, with $\sum \phi_i = 1$ and each ϕ_i vanishing outside of some small region U_i, and replacing A_i by a chain B_i concentrated at $p_i \in U_i$, with $\{B_i\} = \{A_i\}$. In contrast with this, the support of a flat r-chain $A \neq 0$ cannot lie in any plane of dimension $<r$. (We prove this with the help of results from Chapter X.) It would be valuable to find further conditions on the supports of flat chains. For instance, a line segment P can be the support of a flat 1-chain $A \neq 0$, but $\{A\}$ must be a vector along P.

In (IV, C) we showed how cohomology could be introduced into a smooth manifold M in different ways, namely, algebraically (through triangulations), or through differential forms; we showed that the resulting cohomology rings were isomorphic (de Rham's Theorem). (In algebraic topology, further ways of introducing the ring are given.) One may define flat cochains in M (we suppose for simplicity that M is compact); using these gives another definition of the cohomology ring. This definition may in fact be used in a complex (§ 10); we show (§ 12) that the cohomology ring thus determined is isomorphic to the algebraic one. In fact, one may replace the flat cochains by "flat differential forms," thus obtaining a theorem like de Rham's, for complexes; see the introduction to Chapter IX.

We shall use $| \, |^{\odot}$ commonly to denote either $| \, |^{\flat}$ or $| \, |^{\sharp}$.

1. Sharp functions times chains. Let ϕ be a sharp function in E^n (V, 4); we wish to define ϕA for any sharp or flat r-chain A in E^n. This will be the case $s = 0$ of the product $X^s \frown A^r$ in (IX, 16), obtained by setting $\phi = D_X$. The special case is more easily studied, and of especial importance. We will consider ϕA with more general functions ϕ in (XI, 12).

We begin by defining $\phi\sigma$ for an oriented r-simplex σ. Let P be the oriented r-plane of σ, with r-direction α. Using (VI, 7) in P, set

$$\phi\sigma = \tilde{\beta}, \qquad \beta(p) = \phi(p)\alpha \quad \text{in } \sigma, \qquad \beta(p) = 0 \quad \text{in} \quad P - \sigma; \tag{1}$$

this is an r-chain in P, and hence in E^n. By (VI, 7.2), applied to P, and (V, Theorem 16C),

$$| \phi\sigma | = \int_\sigma \langle \phi\alpha \rangle = \int_\sigma \langle \phi \rangle. \tag{2}$$

For any polyhedral r-chain $A = \sum a_i \sigma_i$, set $\phi A = \sum a_i \phi\sigma_i$.

For any sharp r-chain A, write $A = \lim^\sharp A_i$, the A_i polyhedral, and set

$$\phi A = \lim^\sharp \phi A_i; \tag{3}$$

similarly with the flat norm.

The existence and uniqueness of the limit is an immediate consequence of the following inequalities, applied to $A_j - A_i$ (A being an r-chain):

$$\left.\begin{aligned} &| \phi A |^\sharp \leqq N_\phi^{(r)} | A |^\sharp, && (4)\\ &| \phi A |^\flat \leqq N_\phi^{(r)} | A |^\flat, && (5) \end{aligned}\right\} N_\phi^{(r)} = |\phi| + (r+1)\mathfrak{L}_\phi.$$

The last inequality will be strengthened in (IX, 16.19). Note that if ϕ is considered as a sharp 0-form, (V, 10.2) gives

$$| \phi |^\sharp \leqq N_\phi^{(r)} \leqq (r+2) | \phi |^\sharp. \tag{6}$$

We give an inequality that is a consequence of (2) for $A = \sigma$, and hence for A polyhedral:

$$|\phi A| \leqq |\phi| \, |A|. \tag{7}$$

We shall prove these inequalities for sharp or flat r-chains A, and also the following for flat r-chains A:

$$|\phi\partial A - \partial\phi A| \leqq r\mathfrak{L}_\phi |A|, \tag{8}$$

$$|\partial\phi A| \leqq r\mathfrak{L}_\phi |A| + |\phi| \, |\partial A|. \tag{9}$$

(a) One can prove (4) by using $X\cdot\phi A = \phi X\cdot A$ (see § 2) and (VI, 8.5); we give another proof below.

(b) To prove (8) for polyhedral A, we may suppose that A is a simplex σ. Given $\epsilon > 0$, we shall find polyhedral approximations H to $\phi\sigma$ and K to $\phi\partial\sigma$ such that

$$\begin{cases} |H - \phi\sigma| < \epsilon/2, \qquad |K - \phi\partial\sigma| < \epsilon/2, \\ |\partial H - K| < r\mathfrak{L}_\phi |\sigma| + \epsilon; \end{cases} \tag{10}$$

then setting $B = \partial H - K$, we will have

$$|\phi\partial\sigma - \partial\phi\sigma + B|^\flat \leqq |\phi\partial\sigma - K| + |\partial(\phi\sigma - H)|^\flat < \epsilon,$$

which, with the last relation in (10), will give (8) (see (V, 16.1)).

Cut the plane of σ into cubes of side length h, let $\sigma_1, \cdots, \sigma_m$ be the cubes in σ, and let $\sigma_{m+1}, \cdots, \sigma_l$ be the parts of the remaining cubes in σ, as in (V, 7). Let p_i be the center of σ_i, and set $\phi_i = \phi(p_i)$. Set

$$H = \sum_{i=1}^{l} \phi_i \sigma_i;$$

the uniform continuity of ϕ together with (2) shows that the first relation in (10) holds if h is small enough.

Let $\tau_1, \cdots, \tau_s$ be the parts of the $(r-1)$-faces of the σ_i which lie in $\operatorname{int}(\sigma)$, and let $\tau_{s+1}, \cdots, \tau_t$ be the parts in $\partial\sigma$. Then if τ_k is a face of σ_{ν_k} $(k > s)$, setting

$$K = \sum_{k=s+1}^{t} \phi_{\nu_k} \tau_k$$

gives the second relation in (10), for small enough h.

For $k \leqq s$, say τ_k is on σ_{μ_k} positively and on σ_{λ_k} negatively. Now

$$\partial H - K = \sum_{k \leqq s} (\phi_{\mu_k} - \phi_{\lambda_k}) \tau_k.$$

Setting $\epsilon' = \epsilon / r\mathfrak{L}_\phi$, we may suppose h small enough so that

$$lh^r < |\sigma| + \epsilon', \qquad s \leqq rl < r(|\sigma| + \epsilon')/h^r.$$

Also

$$|\phi_{\mu_k} - \phi_{\lambda_k}| \leqq \mathfrak{L}_\phi |p_{\mu_k} - p_{\lambda_k}| \leqq \mathfrak{L}_\phi h;$$

hence

$$|\partial H - K| \leqq \sum_{k \leqq s} \mathfrak{L}_\phi h |\tau_k| \leqq s\mathfrak{L}_\phi h^r < r\mathfrak{L}_\phi |\sigma| + \epsilon,$$

completing the proof.

(c) Using this result gives, for r-simplexes σ,

$$|\phi\partial\sigma|^\flat \leqq r\mathfrak{L}_\phi |\sigma| + |\partial\phi\sigma|^\flat \leqq N_\phi^{(r-1)} |\sigma|.$$

(d) We now prove (5) for polyhedral A. Take any $\epsilon > 0$. Choose D so that $|A - \partial D| + |D| < |A|^\flat + \epsilon$. Then using (8) and (7) for polyhedral A gives

$$\begin{aligned} |\phi A - \partial\phi D| + |\phi D| &\leqq |\phi(A - \partial D)| + |\phi\partial D - \partial\phi D| + |\phi D| \\ &\leqq |\phi|(|A - \partial D| + |D|) + (r+1)\mathfrak{L}_\phi |D| \\ &\leqq [|\phi| + (r+1)\mathfrak{L}_\phi](|A|^\flat + \epsilon), \end{aligned}$$

giving (5).

(e) For a direct proof of (4) (and of (5) again), we use the following inequality:

$$(11) \qquad |\phi T_v\sigma - T_v\phi\sigma| = \int_\sigma |\phi(p+v) - \phi(p)|\,dp \leqq \mathfrak{L}_\phi |v| |\sigma|;$$

hence, by (V, 6.3) (which holds for polyhedral approximations to $\phi\sigma$ and hence for $\phi\sigma$),

$$(12) \qquad |\phi(T_v\sigma - \sigma)|^\sharp \leqq N_\phi^{(r)} |v| |\sigma| / (r+1).$$

We apply this in (f).

(f) Define semi-norms for polyhedral chains by setting (with $\phi \not\equiv 0$ fixed)

$$|A|' = |\phi A|^\flat / N_\phi^{(r)}, \qquad |A|'' = |\phi A|^\sharp / N_\phi^{(r)}.$$

Since, using (c),

$$|\sigma^r|' \leqq |\phi\sigma^r| / N_\phi^{(r)} \leqq |\phi| |\sigma^r| / N_\phi^{(r)} \leqq |\sigma^r|,$$

$$|\partial\sigma^{r+1}|' = |\phi\partial\sigma^{r+1}|^\flat / N_\phi^{(r)} \leqq |\sigma^{r+1}|,$$

(V, Theorem 8A) shows that $|A|' \leqq |A|^\flat$, giving (5); since $|A|'' \leqq |A|'$, (12) and (V, Theorem 8A) give (4). Both relations are proved for polyhedral chains.

(g) The definition of ϕA is therefore satisfactory in either norm; it follows that (4) and (5) hold in the general case. Note that these give ($^\circ$ denoting either $^\flat$ or $^\sharp$)

$$\lim{}^\circ \phi A_k = \phi \lim{}^\circ A_k \quad \text{if} \quad \lim{}^\circ A_k \text{ exists (either norm).} \tag{13}$$

(h) Relations (7) and (8), and hence also (9), hold for flat A. For let $A = \lim^\flat A_i$, $|A| = \lim |A_i|$, the A_i polyhedral. By (13),

$$\phi A_i \xrightarrow{\flat} \phi A, \qquad \phi\partial A_i - \partial\phi A_i \xrightarrow{\flat} \phi\partial A - \partial\phi A,$$

giving (7) and (8).

(i) Similarly, (7) holds for sharp A.

The following properties are evident for simplexes; they hold therefore for polyhedral chains, and hence also for flat or sharp chains:

$$\phi(A + B) = \phi A + \phi B, \qquad (\phi + \psi)A = \phi A + \psi A, \tag{14}$$

$$(\phi\psi)A = \phi(\psi A), \tag{15}$$

$$\phi A = aA \quad \text{if} \quad \phi(p) = a, \text{ all } p. \tag{16}$$

We give a condition under which the mass of a sum equals the sum of the masses (see also (X, 12.11)):

$$|(\phi_1 + \phi_2)A| = |\phi_1 A| + |\phi_2 A| \qquad \text{if} \quad \phi_1(p), \quad \phi_2(p) \geqq 0. \tag{17}$$

As a consequence,

$$|\phi A| \leqq |\psi A| \quad \text{if} \quad 0 \leqq \phi(p) \leqq \psi(p). \tag{18}$$

If $A = \sigma$, (17) is clear from (2); hence it holds for polyhedral A. We prove it for any sharp A (hence for flat A; see (V, Theorem 14B) and (V, Theorem 16A)). We need merely show that for any $\epsilon > 0$,

$$|\phi_1 A| + |\phi_2 A| < |(\phi_1 + \phi_2)A| + \epsilon.$$

We may choose a number M and a sharp function ϕ_3 such that

$$\phi_3(p) = M - \phi_1(p) - \phi_2(p) \geqq 0.$$

By (V, Theorem 16B), there is a $\zeta > 0$ such that

$$|B| \geqq |\phi_i A| - \epsilon/6 \quad \text{if} \quad |B - \phi_i A|^\sharp \leqq \zeta \qquad (i = 1, 2, 3).$$

Set $\eta = \inf\{\zeta/N_{\phi_i}^{(r)}: i = 1, 2, 3\}$, and choose a polyhedral chain A' such that

$$|A' - A|^\sharp < \eta, \qquad |A'| < |A| + \epsilon/2M.$$

Then

$$|\phi_i A' - \phi_i A|^\sharp \leqq N_{\phi_i}^{(r)} |A' - A|^\sharp \leqq \zeta,$$

and hence

$$|\phi_i A'| \geqq |\phi_i A| - \epsilon/6 \qquad (i = 1, 2, 3).$$

Therefore, since A' is polyhedral,

$$\sum |\phi_i A| \leqq \sum |\phi_i A'| + \epsilon/2 = M|A'| + \epsilon/2 < M|A| + \epsilon.$$

Also

$$M|A| \leqq |(\phi_1 + \phi_2)A| + |\phi_3 A|;$$

these inequalities give the required result.

2. Sharp functions times cochains. For any sharp ϕ, we may define ϕX for any sharp or flat X by setting

$$\phi X \cdot A = X \cdot \phi A, \tag{1}$$

A being sharp or flat correspondingly. Then (1.4) and (1.5) give

$$|\phi X|^{\sharp} \leqq N_\phi^{(r)} |X|^{\sharp}, \qquad |\phi X|^{\flat} \leqq N_\phi^{(r)} |X|^{\flat}; \tag{2}$$

see also (IX, 14.25).

From (1.7), (1.8), (1.11), (V, 7.4)and (V, 7.2) (which holds with $A = \phi\sigma$), we find at once, for r-cochains X,

$$|\phi X| \leqq |\phi||X|, \tag{3}$$

$$|\phi\, dX - d\phi X| \leqq (r+1)\,\mathfrak{L}_\phi |X|, \tag{4}$$

$$|d\phi X| \leqq |\phi||dX| + (r+1)\mathfrak{L}_\phi |X|, \tag{5}$$

$$\mathfrak{L}_{\phi X} \leqq |\phi|\mathfrak{L}_X + \mathfrak{L}_\phi |X|. \tag{6}$$

Hence, by (V, 4.8) and (V, 7.8), with either norm,

$$|\phi X|^{\circ} \leqq |\phi||X|^{\circ} + (r+1)\mathfrak{L}_\phi |X|, \tag{7}$$

again giving (2).

We prove that (VI, 8.4) follows for sharp X from the present definition. Take any point p_0 and any r-direction α. Choose $\sigma_1, \sigma_2, \cdots$ to define $D_X(p_0)\cdot\alpha$, as in (V, 10.5). Since, by (1.2),

$$|\phi\sigma_i - \phi(p_0)\sigma_i| \leqq \mathfrak{L}_\phi \operatorname{diam}(\sigma_i)|\sigma_i|,$$

we find

$$\begin{aligned} D_{\phi X}(p_0)\cdot\alpha = \lim (X\cdot\phi\sigma_i)/|\sigma_i| &= \phi(p_0)\lim (X\cdot\sigma_i)/|\sigma_i| \\ &= \phi(p_0)D_X(p_0)\cdot\alpha, \end{aligned}$$

which gives the result.

Clearly (1.14–16) hold for cochains.

We prove that if A is a continuous chain, defined by α, then ϕA is defined by $\phi\alpha$; that is, using (VI, 9.4),

$$\phi(\Phi\alpha) = \Phi(\phi\alpha). \tag{8}$$

For take any sharp X. Then (1), (VI, 7.1) and (VI, 8.4) give

$$X\cdot\phi\Phi\alpha = \phi X\cdot\tilde{\alpha} = \int D_{\phi X}\cdot\alpha = \int D_X\cdot\phi\alpha = X\cdot\Phi\phi\alpha.$$

3. Supports of chains and cochains. We defined the support of a polyhedral chain in (V, 1). We now give more general definitions.

The *support* spt (A) of a flat or sharp chain A is the set of points p such that for each $\epsilon > 0$ there is a sharp cochain X such that

$$X \cdot A \neq 0, \qquad D_X(q) = 0 \quad \text{outside } U_\epsilon(p). \tag{1}$$

Say A is *compact* if spt (A) is. The *support* spt (X) of a flat or sharp cochain X is the set of points p such that for each $\epsilon > 0$ there is a simplex σ such that

$$X \cdot \sigma \neq 0, \qquad \sigma \subset U_\epsilon(p). \tag{2}$$

We prove that the first definition coincides with that in (V, 1) for A polyhedral. Write $A = \sum a_i \sigma_i$, the σ_i non-overlapping and the $a_i \neq 0$; we must show that spt (A) is the union Q of the σ_i under the present definition. Using (V, 10.3), it is clear that spt$(A) \subset Q$. Now take any $p \in Q$; say $p \in \sigma_i$. Take any $\epsilon > 0$. With the method of (V, 12), we find easily an X with spt $(D_X) \subset U_\epsilon(p)$, $X \cdot \sigma_i \neq 0$, $X \cdot \sigma_j = 0$ $(j \neq i)$; now $X \cdot A \neq 0$.

Supports are clearly closed sets. We say a chain or cochain is *in* a set Q if its support is in Q.

Recall the definitions of car (ϕ), spt (ϕ) (App. III). Any sharp function ϕ defines a 0-cochain Z; clearly spt $(\phi) =$ spt (Z). More generally,

$$\text{spt}\,(X) = \text{spt}\,(D_X), \qquad X \text{ sharp}. \tag{3}$$

We shall prove the following properties, all (except (7), if ∂A is not defined) holding in both the flat and sharp norms; ϕ is always sharp:

$$A = 0 \quad \text{if} \quad \text{spt}\,(A) = 0, \tag{4}$$

$$X = 0 \quad \text{if} \quad \text{spt}\,(X) = 0, \tag{5}$$

$$X \cdot A = 0 \quad \text{if} \quad \text{spt}\,(X) \cap \text{spt}\,(A) = 0, \tag{6}$$

$$\text{spt}\,(\partial A) \subset \text{spt}\,(A), \tag{7}$$

$$\text{spt}\,(dX) \subset \text{spt}\,(X), \tag{8}$$

$$\text{spt}\,(\phi A) \subset \text{spt}\,(\phi) \cap \text{spt}\,(A), \tag{9}$$

$$\text{spt}\,(\phi X) \subset \text{spt}\,(\phi) \cap \text{spt}\,(X), \tag{10}$$

$$\phi A = 0 \quad \text{if} \quad \text{car}\,(\phi) \cap \text{spt}\,(A) = 0, \tag{11}$$

$$\phi X = 0 \quad \text{if} \quad \text{car}\,(\phi) \cap \text{spt}\,(X) = 0, \tag{12}$$

$$\phi A = A \quad \text{if} \quad \phi(p) = 1 \text{ in spt}\,(A), \tag{13}$$

$$\phi X = X \quad \text{if} \quad \phi(p) = 1 \text{ in spt}\,(X). \tag{14}$$

To prove (4), it is sufficient to show that $X \cdot A = 0$ for any compact sharp X, by (VI, Lemma 8d). Each $p \in Q = \text{spt}\,(X)$ is in some neighborhood $U(p)$ such that $Y \cdot A = 0$ for any sharp Y with $D_Y = 0$ outside $U(p)$. A finite number of these neighborhoods, say $U_1, \cdots, U_m$, cover Q. Define the partition $\phi_0, \phi_1, \cdots, \phi_m$ of unity as in (App. III, Lemma 2a). Then $\phi_0(p)D_X(p) = 0$ for all p, so that $\phi_0 X = 0$ (see (VI, 8.4)). Now

$$X = \left(\sum \phi_i\right) X = \phi_1 X + \cdots + \phi_m X,$$

and $D_{\phi_i X} = \phi_i D_X = 0$ outside U_i. Hence

$$X \cdot A = \sum (\phi_i X \cdot A) = 0.$$

To prove (5), we show that $X \cdot \sigma = 0$ for all simplexes σ. Each $p \in \sigma$ is in some neighborhood $U(p)$ such that $X \cdot \tau = 0$ for all $\tau \subset U(p)$. We may find a subdivision $\sum \sigma_i$ of σ such that each σ_i is in some $U(p)$; then $X \cdot \sigma = \sum X \cdot \sigma_i = 0$.

To prove (9), take first any p not in spt(ϕ). Then $\phi(q) = 0$ in some $U_\epsilon(p)$. Take any sharp X such that $D_X = 0$ outside $U_\epsilon(p)$. Then $\phi X = 0$, by (VI, 8.4), so that $X \cdot \phi A = \phi X \cdot A = 0$; hence p is not in spt(ϕA). The proof of spt $(\phi A) \subset$ spt (A) is similar.

Relation (10) is clear in the sharp norm. In the flat norm, we prove spt $(\phi X) \subset$ spt (X) as follows. Take any p not in spt (X). Say $X \cdot \tau = 0$ for all $\tau \subset U_\epsilon(p)$. Take any $\sigma \subset U_\epsilon(p)$. Given $\zeta > 0$, using a simplicial subdivision $\sum \sigma_i$ of σ and an approximation ϕ' to ϕ which is constant interior to simplexes gives a polyhedral chain $B = \sum b_i \sigma_i$ such that $| B - \phi\sigma | < \zeta$. Now $X \cdot \sigma_i = 0$, hence $X \cdot B = 0$, and

$$| X \cdot \phi\sigma | = | X \cdot (\phi\sigma - B) | \leqq \zeta | X | ;$$

therefore $\phi X \cdot \sigma = X \cdot \phi\sigma = 0$, and p is not in spt (ϕX). The proof of spt $(\phi X) \subset$ spt (ϕ) is similar but simpler.

We could prove (11) and (12) with spt (ϕ) instead of car (ϕ), then a correspondingly weakened form of (13) and (14), and the remaining properties, without the function ϕ_η to be described; but the stronger properties, as well as the function ϕ_η, are of interest.

Define a numerical function $\gamma_\eta(t)$, for $\eta > 0$:

(15) $$\gamma_\eta(t) = \begin{cases} t - \eta & \text{if } \quad t \geqq \eta, \\ 0 \quad \text{if} & -\eta \leqq t \leqq \eta, \\ t + \eta & \text{if } \quad t \leqq -\eta. \end{cases}$$

For any real function ϕ, define ϕ_η by

(16) $$\phi_\eta(p) = \gamma_\eta(\phi(p)).$$

Clearly

(17) $$\text{spt } (\phi_\eta) \subset \text{car } (\phi) \qquad \text{if } \phi \text{ is continuous.}$$

For any sharp ϕ, sharp or flat X, and simplex σ, we have

$$(18) \qquad \lim_{\eta\to 0} (\phi_\eta X\cdot\sigma) = \lim_{\eta\to 0} (X\cdot\phi_\eta\sigma) = X\cdot\phi\sigma = \phi X\cdot\sigma.$$

To prove (12), use (10) and (17), giving

$$\operatorname{spt}(\phi_\eta X) \subset \operatorname{car}(\phi) \cap \operatorname{spt}(X) = 0;$$

by (5), $\phi_\eta X = 0$. By (18), $\phi X\cdot\sigma = 0$ for all σ, so that $\phi X = 0$.

To prove (14), set $\psi(p) = 1 - \phi(p)$. Then ψ is sharp and $\operatorname{car}(\psi) \cap \operatorname{spt}(X) = 0$, hence $\psi X = 0$, and

$$\phi X = \phi X + \psi X = (\phi + \psi)X = X.$$

Note next that $|\phi_\eta| \leqq |\phi|$, $\mathfrak{L}_{\phi_\eta} \leqq \mathfrak{L}_\phi$, and hence $N^{(r)}_{\phi_\eta} \leqq N^{(r)}_\phi$. Hence, by (2.2), letting $^\odot$ denote either norm,

$$|\phi_\eta X|^\odot \leqq N^{(r)}_{\phi_\eta} |X|^\odot \leqq N^{(r)}_\phi |X|^\odot.$$

This, with (18), shows that

$$(19) \qquad \operatorname*{wkl}^\odot_{\eta\to 0} \phi_\eta X = \phi X.$$

To prove (11), note that, by (9) and (17),

$$\operatorname{spt}(\phi_\eta A) \subset \operatorname{car}(\phi) \cap \operatorname{spt}(A) = 0,$$

and hence $\phi_\eta A = 0$. Because of (19),

$$X\cdot\phi A = \phi X\cdot A = \lim_{\eta\to 0} (\phi_\eta X\cdot A) = \lim_{\eta\to 0} (X\cdot\phi_\eta A) = 0$$

for all X (flat or sharp according as A is); hence $\phi A = 0$.

Relation (13) follows from (11) as (14) did from (12).

To prove (6), suppose first that $Q = \operatorname{spt}(X)$ is compact. Then (App. III, Lemma 1a) there is a sharp function ϕ which equals 1 in Q and vanishes in a neighborhood of $\operatorname{spt}(A)$. By (14), $\phi X = X$, and by (11) or by (9) and (4), $\phi A = 0$; hence

$$X\cdot A = \phi X\cdot A = X\cdot\phi A = 0.$$

For general X and A, choose functions $\phi_1, \phi_2, \cdots$ as in (VI, Lemma 8c). Then because of that lemma and the Remark following it, and what has just been proved,

$$X\cdot A = \lim_{i\to\infty} (\phi_i X\cdot A) = 0.$$

To prove (8), take any p not in $\operatorname{spt}(X)$. Choose $\epsilon > 0$ so that $X\cdot\tau = 0$ for any r-simplex $\tau \subset U_\epsilon(p)$. Then for any $(r+1)$-simplex $\sigma \subset U_\epsilon(p)$, $dX\cdot\sigma = X\cdot\partial\sigma = 0$; hence p is not in $\operatorname{spt}(dX)$.

Finally, to prove (7), take any p not in $\operatorname{spt}(A)$. Choose $\epsilon > 0$ so that $\bar{U}_\epsilon(p) \cap \operatorname{spt}(A) = 0$. Take any sharp $(r-1)$-cochain X such that

$D_X = 0$ outside $U_\varepsilon(p)$. Then by (8), spt $(dX) \cap$ spt $(A) = 0$, and by (6), $X\cdot\partial A = dX\cdot A = 0$; hence p is not in spt (∂A).

We strengthen (6) in the sharp case:

(20) $$X\cdot A = 0 \quad \text{if} \quad \text{car}\,(D_X) \cap \text{spt}\,(A) = 0 \qquad (\text{sharp } X, A).$$

Write $D_X(p) = \omega(p) = \sum_{(\lambda)} \omega_\lambda(p)e^\lambda$. Set (see (16))

$$\omega^\eta_\lambda(p) = \gamma_n(\omega_\lambda(p)), \qquad D_{X^\eta}(p) = \omega^\eta(p) = \sum_{(\lambda)} \omega^\eta_\lambda(p)e^\lambda.$$

Then clearly wkl$^\#$ $X^\eta = X$. Also spt $(X^\eta) \cap$ spt $(A) = 0$, and hence $X\cdot A = \lim\,(X^\eta\cdot A) = 0$.

From the definition of continuous chains (VI, 7) it is clear that

(21) $$\text{spt}\,(\tilde{\alpha}) = \text{spt}\,(\alpha).$$

We prove (compare (XI, 13.6))

(22) $$|A + B| = |A| + |B| \qquad \text{if} \quad \text{spt}\,(A) \cap \text{spt}\,(B) = 0.$$

Suppose first that A and B are compact. Then (App. III, Lemma 1a) there is a sharp ϕ, $0 \leqq \phi(p) \leqq 1$, which is 1 in spt (A) and is 0 in spt (B). Now (1.17), (11) and (13) give, if $\psi(p) = 1 - \phi(p)$,

$$|A + B| = |(\phi + \psi)(A + B)| = |\phi(A + B)| + |\psi(A + B)|$$
$$= |A| + |B|.$$

In the general case, the proof is reduced to the compact case with the help of Theorem 4A below. Note that if one of $|A|$, $|B|$ is infinite, so is $|A + B|$.

We prove (see also (XI, 12.5))

(23) $$|\phi A| = |\langle\phi\rangle A| \quad \text{if} \quad |A| \text{ is finite.}$$

With ϕ_η as in (16), let ϕ^+_η and ϕ^-_η be its positive and negative parts; then $\phi_\eta = \phi^+_\eta - \phi^-_\eta$, $\langle\phi_\eta\rangle = \phi^+_\eta + \phi^-_\eta$. Since spt $(\phi^+_\eta) \cap$ spt $(\phi^-_\eta) = 0$, we have

$$|\phi_\eta A| = |\phi^+_\eta A| + |(-\phi^-_\eta)A| = |\phi^+_\eta A| + |\phi^-_\eta A| = |\langle\phi_\eta\rangle A|.$$

Since $|\phi A - \phi_\eta A| \leqq |\phi - \phi_\eta|\,|A| \leqq \eta\,|A|$ etc., (23) follows at once.

We give the relations (using either norm)

(24) $$\phi X\cdot A = 0 \quad \text{if} \quad \phi(p) = 0 \quad \text{in} \quad \text{spt}\,(X) \cap \text{spt}\,(A),$$

(25) $$|\phi X\cdot A| \leqq N\,|X|\,|A| \quad \text{if} \quad |\phi(p)| \leqq N \quad \text{in} \quad \text{spt}\,(X) \cap \text{spt}\,(A).$$

To prove (24), take any $\eta > 0$. By (10) and (17),

$$\text{spt}\,(\phi_\eta X) \cap \text{spt}\,(A) \subset \text{car}\,(\phi) \cap \text{spt}\,(X) \cap \text{spt}\,(A) = 0;$$

by (6), $\phi_\eta X\cdot A = 0$. Now by (19), $\phi X\cdot A = 0$.

To prove (25), let $\psi(p)$ be $\phi(p)$, N, or $-N$, according as $|\phi(p)| \leqq N$, $\phi(p) > N$, or $\phi(p) < -N$. Then $\theta(p) = \phi(p) - \psi(p) = 0$ in $\operatorname{spt}(X) \cap \operatorname{spt}(A)$, hence $\theta X \cdot A = 0$, and

$$|\phi X \cdot A| = |\psi X \cdot A| \leqq |\psi| \, |X| \, |A| \leqq N |X| \, |A|.$$

Finally (for flat or sharp A and X),

$$(26) \qquad |\phi A| \leqq N |A| \quad \text{if} \quad |\phi(p)| \leqq N \quad \text{in spt}\,(A),$$

$$(27) \qquad |\phi X| \leqq N |X| \quad \text{if} \quad |\phi(p)| \leqq N \quad \text{in spt}\,(X).$$

For take any X; then by (25), $|X \cdot \phi A| \leqq N |X| \, |A|$, and (26) follows, and similarly (27).

4. On non-compact chains. We shall show that most of any chain lies in a compact set. This will generalize (VI, 4.7).

THEOREM 4A. *For any flat chain A and any $\epsilon > 0$ there is a compact flat chain A' such that*

$$(1) \qquad \operatorname{spt}(A') \subset \operatorname{spt}(A), \qquad |A' - A|^\flat < \epsilon,$$

$$(2) \qquad |A' - A| < \epsilon \qquad \textit{if } |A| \textit{ is finite},$$

$$(3) \qquad |\partial(A' - A)| < \epsilon \quad \textit{if} \quad |A| \textit{ and } |\partial A| \textit{ are finite}.$$

This is true in the sharp norm also if we omit (3).

Moreover, if a number N is given, we may find a compact set Q such that for any compact sharp ϕ with

$$(4) \qquad 0 \leqq \phi(p) \leqq 1, \qquad \phi(p) = 1 \text{ in } Q, \qquad \mathfrak{L}_\phi \leqq N,$$

we may use $A' = \phi A$.

To obtain (1), choose B polyhedral so that (if $\dim(A) = r$)

$$|B - A|^\circ < \epsilon/N', \qquad N' = 2 + (r+1)N \qquad \text{(either norm)},$$

and set $Q = \operatorname{spt}(B)$. Then for any ϕ satisfying (4), $\phi B = B$, by (3.13), and (1.4) and (1.5) give

$$|\phi A - A|^\circ \leqq |\phi(A - B)|^\circ + |B - A|^\circ \leqq (N_\phi^{(r)} + 1) |B - A|^\circ < \epsilon.$$

If $|A|$ is finite, choose $\zeta \leqq \epsilon$ so that $|A_1 - A|^\circ < \zeta$ (either norm) implies $|A_1| > |A| - \epsilon$ (V, Theorem 16B). Choose B and Q as above, using ζ in place of ϵ. Then with $A' = \phi A$ as above, (1) holds; also $|\phi A| > |A| - \epsilon$. Setting $\psi(p) = 1 - \phi(p)$ and using (1.17) gives

$$|A' - A| = |\psi A| = |A| - |\phi A| < \epsilon.$$

Suppose both $|A|$ and $|\partial A|$ are finite. By what has just been proved, applied to both A and ∂A, we may find a compact set Q_1 such that for any ϕ' satisfying (4) with Q_1,

$$(5) \qquad |\phi' A - A| < \epsilon/2rN, \qquad |\phi' \partial A - \partial A| < \epsilon/2.$$

Let ϕ_1 be such a function. Set

$$Q = \operatorname{spt}(\phi_1), \qquad \psi_1(p) = 1 - \phi_1(p), \qquad A_1 = \psi_1 A.$$

Now take any ϕ satisfying (4); then (1) and (2) hold with $A' = \phi A$. Set $\psi(p) = 1 - \phi(p)$. By (1.8) and (5) with ϕ_1, since $\mathfrak{L}_\psi \leqq N$,

$$|\psi\partial A_1 - \partial\psi A_1| \leqq r\mathfrak{L}_\psi |A_1| < \epsilon/2.$$

Since car $(\psi) \cap$ spt $(\phi_1) = 0$, (3.7), (3.9) and (3.11) give

$$\psi\partial A = \psi\partial(\phi_1 + \psi_1)A = \psi\partial A_1, \qquad \partial\psi A = \partial\psi A_1.$$

Also

$$\phi\partial A - \partial\phi A = -(\psi\partial A - \partial\psi A),$$

since $\phi + \psi = 1$; hence

$$|\phi\partial A - \partial\phi A| = |\psi\partial A_1 - \partial\psi A_1| < \epsilon/2.$$

Combining this with the last part of (5), with ϕ, gives (3).

REMARK. If $|A|$ is finite, we may omit the hypothesis $\mathfrak{L}_\phi \leqq N$ in (4) in obtaining (1) and (2) (in this connection, see (XI, 12)). For, first find Q_1 so that the first part of (5) holds, using $N = 1$, and choose a corresponding ϕ_1. Set $Q = $ spt (ϕ_1), $\psi_1 = 1 - \phi_1$. Now given ϕ, define ψ as before; then $0 \leqq \psi(p) \leqq \psi_1(p)$, and (with either norm) (1.18) gives

$$|\psi A|^\circ \leqq |\psi A| \leqq |\psi_1 A| < \epsilon.$$

EXAMPLE (a). We cannot omit the hypothesis $\mathfrak{L}_\phi \leqq N$ in obtaining (1). To show this, let B_i be the oriented interval $(i, i + 1/2^i)$ on the real line, and set $B = \sum B_i$, $A = \partial B$. Given any compact Q, we may find a compact ϕ satisfying the first two parts of (4), with $\phi(i) = 1$ and $\phi(i + 1/2^i) = 0$ for some i. Then clearly $|A - \phi A|^\sharp = |A - \phi A|^\flat \geqq 1$.

EXAMPLE (b). We cannot omit the hypothesis that $|A|$ is finite in (3). To show this, construct a 1-cycle A in the plane E^2 as follows. Let Q_i be the square with corners $(\pm 2^i, \pm 2^i)$. Let B_i be the 2-chain formed from a set of 2^{2i} narrow rectangles, each stretching from the bottom of Q_i to the top, evenly distributed over Q_i, of width $1/2^{4i}$; then $|B_i| = 2/2^i$. Let C_i be formed from a similar set of rectangles, crossing Q_i in the other direction. Set

$$B = \sum(B_i + C_i), \qquad A = \partial B.$$

Then $\partial A = 0$. But for any ϕ not identically 0, clearly $|\phi A|$ is infinite, and if ϕ is smooth and not constant, we see easily that $|\partial \phi A|$ is infinite.

Using (1) in the theorem, with either norm, we may give an immediate proof of

(6) $$|A|^\circ = \sup\{X\cdot A\colon\ X \text{ compact}, |X|^\circ \leqq 1\},$$

and hence of (VI, 8.8) and (VI, Lemma 8d), as follows. Choose ϕ, using $\epsilon' = \epsilon/2$, choose Y so that $|Y|^\circ \leqq 1$ and $Y\cdot A \geqq |A|^\circ - \epsilon/2$, and set $X = \phi Y$. We find at once $X\cdot A = Y\cdot\phi A > |A|^\circ - \epsilon$.

5. On polyhedral approximations. We shall prove a theorem similar to Theorem 4A, finding a polyhedral chain. For the proof of (5), we need the second lemma below.

LEMMA 5a. *Given the polyhedral r-chain A, the sharp function ϕ, and $\epsilon > 0$, there is a polyhedral chain B with* spt $(B) \subset$ spt (A), *such that*

(1) $$|B - \phi A| < \epsilon, \qquad |\partial B - \phi\partial A| < r\mathfrak{L}_\phi |A| + \epsilon.$$

Say $A = \sum a_i \sigma_i$. Set $\epsilon' = 2\epsilon \sum |a_i|/3$, and choose H_i and K_i to satisfy (1.10), using σ_i and ϵ'. Set $B = \sum a_i H_i$; then (1) follows at once.

REMARK. The proof shows that we may require B to lie in an arbitrary neighborhood of spt (ϕ).

LEMMA 5b. *Given the polyhedral r-chain A and $\eta > 0$, $\epsilon > 0$, there is a polyhedral chain D such that*

(2) $$D \subset U_\eta(\text{spt}(A)), \quad |A - \partial D| + |D| < [1 + (r+1)/\eta]|A|^\flat + \epsilon.$$

Take $\zeta < \eta$ so that

$$(r+1)|A|^\flat/\zeta < (r+1)|A|^\flat/\eta + \epsilon/4.$$

Choose a polyhedral chain D' such that

$$|A - \partial D'| + |D'| < |A|^\flat + \epsilon', \qquad \epsilon' = \epsilon/[8 + 2(r+1)/\zeta].$$

Define $\phi = \phi_{Q,\zeta}$ as in (V, 12.3), so that $\phi(p) = 1$ in $Q =$ spt (A), $\phi(p) = 0$ outside $U_\zeta(Q)$, and $\mathfrak{L}_\phi = 1/\zeta$. Using D', ϕ and ϵ', choose D by the last lemma and Remark. Then

$$|D - \phi D'| < \epsilon', \qquad |\partial D - \phi\partial D'| < (r+1)\mathfrak{L}_\phi |D'| + \epsilon'.$$

By (3.13), $\phi A = A$. Now

$$\begin{aligned} |A - \partial D| + |D| &\leqq |\phi(A - \partial D')| + |\partial D - \phi\partial D'| \\ &\qquad + |D - \phi D'| + |\phi D'| \\ &\leqq |A - \partial D'| + (r+1)\mathfrak{L}_\phi |D'| + 2\epsilon' + |D'| \\ &\leqq [1 + (r+1)/\zeta](|A - \partial D'| + |D'|) + 2\epsilon' \\ &\leqq [1 + (r+1)/\zeta]|A|^\flat + 3\epsilon/4, \end{aligned}$$

which gives (2).

THEOREM 5A. *For any flat chain A, any neighborhood U of* spt (A), *and any $\epsilon > 0$, there is a polyhedral chain B such that*

(3) $$\text{spt}(B) \subset U, \qquad |B - A|^\flat < \epsilon,$$

(4) $$|B| < |A| + \epsilon \quad \textit{if } |A| \textit{ is finite,}$$

(5) $$|\partial B| < |\partial A| + \epsilon \quad \textit{if } |A| \textit{ and } |\partial A| \textit{ are finite.}$$

The same holds with the sharp norm, omitting (5).

The proof is the same with either norm; we use the flat norm. Because of Theorem 4A, we may clearly suppose that $Q = \text{spt}\,(A)$ is compact. Say $U_{2\eta}(Q) \subset U$.

To prove (3), choose a polyhedral chain B_1 such that

$$| B_1 - A |^\flat < \epsilon' = \epsilon/2[1 + (r + 1)/\eta].$$

Define $\phi = \phi_{Q,\eta}$ as in the proof of the last lemma. Set $B_2 = \phi B_1$. Since $\phi A = A$, we have

$$| B_2 - A |^\flat = | \phi(B_1 - A) |^\flat \leqq N_\phi^{(r)} | B_1 - A |^\flat < \epsilon/2.$$

If $B_1 = \sum b_i \sigma_i$, then $B_2 = \sum b_i \phi \sigma_i$. By approximating to ϕ by a cellwise constant function ϕ' in each σ_i, we obtain a polyhedral chain B in U with $| B - B_2 | < \epsilon/2$; then (3) follows.

Suppose $| A |$ is finite. We may then suppose $| B_1 | < | A | + \epsilon$ in the proof above. Since $| B_2 | \leqq | B_1 |$, and we may clearly take $| B | \leqq | B_2 |$, we obtain $| B | < | A | + \epsilon$.

Suppose finally that $|A|$ and $|\partial A|$ are finite. Choose ϵ' so that $(2 + r/\eta)\cdot 2\epsilon' \leqq \epsilon$. By the proof just given, we may choose polyhedral chains H and K in $U_\eta(Q)$ so that

$$| H - A |^\flat < \epsilon', \qquad | H | < | A | + \epsilon',$$

$$| K - \partial A |^\flat < \epsilon', \qquad | K | < | \partial A | + \epsilon'.$$

Now $| \partial H - K |^\flat < 2\epsilon'$, and by the last lemma, there is a polyhedral chain D in U such that

$$| \partial H - K - \partial D | + | D | < (1 + r/\eta)\cdot 2\epsilon' + \epsilon'.$$

Set $B = H - D$. Then

$$| B - A |^\flat \leqq | H - A |^\flat + | D | < \epsilon,$$

$$| B | \leqq | H | + | D | < | A | + \epsilon,$$

$$| \partial B | \leqq | \partial H - K - \partial D | + | K | < | \partial A | + \epsilon,$$

completing the proof.

6. The r-vector of an r-chain. Recall the definition of the r-vector $\{A\}$ of the cellular (and hence polyhedral) r-chain A in (III, 2). We extend this definition to all sharp chains (and hence flat chains, see (V, Theorem 14B)), by setting

(1) $\{A\} = \lim \{A_i\}$ if $A = \lim^\sharp A_i$, the A_i polyhedral.

We show that the limit exists and is unique, and that

(2) $$| \{A\} |_0 \leqq | A |^\sharp .$$

First we prove (2) for polyhedral A. Given $\epsilon > 0$, write $A = \sum a_i \sigma_i$, and choose the σ_i and v_i and D so that

$$\frac{\sum |a_i| |\sigma_i| |v_i|}{r+1} + \left| \sum a_i T_{v_i} \sigma_i - \partial D \right| + |D| < |A|^\sharp + \epsilon.$$

By (III, Theorem 2B), $\{\partial D\} = 0$. It is clear that $\{T_v \sigma\} = \{\sigma\}$, and $|\{B\}|_0 \leqq |B|$ for any polyhedral B. Hence

$$A = -\sum a_i (T_{v_i} \sigma_i - \sigma_i) + \left(\sum a_i T_{v_i} \sigma_i - \partial D \right) + \partial D,$$

$$|\{A\}|_0 \leqq \left| \sum a_i T_{v_i} \sigma_i - \partial D \right| < |A|^\sharp + \epsilon.$$

Another proof runs as follows. Set $\alpha = \{A\}$, and choose ω_0 so that $|\omega_0|_0 = 1$, $\omega_0 \cdot \alpha = |\alpha|_0$. Define X by $D_X(p) = \omega_0$ (all p); then $|X|^\sharp = |X| = 1$, and (III, 4.1) gives

$$X \cdot A = \omega_0 \cdot \{A\} = |\{A\}|_0.$$

Since $|X \cdot A| \leqq |X|^\sharp |A|^\sharp$, (2) follows.

Because of this, (1) is permissible, and (2) holds for sharp A.

Relation (III, 4.1) now extends to

(3) $$X \cdot A = \omega_0 \cdot \{A\} \quad \text{if} \quad D_X(p) = \omega_0, \quad \text{all } p \qquad (A \text{ sharp}).$$

THEOREM 6A. $\Phi(A) = \{A\}$ *is continuous in* $\mathbf{C}_r^\sharp$.

This follows from (2).

7. Sharp chains at a point. We say the sharp chain $A \neq 0$ is *at the point* p if spt $(A) = p$. Such a chain we call *atomic*; a finite sum of such chains we call *molecular*. There are no flat r-chains $\neq 0$ at p for $r > 0$; see § 9.

THEOREM 7A. *For any point* p *and any* r*-vector* $\alpha \neq 0$ *there is a sharp chain* A *at* p *such that* $\{A\} = \alpha$.

We may suppose α is simple. Let σ_0 be an r-cube with p as a vertex and $\{\sigma_0\} = \alpha$. Let σ_i be the same cube, contracted towards p by the factor $1/2^i$; set $A_i = 2^{ri}\sigma_i$, $A = \lim^\sharp A_i$. Say σ_i has side length $h/2^i$. We may cut σ_i into 2^r cubes σ_{ik}, and find vectors v_{ik}, such that

$$\sigma_{ik} = T_{v_{ik}} \sigma_{i+1}, \qquad |v_{ik}| \leqq d/2^i, \quad d = r^{1/2}h/2.$$

Since $|\sigma_{i+1}| = h^r/2^{r(i+1)}$, (V, 6.3) gives

$$A_{i+1} - A_i = 2^{ri}(2^r \sigma_{i+1} - \sum_k \sigma_{ik}) = 2^{ri} \sum_k (\sigma_{i+1} - T_{v_{ik}} \sigma_{i+1}),$$

$$|A_{i+1} - A_i|^\sharp \leqq \frac{2^{ri+r} |\sigma_{i+1}| d/2^i}{r+1} \leqq \frac{r^{1/2} h^{r+1}}{(r+1)2^{i+1}},$$

proving the existence of A. Since $\{A_i\} = \alpha$, $\{A\} = \alpha$. Clearly spt $(A) = p$.

THEOREM 7B. *There is a one-one correspondence between sharp r-chains at p and r-vectors* $\neq 0$, *defined by* $A \to \{A\}$.

Because of the last theorem, we need merely show that if $\{A\} = 0$, then $A = 0$. Take any sharp X. Define X_0 and Y by

$$D_{X_0}(q) = D_X(p), \quad \text{all } q; \qquad Y = X - X_0.$$

Then $D_Y(p) = 0$, and by (3.20), $Y \cdot A = 0$. Hence, by (6.3),

$$X \cdot A = X_0 \cdot A = D_X(p) \cdot \{A\} = 0, \qquad A = 0.$$

THEOREM 7C. *For any sharp A at p,*

$$X \cdot A = D_X(p) \cdot \{A\}, \qquad |A|^\# = |A| = |\{A\}|_0. \tag{1}$$

The first relation was proved just above. Because of (6.2), we need merely show that for any $\epsilon > 0$, $|A| < |\alpha|_0 + \epsilon$, where $\alpha = \{A\}$. Suppose first that α is simple. By the last theorem, we may express A as in the proof of Theorem 7A. Now

$$|A_i| = |A_0| = |\sigma_0| = |\alpha|, \qquad |A| \leqq |\alpha| = |\alpha|_0.$$

For general α, choose simple r-vectors α_i by (I, 13.1) such that

$$\alpha = \sum \alpha_i, \qquad \sum |\alpha_i| < |\alpha|_0 + \epsilon.$$

Choose A_i at p so that $\{A_i\} = \alpha_i$, by Theorem 7A. Then $\{\sum A_i\} = \sum \alpha_i = \alpha$, and hence $A = \sum A_i$. By the proof above, $|A_i| \leqq |\alpha_i|$. Hence

$$|A| \leqq \sum |A_i| \leqq \sum |\alpha_i| < |\alpha|_0 + \epsilon.$$

We shall give a theorem about the approximation to sharp chains at p. We prove first

$$|A|^\# \leqq |\{A\}|_0 + \epsilon |A|/(r+1) \quad \text{if} \quad \text{spt}(A) \subset U_\epsilon(p). \tag{2}$$

First suppose A is polyhedral. Let N denote the right hand side. It is sufficient to show that $|X \cdot A| \leqq N |X|^\#$ for any sharp X. Set

$$\omega(q) = D_X(q), \qquad \omega_0(q) = \omega(p), \qquad \text{all } q.$$

Using (6.3), we have

$$X \cdot A = \int_A \omega_0(q)\, dq + \int_A [\omega(q) - \omega_0(q)]\, dq,$$

$$\left| \int_A \omega_0(q)\, dq \right| = |\omega(p) \cdot \{A\}| \leqq |\omega(p)|_0 |\{A\}|_0 \leqq |X|^\# |\{A\}|_0,$$

$$\left| \int_A [\omega(q) - \omega_0(q)]\, dq \right| \leqq \mathfrak{L}_0(\omega)\epsilon |A| \leqq |X|^\# \epsilon |A|/(r+1),$$

which gives (2). For general sharp A, (2) follows from the above, using Theorem 5A and (V, 16.1).

THEOREM 7D. *Let $A_1, A_2, \cdots$ be a sequence of chains such that for some N and α,*

$$(3) \qquad |A_i| \leqq N, \qquad \mathrm{spt}(A_i) \subset U_{\epsilon_i}(p), \qquad \epsilon_i \to 0, \quad \{A_i\} \to \alpha.$$

Then

$$(4) \qquad A = \lim{}^{\#} A_i \text{ exists and } \{A\} = \alpha.$$

We may suppose that $\epsilon_j \leqq \epsilon_i$ for $j > i$. By (2),

$$|A_j - A_i|^{\#} \leqq |\{A_j\} - \{A_i\}|_0 + 2N\epsilon_i/(r+1) \quad \text{if} \quad i \leqq j;$$

hence A exists. By Theorem 6A, $\{A\} = \alpha$.

EXAMPLE. We cannot replace $|A_i| \leqq N$ by $|A_i|^{\#} \leqq N$ in (3). For let σ_i be an $(r+1)$-cube of side $1/2^i$, so that $|\sigma_i| = 1/2^{(r+1)i}$, $|\partial\sigma_i| = 2(r+1)/2^{ri}$. Set $A_i = 2^{(r+1)i}\partial\sigma_i$. Then

$$|A_i|^{\#} \leqq 2^{(r+1)i}|\sigma_i| = 1, \qquad \{A_i\} = 0,$$

and if we choose the σ_i to be parallel but with alternating orientations, $\lim^{\#} A_i$ does not exist.

8. Molecular chains are dense.

We prove

THEOREM 8A. *The molecular r-chains are dense in $\mathbf{C}_r^{\#}$.*

We need merely show that given a simplex σ and $\epsilon > 0$, there are chains $A_1, \cdots A_m$ at points $p_1, \cdots, p_m$ such that $|\sigma - \sum A_i|^{\#} \leqq \epsilon$. Write $\sigma = \sum \sigma_i$, the σ_i of diameter $\leqq \zeta = (r+1)\epsilon/2|\sigma|$. Choose $p_i \in \sigma_i$, and let A_i be the chain at p_i with $\{A_i\} = \{\sigma_i\}$. Then (7.2) gives

$$|A_i - \sigma_i|^{\#} \leqq \zeta(|A_i| + |\sigma_i|)/(r+1) = 2\zeta|\sigma_i|/(r+1),$$

and the inequality follows.

We give more precise information for chains of finite mass:

THEOREM 8B. *Let A be a sharp r-chain of finite mass in E^n. Then for each $\epsilon > 0$ there is a compact set Q and a number $\zeta > 0$ with the following property. Let $\phi_0, \phi_1, \cdots, \phi_m$ be sharp functions forming a partition of unity, such that*

$$(1) \qquad \phi_1(p) + \cdots + \phi_m(p) = 1 \quad \text{in } Q, \qquad \mathrm{diam}\,(\mathrm{spt}\,(\phi_i)) \leqq \zeta \quad (i \geqq 1).$$

Let A_i be the r-chain at a point $p_i \in \mathrm{spt}\,(\phi_i)$, with $\{A_i\} = \{\phi_i A\}$. Then

$$(2) \qquad \Big|\sum_{i=1}^{m} A_i - A\Big|^{\#} < \epsilon, \qquad \sum |A_i| = \Big|\sum A_i\Big| \leqq |A|.$$

Choose Q by Theorem 4A and the following Remark, using $\epsilon/2$; set $\zeta = (r+1)\epsilon/4|A|$. Now let the ϕ_i be given. Then (7.2) gives

$$|A_i - \phi_i A|^{\#} \leqq \zeta(|A_i| + |\phi_i A|)/(r+1) \leqq 2\zeta|\phi_i A|/(r+1).$$

By (1.17) and (1.7), $\sum | \phi_i A | \leqq | A |$. By the choice of Q, $| (\phi_1 + \cdots + \phi_m)A - A |^\# < \epsilon/2$. Now

$$\left| \sum_{i=1}^{m} A_i - \sum_{i=1}^{m} \phi_i A \right|^\# \leqq 2\zeta \sum | \phi_i A |/(r+1) \leqq \epsilon/2,$$

and the first part of (2) follows. Also, using (3.22),

$$\left| \sum A_i \right| = \sum | A_i | \leqq \sum | \phi_i A | \leqq | A |.$$

As an application, we show how the mass of a chain may be approximated by sharp norms of "parts" of the chain. See, in this connection, (XI, Theorem 13A).

THEOREM 8C. *In the last theorem, we may choose* Q *and* $\zeta > 0$ *such that, with the* ϕ_i *as before,*

$$\sum_{i=1}^{m} | \phi_i A |^\# > | A | - \epsilon.$$

Choose $\zeta' \leqq \epsilon/2$ such that $| B - A |^\# \leqq \zeta'$ implies $| B | > | A | - \epsilon/2$ (V, Theorem 16B). Now choose Q and ζ by the last theorem, with ζ' in place of ϵ. Then with the ϕ_i as before, some inequalities above give

$$| \phi_i A |^\# \geqq | A_i |^\# - 2\zeta | \phi_i A |/(r+1) = | A_i | - 2\zeta | \phi_i A |/(r+1),$$

and

$$\sum_{i=1}^{m} | \phi_i A |^\# \geqq \left| \sum A_i \right| - 2\zeta | A |/(r+1) > (| A | - \epsilon/2) - \epsilon/2.$$

9. Flat r-chains in E^{r-k} are zero. We prove (using Chapter X), in contrast with the last section,

THEOREM 9A. *Any flat r-chain* A *of* E^n *with* $\operatorname{spt}(A) \subset E^s$ *is zero if* $s < r$.

Suppose E^s is the $(x_1, \cdots, x_s)$-coordinate plane. With γ_η as in (3.15), set

(1) $$f_\eta(x_1, \cdots, x_n) = (x_1, \cdots, x_s, \gamma_\eta(x_{s+1}), \cdots, \gamma_\eta(x_n)).$$

Then

(2) $$\mathfrak{L}_{f_\eta} = 1, \qquad | f_\eta(p) - p | \leqq n\eta, \qquad f_\eta(U_\eta(E^s)) \subset E^s,$$

and f_η maps polyhedral chains into polyhedral chains. (Apply γ_η to each coordinate in turn.) Also

(3) $$| f_\eta(B) |^\flat \leqq | B |^\flat, \qquad \text{any polyhedral } B,$$

which follows at once from the relation (for any polyhedral D)

$$\begin{aligned} | f_\eta(B) - \partial f_\eta(D) | + | f_\eta(D) | &= | f_\eta(B - \partial D) | + | f_\eta(D) | \\ &\leqq | B - \partial D | + | D |. \end{aligned}$$

Suppose $A \neq 0$. Then $|A|^\flat = a > 0$. Let $\epsilon = a/4$. Choose a polyhedral chain B with $|B - A|^\flat < \epsilon$. By (2) and (X, Lemma 5a),

$$|f_\eta(B) - B|^\flat \leqq n\eta(|B| + |\partial B|) < \epsilon$$

for η small enough. By Theorem 5A, we may choose a polyhedral chain B' such that

$$\text{spt}\,(B') \subset U_\eta(E^s), \qquad |B' - A|^\flat < \epsilon.$$

Now $f_\eta(B')$ is a polyhedral r-chain in E^s, and hence is 0. Also $|B - B'|^\flat < 2\epsilon$; applying (3) to $B - B'$ gives

$$|f_\eta B|^\flat = |f_\eta(B - B')|^\flat \leqq |B - B'|^\flat < 2\epsilon.$$

Hence

$$|A|^\flat \leqq |A - B|^\flat + |B - f_\eta B|^\flat + |f_\eta B|^\flat < 4\epsilon = |A|^\flat,$$

a contradiction, proving $A = 0$.

We prove a partial converse of (V, Theorem 16C), from which the last theorem follows at once. (This fails for the sharp norm; see § 7.)

THEOREM 9B. *Let E' be a subspace of E, and let A be a flat chain of E with* spt $(A) \subset E'$. *Then A may be considered as a chain of E'.*

We follow essentially the proof of the last theorem. Given $\epsilon_i = 1/2^i$, find B_i, η_i, B'_i, with $B^*_i = f_{\eta_i} B'_i \subset E'$. We do not have $B^*_i = 0$; but $|A - B^*_i|^\flat_E < 4/2^i$. Hence $\lim^\flat_E B^*_i = A$. Now $|B^*_j - B^*_i|^\flat_{E'} = |B^*_j - B^*_i|^\flat_E \to 0$ (see (V, Theorem 16C)), and $\lim^\flat_{E'} B^*_i = A^*$ exists. But A is precisely the chain A^*, considered as a chain of E (see the theorem quoted).

10. Flat cochains in complexes. Let K be a complex (App. II, 2). Let σ be a cell of K; its space $P(\sigma)$ is an affine space, which may be made Euclidean by the choice of a metric. A *flat r-cochain X in σ* is a linear function $X \cdot A$ of polyhedral r-chains A in σ, such that $|X \cdot A| \leqq N|A|^\flat$ for some N; this condition is independent of the metric chosen. With π as in (V, Lemma 2b), we could extend X through $P(\sigma)$ by setting $X \cdot A = X \cdot \pi A$; then $|X|^\flat$ is unchanged. As noted in (V, 3), we can find $|A|^\flat$ by using polyhedral $(r + 1)$-chains in σ; hence we can find $|X|^\flat$, using chains in σ only.

A *flat r-cochain X in K* is a set of flat r-cochains $X(\sigma)$ in the cells σ of K, such that if σ_1 is a face of σ, then the part of $X(\sigma)$ in σ_1 equals $X(\sigma_1)$; that is, $X(\sigma) \cdot A = X(\sigma_1) \cdot A$ for polyhedral chains A in σ_1. We shall use the same symbol X for all the $X(\sigma)$ when there is no need to distinguish between them. A *polyhedral r-chain A in K* is a sum $\sum A_i$ of polyhedral chains in the cells σ_i of K; set $X \cdot A = \sum X(\sigma_i) \cdot A_i$. The result is independent of the choice of expression of A.

Suppose that K' is a subdivision of K. Let X be a flat cochain of K. For each cell σ' of K', choose a cell σ of K containing it; set $X'(\sigma') = X(\sigma)$ in σ'. This is independent of the choice of σ. Thus a flat cochain X'

in K' is determined. Conversely, let X' in K' be given. Any polyhedral chain A in a cell σ of K may be represented in the form $\sum A_i$, where A_i is in a cell σ_i' of K'; set $X(\sigma)\cdot A = \sum X'(\sigma_i)\cdot A_i$. The result is clearly independent of the representation of A, and determines a flat cochain $X(\sigma)$ in σ; these give a flat cochain X in K. Thus a one-one correspondence between flat cochains X in K and flat cochains X' in K' is determined; we may now speak of a flat cochain X in a polyhedron P. If A is a polyhedral chain in P, we may define $X\cdot A$ with the help of any subdivision of P, and the result is independent of the subdivision.

LEMMA 10a. *A flat cochain in the subcomplex K_1 of K may be extended to be a flat cochain in K.*

It is sufficient to show that if X is defined in the boundary $\partial\sigma$ of a cell σ, it may be extended through σ. Choose $p_0 \in \text{int}(\sigma)$, and set $p_t = (1-t)p_0 + tp$, $p \in \partial\sigma$. Let Q be the subset of σ containing those points p_t for which $t \geqq 1/2$. Set $\pi(p_t) = p$; this maps Q into $\partial\sigma$. For any polyhedral chain A in Q, πA is clearly a polyhedral chain in $\partial\sigma$; setting $X_1\cdot A = X\cdot\pi A$ defines a flat cochain X_1 in Q. Set $\phi(p_t) = 2t - 1$ in Q, and $X = \phi X_1$ in Q; this is a flat cochain in Q (see § 1). Note that X is unchanged in $\partial\sigma$, and $X = 0$ in the subset of Q where $t = 1/2$. Setting $X = 0$ in $\sigma - Q$ completes the definition of X. That X is flat is easy to see.

LEMMA 10b. *Let X be a flat r-cocycle ($r > 0$) in the bounded star shaped subset Q of K. Then X is a coboundary there.*

REMARK. Q might be a cell of K, or the star of a cell, for instance.

Say Q is star shaped from p_0. Set

$$Y\cdot A = X\cdot J(p_0, A), \qquad A \text{ polyhedral in } Q. \tag{2}$$

Using (App. II, 10.3) and the fact that $dX = 0$, we have

$$dY\cdot B = Y\cdot\partial B = X\cdot J(p_0, \partial B) = X\cdot[B - \partial J(p_0, B)] = X\cdot B,$$

and $dY = X$ in Q. Clearly $|Y|$ and $|dY|$ are finite, and hence Y is flat in Q (see the proof of (V, Theorem 4A)).

We now consider *differential forms* in K. An *r-form $\omega(q)$ in K* means a set of r-forms $\omega(\sigma;\ q)$ in the cells σ of K, with the following property. If σ_1 is a face of σ, then for $q \in \sigma_1$, $\omega(\sigma;\ q)\cdot\alpha = \omega(\sigma_1;\ q)\cdot\alpha$ for all r-vectors α in σ_1; thus the part of $\omega(\sigma;\ q)$ in σ_1 equals $\omega(\sigma_1;\ q)$. We can define *flat forms in K* as in (IX, 6); these are needed only in the consideration of products in Theorem 12A below.

11. Elementary flat cochains in a complex. In a complex K, we have flat cochains X as in § 10, and also algebraic cochains x (App. II, 6). Any flat cochain X determines an algebraic cochain ψX, as in (IV, 27.1), by the formula

$$\psi X\cdot\sigma = X\cdot\sigma; \tag{1}$$

thus ψX has the coefficient $X \cdot \sigma$ in the oriented cell σ of K. Clearly

(2) $$d\psi X = \psi dX.$$

Conversely, we shall define a linear mapping ϕ of algebraic cochains x into flat cochains $\phi x = W_x$, such that the following properties hold (with I^0 and $\mathbf{1}$ as in (IV, 27.6)):

(3) $$W_\sigma = 0 \text{ outside St}(\sigma),$$

(4) $$W_{dx} = dW_x,$$

(5) $$\psi\phi x = \psi W_x = x,$$

(6) $$W_{I^0} = \mathbf{1}.$$

To relate this section to (IV, 27), we consider also certain differential forms in K corresponding to the W_σ. The formulas (9) and (12) had been studied by the author in 1947.

In the definition of W_σ, we use the *ratio of two r-vectors* in an affine space, in a certain case:

(7) $$\alpha/\beta = a \quad \text{if} \quad \alpha = a\beta \quad \text{and} \quad \beta \neq 0 \quad \text{is simple.}$$

We must define the flat r-cochain W_σ ($\sigma = \sigma^r$) in each simplex σ' of K. If σ is not a face of σ', set $W_\sigma = 0$ in σ' (i.e. $W_\sigma(\sigma') = 0$ in the notation of § 10). If it is, say

(8) $$\sigma = p_0 \cdots p_r, \quad \sigma' = p_0 \cdots p_r \cdots p_s, \quad \sigma'' = p_{r+1} \cdots p_s \text{ if } s > r;$$

set

(9) $$W_{p_0 \ldots p_r} \cdot (q_0 \cdots q_r) = \frac{\{q_0 \cdots q_r p_{r+1} \cdots p_s\}}{\{p_0 \cdots p_r p_{r+1} \cdots p_s\}}, \qquad q_0 \cdots q_r \subset p_0 \cdots p_s.$$

Using joins, we may write this in the form

(10) $$W_\sigma \cdot \tau = \frac{\{\tau\sigma''\}}{\{\sigma'\}} = \frac{\{\tau\sigma''\}}{\{\sigma\sigma''\}}, \qquad \tau \subset \sigma'.$$

To show that this is a flat cochain in K, take any cells σ_1, σ_2 of K, σ_1 a face of σ_2; we must show that the definitions of W_σ in σ_1 and in σ_2 agree in σ_1. If σ is not a face of σ_2, then $W_\sigma = 0$ in σ_1 and in σ_2. Suppose $\sigma_2 = \sigma'$ as above, but σ is not a face of σ_1. Say $\sigma_1 \subset \sigma^* = p_0 \cdots p_{r-1} p_{r+1} \cdots p_s$. Then for any $\tau = q_0 \cdots q_r \subset \sigma_1$, the numerator in (9) is 0, since $\dim(\sigma^*) < s$ ($\tau\sigma''$ is degenerate); hence $W_\sigma \cdot \tau = 0$, using the definition in σ_2, as it is, using the definition in σ_1. Now suppose σ is a face of σ_1. Then the property follows from the formula

(11) $$\frac{\{q_0 \cdots q_r p_{r+1} \cdots p_{s-1}\}}{\{p_0 \cdots p_r p_{r+1} \cdots p_{s-1}\}} = \frac{\{q_0 \cdots q_r p_{r+1} \cdots p_s\}}{\{p_0 \cdots p_r p_{r+1} \cdots p_s\}},$$
$$q_0 \cdots q_r \subset p_0 \cdots p_{s-1}.$$

To prove this, let a_1 and a_2 denote the two ratios. Then (see (III, 1.2))

$$(q_1 - q_0)\vee\cdots\vee(p_{r+1} - q_r)\vee\cdots\vee(p_{s-1} - p_{s-2}) = a_1(p_1 - p_0)\vee\cdots\vee(p_{s-1} - p_{s-2});$$

multiplying both sides by $p_s - p_{s-1}$ shows that $a_2 = a_1$. (If $s = r + 1$, multiply by $p_{r+1} - p_r$.)

Property (3) holds; since $W_\sigma\cdot\sigma = 1$, (5) holds also. Using (9), (App. II, 10.3) and (III, 2.3) gives, for $q \in \sigma = p_0 \cdots p_s$,

$$[W_{I^0}\cdot q]\{\sigma\} = \sum_{i=0}^{s} (-1)^i\{qp_0 \cdots \hat{p}_i \cdots p_s\} = \{J(q, \partial\sigma)\} = \{\sigma\};$$

hence (6) holds.

We prove (4) in σ', for $x = \sigma$. Take any $\tau = q_0 \cdots q_{r+1}$ in σ'. Since $\{\partial(\tau\sigma'')\} = 0$ (irrespective of $\tau\sigma''$ being degenerate), using the notations above gives

$$\begin{aligned}[dW_\sigma\cdot\tau]\{\sigma'\} &= \left[W_\sigma\cdot \sum_{i=0}^{r+1} (-1)^i(q_0 \cdots \hat{q}_i \cdots q_{r+1})\right]\{\sigma'\} \\ &= \sum_{i=0}^{r+1} (-1)^i\{q_0 \cdots \hat{q}_i \cdots q_{r+1}p_{r+1} \cdots p_s\} \\ &= \sum_{k=r+1}^{s} (-1)^k\{q_0 \cdots q_{r+1}p_{r+1} \cdots \hat{p}_k \cdots p_s\}.\end{aligned}$$

Also, by (App. II, 7.5),

$$W_{d\sigma}\cdot\tau = \sum_{k=r+1}^{s} W_{p_kp_0\cdots p_r}\cdot\tau;$$

comparing these formulas gives (4) for $x = \sigma$ and hence for any x.

Using a metric in σ', (III, 1.3) and (I, 12.14) give

$$\frac{|W_\sigma\cdot\tau|}{|\tau|} \leq \frac{r!(s-r)!}{s!}\frac{|q_r\sigma''|}{|\sigma'|} \leq \frac{r!(s-r-1)!}{s!}\frac{|\sigma''|}{|\sigma'|}\operatorname{diam}(\sigma');$$

hence $|W_\sigma|$ and therefore any $|W_x|$ is finite in σ'. Because of (4), $|dW_x|$ is finite also. Hence W_x is flat in each σ' and therefore also in K.

Let us define differential forms in K as follows. Set $u_{ij} = p_j - p_i$; this is a vector in each simplex of K which has p_ip_j as a face. With σ, σ', σ'' as before, set

$$(12)\qquad \omega_\sigma(q)\cdot\alpha = r!\,\frac{\alpha\vee(p_{r+1} - q)\vee u_{r+1,r+2}\vee\cdots\vee u_{r+1,s}}{u_{01}\vee\cdots\vee u_{0s}} \quad \text{in } \sigma'.$$

Using (III, 1.2), this gives

$$(13)\qquad \omega_0(q)\cdot\{\tau\} = \frac{r!\{\tau\}\vee(s-r)!\{q\sigma''\}}{s!\{\sigma'\}} = \frac{\{\tau\sigma''\}}{\{\sigma'\}}, \qquad q \in \tau \subset \sigma'.$$

Hence, by (III, 4.1) and (10),

$$W_\sigma \cdot \tau = \int_\tau \omega_\sigma, \tag{14}$$

showing that W_σ and ω_σ correspond.

Note that (4) may be proved through the use of the ω_σ, with the help of the formula

$$\sum_{i=0}^{r+1} (-1)^i \{q_0 \cdots \hat{q}_i \cdots q_{r+1} p_{r+1} \cdots p_s\} \tag{15}$$

$$= (-1)^{r+1}(q_1 - q_0) \vee \cdots \vee (q_{r+1} - q_0) \vee u_{r+1,r+2} \vee \cdots \vee u_{r+1,s},$$

whose proof is like that of (III, 1.4).

Finally, we give an explicit formula for ω_σ in K, using the barycentric coordinates μ_i in K:

$$\omega_\sigma(q) = r! \sum_{i=0}^{r} (-1)^i \mu_i(q)\, d\mu_0(q) \vee \cdots \hat{i} \cdots \vee d\mu_r(q), \qquad \sigma = p_0 \cdots p_r. \tag{16}$$

Note that this is exactly (IV, 27.12), using the particular partition $(\mu_1, \mu_2, \cdots)$ of unity in K. We could not use the μ_i (rather, the ν_i) there, since they are not smooth in M.

Let $\bar{\omega}_\sigma$ denote the right hand side of (16); we show that $\bar{\omega}_\sigma = \omega_\sigma$. In any simplex σ' of K (using the notations (8)), μ_i is affine and $d\mu_j$ is constant; hence $\bar{\omega}_\sigma$ is affine. At any vertex of K not in σ, $\bar{\omega}_\sigma$ is 0. The definition of $\bar{\omega}_\sigma$ depends on the oriented simplex σ only, and changes sign with reversal of orientation. The same properties hold for ω_σ. Hence we need only prove that $\bar{\omega}_\sigma(p_0) = \omega_\sigma(p_0)$ in σ'. Using the facts

$$d\mu_i \cdot u_{ji} = 1, \qquad d\mu_i \cdot u_{jk} = 0 \qquad \text{if} \quad i \neq j, \quad i \neq k, \tag{17}$$

and $\mu_i(p_0) = 0$ $(i \neq 0)$, we have

$$\bar{\omega}_\sigma(p_0) \cdot (u_{01} \vee \cdots \vee u_{0r}) = r! \begin{vmatrix} d\mu_1 \cdot u_{01} & \cdots & d\mu_1 \cdot u_{0r} \\ \cdots & \cdots & \cdots \\ d\mu_r \cdot u_{01} & \cdots & d\mu_r \cdot u_{0r} \end{vmatrix} = r!,$$

$$\bar{\omega}_\sigma(p_0) \cdot (u_{0j} \vee \beta) = 0, \qquad j = r+1, \cdots, s;$$

these formulas determine $\bar{\omega}_\sigma(p_0)$. By (12), the same formulas hold for $\omega_\sigma(p_0)$; therefore $\bar{\omega}_\sigma = \omega_\sigma$.

12. The isomorphism theorem. Define the *flat cohomology ring* $\mathbf{H}^\flat$ of K as follows. For each r, $\mathbf{H}^{\flat r}$ is the space of flat r-cocycles in K modulo the space of flat r-coboundaries; $\mathbf{H}^\flat$ is the direct sum of the $\mathbf{H}^{\flat r}$, with multiplication defined through the use of (IX, 14). Let $\mathbf{H}^*$ denote the algebraic cohomology ring of K. Because of (11.2), the mapping ψ of (11.1) defines a linear mapping Ψ of $\mathbf{H}^\flat$ into $\mathbf{H}^*$.

THEOREM 12A. *Ψ is a ring isomorphism of $\mathbf{H}^\flat$ onto $\mathbf{H}^*$.*

First we prove three lemmas.

LEMMA 12a. *Let W be a flat r-cocycle in $\partial\sigma$ ($r \geqq 0$, $\sigma = \sigma^s$, $s \geqq 1$). Suppose*

(1) $$W\cdot\partial\sigma = 0 \quad \text{if} \quad s = r + 1.$$

Then there is a flat r-cocycle W' in σ which equals W in $\partial\sigma$.

LEMMA 12b. *Let W be a flat r-cochain in $\sigma = \sigma^s$ ($r \geqq 1$, $s \geqq 1$), and let X be a flat $(r-1)$-cochain in $\partial\sigma$ such that $dX = W$ there. Suppose*

(2) $$X\cdot\partial\sigma = W\cdot\sigma \quad \text{if} \quad s = r.$$

Then there is a flat cochain X' in σ such that $X' = X$ in $\partial\sigma$ and $dX' = W$ in σ.

Following (IV, 26), we prove (a_0) at once. To prove (b_r), choose X_1 by Lemma 10b so that $dX_1 = W$ in σ. Set $Y = X - X_1$ in $\partial\sigma$. Then $dY = 0$ in $\partial\sigma$, and

$$Y\cdot\partial\sigma = X\cdot\partial\sigma - X_1\cdot\partial\sigma = W\cdot\sigma - dX_1\cdot\sigma = 0 \qquad \text{if} \quad s = r.$$

Hence, by (a_{r-1}), there is a flat cocycle Y' in σ which equals Y in $\partial\sigma$. Set $X' = X_1 + Y'$ in σ.

To prove (a_r), $r > 0$, say $\sigma = p_0 \cdots p_s$; define σ', Q, A as in (IV, 26). Choose the flat cochain X_0 in Q such that $dX_0 = W$ there. If $s - 1 = r$, then

$$W\cdot\sigma' - X_0\cdot\partial\sigma' = W\cdot\sigma' + dX_0\cdot A = W\cdot\partial\sigma = 0;$$

hence, by (b_r), there is a flat cochain X_1 in σ' such that $X_1 = X_0$ in $\partial\sigma'$ and $dX_1 = W$ in σ'. Set $X' = X_0$ in Q and $X' = X_1$ in σ'; then X' is flat in $\partial\sigma$, and $dX' = W$. Let X be an extension of X' through K (Lemma 10a), and set $W' = dX$ in σ; the required properties hold.

LEMMA 12c. *Let W be a flat r-cocycle in K ($r > 0$), and let y be an algebraic $(r-1)$-cochain of K, such that $\psi W = dy$. Then there is a flat $(r-1)$-cochain X in K such that $dX = W$ and $\psi X = y$.*

Let $X = W_y$ (§ 11) in K^{r-1}; then ψX is defined and equals y, and this will hold for any extension of X through K. Note that $dX = W = 0$ here trivially. Take any simplex $\sigma = \sigma^r$ of K. Since

$$X\cdot\partial\sigma = W_y\cdot\partial\sigma = y\cdot\partial\sigma = dy\cdot\sigma = \psi W\cdot\sigma = W\cdot\sigma,$$

we may extend X through σ by Lemma 12b so that $dX = W$ there; thus extend X through K^r. Extend X through the rest of K simplex by simplex, using the same lemma.

The proof of the theorem now goes like that of (IV, Theorem 29A).

REMARK. One consequence of the theorem is that the algebraic cohomology structure (using real coefficients) of a polyhedron P is independent of the (rectilinear) subdivision of P employed.

VIII. Chains and Cochains in Open Sets

So far, we have studied chains and cochains in E^n, with only brief mention of the case that the cochain for instance is defined only in an open set $R \subset E^n$. This case, of obvious importance, is the subject of the present chapter. By defining "R-norms," sharp and flat, of cochains in R, one brings into the theory a wide class of cochains, which need not be extendable through E^n. One may obtain these norms from similar norms of chains in $\dot{R}$. The general study of these norms occupies the first three sections. In the last sections we give expressions for flat and for sharp chains that are of equal importance for chains in E^n.

The definitions of the R-norms are obvious modifications of the definitions in E^n. But one must be careful what chains are called "chains of R". For instance, if R is the interior of a circle C in E^2, we may orient it and assign it the coefficient 1, forming a 2-chain A. It is doubtful whether one should consider A to be a chain of R. If so, it should also be considered as a chain of R', formed from R for instance by cutting out a line segment. Still less would one think of ∂A as a chain of R (its support is C). It is certainly reasonable to require the support spt (A) to lie in R. But even this does not suffice for $X \cdot A$ to be defined for reasonable X; see an example at the beginning of § 1. It is sufficient, however, if A is compact (Theorem 2D) or of finite mass (Theorem 3D), or if A is the limit in the R-norm of a sequence of polyhedral chains of R (Theorem 4A). If A is a chain of R (either norm), it is also a chain of any larger open set (as will be obvious), and is a chain of some smaller open set (Theorems 5B and 6B).

The formulas for R-norms are essentially as in E^n. If R is convex, the norms are the same as in E^n; moreover, cochains in R are extendable through E^n (we use Chapter X in the proof of this fact). Less elementary properties of chains and cochains depend on extending to the present situation the theory of sharp functions as operators and of supports (§ 2). Properties of mass (§ 3) then follow, as in Chapter V.

Recall (V, 3.1) that for polyhedral A, $|A|^\flat$ was defined as the lower bound of $|A - \partial D| + |D|$, with D polyhedral. In § 5 we show that the same formula holds for any flat A, allowing D to be flat. A similar theorem in the sharp case (§ 6) is more difficult; we make use of the "translation norm" of a chain, larger in general than the sharp norm (and existing for instance if $|A|$ is finite).

REMARK. For a more complete theory, one should consider cochains in R which are flat or sharp in each compact subset of R; this corresponds to the use of improper integrals as in (V, 6).

Given the point set Q and the vector v, define the point set $\mathscr{D}_v(Q)$ as follows:

$$\mathscr{D}_v(Q) = \text{all points } q + tv, \qquad q \in Q, \qquad 0 \leqq t \leqq 1.$$

1. Chains and cochains in open sets, elementary properties. In applications, one may have a differential form with bounded first partial derivatives in an open set R, which is not extendable through E^n to have this property. To work with these, one needs a new norm for chains and cochains, depending on the set R. For instance, in E^1, if R is E^1 minus the origin, $r = 0$, and $D_X(t) = t/|t|$, then $X \cdot \partial\sigma = 0$ for any 1-simplex σ in R; but X is not extendable to be a cochain in E^1. If we cut out a small neighborhood of 0, then the X thus formed is extendable to be a cochain Y, but $|dY| > |dX|$. In E^2, if we cut out the closed positive x-axis, or the set of points (x, y) with $y^2 \leqq x^3$, we may set $D_X(x, y) = \inf\{x, 1\}$ for $x \geqq 0$, $y > 0$, and $D_X = 0$ elsewhere, giving a non-extendable 0-cochain; similarly for $r = 1$.

The non-extendable cochains X may not give values $X \cdot A$ for certain chains A lying in R. For instance, if R is E^2 minus the closed positive x-axis, and

$$p_i = (i, 1/2^i), \qquad q_i = (i, -1/2^i), \qquad A = \sum (p_i - q_i),$$

then $|A|^\flat$ is finite, but with X as above, $X \cdot A$ does not exist. With the new norm $|A|_R^\flat$ to be defined, A would not exist, even though $\operatorname{spt}(A) \subset R$. Again, if

$$p_i' = (1/2^i, 1/4^i), \qquad q_i' = (1/2^i, -1/4^i), \qquad A_i = 2^i(p_i' - q_i'),$$

then $|A_i|^\flat = 2/2^i$, $X \cdot A_i = 1$, and $|X \cdot A_i|/|A_i|^\flat$ is unbounded.

We shall define new norms relative to R, and show that in terms of these, practically all the results of chapters V and VII go through.

(a) FLAT CHAINS OF R. For any polyhedral r-chain A in the open set R, $|A|$ is defined as before. Set

(1) $$|A|_R^\flat = \inf\{|A - \partial D| + |D| : \text{polyhedral } D \text{ in } R\}.$$

This is a norm, the *flat R-norm*, and

(2) $$|A|_{R'}^\flat \geqq |A|_R^\flat \geqq |A|^\flat, \qquad A \subset R' \subset R.$$

As before,

(3) $$|\partial A|_R^\flat \leqq |A|_R^\flat.$$

The space of polyhedral chains in R, completed in this norm, is a Banach space $\mathbf{C}_r'^{\flat}(R)$. We shall use only a subset $\mathbf{C}_r^{\flat}(R)$ of this space, consisting of those chains A (in E^n) with the following property. There is a closed set $Q \subset R$, and there is a Cauchy sequence $A_1, A_2, \cdots$ (in the R-norm) of polyhedral chains in Q, such that $A = \lim^{\flat} A_i$. Then $A = \lim^{\flat}_R A_i$ also. Such chains we say are flat chains *of* R, lying *in* Q. We will define spt (A) in § 2; this is the intersection of such sets Q (Theorem 2A). We will then say, more generally, that A is *in* spt (A). An example above shows that for chains of E^n, we can have spt $(A) \subset R$, while A is not a chain of R; this is why we speak of chains "of" R, not "in" R.

Clearly for any flat chain A of R, ∂A is a flat chain of R.

Suppose $r = n$, and R is bounded. Let $A_1, A_2, \cdots$ be similarly oriented polyhedral regions in R, whose union equals R. Though this is a Cauchy sequence in the flat R-norm, the limit A is not in $\mathbf{C}_n^{\flat}(R)$. Nor is ∂A a chain of R (which is quite natural).

If R is E^1 minus the origin, and $A = \partial\sigma$, σ being a 1-cell containing the origin, then $|A|_R^{\flat} = 2$. The reader may examine the R-norm for the other examples above.

The former proofs (V, 3) go through for the following facts, at present for A polyhedral:

$$(4) \qquad |T_v A - A|_R^{\flat} \leqq |v|(|A| + |\partial A|) \quad \text{if} \quad \mathscr{D}_v(\text{spt}\,(A)) \subset R,$$

$$(5) \qquad |\sigma|_R^{\flat} = |\sigma|^{\flat} = |\sigma| \qquad \text{for simplexes } \sigma,$$

$$(6) \qquad |A|_R^{\flat} = \inf\{|A - \partial D|_R^{\flat} + |D|_R^{\flat}\colon D \text{ polyhedral in } R\}.$$

We compare $|A|_R^{\flat}$ with $|A|^{\flat}$ in § 2 below. Because of (V, Lemma 2b) (see also (h) below),

$$(7) \qquad |A|_R^{\flat} = |A|^{\flat} \qquad (\text{all } A) \quad \text{if} \quad R \text{ is convex.}$$

(b) FLAT COCHAINS IN R. A *flat r-cochain* in R is an element of the conjugate space $\mathbf{C}^{\flat r}(R)$ of $\mathbf{C}_r^{\flat}(R)$. The *flat R-norm* $|X|_R^{\flat}$ and *comass* $|X|$ are, as before, the smallest numbers satisfying

$$(8) \quad |X\cdot A| \leqq |X|_R^{\flat}|A|_R^{\flat}, \qquad |X\cdot A| \leqq |X||A| \quad (\text{polyhedral } A \text{ in } R).$$

As before,

$$(9) \qquad |A|_R^{\flat} = \sup\{|X\cdot A|\colon X \text{ in } R, |X|_R^{\flat} = 1\}.$$

If X is in R, so is dX, defined as before; $ddX = 0$. We could define $dX\cdot A$ if simply ∂A is a chain of R; but we shall not use this. Again we prove (considering X in R alone, even if it is defined in a larger set)

$$(10) \qquad |X|_R^{\flat} = \sup\{|X|, |dX|\}.$$

Note that if X_S denotes X, considered in S only, then

$$(11) \qquad |X_{R'}|^\flat_{R'} \leqq |X_R|^\flat_R \leqq |X|^\flat, \qquad R' \subset R,$$

for X defined in E^n, either because of (8) and (2) or because of (10). If X is defined just in R, the first part holds.

As before, a linear function $X \cdot A$ of polyhedral r-chains A of R with $|X|$ and $|dX|$ finite defines a flat r-cochain in R.

Again we find

$$(12) \qquad |X| = \sup |X \cdot \sigma| / |\sigma|, \qquad |dX| = \sup |X \cdot \partial\sigma| / |\sigma|.$$

(c) Sharp chains of R. The *sharp R-norm* $|A|^\sharp_R$ of a polyhedral r-chain A in R is defined by (V, 6.1), using $|\ |^\flat_R$ in place of $|\ |^\flat$, and with the requirement that

$$(13) \qquad \text{each } \mathscr{D}_{v_i}(\sigma_i) \text{ lie in } R;$$

compare (2.8). This is a norm, and

$$(14) \quad |A|^\sharp_R \leqq |A|^\flat_R, \qquad |A|^\sharp_{R'} \geqq |A|^\sharp_R \geqq |A|^\sharp, \qquad A \subset R' \subset R.$$

Clearly (5) holds in this norm. Also

$$(15) \qquad |T_v A - A|^\sharp_R \leqq |v| \, |A| / (r+1) \quad \text{if} \quad \mathscr{D}_v(\text{spt}(A)) \subset R.$$

Sharp chains of R are defined as in the flat case.

For $r = 0$, the sharp and flat R-norms coincide.

(d) Sharp cochains in R. Define $\mathbf{C}^{\sharp r}(R)$ like $\mathbf{C}^{\flat r}(R)$. Then

$$(16) \qquad |X \cdot A| \leqq |X|^\sharp_R |A|^\sharp_R, \qquad |X|^\flat_R \leqq |X|^\sharp_R,$$

$$(17) \qquad |X_{R'}|^\sharp_{R'} \leqq |X_R|^\sharp_R \qquad (X_R \text{ in } R,\ R' \subset R).$$

Every sharp cochain in R is a flat cochain in R. Set

$$(18)\ \mathfrak{L}_{X,R} = \sup \left\{ \frac{|X \cdot (T_v A - A)|}{|A| \, |v|} : A \text{ polyhedral, } \mathscr{D}_v(\text{spt}(A)) \subset R \right\}.$$

Again

$$(19) \qquad \mathfrak{L}_{X,R} = \sup \left\{ \frac{|X \cdot (T_v \sigma - \sigma)|}{|\sigma| \, |v|} : \text{simplexes } \sigma,\ \mathscr{D}_v(\sigma) \subset R \right\}.$$

As before, requiring that all deformations of chains by vectors lie in R, we prove

$$(20) \qquad |dX| \leqq (r+1)\mathfrak{L}_{X,R} \quad \text{if} \quad X \text{ is sharp in } R,$$

$$(21) \quad |X|^\sharp_R = \sup \{|X|^\flat_R, (r+1)\mathfrak{L}_{X,R}\} = \sup \{|X|, (r+1)\mathfrak{L}_{X,R}\}.$$

Any flat cochain in R with $\mathfrak{L}_{X,R}$ finite is sharp in R.

A *sharp function ϕ in R* is a bounded real function with

$$(22) \qquad \mathfrak{L}_{\phi,R} = \sup \left\{ \frac{|\phi(q) - \phi(p)|}{|q - p|} : \text{segment } pq \subset R \right\}$$

finite; sharp functions in R correspond exactly to sharp 0-cochains in R. For $r = n$, sharp functions also correspond to sharp n-cochains, the correspondence depending on the metric.

(e) For each r, the mass, flat R-norm, and sharp R-norm for polyhedral chains in R are *characterized* as the largest semi-norms satisfying the first, the first two, and all of the relations

$$(23) \qquad \begin{cases} |\sigma^r|' \leqq |\sigma^r| \quad (\sigma^r \subset R), \quad |\partial\sigma^{r+1}|' \leqq |\sigma^{r+1}| \quad (\sigma^{r+1} \subset R), \\ |T_v\sigma^r - \sigma^r|' \leqq |\sigma^r|\,|v|/(r+1) \qquad (\mathscr{D}_v(\sigma^r) \subset R) \end{cases}$$

respectively. The proof in (V, 8) goes through, requiring all chains and deformations of chains to lie in R.

(f) SHARP r-FORMS IN R. The *Lipschitz comass constant in R* of the differential r-form ω, defined in R, is

$$(24) \qquad \mathfrak{L}_0(\omega, R) = \sup\{|\omega(q) - \omega(p)|_0/|q - p| : pq \subset R\}.$$

We say ω (defined in R) is *sharp in R* if $|\omega|_0$ and $\mathfrak{L}_0(\omega, R)$ are finite; the *sharp R-norm* of ω is

$$(25) \qquad |\omega|_R^\sharp = \sup\{|\omega|_0, (r+1)\mathfrak{L}_0(\omega, R)\}.$$

As in (V, Theorem 10A), the relation $X\cdot\sigma = \int_\sigma D_X$ $(\sigma \subset R)$ defines a one-one correspondence between sharp cochains and sharp forms in R, and

$$(26) \qquad |D_X|_0 = |X|, \qquad \mathfrak{L}_0(D_X, R) = \mathfrak{L}_{X,R}, \qquad |D_X|_R^\sharp = |X|_R^\sharp.$$

(g) WEAK CONVERGENCE. Letting $\odot$ denote either norm, write

$$(27) \qquad X = \mathrm{wkl}_R^\odot X_i \quad \text{if} \quad |X_i|_R^\odot \leqq N, \qquad \lim (X_i\cdot A) = X\cdot A,$$

for some N, and all chains A of R. As in (V, 13), the last relation may be replaced by $\lim (X_i\cdot\sigma) = X\cdot\sigma$ (all σ in R). If X has these properties, it is a cochain in the corresponding norm.

Given the flat or sharp X in R, we may smooth it in $\mathrm{int}_\zeta(R)$:

$$(28) \qquad X_\eta\cdot A = \int_V \kappa_\eta(v)(X\cdot T_vA)\,dv, \qquad A \text{ in } \mathrm{int}_\zeta(R), \qquad \eta < \zeta.$$

For we may write $A = \lim_R^\odot A_i$, the A_i being polyhedral chains in $Q = \overline{\mathrm{int}_\zeta(R)}$; then $T_vA_i \subset Q_\zeta = \bar{U}_\zeta(Q)$, Q_ζ is closed and is in R, and hence $T_vA = \lim_R^\odot T_vA_i$ exists for $|v| \leqq \eta$, so that (28) has meaning. (By (X, Theorem 7B), T_vA is continuous.)

The properties of (V, 13) hold in $\mathrm{int}_\zeta(R)$.

PROBLEM. Given the flat r-cochain X in R, does there exist a sequence of sharp r-cochains $X_1, X_2, \cdots$ in R such that $\mathrm{wkl}_R^\flat X_i = X$? The difficulty is in defining the X_i so that the $|dX_i|$ are uniformly bounded.

Because the problem is unanswered, we cannot at once extend the results of (V, 14) to the present case. In this connection, see the Remark in the introductory pages.

(h) CONVEX OPEN SETS. We shall show that, in either norm,

(29) $$| A |_R^{\odot} = | A |^{\odot} \qquad \textit{if } R \textit{ is convex,}$$

and also that *any cochain X in R can be extended to a cochain Y in E so that*

(30) $$| Y |^{\odot} = | X |_R^{\odot} \qquad \textit{if } R \textit{ is convex.}$$

In the flat norm, (29) is (7). Next, let π be the projection of E^n onto $\bar{R}$: $\pi(p) =$ nearest point of $\bar{R}$ to p. Then $\mathfrak{L}_\pi = 1$. For, given p, q, set $p' = \pi(p)$, $q' = \pi(q)$, and let P, Q be the $(n - 1)$-planes through p', q', respectively, perpendicular to $p'q'$. Since $p'q' \subset \bar{R}$, neither p nor q lie between these planes; hence $| q - p | \geqq | q' - p' |$. Now suppose X is a sharp cochain in R, so that $| D_X |$ and $\mathfrak{L}_0(D_X)$ are finite. Extend D_X so as to be continuous in $\bar{R}$, and set $D_Y(p) = D_X(\pi(p))$ in E^n. Then $| D_Y | = | D_X |$, $\mathfrak{L}_0(D_Y) = \mathfrak{L}_0(D_X)$, so that Y, defined by D_Y, is the required extension of X. Because of this fact and (9) with $^\sharp$, (29) holds in the sharp norm.

Finally, given the flat cochain X in R, we find the extension Y, with the help of Chapter X, as follows. For any simplex $\sigma = p_0 \cdots p_r$ with vertices in $\bar{R}$, choose a sequence $p_{i1}, p_{i2}, \cdots \to p_i$ of points of R, and set

$$Y\cdot(p_0 \cdots p_r) = \lim_{k \to \infty} X\cdot(p_{0k} \cdots p_{rk});$$

the existence and properties of Y so far follow from (X, Lemma 5a). Next, for any Lipschitz chain A in $\bar{R}$ (X, 6), let $A = \lim^\flat B_k$, as in (X, 6.1), and set $Y\cdot A = \lim Y\cdot B_k$. Finally, for any polyhedral chain A in E^n, πA is a Lipschitz chain in $\bar{R}$, and we may set $Y\cdot A = Y\cdot\pi A$; (30) follows from (X, 7.10).

2. Chains and cochains in open sets, further properties. The remaining parts of Chapters V and VII which concern us, applied to an open set R, are less easy to obtain. Some rearrangement and some new material is necessary.

(a) SHARP FUNCTIONS TIMES CHAINS AND COCHAINS. Let ϕ be sharp in R. (Usually ϕ will be the part in R of a sharp function in E^n.) We define ϕA for polyhedral A in R as before. With

(1) $$N_{\phi,R}^{(r)} = | \phi | + (r + 1)\mathfrak{L}_{\phi,R},$$

we prove, exactly as before, letting $^\odot$ denote either norm,

(2) $$| \phi A |_R^{\odot} \leqq N_{\phi,R}^{(r)} | A |_R^{\odot},$$

(3) $$| \phi\partial A - \partial\phi A |_R^{\flat} \leqq r\mathfrak{L}_{\phi,R} | A |,$$

(4) $$| \phi(T_v\sigma - \sigma) |_R^{\sharp} \leqq N_{\phi,R}^{(r)} | v | | \sigma |/(r + 1) \quad \text{if} \quad \mathscr{D}_v(\sigma) \subset R,$$

at first for polyhedral A in R. Since we have not defined mass for chains of R yet, we give the weakened form (3) of (VII, 1.8); its proof clearly gives (3). See (3.10) below for the strong inequality. Because of (2), we may define ϕA for flat or sharp chains A of R, and (2) still holds. We consider proofs involving mass later.

The definition and properties of ϕX go as before.

Lemma 8c of Chapter VI holds, in either norm:

$$\text{(5)} \quad \text{wkl}_R^{\circ}\, \phi_i X = X \quad \text{if} \quad |\phi_i|_R^{\circ} \leqq N, \qquad \phi_i = 1 \text{ in } R_i, \qquad R_i \subset R_{i+1}, \qquad \bigcup R_i = R.$$

(b) SUPPORTS. The *support* of a chain or cochain of R is defined just as before; take the neighborhoods $U_\epsilon(p)$ in R. Clearly spt (A) *is the same as if we consider A in E^n*. If A is in the closed set Q, clearly spt $(A) \subset Q$; see also Theorem 2A below. Supports of chains of R are closed sets which lie in R; supports of cochains in R are closed subsets of R (which need not be closed in E^n).

Say A is *compact* if we can write $A = \lim_R^{\circ} A_i$, the A_i being polyhedral chains in a compact set in R. (Because of Theorem 2A below, this is equivalent to requiring spt (A) to be compact.) Many properties can be best proved for the compact case, and then extended to the general case by means of the following lemma:

LEMMA 2a. *Given the chain A of R (either norm) and $\epsilon > 0$, there is a compact sharp ϕ in E^n such that*

$$\text{(6)} \qquad 0 \leqq \phi(p) \leqq 1, \qquad \mathfrak{L}_\phi \leqq 1, \qquad |\phi A - A|_R^{\circ} < \epsilon.$$

The proof of (VII, 4.1) applies, using (2).

The next lemma becomes trivial after properties of supports have been obtained.

LEMMA 2b. *For any chain A of R (either norm), ϕA is compact if ϕ is.*

Write $A = \lim_R^{\circ} A_i$, the A_i being polyhedral chains in the closed set $Q \subset R$. We may approximate to ϕA_i by a polyhedral chain B_i with spt $(B_i) \subset$ spt (ϕA_i). Now $\phi A = \lim_R^{\circ} B_i$, the B_i lying in spt $(\phi) \cap Q$.

We prove that spt $(A) = 0$ *implies that* $A = 0$, first under the assumption that A is compact (as defined above). Say $A = \lim_R^{\circ} A_i$, the A_i being polyhedral chains in Q, $Q \subset R$ compact. Each $p \in Q$ is in a neighborhood $U(p)$ such that for any sharp Y with $D_Y = 0$ outside $U(p)$, $Y \cdot A = 0$. A finite number of these neighborhoods, say $U_1, \cdots, U_m$, cover Q. Let $\phi_0, \phi_1, \cdots, \phi_m$ be a corresponding partition of unity (App. III, Lemma 2a). Now take any sharp X in R. Then $\phi_i X \cdot A = 0$ $(i > 0)$, since $D_{\phi_i X} = \phi_i D_X = 0$ outside U_i, and $\phi_0 X \cdot A = 0$, since $\phi_0 X \cdot A_i = 0$. Hence $X \cdot A = \sum(\phi_i X \cdot A) = 0$. In the sharp case, this proves $A = 0$. In the flat case, let the above X be flat. Then $\phi_0 X \cdot A = X \cdot \lim \phi_0 A_k = 0$. For $i > 0$, we may extend $\phi_i X$ to be a flat cochain Y_i in E, with $Y_i \cdot \sigma = 0$

for σ outside R (see (d) below, with $\rho = 1$). Let $(Y_i)_\eta$ be the average of Y_i (V, Theorem 13A). For η small enough, clearly $D_{(Y_i)_\eta} = 0$ outside U_i; hence $\phi_i X \cdot A = \lim (Y_i)_\eta \cdot A = 0$. Again $X \cdot A = 0$, and $A = 0$.

For general A in R, we may choose a sequence $\phi_1, \phi_2, \cdots$ of functions as in Lemma 2a such that $\lim_R^\circ \phi_i A = A$. Clearly $\operatorname{spt}(\phi_i A) \subset \operatorname{spt}(A) = 0$. By Lemma 2b, $\phi_i A$ is compact. By the proof above, $\phi_i A = 0$. Hence $A = 0$.

All the other properties of supports, (VII, 3.5-14), (VII, 3.19-20), go through with no essential change, except the following. To prove (VII, 3.6), suppose first that A is compact (as defined above). Then dist (spt (A), spt (X)) > 0 ,and we may construct ϕ as in the former proof, and find $X \cdot A = 0$. For general A, use Lemma 2a.

The following lemma will be extended in Theorem 3C below.

LEMMA 2c. *For any chain A of R (either norm), any neighborhood U of* spt (A), *and any $\epsilon > 0$, there is a polyhedral chain B such that*

$$\operatorname{spt}(B) \subset U, \qquad |B - A|_R^\circ < \epsilon.$$

The proof is the same as that of (VII, 5.3), using Lemma 2a.

THEOREM 2A. *For any chain A of R (either norm),* spt (A) *is the intersection of all closed sets $Q \subset R$ such that A is in Q, in the sense of* § 1, (a).

Clearly spt $(A) \subset Q$ for any such Q. Conversely, let $Q_1, Q_2, \cdots$ be a sequence of closed sets in R, each containing a neighborhood of spt (A), such that their intersection is spt (A). The last lemma shows that A lies in each Q_i.

Of course spt (A) need not be such a set Q, for instance, if A is a 1-chain formed by a curved arc.

(c) THE R-ρ-NORMS. We extend the definitions of (V, 15) to the case of open sets. We need various properties in studying mass, and other properties, with $\rho = 1$, for general purposes.

The *flat R-ρ-norm* of the polyhedral chain A of R is

$$(7) \qquad |A|^\flat_{R,\rho} = \inf \{|A - \partial D| + |D|/\rho\colon \text{ polyhedral } D \text{ in } R\}.$$

The *sharp R-ρ-norm is*

$$(8) \qquad |A|^\sharp_{R,\rho} = \inf \left\{ \frac{\sum |a_i|\,|\sigma_i|\,|v_i|}{(r+1)\rho} + \left| \sum a_i T_{v_i} \sigma_i \right|^\flat_{R,\rho} : \right.$$
$$\left. A = \sum a_i \sigma_i, \text{ each } \mathscr{D}_{v_i}(\sigma_i) \subset R \right\}.$$

Note that each σ_i is in R. Following former proofs, we find at once, at first in the polyhedral case,

$$(9) \qquad |\partial A|^\flat_{R,\rho} \leqq |A|^\flat_{R,\rho}/\rho,$$

$$(10) \qquad |T_v A - A|^\sharp_{R,\rho} \leqq |A|\,|v|/[(r+1)\rho] \quad \text{if} \quad \mathscr{D}_v(\operatorname{spt}(A)) \subset R.$$

The first extends at once to any flat chain A of R (in the R-ρ-norm). See also (3.7), (3.8). We find also, for X in R,

$$(11)\qquad |X|^{\flat}_{R,\rho} = \sup\{|X|, \rho|dX|\},$$

$$(12)\qquad |X|^{\sharp}_{R,\rho} = \sup\{|X|, (r+1)\rho\mathfrak{L}_{X,R}\}.$$

The remaining formulas and theorems of (V, 15) go through.

As in § 1, (e), the norms are characterized by two or three of the relations (taking σ and $\mathscr{D}_v(\sigma)$ in R)

$$(13)\qquad |\sigma|' \leqq |\sigma|, \quad |\partial\sigma|' \leqq |\sigma|/\rho, \quad |T_v\sigma - \sigma|' \leqq |\sigma||v|/[(r+1)\rho].$$

With

$$(14)\qquad N^{(r)}_{\phi,R,\rho} = |\phi| + (r+1)\rho\mathfrak{L}_{\phi,R},$$

we find, in either norm,

$$(15)\qquad |\phi A|^{\odot}_{R,\rho} \leqq N^{(r)}_{\phi,R,\rho}|A|^{\odot}_{R,\rho}, \qquad |\phi X|^{\odot}_{R,\rho} \leqq N^{(r)}_{\phi,R,\rho}|X|^{\odot}_{R,\rho}.$$

(d) On the extension of cochains. (See also Lemma 4b.) Let X be a flat or sharp cochain in R such that spt (X) is a closed set lying in R. Then we may define an extension Y of X in E, $Y = 0$ outside R, and will have

$$(16)\qquad |Y| = |X|, \quad |dY| = |dX|, \quad \mathfrak{L}_Y = \mathfrak{L}_{X,R}, \quad |Y|^{\odot}_{\rho} = |X|^{\odot}_{R,\rho}.$$

To define Y, take any σ, and write it as a polyhedral chain $\sum\sigma_i$ such that each σ_i either lies in R or has no points in spt (X). Write $Y\cdot\sigma_i = X\cdot\sigma_i$ in the first case and $=0$ in the second, and set $Y\cdot\sigma = \sum Y\cdot\sigma_i$. The result is clearly independent of the expression $\sum\sigma_i$ used (see (App. II, Lemma 3b)).

The first two relations in (16) are evident. The next will follow if we prove the proper inequality on $Y\cdot(T_v\sigma - \sigma)$ for any sufficiently small σ. This will follow from the inequalities

$$|Y\cdot(T_{kv/m}\sigma - T_{(k-1)v/m}\sigma)| \leqq \mathfrak{L}_{X,R}|v||\sigma|/m, \qquad k = 1, \cdots, m.$$

With m large enough, the deformation chain concerned either lies in R or has no points in spt (X), which makes the inequality clear. The last inequality in (16) now follows also.

(e) On the norms of compact chains. Let A be a chain (sharp or flat) of R. Because of (1.2) and (1.14), it may also be considered as a chain of R', if $R \subset R'$; see (f) below. The converse need not be true; see the beginning of § 1. But it is true if A is compact. This is based on the following theorem.

Theorem 2B. *Let A be a chain (either norm) of R; then*

$$(17)\qquad |A|^{\odot}_{R,\rho} \leqq [1 + (r+1)\rho/\eta]|A|^{\odot}_{\rho} \quad \text{if} \quad U_\eta(\text{spt}(A)) \subset R.$$

Take any $\zeta < \eta$, and any $\epsilon > 0$. Choose a cochain X in R (same norm) such that

$$|X|^{\odot}_{R,\rho} \leqq 1, \qquad X\cdot A \geqq |A|^{\odot}_{R,\rho} - \epsilon.$$

Define $\phi = \phi_{Q,\eta}$, $Q = \operatorname{spt}(A)$, as in (V, 12.3). By (d), (15), (14) and (VII, 3.13), we may extend ϕX to give a cochain Y in E^n, with

$$|Y|^{\odot}_{\rho} = |\phi X|^{\odot}_{R,\rho} \leqq 1 + (r+1)\rho/\zeta,$$

$$Y\cdot A = \phi X\cdot A = X\cdot\phi A = X\cdot A \geqq |A|^{\odot}_{R,\rho} - \epsilon.$$

These relations prove

$$|A|^{\odot}_{\rho} \geqq (|A|^{\odot}_{R,\rho} - \epsilon)/[1 + (r+1)\rho/\zeta],$$

which gives (17).

Note that with the flat norm, (17) is an immediate consequence of (VII, Lemma 5b) if A is polyhedral; obvious changes in the proof are required for $\rho \neq 1$. The case of general flat A may be reduced to this with the help of Lemma 2c.

(f) RELATIONS BETWEEN SPACES OF CHAINS AND COCHAINS. Suppose $R' \subset R$. Because of (1.11) and (1.17), every cochain X (flat or sharp) in R defines a cochain $X_{R'}$ in R', the part of X in R'. As noted at the beginning of § 1, a cochain in R' need not be extendable to a cochain in R. We recall that we do not know if $\mathbf{C}^{\sharp r}_R$ is weakly dense in $\mathbf{C}^{\flat r}_R$.

Theorem 14B of Chapter V holds: flat chains of R may be considered as sharp chains of R. We must show that if $A \neq 0$ is a flat chain of R, then the corresponding sharp chain A' is $\neq 0$. By (b), $\operatorname{spt}(A) \neq 0$; hence there is a point $p \in R$ and a sharp chain X in R with $D_X = 0$ outside some $U_\epsilon(p)$, such that $X\cdot A \neq 0$. If $A = \lim^{\flat}_R A_i$, the A_i polyhedral and in the closed set $Q \subset R$, then $X\cdot A = \lim X\cdot A_i$. But also $A' = \lim^{\sharp}_R A_i$; hence $X\cdot A' \neq 0$, and $A' \neq 0$.

THEOREM 2C. *Suppose* $R' \subset R$. *Let* A *be a chain of* R' (*either norm*). *Then it may also be considered as a chain of* R, *in the* R-ρ-*norm, for any* ρ. *The point set* spt (A) *is the same in all cases. If* $A = 0$ *in one norm, the same is true in all; hence the mapping of* $\mathbf{C}^{\odot}_r(R')$ *into* $\mathbf{C}^{\odot}_r(R)$ *is one-one.*

Say $A = \lim^{\odot}_{R'} A_i$, the A_i being polyhedral chains in the closed set $Q \subset R'$. Because of (1.2), (1.14), and (V, 15.7) for R, $\lim^{\odot}_{R,\rho} A_i$ exists for all ρ; hence a chain A' of R in the R-ρ-norm is defined, which may be identified with A as soon as we prove that $A' \neq 0$ if $A \neq 0$. The definition of spt (A) shows that it is independent of R, provided $R' \subset R$; as in (V, Theorem 15A), we see that it is independent of ρ. If $A \neq 0$, then $\operatorname{spt}(A') = \operatorname{spt}(A) \neq 0$, and $A' \neq 0$, completing the proof.

THEOREM 2D. *Let* A *be a chain of* R (*either norm*), *and suppose* $U_\eta(\operatorname{spt}(A)) \subset R'$. *Then* A *may be considered as a chain of* R', *in the* R'-ρ-*norm, for any* ρ. *In particular, if* A *is compact, this holds for any* R' *containing* spt (A).

This is proved like the last theorem, using Lemma 2C and Theorem 2B.

3. Properties of mass. We shall show that the mass of a chain of R is definable as in (V, 16), and is independent of R; former properties of mass continue to hold.

Suppose A is a flat chain of R. Then we define

$$(1) \qquad |A|_{R,\flat} = \inf \{\liminf_{i\to\infty} |A_i| : \lim{}_R^\flat A_i = A, \quad A_i \text{ polyhedral}, \quad A_i \text{ in } Q \text{ (all } i), Q \text{ closed}, Q \subset R\},$$

$$(2) \qquad |A|_{R,(\flat)} = \lim_{\rho\to 0} |A|_{R,\rho}^{\flat},$$

$$(3) \qquad |A|_{R,[\flat]} = \sup \{X \cdot A : X \text{ flat in } R, |X| \leqq 1\}.$$

Define sharp masses similarly; for sharp A, these three at least are defined. Because of the following theorem, we may let $|A|$ denote any of these.

THEOREM 3A. *For flat chains, all six masses are equal; for sharp chains, the sharp masses are equal. For polyhedral A, they all equal $|A|$. They are independent of R, and equal $|A|$, defined by considering A as lying in E^n.*

First we prove $|A|_{R,[\odot]} = |A|_{R,(\odot)}$ (both norms) as before; also $|A|_{R,(\odot)} \leqq |A|_{R,\odot}$.

Next we prove

$$(4) \qquad |A|_{R,\odot} \leqq |A|.$$

We may suppose the latter is finite. Suppose first that A is compact; say $U_{2\eta}(Q_0) \subset R$, $Q_0 = \operatorname{spt}(A)$. Set $R' = U_\eta(Q_0)$, $Q = \bar{R}'$. Given $\epsilon > 0$, we need merely find a polyhedral B in Q such that

$$(5) \qquad |B - A|_R^{\odot} < \epsilon, \qquad |B| < |A| + \epsilon.$$

Taking first the flat norm, choose $\rho \leqq 1$ so that

$$(r+1)\rho |A|/\eta < \epsilon/2, \qquad \rho(|A| + \epsilon) < \epsilon/2.$$

Choose a polyhedral A' in R' so that (see Theorem 2D)

$$|A' - A|_{R',\rho}^{\flat} < \epsilon/2.$$

Then, since $|A|_\rho^\flat \leqq |A|$, (2.17) gives

$$|A'|_{R',\rho}^{\flat} < |A|_{R',\rho}^{\flat} + \epsilon/2 \leqq [1 + (r+1)\rho/\eta] |A|_\rho^\flat + \epsilon/2 < |A| + \epsilon.$$

Hence we may choose a polyhedral D in R' such that

$$|A' - \partial D| + |D|/\rho < |A| + \epsilon.$$

Set $B = A' - \partial D$. Then $|B| < |A| + \epsilon$, and

$$|B - A|_R^\flat \leqq |A' - A|_R^\flat + |\partial D|_R^\flat \leqq |A' - A|_{R',\rho}^\flat + |D|$$
$$< \epsilon/2 + \rho(|A| + \epsilon) < \epsilon.$$

With the sharp norm, choose ρ and A' as before. Then $|A'|^\sharp_{R',\rho} < |A| + \epsilon$, and we may write $A' = \sum a_i\sigma_i$, and find vectors v_i and a polyhedral chain D in R', such that $\mathscr{D}_{v_i}(\sigma_i) \subset R'$, and

$$\frac{\sum|a_i||\sigma_i||v_i|}{(r+1)\rho} \qquad \left|\sum a_i T_{v_i}\sigma_i - \partial D\right| + \frac{|D|}{\rho} < |A| + \epsilon.$$

Set $B = \sum a_i T_{v_i}\sigma_i - \partial D$. Then $|B| < |A| + \epsilon$, and

$$|B - A|^\sharp_R \leqq |A - A'|^\sharp_{R',\rho} + \left|\sum a_i(T_{v_i}\sigma_i - \sigma_i)\right|^\sharp_R + |\partial D|^\sharp_R$$

$$< \frac{\epsilon}{2} + \frac{\sum|a_i||\sigma_i||v_i|}{r+1} + |D| < \frac{\epsilon}{2} + \rho(|A| + \epsilon) < \epsilon.$$

If A is not compact, choose a compact ϕ by Lemma 2a so that $|\phi A - A|^{\odot}_R < \epsilon/2$. Choose a closed neighborhood Q of spt (A), $Q \subset R$. Then $U_\eta(\text{spt}(\phi A)) \subset Q$ for some $\eta > 0$. Choose B in Q, by the proof above, so that $|B - \phi A|^{\odot}_R < \epsilon/2$, $|B| < |\phi A| + \epsilon$. Since $|\phi A| \leqq |A|$, by (VII, 1.7), this gives (5) and hence (4).

Since $|A|^{\odot}_\rho \leqq |A|^{\odot}_{R,\rho}$, letting $\rho \to 0$ gives

$$|A| = |A|_{\odot} \leqq |A|_{R,(\odot)}.$$

Hence all masses (whenever defined) equal $|A|$, completing the proof.

THEOREM 3B. *Mass is lower semi-continuous in any of the spaces considered.*

This is an immediate consequence of (V, Theorem 16B) and the fact that if a chain of R is considered as a chain of E^n, its norm is thereby not increased.

THEOREM 3C. *Theorems* 4A *and* 5A *of Chapter* VII *hold for chains of* R.

The former proofs holds. We obtain $|A' - A|^{\odot}_R < \epsilon$, $|B - A|^{\odot}_R < \epsilon$ (either norm).

THEOREM 3D. *For any chain A of R (either norm) of finite mass, we may consider A as a chain of any open set R' which contains* spt (A).

Choose a neighborhood U of spt (A) whose closure Q is in R'. Choose a compact ϕ_i by the last theorem, so that

$$|A - \phi_i A| < 1/2^{i+1}, \qquad i = 1, 2, \cdots.$$

Say $Q_i = \text{spt}(\phi_i A)$, $U_{2\eta_i}(Q_i) \subset U$. Choose a polyhedral chain B_i in $U_{\eta_i}(Q_i)$ by the last theorem, so that

$$|B_i - \phi_i A|^{\odot}_R < \epsilon_i, \qquad [1 + (r+1)/\eta_i]\epsilon_i < 1/2^{i+1}.$$

Then, by (2.17),

$$| B_i - \phi_i A |_{R'}^{\odot} \leqq [1 + (r+1)/\eta_i] | B_i - \phi_i A |^{\odot} < 1/2^{i+1}.$$

Hence $B_1, B_2, \cdots$ is a Cauchy sequence, in the R'-norm, of chains in Q, defining therefore a chain A' of R'. Using $B_1, B_2, \cdots$ in the R-norm shows that A equals A', considered as a chain of R.

The remaining elementary properties of mass hold, as before; for instance,

$$(6) \qquad | X \cdot A | \leqq | X | \, | A |,$$

$$\left.\begin{aligned} &(7) \quad | T_v A - A |_{R,\rho}^{\flat} \leqq | v | (| A |/\rho + | \partial A |) \\ &(8) \quad | T_v A - A |_{R,\rho}^{\sharp} \leqq | v | \, | A |/(r+1)\rho \end{aligned}\right\} \quad \text{if} \quad \mathscr{D}_v(\text{spt}\,(A)) \subset R,$$

$$(9) \qquad | \phi A | \leqq | \phi | \, | A |, \qquad | \partial \phi A | \leqq r \mathfrak{L}_{\phi,R} | A | + | \phi | \, | \partial A |,$$

$$(10) \qquad | \phi \partial A - \partial \phi A | \leqq r \mathfrak{L}_{\phi,R} | A |.$$

In deducing (10) for flat A from the case of polyhedral A, we use a sequence of polyhedral chains $B_1, B_2, \cdots$ in a closed set $Q \subset R$, such that $\lim_R^{\flat} B_i = A$, $\lim | B_i | = | A |$, found by Theorem 3C; similarly for (7) and (8), requiring $\mathscr{D}_v(\text{spt}\,(B_i)) \subset Q$.

THEOREM 3E. *Let $\alpha(p)$ be a cellwise continuous summable r-vector valued function in E^n. Set $Q = \text{spt}\,(\alpha)$. Then for any open set $R \supset Q$, α defines a flat chain $\tilde{\alpha}$ of R, and $| \tilde{\alpha} | = \int \langle \alpha \rangle_0$, $\text{spt}\,(\tilde{\alpha}) = Q$.*

We may apply (VI, 7.2), Theorems 3D and 3A, and (VII, 3.21).

4. On the open sets to which a chain belongs. In § 1, we defined a chain of R (either norm) as an element of the completion of the linear space of polyhedral chains in the R-norm, provided a certain condition was satisfied. Theorem 4A shows that an equivalent condition is simply $\text{spt}\,(A) \subset R$. The next theorem shows that if A is a chain of R, then it is also a chain of certain smaller open sets. See also Theorems 5B and 6B.

We shall say the sets $Q_1, Q_2, \cdots \to \infty$ if each compact set touches but a finite number of them. We say the functions $\phi_1, \phi_2, \cdots$ form a *spreading sequence* if $0 \leqq \phi_i(p) \leqq 1$, the sets $\text{spt}\,(\phi_i)$ are compact, the sets $\text{spt}\,(1 - \phi_i) \to \infty$, and for some N, $\mathfrak{L}_{\phi_i} \leqq N$ (all i).

LEMMA 4a. *With either norm, let $A_1, A_2, \cdots$ be chains of R, such that $\sum | A_i |_{R,\rho}^{\odot}$ converges, and the sets $\text{spt}\,(A_i) \to \infty$. Then $A = \sum^{\odot} A_i$ is a chain of R, and $A = \sum_{R,\rho}^{\odot} A_i$.*

Choose neighborhoods U_i of $\text{spt}\,(A_i)$ so that $\bar{U}_i \subset R$ and $\bar{U}_i \to \infty$. For each i and k, choose a polyhedral chain A_{ik} in U_i so that $| A_i - A_{ik} |_{R,\rho}^{\odot} < 1/2^{i+k}$ (Lemma 2c, 2(c)); set

$$B_k = A_{1k} + \cdots + A_{kk}.$$

Then if $k' > k$, we see easily that

$$|B_{k'} - B_k|^{\odot}_{R,\rho} \leqq \sum_{j=1}^{k} |A_{jk'} - A_{jk}|^{\odot}_{R,\rho} + \sum_{j=k+1}^{k'} |A_{jk'}|^{\odot}_{R,\rho}$$

$$< \sum_{j=k+1}^{k'} |A_j|^{\odot}_{R,\rho} + 2/2^k;$$

hence $B = \lim^{\odot} B_k$ exists. Since

$$|B_k - (A_1 + \cdots + A_k)|^{\odot} < 1/2^k,$$

we have $B = A$. Now $Q = \bigcup \bar{U}_i$ is a closed set in R, the B_k are polyhedral chains in Q which form a Cauchy sequence in the R-ρ-norm, and $A = \lim^{\odot} B_k$; therefore A is a chain of R, and $A = \lim^{\odot}_{R,\rho} B_k = \sum^{\odot}_{R,\rho} A_k$.

THEOREM 4A. *With either norm, let $A_1, A_2, \cdots$ be chains of R, let A be a chain of E, and suppose*

(1) $$\lim_{i,j\to\infty} |A_j - A_i|^{\odot}_{R,\rho} = 0, \qquad \lim_{i\to\infty}{}^{\odot} A_i = A, \qquad \operatorname{spt}(A) \subset R.$$

Then A is a chain of R, and $A = \lim^{\odot}_{R,\rho} A_i$.

Recall that, as we showed by an example near the beginning of § 1, the condition $\operatorname{spt}(A) \subset R$ alone is not sufficient; but see Theorems 2D and 3D.

First, choosing a polyhedral chain A'_i in R with $|A'_i - A_i|^{\odot}_{R,\rho} < 1/2^i$, we see that the conditions hold with the A'_i in place of the A_i; hence we may suppose the A_i are polyhedral. Also, taking a subsequence if necessary, we may suppose that

(2) $$|A_j - A_i|^{\odot}_{R,\rho} < 1/(r+3)2^i \quad \text{if} \quad j \geqq i.$$

Let $\phi_1, \phi_2, \cdots$ be a spreading sequence such that $\phi_i = 1$ in $\operatorname{spt}(A_i)$ and $\mathfrak{L}_{\phi_i} \leqq 1/\rho$. Then using (VII, 3.13), (2.15) and (2.14) gives

(3) $$|A_j - \phi_i A_j|^{\odot}_{R,\rho} \leqq |A_j - A_i|^{\odot}_{R,\rho} + |\phi_i(A_j - A_i)|^{\odot}_{R,\rho}$$
$$\leqq (1 + N^{(r)}_{\phi_i,R,\rho})/(r+3)2^i \leqq 1/2^i, \qquad j \geqq i.$$

It follows that

(4) $$|(\phi_i - \phi_{i-1})A_j|^{\odot}_{R,\rho} \leqq 3/2^i, \qquad j \geqq i.$$

By Theorem 2D, $\phi_i A$ and $(\phi_i - \phi_{i-1})A$ are chains of R. We now show that

(5) $$\lim_{j\to\infty}{}^{\odot}_{R,\rho} \phi_i A_j = \phi_i A, \qquad \lim_{j\to\infty}{}^{\odot}_{R,\rho} (\phi_i - \phi_{i-1})A_j = (\phi_i - \phi_{i-1})A.$$

Since $\lim^{\odot}_{\rho} A_j = A$ (V, 15), these follow from (2.15), if we omit the R. Since $\operatorname{spt}(\phi_i)$ and $\operatorname{spt}(\phi_i - \phi_{i-1})$ are compact, using (2.17) gives them with R also.

By (VII, Theorem 4A),

$$\phi_0 A + \sum_{i=1}^{\infty}{}^{\odot} (\phi_i - \phi_{i-1})A = \lim_{i\to\infty}{}^{\odot} \phi_i A = A.$$

Also (4) and (5) show that $|(\phi_i - \phi_{i-1})A|_{R,\rho}^{\odot} \leq 3/2^i$. Since spt $((\phi_i - \phi_{i-1})A) \to \infty$, Lemma 4a shows that A is a chain of R, and

$$A = \lim_{i\to\infty}{}_{R,\rho}^{\odot} \phi_i A. \tag{6}$$

Given $\epsilon > 0$, choose k so that

$$|A - \phi_k A|_{R,\rho}^{\odot} < \epsilon/3, \qquad 1/2^k < \epsilon/3.$$

Then by (3), $|A_j - \phi_k A_j|_{R,\rho}^{\odot} \leqq \epsilon/3$, $j \geqq k$. By (5), we may choose $j_0 \geqq k$ so that

$$|\phi_k A - \phi_k A_j|_{R,\rho}^{\odot} < \epsilon/3, \qquad j \geqq j_0.$$

These give $|A - A_j|_{R,\rho}^{\odot} < \epsilon$, $j \geqq j_0$, proving $A = \lim_{R,\rho}^{\odot} A_j$. The theorem is now proved.

In the proof of the next theorem, we need the following lemma.

LEMMA 4b. *Let X be a cochain (either norm) in the open set R. Set*

$$S = \overline{\text{spt}(X) - R}, \qquad R^* = E - S. \tag{7}$$

Then there is a cochain Y in R^* which equals X in R and equals zero in $E - \bar{R}$, such that

$$|Y| = |X|, \qquad |dY| = |dX|, \qquad \mathfrak{L}_{Y,R^*} = \mathfrak{L}_{X,R}, \tag{8}$$

$$|Y|_{R^*,\rho}^{\odot} = |X|_{R,\rho}^{\odot}.$$

The proof in (d) of § 2 applies.

THEOREM 4B. *Let A be a chain of R (either norm). Set*

$$R' = R \cap U_\eta(Q), \qquad Q = \text{spt}(A). \tag{9}$$

Then A is a chain of R', and

$$|A|_{R',\rho}^{\odot} \leqq [1 + (r+1)\rho/\eta]\, |A|_{R,\rho}^{\odot}. \tag{10}$$

Suppose first that A is polyhedral; then A is a chain of R'. Given $\epsilon < \eta$, choose X in R' so that

$$|X|_{R',\rho}^{\odot} \leqq 1, \qquad X\cdot A > |A|_{R',\rho}^{\odot} - \epsilon.$$

Set $\zeta = \eta - \epsilon$, $\phi = \phi_{Q,\zeta}$ (V, 12.3), $Y = \phi X$; extend Y as in the last lemma, using ϕX and R' in place of X and R. Take any $p \in R$. If $p \in R'$, then p is not in S; otherwise, dist $(p, Q) \geqq \eta > \zeta$, p is not in spt (ϕX), and again p is not in S. This shows that $R \subset R^*$, and Y is defined in R. By (8), (2.15) and (2.14),

$$|Y|_{R,\rho}^{\odot} = |\phi X|_{R',\rho}^{\odot} \leqq 1 + (r+1)\rho/\zeta.$$

Also, using (VII, 3.13),

$$Y\cdot A = \phi X\cdot A = X\cdot\phi A = X\cdot A > |A|^{\odot}_{R',\rho} - \epsilon.$$

Since $Y\cdot A \leqq |Y|^{\odot}_{R,\rho} |A|^{\odot}_{R,\rho}$, (10) follows.

Now consider the general case. Given $\epsilon < \eta$, set $\zeta = \eta - \epsilon$, and choose polyhedral chains $A_0, A_1, \cdots$ in $R'' = R \cap U_\epsilon(Q)$ (Lemma 2c) so that

$$|A_i - A|^{\odot}_{R,\rho} < \epsilon/2^i.$$

Set $B_i = A_i - A_{i-1}$ $(i \geqq 1)$; then

$$R_i = R \cap U_\zeta(\text{spt}\,(B_i)) \subset R',$$

and hence, by the proof above, if $N_\zeta = 1 + (r+1)\rho/\zeta$,

$$|B_i|^{\odot}_{R',\rho} \leqq |B_i|^{\odot}_{R_i,\rho} \leqq N_\zeta |B_i|^{\odot}_{R,\rho} \leqq 3N_\zeta\epsilon/2^i.$$

Also

$$|A_0|^{\odot}_{R',\rho} \leqq N_\zeta |A_0|^{\odot}_{R,\rho} \leqq N_\zeta(|A|^{\odot}_{R,\rho} + \epsilon).$$

Now the A_i are chains of R', $\lim |A_j - A_i|^{\odot}_{R',\rho} = 0$, $\lim^{\odot} A_i = A$, and spt $(A) \subset R'$; hence, by Theorem 4A, A is a chain of R', and $A = \lim^{\odot}_{R',\rho} A_i$. Also

$$|A|^{\odot}_{R',\rho} \leqq |A_0|^{\odot}_{R',\rho} + \sum |B_i|^{\odot}_{R',\rho} \leqq N_\zeta(|A|^{\odot}_{R,\rho} + 4\epsilon),$$

and (10) follows.

5. An expression for flat chains. We shall show that any flat chain A of R is the sum of two chains, one, $A - \partial D$, of finite mass, the other, ∂D, the boundary of a chain of finite mass; moreover, as in (1.1), the sum of these masses is arbitrarily near the R-norm of the chain. We shall use the R-ρ-norm, and prove:

THEOREM 5A. *Let A be a flat chain of R. Then for each $\epsilon > 0$ there is a flat chain D of R such that*

(1) $$|A - \partial D| + |D|/\rho < |A|^{\flat}_{R,\rho} + \epsilon.$$

First suppose A is compact. Say

$$U_{3\eta}(Q) \subset R, \qquad Q = \text{spt}\,(A).$$

Set $N_\eta = 1 + (r+1)\rho/\eta$. Given $\epsilon > 0$, set $\epsilon' = \epsilon/(1 + 3N_\eta)$, and choose polyhedral chains $A_0, A_1, \cdots$ in $U_\eta(Q)$ (Lemma 2c) so that

$$|A_i - A|^{\flat}_{R,\rho} < \epsilon'/2^i.$$

We may choose a polyhedral chain D_0 in R so that

$$|A_0 - \partial D_0| + |D_0|/\rho < |A|^{\flat}_{R,\rho} + \epsilon'.$$

For $i \geqq 1$, setting $R' = U_{2\eta}(Q)$, Theorem 2B gives

$$| A_i - A_{i-1} |^{\flat}_{R',\rho} \leqq N_\eta | A_i - A_{i-1} |^{\flat}_{\rho} < 3N_\eta \epsilon'/2^i;$$

hence we may choose a polyhedral chain D_i in $U_{2\eta}(Q)$ so that

$$| A_i - A_{i-1} - \partial D_i | + | D_i | /\rho < 3N_\eta \epsilon'/2^i.$$

Now $D = \sum D_i$ is a chain which lies in a compact set in R; hence D is a chain of R. Also $A = A_0 + \sum(A_i - A_{i-1})$, the sum being convergent in the flat R-ρ-norm; hence

$$A - \partial D = (A_0 - \partial D_0) + (A_1 - A_0 - \partial D_1) + \cdots,$$

and

$$| A - \partial D | + | D |/\rho < | A |^{\flat}_{R,\rho} + \epsilon' + \sum 3N_\eta \epsilon'/2^i = | A |^{\flat}_{R,\rho} + \epsilon.$$

Now take any flat chain A of R. Given $\epsilon > 0$, define N_1 ($\eta = 1$) and ϵ' as before. Choose a spreading sequence $\phi_0, \phi_1, \cdots$ (§ 4) such that (Lemma 2a)

$$| \phi_i A - A |^{\flat}_{R,\rho} < \epsilon'/2^i.$$

By the proof above, there is a flat chain D_0 of R such that

$$| \phi_0 A - \partial D_0 | + | D_0 |/\rho < | A |^{\flat}_{R,\rho} + \epsilon'.$$

Set

$$R_i = R \cap U_1(\operatorname{spt}(\phi_i A - \phi_{i-1} A)).$$

By Theorem 4B,

$$| (\phi_i - \phi_{i-1})A |^{\flat}_{R_i,\rho} \leqq N_1 | (\phi_i - \phi_{i-1})A |^{\flat}_{R,\rho} < 3N_1 \epsilon'/2^i;$$

hence there is a flat chain D_i of R_i such that

$$| (\phi_i - \phi_{i-1})A - \partial D_i | + | D_i |/\rho < 3N_1 \epsilon'/2^i.$$

Since $\sum | D_i |$ converges and $\operatorname{spt}(D_i) \to \infty$, $D = \sum D_i$ is a chain of R (Lemma 4a, or Theorem 3D). Just as before, (1) holds.

As a consequence of the last theorem, we prove a continuity theorem:

THEOREM 5B. *For any flat chain A of R, and ρ, $\epsilon > 0$, there is an open set R' such that $\bar{R}' \subset R$, A is a flat chain of R', and*

$$| A |^{\flat}_{R',\rho} < | A |^{\flat}_{R,\rho} + \epsilon. \tag{2}$$

Choose D by the last theorem, and let R' be a neighborhood of $\operatorname{spt}(A) \cup \operatorname{spt}(D)$ such that $\bar{R}' \subset R$. By Theorem 3D, $A - \partial D$ and D are flat chains of R'; hence so are ∂D and A. Also

$$A |^{\flat}_{R',\rho} \leqq | A - \partial D |^{\flat}_{R',\rho} + | \partial D |^{\flat}_{R',\rho} \leqq | A - \partial D | + | D |/\rho < | A |^{\flat}_{R,\rho} + \epsilon.$$

6. An expression for sharp chains. We shall prove a theorem bearing the same relation to (2.8) that Theorem 5A does to (2.7). First we introduce a new norm.

Given an open set R and $\rho > 0$, the *translation R-ρ-norm* $|A|^T_{R,\rho}$ of a polyhedral chain A in R is given by (2.8), with the flat R-ρ-norm replaced by mass. Clearly

(1) $$|A|^\sharp_{R,\rho} \leqq |A|^T_{R,\rho} \leqq |A|.$$

This is a norm in the linear space of polyhedral chains in R. The elements of the completion of this space, using sequences of polyhedral chains in a closed set in R, form the space of *translation chains of R*. Translation chains may be considered also as sharp chains, as in the case for flat chains (2 (f)).

As in (2.10), we find at once (see (V, Theorem 6A); we need the polyhedral case only)

(2) $$|T_v A - A|^T_{R,\rho} \leqq |A|\,|v|/(r+1)\rho \quad \text{if} \quad \mathscr{D}_v(\text{spt}\,(A)) \subset R.$$

THEOREM 6A. *Let A be a sharp chain of R. Then given $\epsilon > 0$, there is a flat chain C of R and a flat chain of finite mass D of R, such that $A - C$ is a translation chain of R, and*

(3) $$|A - C|^T_{R,\rho} + |C - \partial D| + |D|/\rho < |A|^\sharp_{R,\rho} + \epsilon.$$

Suppose first that A is compact. Say $U_{3\eta}(Q) \subset R$, $Q = \text{spt}\,(A)$. Set $N_\eta = 1 + (r+1)\rho/\eta$, $\epsilon' = \epsilon/(1 + 3N_\eta)$, and choose polyhedral chains $A_0, A_1, \cdots$ in $U_\eta(Q)$ such that

$$|A_i - A|^\sharp_{R,\rho} < \epsilon'/2^i, \qquad i = 0, 1, \cdots.$$

We may write $A_0 = \sum_k a_{0k}\sigma_{0k}$, $C_0 = \sum_k a_{0k} T_{v_{0k}} \sigma_{0k}$, such that $\mathscr{D}_{v_{0k}}(\sigma_{0k}) \subset R$, and find a polyhedral chain D_0 in R, such that

$$|A_0 - C_0|^T_{R,\rho} + |C_0 - \partial D_0| + |D_0|/\rho < |A|^\sharp_{R,\rho} + \epsilon'.$$

Since $U_\eta(\text{spt}\,(A_i - A_{i-1})) \subset R' = U_{2\eta}(Q)$,

$$|A_i - A_{i-1}|^\sharp_{R',\rho} \leqq N_\eta |A_i - A_{i-1}|^\sharp_\rho < 3N_\eta \epsilon'/2^i,$$

and we may write $A_i - A_{i-1} = \sum_k a_{ik}\sigma_{ik}$, $C_i = \sum_k a_{ik} T_{v_{ik}} \sigma_{ik}$, such that $\mathscr{D}_{v_{ik}}(\sigma_{ik}) \subset R'$, and find a polyhedral chain D_i in R', such that

$$|A_i - A_{i-1} - C_i|^T_{R',\rho} + |C_i - \partial D_i| + |D_i|/\rho < 3N_\eta \epsilon'/2^i.$$

Now $A = A_0 + \sum(A_i - A_{i-1})$, and

$$C = \sum C_i, \qquad D = \sum D_i$$

are defined. The last inequality shows that $\sum |C_i|^\flat_{R,\rho}$ and $\sum |D_i|$ converge; hence C is flat, and D is of finite mass. Since

$$A - C = (A_0 - C_0) + \sum (A_i - A_{i-1} - C_i),$$

we see also that $A - C$ is a translation chain. Since the approximating chains all lie in a closed set in R, all are chains of R. The inequality (3) is immediate.

The general case is carried out as in the similar proof in Theorem 5A. We obtain $|\phi_i A - A|^\sharp_{R,\rho} < \epsilon'/2^i$, and

$$|\phi_0 A - C_0|^T_{R,\rho} + |C_0 - \partial D_0| + |D_0|/\rho < |A|^\sharp_{R,\rho} + \epsilon',$$

$$|(\phi_i - \phi_{i-1})A - C_i|^T_{R,\rho} + |C_i - \partial D_i| + |D_i|/\rho < 3N_1\epsilon'/2^i,$$

etc.

THEOREM 6B. *Theorem* 5B *holds also in the sharp norm.*

Using (3), let R' be an open set whose closure is in R, which contains the supports of A, C, and D, and also all the deformation chains $\mathscr{D}_{v_{ik}}(\sigma_{ik})$ used; this is possible for each $(\phi_i - \phi_{i-1})A$ by construction, and in the general case, since spt $((\phi_i - \phi_{i-1})A) \to \infty$. The last inequalities in the proof of the last theorem now hold with R' in place of R on the left, and using Lemma 4a shows that $A - C$ is a sharp chain (in fact, a translation chain) of R'. As in the proof of Theorem 5B, D and $C - \partial D$ are flat chains of R'; hence so are ∂D and C; now $A - C$ and C are sharp chains of R', and hence so is A. Finally, (3) holds with R' in place of R on the left, and gives

$$|A|^\sharp_{R',\rho} \leqq |A - C|^\sharp_{R',\rho} + |C - \partial D|^\sharp_{R',\rho} + |\partial D|^\sharp_{R,\rho} < |A|^\sharp_{R,\rho} + \epsilon.$$

PART III

LEBESGUE THEORY

IX. Flat Cochains and Differential Forms

In Chapter V we set up an abstract definition of r-dimensional integration in E^n; an integrand was called a flat or a sharp cochain, according to which of two continuity assumptions were made. It was proved in (V, 10) that a sharp cochain corresponds to a differential form which is bounded and satisfies a Lipschitz condition. The more general integral, and in many ways the more important, is given by the flat cochains. The principal object of this chapter is to prove Wolfe's Theorem (Theorem 7C), that a flat r-cochain in E^n corresponds to a differential form, a "flat" r-form, and conversely. Flat forms are measurable, but not continuous in general. ("Measurable" will always mean "Lebesgue measurable" in this chapter.) With the help of flat forms, the theory of products ("cup" and "cap" products in algebraic topology) is derived.

Besides the results of Chapter V, we need the theory of (VI, 7). In one place (Theorem 12A) we use (VII, Lemma 10b); the proof of this lemma is simple. Chapter VIII is used only in so far as we use cochains defined in open sets; only elementary facts about such cochains are needed. However, the reader is expected to have a good understanding of Lebesgue theory (see Appendix III).

We consider some extreme cases. For $r = n$, Wolfe's Theorem becomes the differentiation theorem about additive set functions $\Phi(Q)$ satisfying $|\Phi(Q)| \leqq N|Q|$ for some N; the flat n-forms are the bounded measurable functions. In this case, it is easy to remove the boundedness restriction; in the general case, this is a very difficult problem. For $r = 0$, Wolfe's Theorem is elementary. For $r = 1$, using 1-cochains which are coboundaries, the theorem, together with the approximation theorem of § 10, gives Rademacher's Theorem (§ 11) on the total differentiability of Lipschitz functions a.e. (almost everywhere). For other cases, the theorem is new.

The flat form $D_X(p)$ corresponding to the flat cochain X is defined by its values $D_X(p)\cdot\alpha = D_X(p, \alpha)$ for r-directions α, by the same definition (4.1) as in (V, 10.5). However, as is usual in Lebesgue theory, there is a restriction on the sequence of simplexes used; it must be "full", in that the simplexes must not have too small volumes compared to their diameters. This concept, used already in (IV, 14), is studied further in § 2. We study projections of simplexes in § 3. The next section is devoted to proving that for fixed α, $D_X(p, \alpha)$ exists a.e. and is measurable, and for fixed p, $D_X(p, \alpha)$ is defined over a closed set of values of α and is continuous there. To find further properties of $D_X(p)$ (in particular, to prove linearity), we smooth X (V, 13), forming X_ρ, find the corresponding sharp form D_{X_ρ} (V, 10), and let $\rho \to 0$. (Wolfe's proof did not use smoothing.) The form D_X corresponds to X in that

$$\int_\sigma D_X = X \cdot \sigma, \qquad \text{all } r\text{-simplexes } \sigma,$$

as in (V, 10.3). Hence it satisfies two inequalities:

$$\left| \int_{\sigma^r} D_X \right| \leqq |X| \, |\sigma^r|, \qquad \left| \int_{\partial \sigma^{r+1}} D_X \right| \leqq |dX| \, |\sigma^{r+1}|.$$

The converse theorem is as follows. Let ω be measurable. For most simplexes σ^r and σ^{r+1} (in particular, ω must be measurable in them, which need not be true in general if their dimensions are $<n$), assume the above inequalities are satisfied. (The first inequality states simply that ω is essentially bounded.) Such a form is called "flat". Then the corresponding flat cochain X exists, $\int_\sigma \omega = X \cdot \sigma$ for most simplexes, and ω is equivalent to D_X, i.e. $\omega = D_X$ a.e. The flat cochains correspond exactly to the equivalence classes of flat forms.

If we start with a flat form ω, find the corresponding X, and construct D_X, this D_X is in general an improvement over ω; see for instance (5.2), Theorem 17B and (X, Theorem 9B). That $\omega^* = \lim_{\rho \to 0} D_{X_\rho}$ need not be as good as D_X is shown by an example in § 13.

Though $\int_\sigma D_X = X \cdot \sigma$ for all r-simplexes σ, this does not mean, however, that $D_X(p) \cdot \alpha$ exists in σ for other r-directions α than that of σ. For example, define the flat 0-cochain Y and the flat 1-cochain X in the plane by

$$Y \cdot (x, y) = x + |y|, \qquad X = dY.$$

With the unit vectors e_1, e_2, let σ_t be any 1-simplex from a point p to a point $p + te_1$; then

$$X \cdot \sigma_t = Y \cdot \partial \sigma_t = Y \cdot (p + te_1) - Y \cdot p = t;$$

hence $X \cdot \sigma_t / |\sigma_t| = 1$, and $D_X(p) \cdot e_1 = 1$. Therefore $X \cdot \sigma_t = \int_{\sigma_t} D_X$ for all such simplexes σ_t. Yet $D_X(p) \cdot e_2$ exists for no p on the x-axis.

In §§ 8 and 9 we find assumptions on the values $\omega(p) \cdot \alpha$ for r-directions α and on the components of ω sufficient to insure that ω be flat. Though the definition of $D_X(p, \alpha)$ uses only sequences of simplexes with p as a vertex, the approximation theorem of § 10 shows that more general sequences could be used. (Part of this theorem was found by Wolfe.)

Though the exterior differential $d\omega$ of a form ω which is not continuous hardly seems definable, we can define it to be D_{dX}, X being the cochain corresponding to ω. The requisite properties, such as $dd\omega = 0$ and the fact given in Theorem 12A, hold; see also (X, Theorem 9C). The analytical formula for $d\omega$ holds provided ω satisfies a Lipschitz condition; see Theorem 12B.

Suppose ξ and η are flat forms. Then we can show (in part, with the help of cochains) that the product $\xi \vee \eta$ is flat. Given the flat cochains X and Y, we can now define their product as the cochain corresponding to the form $D_X \vee D_Y$. The usual properties of products, in particular, the formulas for $d(X \smile Y)$ and $d(\xi \vee \eta)$ and inequalities on norms of products (in both the flat and sharp cases), are now easily derived. This is an example of where the theories of flat cochains and of flat forms are, together, helpful to each other. (The theory of products of cochains could be derived independently of forms; see the end of § 18.)

In § 15 an analytical representation of flat n-chains in E^n is given; these chains correspond to the measurable summable functions. The measurable summable r-vector valued functions correspond to a subset of the r-chains. We use this correspondence in § 16 to define the product $X \frown A$ of a cochain and a chain; the expected properties and inequalities on norms hold. A theorem on weak limits of products is given in § 17; this is used in the proof of the formula (X, 11.1) for $f^*(X \smile Y)$. The cup products are characterized in § 18.

Recall that in (VII, 12), we showed that the algebraic cohomology ring of a complex K (with real coefficients) was the same as that determined by flat cochains. Using Wolfe's Theorem and § 12, we see that the ring is equally well determined by the flat forms in K. One may clearly define flat forms in a compact smooth manifold M (compare (VII, 10); in the non-compact case, use "locally flat" forms); using these, we have de Rham's Theorem on the cohomology ring of M, using flat forms in place of smooth forms.

1. n-cochains in E^n. We show here that flat n-cochains in an open subset R of E^n correspond to real valued bounded measurable functions in R, or rather, to equivalence classes of such functions. As usual, we let E^n be metric and oriented. Recall that two functions, each defined a.e. in R, are *equivalent* if they are the same except on a set of measure 0.

THEOREM 1A. *The flat n-cochains X in an open set $R \subset E^n$ are in one-one correspondence with the equivalence classes of bounded measurable functions ϕ in R, as follows.*

(a) *Given ϕ, define X as follows. For any n-simplex $\sigma \subset R$, oriented like E^n, set*

$$X \cdot \sigma = \int_\sigma \phi(p)\, dp. \tag{1}$$

(b) *Given X, define ϕ as follows. For any $p \in R$, choose a sequence $\sigma_1, \sigma_2, \cdots$ of n-simplexes oriented like E^n, each with p as a vertex, of diameters $\to 0$, and such that for some $\eta > 0$, $|\sigma_i|/[\text{diam}(\sigma_i)]^n \geqq \eta$ for all i. Set*

$$(2) \qquad \phi(p) = \lim_{i\to\infty} X\cdot\sigma_i/|\sigma_i|,$$

provided that the limit exists, independently of the sequence.

In (b), instead of requiring that p be a vertex of each σ_i, one could require simply that $p \in \sigma_i$. One could use more general sets σ_i (for instance, the union of a simplex not containing p and the point p), with the corresponding requirements; it is well known from Lebesgue theory that any such resulting function ϕ' is defined a.e. and equals ϕ a.e.

To prove the theorem, first suppose that ϕ is given. Define $X\cdot\sigma$ by (1), and for any polyhedral n-chain $A = \sum a_i\sigma_i$, set $X\cdot A = \sum a_i X\cdot\sigma_i$. Clearly $X\cdot A$ is uniquely defined, and is linear in A. With $|\phi|_0 = \text{ess sup}\,|\phi(p)|$ (App. III, 5.1), we have $|X\cdot\sigma| \leqq |\phi|_0 |\sigma|$; hence, by (V, 4.9), $|X| \leqq |\phi|_0$. Since there are no polyhedral $(n+1)$-chains $\neq 0$ in E^n, $dX = 0$; hence by (V, Theorem 4A), X is a flat n-cochain. Clearly if $\phi = \phi'$ a.e., the same cochain is defined.

Given X, any polyhedral region P (oriented like E^n) is expressible as an n-chain A; now $\Phi(P) = X\cdot A$ defines an additive set function Φ. Since $|\Phi(P)| \leqq |X|\,|P|$, standard theory (App. III, 5) shows that $\phi(p)$, defined by (2), exists a.e. and is measurable, $|\phi|_0 \leqq |X|$, and (1) holds. The theorem is now proved.

2. Some properties of fullness. We discuss some properties of simplexes, some of which may be found in (IV, 14). Given the point p and the simplex σ, set

$$(1) \qquad \delta_\sigma = \text{diam}(\sigma), \qquad \delta_{p\sigma} = \text{diam}(p \cup \sigma).$$

The *fullness* $\Theta(\sigma^r)$ and *p-fullness* $\Theta_p(\sigma^r)$ of σ^r are

$$(2) \qquad \Theta(\sigma^r) = |\sigma^r|/\delta^r_\sigma, \qquad \Theta_p(\sigma^r) = |\sigma^r|/\delta^r_{p\sigma};$$

we assume $r > 0$. (We could set $\Theta(\sigma^0) = 1$.) Note that

$$(3) \qquad \Theta_p(\sigma) \leqq \Theta(\sigma), \qquad \Theta_p(\sigma) = \Theta(\sigma) \quad \text{if} \quad p \in \sigma.$$

Recall, from (IV, 14), that if $v_1, \cdots, v_r$ is a defining set of vectors for σ^r,

$$(4) \qquad |\sigma^r| = |v_1 \vee \cdots \vee v_r|/r! \leqq \delta^r_\sigma/r!, \qquad \Theta(\sigma^r) \leqq 1/r!,$$

$$(5) \qquad k!\Theta(\sigma^k) \geqq r!\Theta(\sigma^r) \qquad \text{if} \quad \sigma^k \text{ is a face of } \sigma^r.$$

Using (4) gives

$$(6) \qquad \delta_{p\sigma} = [|\sigma|/\Theta_p(\sigma)]^{1/r} \leqq \delta_\sigma/[r!\Theta_p(\sigma)]^{1/r}.$$

If $\sigma_0, \cdots, \sigma_r$ are the $(r-1)$-faces of σ, then

$$|\sigma_i| \leqq \delta_\sigma^{r-1}/(r-1)! = |\sigma|/(r-1)!\Theta(\sigma)\delta_\sigma;$$

hence

$$|\partial\sigma| \leqq \frac{(r+1)|\sigma|}{(r-1)!\Theta(\sigma)\delta_\sigma}. \tag{7}$$

We say a sequence of simplexes $\sigma_1, \sigma_2, \cdots$ is a *full p-sequence* if p is a vertex of each σ_i, $\Theta(\sigma_i) \geqq \eta > 0$ for some η and all i, and $\delta_{\sigma_i} \to 0$; it is a *p-full sequence* if $\Theta_p(\sigma_i) \geqq \eta > 0$ and $\delta_{p\sigma_i} \to 0$ (equivalently, $\delta_{\sigma_i} \to 0$).

3. Properties of projections. Let P and P' be r-planes in E^n, and let π be the orthogonal projection of E^n (hence of P') into P. Let h be the distance (I, 15) between the r-directions of P and P'; we suppose $h < 1$. Then for any r-simplex σ in P', (I, 15.6) and (I, 15.3) give

$$|\pi\sigma| = |\cos\theta||\sigma| = (1 - h^2/2)|\sigma| \leqq |\sigma|, \tag{1}$$

$$|\sigma| - |\pi\sigma| = h^2|\sigma|/2. \tag{2}$$

Let P and P' have the point p in common. For any $q \in P'$, using (I, 15.7) with $v = q - p$ gives

$$|q - \pi q| \leqq h|q - p| \qquad (p \in P \cap P', \quad q \in P'). \tag{3}$$

Given $\sigma = p_0 \cdots p_s \in P'$, set $q_i = \pi(p_i)$, and

$$H(\sigma) = \sum_{i=0}^{s} (-1)^i \tau_i, \qquad \tau_i = p_0 \cdots p_i q_i \cdots q_s. \tag{4}$$

Then

$$|H(\sigma)| \leqq \zeta|\sigma|/s!\Theta(\sigma) \quad \text{if} \quad |q_i - p_i| \leqq \zeta. \tag{5}$$

For $|q_k - q_j| \leqq |p_k - p_j| \leqq \delta_\sigma$, and hence

$$(s+1)!|\tau_i| \leqq |p_1 - p_0| \cdots |q_i - p_i| \cdots |q_s - q_{s-1}| \leqq \zeta\, \delta_\sigma^s;$$

summing over i and applying the definition of $\Theta(\sigma)$ gives (5).

We prove an inequality (from which the denominator could be omitted; see (X, Lemma 5a)): For an r-simplex σ,

$$|\pi\sigma - \sigma|^\flat \leqq \frac{h\,\delta_{p\sigma}(|\sigma| + |\partial\sigma|)}{r!\Theta(\sigma)}. \tag{6}$$

By (App. II, 12.4), $\partial H(\sigma) = \pi\sigma - \sigma - H(\partial\sigma)$. Say $\partial\sigma = \sum \sigma_i$. Applying (5) and (3) gives

$$|\pi\sigma - \sigma|^\flat = |\partial H(\sigma) + H(\partial\sigma)|^\flat \leqq |H(\sigma)| + |H(\partial\sigma)|$$

$$\leqq \frac{h\,\delta_{p\sigma}|\sigma|}{r!\Theta(\sigma)} + \sum_i \frac{h\,\delta_{p\sigma}|\sigma_i|}{(r-1)!\Theta(\sigma_i)};$$

using (2.5) gives (6).

By (2.6), (2.7) and (2.3),

$$(7)\qquad \delta_{p\sigma}|\,\partial\sigma\,| \leqq N_{r,\Theta_p(\sigma)}|\,\sigma\,|, \quad N_{r,\eta} = \frac{r+1}{(r-1)!(r!)^{1/r}\eta^{1+1/r}}.$$

Hence (6) gives

$$(8)\qquad |\,\pi\sigma - \sigma\,|^\flat \leqq \frac{h(\delta_{p\sigma} + N_{r,\Theta_p(\sigma)})\,|\,\sigma\,|}{r!\Theta(\sigma)}.$$

Because of (2),

$$\left|\frac{\pi\sigma}{|\,\pi\sigma\,|} - \frac{\pi\sigma}{|\,\sigma\,|}\right| = \frac{|\,\sigma\,| - |\,\pi\sigma\,|}{|\,\sigma\,|} = \frac{h^2}{2};$$

combining this with the last inequality gives

$$\left|\frac{\pi\sigma}{|\,\pi\sigma\,|} - \frac{\sigma}{|\,\sigma\,|}\right|^\flat \leqq \frac{h^2}{2} + \frac{h(\delta_{p\sigma} + N_{r,\Theta_p(\sigma)})}{r!\Theta(\sigma)}.$$

Hence, if we set

$$(9)\qquad N'_{r,\eta} = \frac{1}{2} + \frac{1+N_{r,\eta}}{r!\eta},$$

we have

$$(10)\qquad \left|\frac{\pi\sigma}{|\,\pi\sigma\,|} - \frac{\sigma}{|\,\sigma\,|}\right|^\flat \leqq N'_{r,\eta}h \quad \text{if} \quad h \leqq 1, \quad \delta_{p\sigma} \leqq 1,\ \Theta_p(\sigma) \geqq \eta.$$

4. Elementary properties of $D_X(p, \alpha)$. We first introduce some notations. For any oriented r-simplex σ, let $P(\sigma)$ and $\alpha(\sigma) = \{\sigma\}/|\,\sigma\,|$ denote the oriented r-plane through σ and the r-direction of σ respectively. Let $\alpha(P)$ denote the r-direction of the oriented r-plane P. Let $P(p, \alpha)$ be the oriented r-plane through p with the r-direction α.

Call σ a *p-α-simplex* if it has p as a vertex and has r-direction α. The p-α-simplexes $\sigma_1, \sigma_2, \cdots$ form a *p-α-sequence* if $\delta_{\sigma_i} \to 0$.

Let X be a flat r-cochain in the open set $R \subset E^n$. For any point $p \in R$ and any r-direction α, set

$$(1)\qquad D_X(p, \alpha) = \lim_{i\to\infty} X\cdot\sigma_i/|\,\sigma_i\,|,$$

where $\sigma_1, \sigma_2, \cdots$ is any full p-α-sequence of simplexes, provided that the limit exists and is independent of the sequence.

Suppose $D_X(p, \alpha)$ exists. Then it is clear that for each $\epsilon > 0$ and $\eta > 0$ there is a $\zeta > 0$ such that

$$(2)\qquad \left|D_X(p, \alpha) - \frac{X\cdot\sigma}{|\,\sigma\,|}\right| < \epsilon \quad \text{if} \quad \sigma \subset U_\zeta(p), \quad \Theta(\sigma) \geqq \eta,$$

and σ has p as a vertex and $\alpha(\sigma) = \alpha$.

We could change the definition, allowing for instance p-full sequences of simplexes in $P(p, \alpha)$; this would not materially affect matters, however, as is shown by Theorem 10A below.

For $r = n$, $\phi(p) = D_X(p, \alpha_0)$ ($\alpha_0 = n$-direction of E^n) is the function given by Theorem 1A. For $r = 0$, $D_X(p)$ is a sharp function; see (V, 10). We henceforth assume $0 < r < n$.

The definition (1) shows that

$$D_X(p, -\alpha) = -D_X(p, \alpha). \tag{3}$$

Let P be any oriented r-plane. Then, by Theorem 1A, $D_X(p, \alpha(P))$ exists a.e. in $P \cap R$, and

$$X \cdot \sigma = \int_\sigma D_X(p, \alpha(P))\, dp, \qquad \sigma \subset P \cap R, \quad \sigma \text{ oriented like } P. \tag{4}$$

Lemma 4a. *For a fixed r-direction* α, $D_X(p, \alpha)$ *exists a.e. in* R *and is measurable in* R; *in fact this holds in* $P^s \cap R$, *for any s-plane* P^s *containing* α.

Let $\phi(p, \eta, \zeta, \zeta')$ be the least upper bound of numbers a such that for some σ with p as vertex and with $\alpha(\sigma) = \alpha$, we have

$$\Theta(\sigma) \geqq \eta, \qquad \zeta' \leqq \delta_\sigma \leqq \zeta, \qquad X \cdot \sigma / |\sigma| = a.$$

For fixed η, ζ, ζ', the function ϕ is continuous. For, given $\epsilon > 0$, choose σ so that

$$X \cdot \sigma / |\sigma| > \phi(p, \eta, \zeta, \zeta') - \epsilon.$$

Using (V, 3.6) and (2.7) gives

$$\left| \frac{X \cdot T_v \sigma}{|\sigma|} - \frac{X \cdot \sigma}{|\sigma|} \right| \leqq |X|^\flat |v| \left(1 + \frac{|\partial \sigma|}{|\sigma|}\right) \leqq |X|^\flat |v| \left[1 + \frac{r+1}{(r-1)!\eta\zeta'}\right],$$

which is $< \epsilon$ for $|v|$ small enough. Thus $\phi(p + v, \cdots) > \phi(p, \cdots) - \epsilon$ for v small enough. The same holds with p and $p + v$ interchanged.

Set

$$\phi(p, \eta) = \lim_{\zeta \to 0} \lim_{\zeta' \to 0} \phi(p, \eta, \zeta, \zeta'), \qquad \phi(p) = \lim_{\eta \to 0} \phi(p, \eta);$$

these are measurable. Define similar functions ψ, using greatest lower bounds. We now show that

$$\psi(p) = D_X(p, \alpha) = \phi(p) \qquad \text{if} \quad D_X(p, \alpha) \text{ exists},$$

while $\psi(p) < \phi(p)$ otherwise. If $D_X(p, \alpha)$ exists, its definition shows that it equals $\phi(p, \eta)$ and $\psi(p, \eta)$; this being true for all η, $D_X(p, \alpha) = \phi(p) = \psi(p)$. If $D_X(p, \alpha)$ does not exist, there are full p-α-sequences $\sigma_1, \sigma_2, \cdots$ and $\tau_1, \tau_2, \cdots$ with

$$a = \lim X \cdot \sigma_i / |\sigma_i| < \lim X \cdot \tau_i / |\tau_i| = b.$$

Say $\Theta(\sigma_i), \Theta(\tau_i) \geqq \eta > 0$. Then

$$\psi(p) \leqq \psi(p, \eta) \leqq a < b \leqq \phi(p, \eta) \leqq \phi(p).$$

Note that ϕ and ψ are defined throughout R, though $\phi(p, \eta, \zeta, \zeta')$ and $\psi(\cdots)$ need not be.

Let Q be the subset of R where $\phi = \psi$. Choose an $(n - r)$-plane P^{n-r} orthogonal to α. For each $q \in P^{n-r}$, § 1 shows that (letting $|H|_s$ denote s-dimensional Lebesgue measure in an s-plane)

$$|P(q, \alpha) \cap R - Q|_r = 0.$$

Since ϕ and ψ are measurable, Fubini's Theorem shows that $|R - Q| = 0$. Hence $D_X(p, \alpha)$ equals the measurable function $\phi(p)$ a.e. in R, and the lemma is proved, using R. Given P^s, we need merely apply the proof with E^n replaced by P^s.

LEMMA 4b. *For $p \in R$, let $H(p)$ be the set of r-directions α such that $D_X(p, \alpha)$ exists. Then $H(p)$ is closed, and $D_X(p, \alpha)$ is continuous in $H(p)$.*

First, let α, β be r-directions with $h = |\beta - \alpha| \leqq 1$. Let π be the projection onto $P(p, \alpha)$. For any p-β-simplex σ, (3.1) gives

$$\Theta(\pi\sigma) = \frac{|\pi\sigma|}{\delta^r_{\pi\sigma}} \geqq \frac{(1 - h^2/2)|\sigma|}{\delta^r_\sigma} \geqq \frac{1}{2}\Theta(\sigma). \tag{5}$$

Next, let $\sigma_1, \sigma_2, \cdots$ be a full p-β-sequence of simplexes; say $\Theta(\sigma_i) \geqq 2\eta$. Then $\pi\sigma_1, \pi\sigma_2, \cdots$ is a p-α-sequence, and $\Theta(\pi\sigma_i) \geqq \eta$. If $\alpha, \beta \in H(p)$, then (3.10) shows that

$$|D_X(p, \beta) - D_X(p, \alpha)| \leqq |X|^\flat N'_{r,\eta} h.$$

Since we may use a fixed η, this shows that $D_X(p, \alpha)$ is continuous in $H(p)$.

Finally suppose $\alpha \notin H(p)$. Then there are p-α-sequences $\sigma_1, \sigma_2, \cdots$ and $\tau_1, \tau_2, \cdots$ with $\Theta(\sigma_i), \Theta(\tau_i) \geqq \eta > 0$, and

$$X \cdot \sigma_i / |\sigma_i| \leqq a < b \leqq X \cdot \tau_i / |\tau_i|.$$

Set $\epsilon = (b - a)/3$, $h = \epsilon / N'_{r,\eta} |X|^\flat$. Now take any β, $|\beta - \alpha| < h$. Then if π is the projection onto $P(p, \beta)$, (3.10) gives

$$X \cdot \pi\sigma_i / |\pi\sigma_i| \leqq a + \epsilon, \qquad X \cdot \pi\tau_i / |\pi\tau_i| \geqq b - \epsilon,$$

showing that $\beta \notin H(p)$. Hence $H(p)$ is closed.

5. The r-form defined by a flat r-cochain. We shall now prove that to each flat cochain there corresponds a differential form.

Let P^s be an s-plane in E^n. Let $\xi \cdot \alpha$ be defined for certain r-vectors α. We say ξ is an *r-covector relative to P^s* if $\xi \cdot \alpha$ is defined for all α lying in P^s, and is linear in these α. Let S be a measurable subset of P^s. We say ω is *a measurable r-form in S relative to P^s* if (a) for some Q with $|S - Q|_s = 0$, $\omega(p)$ is an r-covector relative to P^s for each $p \in Q$, and (b) for each α in P^s, $\omega(p) \cdot \alpha$ is measurable in S (as a subset of P^s); equivalently, each component $\omega_\lambda(p)$ (in some coordinate system for P^s) is measurable. In particular, ω is a *measurable r-form in S* if this holds with $P^s = E^n$.

If ω is a bounded measurable r-form in the oriented r-simplex $\sigma \subset P^r$ relative to P^r, we may define the Lebesgue integral

$$\int_\sigma \omega = \int_\sigma \omega(p)\cdot\alpha(\sigma)\,dp. \tag{1}$$

If $\omega(p, \alpha)$ is a function of r-directions α, with the same properties, we define similarly $\int_\sigma \omega = \int_\sigma \omega(p, \alpha(\sigma))\,dp$.

THEOREM 5A. *Let X be a flat r-cochain in the open set $R \subset E^n$. Then there is a set $Q \subset R$ with $|R - Q| = 0$, such that for each $p \in Q$, $D_X(p, \alpha)$ is defined for all r-directions α, and is extendable to all r-vectors α, giving an r-covector $D_X(p)$; D_X is a bounded and measurable r-form in R. For any r-simplex σ in R, D_X is a measurable r-form in σ relative to $P(\sigma)$, and*

$$X\cdot\sigma = \int_\sigma D_X. \tag{2}$$

The same facts are true for dX. Also

$$|D_X|_0 = |X|, \qquad |D_{dX}|_0 = |dX|. \tag{3}$$

To begin with, smooth X by taking averages, as in (V, 13):

$$X_\rho\cdot A = \int_V \kappa_\rho(v)(X\cdot T_v A)\,dv, \qquad A \text{ polyhedral}. \tag{4}$$

Then X_ρ is defined in $\operatorname{int}_\rho(R)$ (see (App. III, 3)), and is a sharp r-cochain (V, Theorem 13A). Recall that

$$dX_\rho = (dX)_\rho, \qquad |X_\rho| \leqq |X|, \qquad |dX_\rho| \leqq |dX|. \tag{5}$$

We now show that D_{X_ρ} is the ρ-average of D_X:

$$D_{X_\rho}(p, \alpha) = \int_V \kappa_\rho(v) D_X(p + v, \alpha)\,dv, \qquad p \in \operatorname{int}_\rho(R). \tag{6}$$

With α fixed, let $\phi(p)$ denote the right hand side. Take any r-plane P with $\alpha(P) = \alpha$, and any r-simplex $\sigma \subset P$. Applying (4), (4.4) and Fubini's Theorem gives

$$X_\rho\cdot\sigma = \int_V \kappa_\rho(v)\int_\sigma D_X(p + v, \alpha(\sigma))\,dp\,dv = \int_\sigma \phi(p)\,dp.$$

That Fubini's Theorem is applicable may be seen as follows. Write V as a direct sum $V_1 \oplus V_2$, V_1 parallel to σ. Then

$$\int_V\int_\sigma = \int_{V_2}\int_{V_1}\int_\sigma = \int_{V_2}\int_\sigma\int_{V_1} = \int_\sigma\int_{V_2}\int_{V_1} = \int_\sigma\int_V,$$

the function $D_X(p, \alpha(\sigma))$ having clearly the required measurability property in each case. Since ϕ is continuous, (6) follows for $p \in P \cap \operatorname{int}_\rho(R)$, and hence for $p \in \operatorname{int}_\rho(R)$.

Because of (6), (App. III, Lemma 3b) gives

$$\lim_{\rho\to 0} D_{X_\rho}(p, \alpha) = D_X(p, \alpha) \qquad \text{a.e. in } R, \text{ for each } \alpha. \tag{7}$$

Let $\alpha_1, \alpha_2, \cdots$ be a sequence of r-directions, which is dense in the set of all r-directions. For each i, let Q_i be the set of points p such that $D_X(p, \alpha_i)$ exists and (7) holds with α_i; then $|R - Q_i| = 0$. Set $Q = Q_1 \cap Q_2 \cap \cdots$. Now take any $p \in Q$. Then $D_X(p, \alpha_i)$ exists for all i, and by Lemma 4b, $D_X(p, \alpha)$ exists for all r-directions α and is continuous in α. By Lemma 4a, $D_X(p, \alpha)$ is measurable in R for each α.

Let the first of the α_i consist of the r-directions e_λ $(\lambda_1 < \cdots < \lambda_r)$. Take any $p \in Q$, and any r-vector $\alpha = \sum \alpha^\lambda e_\lambda$. Since (V, Theorem 10A) $D_{X_\rho}(p)\cdot\alpha$ is linear in α (taking ρ so small that $p \in \operatorname{int}_\rho(R)$), and (7) holds at p for each α_i,

$$\lim_{\rho\to 0} [D_{X_\rho}(p)\cdot\alpha] = \lim_{\rho\to 0} \sum_{(\lambda)} \alpha^\lambda D_{X_\rho}(p, e_\lambda) = \sum_{(\lambda)} \alpha^\lambda D_X(p, e_\lambda),$$

which is linear in α. Hence we may define

$$D_X(p)\cdot\alpha = \lim_{\rho\to 0} [D_{X_\rho}(p)\cdot\alpha] \qquad (p \in Q), \tag{8}$$

and $D_X(p)$ is an r-covector, $p \in Q$. Now for $p \in Q$,

$$D_X(p, \alpha_i) = \lim_{\rho\to 0} D_{X_\rho}(p, \alpha_i) = \lim_{\rho\to 0} [D_{X_\rho}(p)\cdot\alpha_i] = D_X(p)\cdot\alpha_i$$

for each i. Since $D_X(p, \alpha)$ and $D_X(p)\cdot\alpha$ are both continuous in α (see Lemma 4b), this proves

$$D_X(p)\cdot\alpha = D_X(p, \alpha) \qquad (r\text{-directions } \alpha, p \in Q). \tag{9}$$

Thus the r-form $D_X(p)$ is an extension of $D_X(p, \alpha)$ in Q.

Since dX is a flat cochain in R, the same facts hold for it. Relation (2) is clear from the definition of D_X. The proof of (3) is like the corresponding proof in (V, 10). The theorem is now proved.

6. Flat r-forms. We turn now to the problem of determining a flat cochain from a differential form satisfying certain conditions; we take these conditions from the conclusion of Theorem 5A.

Given ω and Q, we say an s-plane P^s $(s \geqq r)$ is *Q-good* (for ω) if $|P^s \cap R - Q|_s = 0$ and ω (that is, each $\omega(p)\cdot\alpha$) is measurable in $P^s \cap R$. (We use all α, not simply those α in P^s, for the purposes of the proof of Lemma 6b.) We say the s-simplex σ^s in R $(s \geqq r)$ is *Q-good* if $|\sigma^s - Q|_s = 0$ and ω is measurable in σ^s. We say σ^s is *Q-excellent* if σ^s and each of its faces of dimension $\geqq r$ is Q-good.

We say ω is a *flat r-form* in the open set $R \subset E^n$ if there is a measurable subset Q of R such that $|R - Q| = 0$, and

(a) ω is a measurable r-form in R,

(b) $|\omega|_0 \leqq N$ for some N (see (App. III, 5.1)),

(c) there is an N' such that for any Q-excellent simplex σ^{r+1} in R,

$$\left| \int_{\partial \sigma^{r+1}} \omega \right| \leqq N' \, |\sigma^{r+1}|. \tag{1}$$

If ω is flat in terms of Q, then it is flat in terms of any $Q' \subset Q$ with $|R - Q'| = 0$.

We shall need some lemmas in the nature of Fubini's Theorem, showing that most planes and simplexes are Q-good; ω is assumed flat in R in terms of Q.

LEMMA 6a. *Let $P^s(p)$ denote the plane orthogonal to a fixed P^{n-s} and containing $p \in P^{n-s}$ $(s \geqq r)$. Then for almost all $p \in P^{n-s}$, $P^s(p)$ is Q-good.*

For each $\lambda = (\lambda_1, \cdots, \lambda_r)$, $\lambda_1 < \cdots < \lambda_r$, Fubini's Theorem shows that there is a set $H_\lambda \in P^{n-s}$ with $|P^{n-s} - H_\lambda|_{n-s} = 0$, such that for $p \in H_\lambda$, $|P^s(p) \cap R - Q|_s = 0$ and $\omega_\lambda(p)$ is measurable in $P^s(p) \cap R$. Set $H = \bigcap_\lambda H_\lambda$; then $P^s(p)$ is Q-good for $p \in H$.

LEMMA 6b. *Let $\sigma^s = p_0 p_1 \cdots p_s$ $(s \geqq r)$ be a simplex in R, and let K be the set of points p such that $p\sigma' = pp_1 \cdots p_s$ is a (non-degenerate) simplex in R. Let P^m be a plane through p_0 such that $\sigma^s \cup P^m$ spans E^n. Then for almost all $p \in P^m \cap K$, $p\sigma'$ is Q-good.*

Let P^s and P^{s-1} be the planes of σ^s and $\sigma' = p_1 \cdots p_s$ respectively. Let P_0^{n-s} be a plane in P^m with only p_0 in common with P^s; then P^s and P_0^{n-s} span E^n. Let H be the set of $p \in K$ such that $p\sigma'$ is Q-good; H is measurable. For any $q \in P^m \cap P^s$, let $P^{n-s}(q)$ be the plane through q parallel with P_0^{n-s}. For $q \in P^m \cap P^s - P^{s-1}$, and $p \in P^{n-s}(q)$, let $P^s(p)$ be the plane through P^{s-1} and p. Now (App. III, 5) for each λ, $|P^s(p) \cap R - Q|_s = 0$ and $\omega_\lambda(p)$ is measurable in $P^s(p) \cap R$ for almost all $p \in P^{n-s}(q)$; hence $P^s(p)$ is Q-good for almost all p in $P^{n-s}(q)$, and $|P^{n-s}(q) \cap K - H|_{n-s} = 0$. Hence, by Fubini's Theorem, $|P^m \cap K - H|_m = 0$.

7. Flat r-forms and flat r-cochains. Let ω be a flat r-form in the open set $R \subset E^n$, and let X be a flat r-cochain in R. We say ω and X are *associated* if there is a set $Q \subset R$ with $|R - Q| = 0$ such that ω is a flat form in terms of Q, and

$$\int_\sigma \omega = X \cdot \sigma \qquad \text{for any } Q\text{-good } r\text{-simplex } \sigma. \tag{1}$$

We now prove a converse of Theorem 5A.

THEOREM 7A. *Any flat r-form in the open set $R \subset E^n$ is associated with a unique flat r-cochain in R. With N and N' as in (b) and (c) of § 6, $|X| \leqq N$ and $|dX| \leqq N'$.*

Given ω, define $X\cdot\sigma$ by (1) for all Q-good σ.

LEMMA 7a. *Let σ and σ' be Q-good r-simplexes, contained in an open convex subset S of R. Say $\sigma = p_0 \cdots p_r$, $\sigma' = p_0' \cdots p_r'$, $\delta_\sigma =$* diam (σ), *and*

$$\delta_\sigma + 2\zeta \leqq \delta, \qquad |p_i' - p_i| \leqq \zeta. \tag{2}$$

Then

$$\left|\int_{\sigma'} \omega - \int_\sigma \omega\right| \leqq \zeta\left[\frac{N'\delta^r}{r!} + \frac{(r+1)N\delta^{r-1}}{(r-1)!}\right]. \tag{3}$$

Given $\zeta' > 0$, we shall find simplexes $\sigma'' = p_0'' \cdots p_r''$ and

$$\tau_i = p_0 \cdots p_i p_i'' \cdots p_r'', \qquad \tau_i' = p_0' \cdots p_i' p_i'' \cdots p_r'',$$

all in S, such that

$$|p_i'' - p_i| \leqq \zeta, \qquad |p_i'' - p_i'| \leqq \zeta',$$

and the τ_i and τ_i' are Q-excellent. We choose $p_r'', p_{r-1}'', \cdots, p_0''$ in succession. Having found $p_r'', \cdots, p_{i+1}''$, we must find p_i'' so that τ_i and τ_i' are Q-excellent. Since

$$p_0 \cdots p_i p_{i+1}'' \cdots p_r'', \qquad p_0' \cdots p_i' p_{i+1}'' \cdots p_r''$$

are good, we need merely make τ_i and τ_i' and their r-faces which contain p_i'' good. That this can be done follows at once from Lemma 6b, using $P^m = E^n$.

By (App. II, 12.4), we have

$$\partial \sum_i (-1)^i \tau_i = \sigma'' - \sigma + \sum_j \tau_j^*,$$

the τ_j^* being faces of the τ_i. Hence

$$\int_{\sigma''} \omega - \int_\sigma \omega = \sum_i (-1)^i \int_{\partial\tau_i} \omega - \sum_j \int_{\tau_j^*} \omega.$$

Using (2.4) and (b) and (c) of § 6 gives

$$\left|\int_{\partial\tau_i} \omega\right| \leqq N' \,|\tau_i| \leqq N'\delta^r\zeta/(r+1)!,$$

$$\left|\int_{\tau_j^*} \omega\right| \leqq N\,|\tau_j^*| \leqq N\delta^{r-1}\zeta/r!.$$

The two sums contain $r + 1$ and $(r + 1)r$ terms respectively, and (3) follows with σ' replaced by σ''. Also (3) holds with σ' and σ'', with ζ replaced by ζ'. Since ζ' is arbitrary, (3) follows.

To return to the theorem, let $\sigma = p_0 \cdots p_r$ be any r-simplex in R which is not Q-good. By Lemma 6b, there is a sequence of points p_{01}, $p_{02}, \cdots \to p_0$ such that each $\sigma_k = p_{0k}p_1 \cdots p_r$ is Q-good. Set

$$X\cdot\sigma = \lim_{k\to\infty} X\cdot\sigma_k. \tag{4}$$

(We could in fact use $\sigma_k = p_{0k} \cdots p_{rk}$, $p_{ik} \to p_i$.) The lemma above shows that the limit exists and is independent of the sequence chosen.

If we prove $|X\cdot\sigma| \leqq N|\sigma|$, $|X\cdot\partial\sigma^{r+1}| \leqq N'|\sigma^{r+1}|$, it will follow from (V, Theorem 4A), or rather, (VIII, 1(b)), that X is a flat r-cochain in R. The first inequality follows from (1), (4) and § 6, (b). To prove the second, say $\sigma^{r+1} = p_0 \cdots p_{r+1}$. Given $\zeta > 0$, we see from Lemma 6b that there is a Q-excellent $\sigma' = p_0' \cdots p_{r+1}'$ such that $|p_i' - p_i| < \zeta$. By (1) and (6.1), $|X\cdot\partial\sigma'| \leqq N'|\sigma'|$. Using (4) gives the required inequality.

The uniqueness of X is clear, and the theorem is proved.

Theorem 7B. *If ω and ω' are both associated with X, then $\omega(p) = \omega'(p)$ a.e. in R.*

It is sufficient to show that $\omega_\lambda(p) = \omega_\lambda'(p)$ a.e. for each λ. There is a set Q with $|R - Q| = 0$ such that (1) holds for both ω and ω'. Let P^{n-r} be a plane orthogonal to e_λ. Lemma 6a shows that $P(q, e_\lambda)$ is Q-good, with both ω and ω', for almost all $q \in P^{n-r}$. For any such q, (1) shows that $\int_\sigma \omega = \int_\sigma \omega'$ for all σ in $P(q, e_\lambda) \cap R$. It follows that $\omega_\lambda(p) = \omega_\lambda'(p)$ a.e. in $P(q, e_\lambda) \cap R$. By Fubini's Theorem, this holds a.e. in R.

Say the r-forms ω, ω' in R are *equivalent* if $\omega = \omega'$ a.e. in R. Theorems 5A, 7A and 7B give

Theorem 7C. *With the correspondence* (1), *the flat r-cochains in the open set R correspond in a one-one manner to the equivalence classes of flat r-forms in R. The norms agree; see* (5.3) *and* (12.6) *below.*

It follows, incidentally, that the space of equivalence classes of flat forms is complete, and hence is a Banach space.

Lemma 7b. *Any form ω' equivalent to a flat form ω is flat.*

Suppose ω is flat in terms of Q, and $\omega' = \omega$ in Q_1, $|R - Q_1| = 0$. Set $Q' = Q \cap Q_1$; we show that ω' is flat, in terms of Q'. We have $|R - Q'| = 0$, and (a) and (b) of § 6 hold. We must prove (c); we use the N' given by ω. Take any simplex $\sigma = \sigma^{r+1}$ which is Q'-excellent for ω'. Let $\sigma_0, \cdots, \sigma_{r+1}$ be the r-faces of σ. Since $Q' \subset Q$, $|\sigma - Q|_{r+1} = 0$. Also $|\sigma - Q_1|_{r+1} = 0$, hence $\omega = \omega'$ a.e. in σ, and ω is measurable in σ. Therefore σ is Q-good for ω. Similarly each σ_i is Q-good for ω; hence σ is Q-excellent for ω, and (6.1) holds. Also $|\partial\sigma - Q_1|_r = 0$; hence $\omega' = \omega$ a.e. in $\partial\sigma$, and (6.1) holds for ω', proving (c) for ω'. This completes the proof.

The next lemma will be used in § 14. Let ϕ be a function with values in a vector space, which is measurable in R (i.e. each component, in some coordinate system, is measurable). We say σ^s is *(Q, ϕ)-good* for ω if $|\sigma^s - Q|_s = 0$ and both ω and ϕ are measurable in σ^s; similarly for *(Q, ϕ)-excellence*. We say ω is *weakly flat* in R if, for some Q and ϕ, the conditions for flatness of ω hold, except that (c) is required only for (Q, ϕ)-excellent simplexes.

LEMMA 7c. *Any weakly flat form is flat.*

The proof of Lemmas 6a and 6b and Theorem 7A go as before, using ϕ throughout; this proves the existence of the flat cochain X such that

$$(5) \qquad \int_\sigma \omega = X \cdot \sigma, \qquad \text{all } (Q, \phi)\text{-good } \sigma.$$

The proof of Theorem 7B now shows that $\omega = D_X$ a.e. By the last lemma, ω is flat.

We extend (VI, 8.4) to flat forms:

THEOREM 7D. *For any flat cochain X and sharp function ϕ in R,*

$$(6) \qquad D_{\phi X}(p) = \phi(p) D_X(p) \qquad \text{wherever } D_X(p) \text{ is defined.}$$

Take any p such that $D_X(p)$ is defined; set $a = \phi(p)$. Take any r-direction α; let $\sigma_1, \sigma_2, \cdots$ be a full p-α-sequence. There is a sequence $\epsilon_1, \epsilon_2, \cdots \to 0$ such that

$$|\phi(q) - a| \leqq \epsilon_i, \qquad q \in \sigma_i.$$

Now

$$\left| \frac{\phi X \cdot \sigma_i}{|\sigma_i|} - a \frac{X \cdot \sigma_i}{|\sigma_i|} \right| = \frac{|X \cdot (\phi \sigma_i - a \sigma_i)|}{|\sigma_i|} \leqq |X| \, \epsilon_i,$$

which $\to 0$ as $i \to \infty$, proving (6).

8. Flat r-direction functions. In the proof of Theorem 7A, no direct use was made of the fact that $\omega(p)$ was an r-covector for $p \in Q$. We show here how weaker assumptions on ω are enough to insure that the corresponding X exists; this in turn shows that ω is extendable to be a flat r-form a.e. in R.

We say ω is a *bounded measurable r-direction function* in the open set $R \subset E^n$ if there is a measurable set $Q \subset R$ with $|R - Q| = 0$, with the following properties:

(a′) For each $p \in Q$, $\omega(p, \alpha)$ is defined for all r-directions α and is continuous in α.

(b′) $\omega(p, -\alpha) = -\omega(p, \alpha)$.

(c′) For each α, $\omega(p, \alpha)$ is measurable in R.

(d′) $|\omega(p, \alpha)| \leqq N$ for some N.

We now give a condition from which (c) of § 6 may be deduced. Define Q-goodness and Q-excellence of simplexes as in § 6. Given $p \in R$ and an $(r + 1)$-direction β, set

$$(1) \qquad \bar{d}\omega(p, \beta) = \sup \left\{ \limsup_{i \to \infty} \frac{1}{|\sigma_i|} \left| \int_{\partial \sigma_i} \omega \right| \right\},$$

using all full sequences $\sigma_1, \sigma_2, \cdots$ of Q-excellent $(r + 1)$-simplexes in $P(p, \beta)$ containing p and converging to p.

We say ω is a *flat r-direction function* in R if it is a bounded measurable r-direction function, and

(e′) for some N', $\bar{d}\omega(p, \beta) \leqq N'$ for all (p, β).

The definition of ω and X being associated is the same as in § 7; also the definition of equivalence of ω and ω'.

THEOREM 8A. *Any flat r-direction function ω in R is associated with a flat r-cochain X in R, by* (7.1); *this establishes a one-one correspondence between equivalence classes of flat r-direction functions and flat r-cochains. Given ω, there is a set $Q' \subset R$ with $|R - Q'| = 0$ and the property that for $p \in Q'$, there is an r-covector $\bar{\omega}(p)$ such that $\bar{\omega}(p)\cdot\alpha = \omega(p, \alpha)$ for all r-directions α; $\bar{\omega}$ is a flat r-form in R, defining the same X that ω does.*

First we note that Lemma 6b continues to hold. For, let $\alpha_1, \alpha_2, \cdots$ be a sequence of r-directions which is dense in the set of all r-directions. Following the proof of the lemma, let $H_k(q)$ be the set of points $p \in P^{n-s}(q)$ such that $|P^s(p) \cap R - Q|_s = 0$ and $\omega(p, \alpha_k)$ is measurable in $P^s(p) \cap R$; then $|P^{n-s}(q) - H_k(q)|_{n-s} = 0$. Set $H(q) = \bigcap_k H_k(q)$. Then for $p \in H(q)$, each $\omega(p', \alpha_k)$ is measurable in $P^s(p) \cap R$. Take any r-direction α; say $\alpha_{\mu_1}, \alpha_{\mu_2}, \cdots \to \alpha$. Since $\omega(p', \alpha_{\mu_j}) \to \omega(p', \alpha)$ for $p' \in Q$, by (a'), $\omega(p', \alpha)$ is the limit a.e. in $P^s(p) \cap R$ of a sequence of measurable functions, and hence is measurable. The proof is now completed as before.

Next we prove (c) of § 6 (which has meaning here). Let $\sigma = \sigma^{r+1}$ be Q-excellent. Consider $\omega(p, \alpha)$ simply in the plane P^{r+1} of σ, α being in P^{r+1}; σ is also Q-excellent in this restricted sense. Say $\sigma = p_0 \cdots p_{r+1}$. Take the standard subdivision $\mathfrak{S}\sigma$ of σ, and consider any r-simplex τ of $\mathfrak{S}\sigma$ not in $\partial\sigma$. By (App. II, Lemma 4c), the plane P^r of τ contains the mid point $p_{0,r+1}$ of p_0p_{r+1}, but no other points of p_0p_{r+1}. Say $\tau = p_{0,r+1}\tau'$; let P^{r-1} be the plane of τ'. Then P^{r-1} and the line P^1 of p_0p_{r+1} span P^{r+1}. Set $\tau(p) = p\tau'$. By Lemma 6b, applied to the space P^{r+1}, $\tau(p)$ is Q-good for almost all p in P^1. The same is true of each other r-simplex of $\mathfrak{S}\sigma$ not in $\partial\sigma$. Since the simplexes in $\partial\sigma$ are Q-good, we can find a point p' in p_0p_{r+1} arbitrarily near $p_{0,r+1}$ such that if $\mathfrak{S}_1\sigma$ is formed with p' in place of $p_{0,r+1}$, all the r-simplexes of $\mathfrak{S}_1\sigma$ are Q-good, and hence all the $(r + 1)$-, simplexes are Q-excellent (always using P^{r+1} in place of E^n). Continuing, we find a sequence $\mathfrak{S}_1\sigma, \mathfrak{S}_2\sigma, \cdots$ of subdivisions, like the sequence of standard subdivisions, such that all r-simplexes are Q-good, and all $(r + 1)$-simplexes are of fullness $\geqq \eta$ for some $\eta > 0$.

Suppose $\left|\int_{\partial\sigma} \omega\right| = a\,|\sigma|$, $a > N'$. Let $\tau_{11}, \cdots, \tau_{1m}$ be the $(r + 1)$-simplexes of $\mathfrak{S}_1\sigma$. If $\left|\int_{\partial\tau_{1i}} \omega\right| < a\,|\tau_{1i}|$ were true for each i, then adding these inequalities would give a contradiction to the above equality. Hence $\left|\int_{\partial\tau_{1i}} \omega\right| \geqq a\,|\tau_{1i}|$ for some i. If $\tau_{21}, \cdots, \tau_{2m}$ are the $(r + 1)$-simplexes of $\mathfrak{S}_2\sigma$ lying in τ_{1i}, then $\left|\int_{\partial\tau_{2j}} \omega\right| \geqq a\,|\tau_{2j}|$ for some j, by the same reasoning. Continuing gives a full sequence $\tau_{1i}, \tau_{2j}, \cdots$ of Q-excellent simplexes, with a common point $\bar{p}$, showing that $\bar{d}\omega(\bar{p}, \alpha(\sigma)) \geqq a > N'$, a contradiction; hence (c) of § 6 holds.

The determination of X from ω now proceeds exactly as in the proof of Theorem 7A. Again X is unique.

Suppose ω and ω' are both associated with X. Let $\alpha_1, \alpha_2, \cdots$ be a dense sequence of r-directions, as above. For each i, the proof of Theorem 7B shows that $\omega(p, \alpha_i) = \omega'(p, \alpha_i)$ a.e. Because of (a'), $\omega(p, \alpha) = \omega'(p, \alpha)$ for all α, a.e. in R.

Given ω, determine X, and then the flat r-form $\bar{\omega} = D_X$ (Theorem 5A). Then setting $\omega'(p, \alpha) = \bar{\omega}(p)\cdot\alpha$ for r-directions α gives a flat r-direction function ω', associated with X. Hence there is a set Q, $|R - Q| = 0$, such that ω, $\bar{\omega}$ and ω' are all flat in terms of Q, and $\omega(p, \alpha) = \omega'(p, \alpha)$ for all α and all $p \in Q$. Hence $\omega(p, \alpha) = \bar{\omega}(p)\cdot\alpha$ for all α and all $p \in Q$, and the theorem is proved.

9. Flat forms defined by components. With a rectangular coordinate system, we give here conditions on the components ω_λ of a form ω sufficient to insure that ω be a flat form. The condition $\bar{d}\omega \leqq N'$ of § 8 will appear in a form using coordinate intervals instead of simplexes.

A *coordinate interval of type* $\mu = (\mu_1, \cdots, \mu_{r+1})$ is defined by a set of relations

$$a_{\mu_i} \leqq x^{\mu_i} \leqq b_{\mu_i} \quad (i = 1, \cdots, r+1), \qquad x^k = a_k \quad (\text{other } k).$$

Say an s-interval A $(s \geqq r)$ is *Q-good* if $|A - Q|_s = 0$ and each ω_λ is measurable in A; it is *Q-excellent* if it and each of its faces of dimension $\geqq r$ is Q-good.

For any μ, set

$$\tilde{d}_\mu\omega(p) = \sup\left\{\limsup_{i\to\infty} \frac{1}{|A_i|}\left|\int_{\partial A_i} \omega\right|\right\}, \tag{1}$$

using all full sequences $A_1, A_2, \cdots$ of Q-excellent intervals of type μ, containing p and converging to p. The integral over each face of A_i uses the corresponding component ω_λ.

THEOREM 9A. *Let the ω_λ $(\lambda_1 < \cdots < \lambda_r)$ be bounded measurable functions, defined a.e. in the open set $R \subset E^n$. Let $\tilde{d}_\mu\omega$ be bounded in R. Then the ω_λ are components of a flat r-form in R, and hence define uniquely a flat r-cochain in R. The flat r-cochains in R correspond to the equivalence classes of sets of ω_λ in R.*

Say $|\omega_\lambda| \leqq N_0$, $\tilde{d}_\mu\omega \leqq N_0'$. We show first that

$$\left|\int_{\partial A} \omega\right| \leqq N_0'|A| \qquad \text{for } Q\text{-excellent coordinate intervals } A. \tag{2}$$

Suppose not; say $\left|\int_{\partial A}\omega\right| = a|A|$, $a > N_0'$, A of type μ. We may find a coordinate r-interval B_i cutting A approximately in half, so that B_i is Q-good, B_i being orthogonal to e_{μ_i}, for $i = 1, \cdots, r + 1$. Thus A is cut into $m = 2^{r+1}$ Q-excellent intervals $A_{11}, \cdots, A_{1m}$, each approximately of the same shape as A. As in the proof of Theorem 8A, $\left| \int_{\partial A_{1i}} \omega \right| \geqq a |A_{1i}|$ for some i. Cut A_{1i} into Q-excellent intervals $A_{21}, \cdots, A_{2m}$, etc. As in the proof of Theorem 8A, we are led to a contradiction with (1).

For any $\rho_0 > 0$, we shall show that $\omega = \sum \omega_\lambda e^\lambda$ is a flat r-form in $R' = \text{int}_{\rho_0}(R)$, with fixed bounds N, N' for $|\omega|_0$, $|d\omega|_0$; then the same is true in R itself.

For each ρ, $0 < \rho \leqq \rho_0$, we may define the ρ-average ω_λ^ρ of ω_λ (App. III, 3). Set $\omega^\rho = \sum \omega_\lambda^\rho e^\lambda$; this is a smooth r-form in R'. There is a set $Q \subset R'$ with $|R' - Q| = 0$ and

$$\lim_{\rho \to 0} \omega^\rho(p) = \omega(p), \qquad p \in Q. \tag{3}$$

Let X_ρ correspond to ω^ρ (V, Theorem 10A); then dX_ρ corresponds to $\xi^\rho = d\omega^\rho$ (V, Theorem 10B). We shall find a bound for $|dX_\rho|$.

First, take any coordinate $(r + 1)$-interval A. With slightly condensed notation, we have

$$\begin{aligned} dX_\rho \cdot A = \int_{\partial A} \omega^\rho(p)\, dp &= \int_{\partial A} \int_V \kappa_\rho(v) \omega(p + v)\, dv\, dp \\ &= \int_V \kappa_\rho(v) \int_{\partial A} \omega(p + v)\, dp\, dv; \end{aligned}$$

compare the proof of (5.6). Applying (2) gives (since almost all the $T_v\, \partial A$ are Q-excellent)

$$|dX_\rho \cdot A| \leqq N_0' |A| \int_V \kappa_\rho(v)\, dv = N_0' |A|.$$

Given $p \in R'$ and $\mu = (\mu_1, \cdots, \mu_{r+1})$, using a full sequence of intervals $A_1, A_2, \cdots$ of type μ shows that

$$|\xi_\mu^\rho(p)| = |\lim_{k \to \infty} dX_\rho \cdot A_k| / |A_k| \leqq N_0'.$$

Hence by (I, 13.3), (I, 12.18) (for covectors) and (V, 10.4),

$$|\xi^\rho(p)|_0 \leqq \binom{n}{r}^{1/2} N_0' = N', \qquad |dX_\rho| \leqq N'. \tag{4}$$

We now prove (c) of § 6, using Q. Since $|\omega_\lambda^\rho| \leqq |\omega_\lambda| \leqq N_0$, using (3) gives, for any Q-excellent σ,

$$\left| \int_{\partial \sigma} \omega \right| = \left| \lim_{\rho \to 0} \int_{\partial \sigma} \omega^\rho \right| = |\lim_{\rho \to 0} dX_\rho \cdot \sigma| \leqq N' |\sigma|.$$

Hence ω is a flat r-form in R', and hence in R. The rest of the theorem is clear.

EXAMPLE. In the plane, set

$$\omega(x, y) = \begin{cases} e^1 + e^2, & x + y < 0, \\ 0, & x + y > 0. \end{cases}$$

Then $\omega = \omega_1 e^1 + \omega_2 e^2$ is flat, but $\omega_1 e^1$ is not, as we see by considering small squares with diagonal on the line $x + y = 0$.

10. Approximation to $D_X(p)$ by $X \cdot \sigma / |\sigma|$. In the definition (4.1) of $D_X(p, \alpha)$, sequences $\sigma_1, \sigma_2, \cdots$ of simplexes were used, each simplex having p as a vertex. The limit might not exist if more general p-full sequences were used. However, for given p, if the covector $D_X(p)$ exists, then $D_X(p, \alpha)$ exists, for all α, using p-full sequences, as the following theorem shows. For $r = n$, the theorem is standard (App. III, 5); for $r = 0$, it is trivial. As an application of the theorem, we find the theorem on total differentiability of the next section.

THEOREM 10A. *Let X be a flat r-cochain in the open set $R \subset E^n$. Let $p \in R$ be such that $D_X(p, \alpha)$ exists for all r-directions α, and is extendable to all r-vectors α, defining an r-covector $D_X(p)$. Then for each $\epsilon > 0$ and $\eta > 0$ there is a $\zeta > 0$ with the following property. For any r-simplex σ,*

$$(1) \qquad |X \cdot \sigma - D_X(p) \cdot \{\sigma\}| < \epsilon |\sigma| \quad \text{if} \quad \sigma \subset U_\zeta(p), \quad \Theta_p(\sigma) \geqq \eta.$$

Dividing by $|\sigma|$ gives

$$(2) \qquad |X \cdot \sigma / |\sigma| - D_X(p) \cdot \alpha(\sigma)| < \epsilon.$$

REMARK. Suppose the hypothesis holds, restricting α to lie in a plane P^s containing p. Then the conclusion holds, restricting σ to lie in P^s. For we need merely apply the theorem to X, restricted to the open subset $P^s \cap R$ of the space P^s.

We first define the *affine approximation* W to X at p, by

$$(3) \qquad W \cdot A = D_X(p) \cdot \{A\}, \qquad \text{any flat } r\text{-chain } A \text{ in } E^n.$$

Using (III, 2.3), we find

$$(4) \qquad |W| \leqq |X|, \qquad dW = 0, \qquad \mathfrak{L}(W) = 0.$$

Because of (4), W is a sharp cochain. (It is a coboundary; see (VII, Lemma 10b).)

We start by proving a special case of the theorem:

LEMMA 10a. *The theorem holds if we require that σ have p as a vertex.*

With $N'_{r,\eta}$ as in (3.9), choose r-directions $\alpha_1, \cdots, \alpha_m$ such that any r-direction is within $\epsilon_1 = \epsilon / 4N'_{r\eta} |X|^\flat$ of one of these. Using α_i, $\epsilon/2$ and $\eta/2$, choose $\zeta_i > 0$ to satisfy (4.2). Let ζ be the smallest of $\zeta_1, \cdots, \zeta_m$.

Now take any σ with p as vertex, satisfying the last relations in (1). Choose i so that

$$|\alpha(\sigma) - \alpha_i| \leqq \epsilon_1,$$

and let π denote the projection onto $P(p, \alpha_\iota)$. Supposing $\epsilon_1 \leqq 1$, $\zeta \leqq \frac{1}{2}$, (3.10) gives, if $\tau = \pi\sigma$,

$$\left| X\cdot\tau/ \left|\tau\right| - X\cdot\sigma/ \left|\sigma\right| \right| \leqq N'_{r,\eta}\epsilon_1 \left| X \right|^\flat = \epsilon/4;$$

the same inequality holds with W, because of (4). By (4.5), $\Theta(\tau) \geqq \eta/2$. Therefore, by the choice of ζ_i and the definition of W,

$$\left| W\cdot\tau/ \left|\tau\right| - X\cdot\tau/ \left|\tau\right| \right| < \epsilon/2.$$

These inequalities give (2).

We now prove the theorem. Set

$$\epsilon_1 = \frac{\epsilon}{4(r+1)\left| X \right|}, \quad \epsilon' = \frac{r!\eta\epsilon}{2(r+1)}, \quad \eta' = \epsilon_1\eta, \quad \zeta_0 = \frac{(r+1)\epsilon}{2\left| dX \right|}.$$

Choose $\zeta \leqq \zeta_0$ by the lemma, with ϵ and η replaced by ϵ' and η' respectively. Now take any $\sigma \in U_\zeta(p)$ with $\Theta_p(\sigma) \geqq \eta$. Let $\sigma_0, \cdots, \sigma_r$ denote the $(r-1)$-faces of σ, and define the simplexes $\tau = p\sigma$, $\tau_i = p\sigma_i$ (some may be degenerate). We shall show that

$$(5) \qquad \left| X\cdot\tau_i - W\cdot\tau_i \right| < \epsilon \left| \sigma \right| /2(r+1), \qquad i = 0, \cdots, r.$$

Given i, suppose first that $\Theta(\tau_i) \geqq \eta'$. Then the lemma gives

$$\left| X\cdot\tau_i - W\cdot\tau_i \right| < \epsilon' \left| \tau_i \right|.$$

Moreover, using (2.4),

$$\left| \tau_i \right| \leqq \delta^r_{\tau_i}/r! \leqq \delta^r_{p\sigma}/r! = \left| \sigma \right| /r!\Theta_p(\sigma) \leqq \left| \sigma \right| /r!\eta;$$

these inequalities give (5). Suppose next that $\Theta(\tau_i) < \eta'$. Then

$$\left| \tau_i \right| < \eta'\delta^r_{\tau_i} \leqq \eta'\delta^r_{p\sigma} = \eta' \left| \sigma \right| /\Theta_p(\sigma) \leqq \eta' \left| \sigma \right| /\eta = \epsilon_1 \left| \sigma \right|,$$

and hence

$$\left| X\cdot\tau_i \right| < \epsilon_1 \left| \sigma \right| \left| X \right|, \qquad \left| W\cdot\tau_i \right| < \epsilon_1 \left| \sigma \right| \left| X \right|;$$

these give (5) again.

We have, furthermore,

$$\left| dX\cdot\tau \right| \leqq \left| dX \right| \left| \tau \right| \leqq \left| dX \right| \left| \sigma \right| \zeta/(r+1) \leqq \epsilon \left| \sigma \right| /2.$$

Since (App. II, 10.3) $\partial\tau = \sigma - \sum(-1)^i\tau_i$ and $dW = 0$,

$$X\cdot\sigma - W\cdot\sigma = dX\cdot\tau + \sum(-1)^i(X\cdot\tau_i - W\cdot\tau_i).$$

Combining this with the above inequalities gives (1), completing the proof.

11. Total differentiability of Lipschitz functions. As an application of Theorem 10A, we shall prove Rademacher's Theorem:*

* H. Rademacher, Ueber partielle und totale Differenzierbarkeit I, *Math. Ann.* **79** (1919) 340–359. See S. Saks, p. 311.

THEOREM 11A. *Let $f(p)$ be a real valued function in an open set $R \subset E^n$ which satisfies a Lipschitz condition. Then there is a measurable set $Q \subset R$ with $|R - Q| = 0$, such that:*

(a) *For each $p \in Q$, $\nabla f(p, v)$ exists* (II, 1.1) *and is linear in v.*

(b) *$\nabla f(p)$ is measurable in R; that is, each $\nabla f(p, v)$ is.*

(c) *For each $p_0 \in Q$, $\epsilon > 0$ and $\eta > 0$ there is a $\zeta > 0$ with the following property:*

(1) $$|f(q) - f(p) - \nabla f(p_0, q - p)| \leqq \epsilon |q - p|$$
$$\text{if} \quad p, q \in U_\zeta(p_0), \quad |q - p| \geqq \eta |p - p_0|.$$

If we prove the theorem in any open bounded part R' of R, it will follow for R. Set $Y \cdot p = f(p)$ in R'; this defines a flat 0-cochain Y in R' (V, Theorem 4B). Set $X = dY$. Let Q be the set where $D_X(p)$ exists; by Theorem 5A, $|R - Q| = 0$. We shall show that

(2) $$\nabla f(p, v) = D_X(p) \cdot v, \qquad p \in Q.$$

Set $p_t = p + tv$, $\sigma_t = pp_t$ (if $v \neq 0$). Then if t is small enough so that pp_t is in R',

$$f(p_t) - f(p) = Y \cdot \partial(pp_t) = X \cdot \sigma_t,$$

$$\nabla f(p, v) = \lim_{t \to 0+} \frac{f(p_t) - f(p)}{t} = \lim_{t \to 0+} \frac{X \cdot \sigma_t}{|\sigma_t|} |v| = D_X(p) \cdot v.$$

Property (a) now holds; (b) follows from Theorem 5A.

To prove (c), given $p_0 \in Q$, $\epsilon > 0$, $\eta > 0$, set $\eta' = \eta/(1 + \eta)$, and choose $\zeta > 0$ by Theorem 10A, using η'. Now take any $p, q \in R'$ as in (1). We may suppose $\eta \leqq 1$; then

$$|q - p_0| \leqq |q - p| + |p - p_0| \leqq (1 + 1/\eta)|q - p| = |q - p|/\eta',$$

and hence $\operatorname{diam}(p_0pq) \leqq |q - p|/\eta'$. Set $\sigma = pq$. Then $\sigma \subset U_\zeta(p_0)$, and the above inequality gives $\Theta_{p_0}(\sigma) \geqq \eta'$. Hence we may apply (10.1), which gives (1), completing the proof.

THEOREM 11B. *Let f be a mapping of an open set $R \subset E^n$ into E^m, which satisfies a Lipschitz condition. Then for some Q with $|R - Q| = 0$, the following is true. $\nabla f(p, v)$ exists and is linear in Q; ∇f is measurable in R; property* (c) *of Theorem* 11A *holds.*

On writing $f(p)$ in terms of its components f^i in an affine coordinate system, as in (II, 5.20), the theorem follows at once by applying the last theorem to each f^i.

12. On the exterior differential of r-forms. Let ω be a flat r-form in the open set $R \subset E^n$. Let $X = \Psi\omega$ be the corresponding flat r-cochain in R. Write ΦX for D_X. By Theorem 7C, we have

(1) $$\Psi\Phi X = X, \qquad \Phi\Psi\omega = \omega \quad \text{a.e.}$$

We can define $d\omega$ by the formula

$$d\omega = \Phi\, d\Psi\omega; \tag{2}$$

it is a flat $(r+1)$-form, defined a.e. in R, and $d\omega = d\omega'$ *if* $\omega = \omega'$ a.e. Because of (1) and (2).

$$\begin{cases} \Psi\, d\omega = d\Psi\omega, & \Phi\, dX = d\Phi X, \\ dd\omega = \Phi\, d\Psi\Phi\, d\Psi\omega = \Phi\, dd\Psi\omega = 0. \end{cases} \tag{3}$$

Because of Theorem 5A and (2), we have Stokes' Theorem for the flat forms D_X:

$$\int_{\partial\sigma} \omega = \int_\sigma d\omega, \quad \text{all simplexes } \sigma, \quad \text{if} \quad \omega = D_X. \tag{4}$$

We could replace σ by more general regions by the methods of Chapters III and X.

For flat forms ω, define (see (App. III, 5.1))

$$|\omega|^\flat = \sup\{|\omega|_0, |d\omega|_0\}. \tag{5}$$

Then, by (5.3),

$$|X|^\flat = |\omega|^\flat \qquad \text{if } X \text{ and } \omega \text{ are associated.} \tag{6}$$

THEOREM 12A. *Let ω be a flat r-form $(r > 0)$ in the convex open set $R \subset E^n$, and let $d\omega = 0$ a.e. in R. Then there is a flat $(r-1)$-form ξ in R such that $d\xi = \omega$ a.e. in R.*

Set $X = \Psi\omega$; then $dX = 0$. By (VII, Lemma 10b), there is a flat $(r-1)$-cochain Y in R with $dY = X$. Set $\xi = \Phi Y$. Then by (3) and (1),

$$d\xi = \Phi\, dY = \Phi X = \Phi\Psi\omega = \omega \quad \text{a.e.}$$

Recall the analytical formula (II, 8.1) for $d\omega$, which we call $d'\omega$ for the moment:

$$d'\omega(p)\cdot(v_1\vee\cdots\vee v_{r+1}) = \sum_{i=1}^{r+1}(-1)^{i-1}\nabla_{v_i}\omega(p)\cdot(v_1\vee\cdots\hat{v}_i\cdots\vee v_{r+1}). \tag{7}$$

We shall show that for sharp forms (V, 10), this has meaning and equals $d\omega$ a.e.

THEOREM 12B. *Let ω be a sharp r-form in the open set $R \subset E^n$. Then $d'\omega$ is a flat $(r+1)$-form, and equals $d\omega$ a.e. in R.*

Let $\Psi\omega = X$; then $d\omega$ is the $(r+1)$-form corresponding to dX. By Theorem 11B, there is a set $Q \subset R$ with $|R - Q| = 0$, and with the following property: For the mapping ω of R into $V^{[r]}$, $\nabla_v\omega(p)$ exists and is linear in v, for $p \in Q$; it is measurable, and satisfies (c) of Theorem 11A in Q.

Now $\nabla_v\omega(p)\cdot(u_1\vee\cdots\vee u_r)$ is linear in all the vectors for $p \in Q$; hence $d'\omega(p)$ is an $(r+1)$-covector $(p \in Q)$. It is measurable in R. Let $d_\mu\omega$

denote the component $(d\omega)_\mu$. If we show that for each $\mu = (\mu_1, \cdots, \mu_{r+1})$, $d'_\mu\omega = d_\mu\omega$ a.e. in R, the theorem will follow.

First we show the following. Let P^{r+1} be any plane with the direction of e_μ, and with $|P^{r+1} \cap R - Q|_{r+1} = 0$, such that $d'\omega$ is measurable in $P^{r+1} \cap R$. Then for any $p_0 \in P^{r+1} \cap Q$, $\epsilon > 0$, and $\eta > 0$, there is a $\zeta > 0$ with the following property. Let A be any coordinate interval in P^{r+1}. Then

$$(8) \qquad \left| \int_{\partial A} \omega - d'\omega(p_0)\cdot\{A\} \right| \leqq \epsilon |A| \quad \text{if} \quad p_0 \in A \subset U_\zeta(p_0), \quad \Theta(A) \geqq \eta.$$

Set $\epsilon_1 = \epsilon/(r+1)$, and choose ζ for ω, p_0, ϵ_1, η, by Theorem 11B. Let $v_1 = c_1 e_{\mu_1}, \cdots, v_{r+1} = c_{r+1}e_{\mu_{r+1}}$ be the edge vectors of A. Let A_i^-, A_i^+ be the faces of A; then

$$\{A\} = v_1 \vee \cdots \vee v_{r+1}, \qquad \{A_i^-\} = \{A_i^+\} = v_1 \vee \cdots \hat{v}_i \cdots \vee v_{r+1}.$$

Take any $p \in A_i^-$; then $q = p + v_i \in A_i^+$. Set $\delta = \operatorname{diam}(A)$. Then

$$|v_i|\,|A_i^-| = |A| = \Theta(A)\delta^{r+1} \geqq \eta\delta^{r+1}.$$

Also $|v_j| \leqq \delta$, hence $|A_i^-| \leqq \delta^r$, and

$$|q - p| = |v_i| \geqq \eta\delta \geqq \eta\,|p - p_0|.$$

Therefore, applying (11.1) to ω,

$$|\omega(q) - \omega(p) - \nabla_{v_i}\omega(p_0)| \leqq \epsilon_1 |v_i|,$$

and hence

$$|\omega(q)\cdot\{A_i^+\} - \omega(p)\cdot\{A_i^-\} - \nabla_{v_i}\omega(p_0)\cdot(v_1 \vee \cdots \hat{v}_i \cdots \vee v_{r+1})| \leqq \epsilon_1 |A|.$$

Letting p run over A_i^- and integrating gives

$$\left| \int_{A_i^+} \omega - \int_{A_i^-} \omega - \nabla_{v_i}\omega(p_0)\cdot(v_1 \vee \cdots \hat{v}_i \cdots \vee v_{r+1}) \right| \leqq \epsilon_1 |A|;$$

summing over i gives (8), since $\partial A = \sum (-1)^{i-1}(A_i^+ - A_i^-)$.

Take a plane $P' = P^{n-r-1}$ orthogonal to e_μ. For almost all $q \in P'$, $P(q, e_\mu)$ has the properties of P^{r+1} above. Take any $p_0 \in Q \cap P(q, e_\mu)$, and let $A_1, A_2, \cdots$ be a full sequence of intervals as above. Then (10.1) (applied to the regular subdivision of each A_i) and (8) give

$$d_\mu\omega(p_0) = d\omega(p_0)\cdot e_\mu = \lim_{i\to\infty} \frac{dX\cdot A_i}{|A_i|}$$

$$= \lim_{i\to\infty} \frac{1}{|A_i|} \int_{\partial A_i} \omega = d'\omega(p_0)\cdot e_\mu = d'_\mu\omega(p_0).$$

Hence $d_\mu\omega = d'_\mu\omega$ a.e. in $P(q, e_\mu)$, and consequently a.e. in R. This completes the proof.

13. On averages of r-forms. Given the flat r-form ω, let ω^ρ be its ρ-average in $\operatorname{int}_\rho(R)$ (§ 9). Let $X = \Psi\omega$. Since $D_X = \omega$ a.e., (5.6) gives $\omega^\rho = (D_X)^\rho = D_{X_\rho}$. Therefore also, by (12.1), $X_\rho = \Psi\Phi X_\rho = \Psi\omega^\rho$. Thus

$$(1) \qquad (\Phi X)_\rho = \Phi X_\rho, \qquad (\Psi\omega)_\rho = \Psi\omega^\rho.$$

These relations, with (12.2) and (5.5), give

$$(2) \qquad (d\omega)^\rho = d\omega^\rho.$$

Given ω and $X = \Psi\omega$, set $\bar\omega = D_X$. In general $\bar\omega$ has better properties than ω; see Theorems 5A and 17B, and (X, Theorem 9B). We may find $\bar\omega$ directly from ω by a double limiting process, as follows. Given p and the r-direction α, choose vectors $v_1, v_2, \cdots \to 0$ such that for each j, $|P(p + v_j, \alpha) \cap R - Q|_r = 0$ and $\omega(q, \alpha')$ is measurable in $P(p + v_j, \alpha) \cap R$ for all α', Q being a set in terms of which the conditions of § 6 hold. Then (see (7.1))

$$\lim_{j\to\infty} \int_{T_{v_j}\sigma} \omega = \lim_{j\to\infty} X\cdot T_{v_j}\sigma = X\cdot\sigma, \qquad \sigma \subset P(p, \alpha).$$

Choose a p-full sequence $\sigma_1, \sigma_2, \cdots$ in $P(p, \alpha)$, $\alpha(\sigma_i) = \alpha$; then

$$(3) \qquad \bar\omega(p)\cdot\alpha = \lim_{i\to\infty} \frac{X\cdot\sigma_i}{|\sigma_i|} = \lim_{i\to\infty}\lim_{j\to\infty} \frac{1}{|\sigma_i|}\int_{T_{v_j}\sigma} \omega.$$

An obvious method of improving ω is as follows: First take the ρ-average ω^ρ, then let $\rho \to 0$:

$$(4) \qquad \omega^*(p, \alpha) = \lim_{\rho\to 0} \int_V \kappa_\rho(v)\omega(p + v, \alpha)\, dv.$$

This is not as satisfactory as $\bar\omega$; see the example below. So far as ω^* exists, it has satisfactory values:

THEOREM 13A. *Given the flat r-form ω in R, ω^* exists a.e. in R, and is a flat r-form; $\omega^* = \omega$ a.e. in R. For any r-simplex σ such that $\omega^*(p, \alpha(\sigma))$ exists a.e. in σ, $\int_\sigma \omega^* = X\cdot\sigma$ $(X = \Psi\omega)$.*

Define the average X_ρ as in § 5. Then $D_{X_\rho} = \omega^\rho$; hence, for any α,

$$(5) \qquad \omega^*(p, \alpha) = \lim_{\rho\to 0} D_{X_\rho}(p, \alpha) \qquad \text{if } \omega^*(p, \alpha) \text{ exists.}$$

By (5.7), this holds for all α a.e. in R, and $\omega^* = D_X$ a.e.; hence ω^* is a flat r-form equivalent to ω.

Now take any σ as described. Then using (V, 13.12) gives

$$\int_\sigma \omega^* = \int_\sigma \lim_{\rho\to 0} D_{X_\rho} = \lim_{\rho\to 0}\int_\sigma D_{X_\rho} = \lim_{\rho\to 0} X_\rho\cdot\sigma = X\cdot\sigma.$$

EXAMPLE. We shall define a flat 1-form ω in the x-y-plane such that

$\omega^*(p, e_1)$ exists at almost no points of the x-axis. The key to the construction is (V, 13, Example (a)).

Define the strips

$$R_0:\ 3/16 \leqq y \leqq 4/16, \qquad R_0':\ 2/16 \leqq y \leqq 5/16.$$

Cut R_0 into rectangles of length 4, which we call alternately "even" and "odd". Let $\phi(p)$ be 0 at the left hand end of each even interval, let it climb with the slope 1 to the value 4 at the right hand end, and let it go back to 0 in the next odd interval. Let ϕ be 0 on the edges of R_0', and let it be linear on the vertical lines of $R_0' - R_0$, making it continuous in R_0'.

By contracting the plane towards the origin with the factor 1/16, R_0 and R_0' go into strips R_1 and R_1'; R_1 is cut into rectangles of length 1/4. Define ϕ in R_1' as in R_0', ϕ having slopes 1 and -1 in the rectangles of R_1, and having the maximum value 1/4. Continue with R_2, R_2', etc. Set $\phi(p) = 0$ outside the strips R_i'.

Take intervals of length 2 on the x-axis E^1, each with its center below the center of an even rectangle of R_0; let A_0 be the union of all such intervals. Let B_0 be the similar intervals below the odd rectangles of R_0. Define A_i and B_i similarly, using R_i. Set

$$A_i' = A_i \cup A_{i+1} \cup \cdots, \qquad A = A_0' \cap A_1' \cap \cdots,$$

and define B_i', B, similarly. Clearly $|E^1 - A_i'| = 0$; hence

$$|E^1 - A| = 0, \quad |E^1 - B| = 0, \quad |E^1 - A \cap B| = 0.$$

The function $\phi(p)$ defines a 0-cochain Y; set $X = dY$, $\omega = D_X$. Define $\omega^\rho = D_{X_\rho} = \rho$-average of ω, and ω^* as above. We shall sketch the proof that $\omega^*(p, e_1)$ exists at almost no points of the x-axis, *under the assumption* that $\kappa_\rho(v)$ equals its maximum value in most of $U_\rho(0)$.

Take any $p \in A_0$, and consider the average ω^1. Since $U_1(p)$ contains a portion H of an even rectangle of R_0 but no points of any odd rectangle of R_0, and $\omega(p, e_1) = 1$ in H, (5.6) shows that the integral over R_0' contributes an amount $a' > 0$ to $\omega^1(p, e_1)$. For $i > 0$, there are approximately as many even as odd rectangles of R_i in $U_1(p)$; hence the contribution of R_i' to $\omega^1(p, e_1)$ is small. Thus we see that $\omega^1(p, e_1) \geqq a > 0$ for $p \in A_0$. Since $U_{1/16^i}(p) \cap R_j' = 0$ for $j < i$ $(p \in E^1)$, we see similarly that $\omega^{1/16^i}(p, e_1) \geqq a$, $p \in A_i$. Hence for each j and for each $p \in A_j'$ there is an $i \geqq j$ such that $\omega^{1/16^i}(p, e_1) \geqq a$. Consequently, for $p \in A$, there is a sequence $i_1, i_2, \cdots$ such that for each $i_s = k$, $\omega^{1/16^k}(p, e_1) \geqq a$. For $p \in B$, we find similarly $\omega^{1/16^l}(p, e_1) \leqq -a$. Therefore in $A \cap B$, $\omega^*(p, e_1)$ does not exist.

14. Products of cochains. We first define products of forms; this will give products of cochains.

Let $\xi(p)$ and $\eta(p)$ be flat forms in the open set $R \subset E^n$, of degrees r and s respectively. Their *product* is the $(r + s)$-form $\xi \vee \eta$:

$$(1) \qquad (\xi\vee\eta)(p) = \xi(p)\vee\eta(p), \qquad \text{defined a.e. in } R.$$

By (I, 14.2) and (I, 14.3),

$$(2) \qquad |\xi\vee\eta|_0 \leqq \binom{r+s}{r} |\xi|_0 |\eta|_0,$$

$$(3) \qquad |\xi\vee\eta|_0 \leqq |\xi|_0 |\eta|_0 \qquad \text{if } \xi \text{ or } \eta \text{ is simple a.e.,}$$

the latter being true if either r or s equals 0, 1, $n - 1$ or n; see (I, 9).

To prove that $\xi\vee\eta$ is a flat $(r + s)$-form, we show that it is weakly flat (§ 7), letting ϕ be the set of functions $(\xi, \eta, d\xi, d\eta)$. We shall use the averages ξ^ρ, η^ρ of ξ and η. By Theorem 13A, $\lim_{\rho\to 0} \omega^\rho = \omega$ a.e. for any flat ω. Hence, using (13.2),

$$(4) \qquad \xi^\rho \to \xi, \qquad \eta^\rho \to \eta, \qquad d\xi^\rho \to d\xi, \qquad d\eta^\rho \to d\eta,$$

a.e. in R as $\rho \to 0$. Let these hold in Q, $|R - Q| = 0$. We must prove (c) of § 6, using (Q, ϕ)-excellence. Take any $\sigma = \sigma^{r+s+1}$ such that $|\sigma - Q|_{r+s+1} = 0$, $|\partial\sigma - Q|_{r+s} = 0$, and all the above functions are measurable in σ and in $\partial\sigma$. Since the functions are bounded, and ξ^ρ and η^ρ are smooth, we find

$$\begin{aligned}\int_{\partial\sigma} \xi\vee\eta &= \int_{\partial\sigma} \lim\,(\xi^\rho\vee\eta^\rho) = \lim \int_{\partial\sigma} \xi^\rho\vee\eta^\rho \\ &= \lim \int_\sigma d(\xi^\rho\vee\eta^\rho) = \lim \int_\sigma (d\xi^\rho\vee\eta^\rho \pm \xi^\rho\vee d\eta^\rho) \\ &= \int_\sigma (\lim d\xi^\rho \vee \lim \eta^\rho \pm \lim \xi^\rho \vee \lim d\eta^\rho) = \int_\sigma (d\xi\vee\eta \pm \xi\vee d\eta).\end{aligned}$$

Hence, using (2),

$$(5) \qquad \left|\int_{\partial\sigma} \xi\vee\eta\right| \leqq N' |\sigma|, \qquad N' = c_{rs}(|d\xi|_0 |\eta|_0 + |\xi|_0 |d\eta|_0),$$

with c_{rs} defined by (15), as required. Hence $\xi\vee\eta$ is weakly flat, and by Lemma 7c, it is flat.

Let $\omega = \xi\vee\eta$ correspond to W. Then the definition (12.2) of $d\omega$ gives, with σ as above,

$$\int_\sigma d(\xi\vee\eta) = dW\cdot\sigma = W\cdot\partial\sigma = \int_{\partial\sigma} \xi\vee\eta;$$

using the above equality proves (compare the proof of Theorem 7B)

$$(6) \qquad d(\xi\vee\eta) = d\xi\vee\eta + (-1)^r \xi\vee d\eta \quad \text{a.e.} \qquad (\deg(\xi) = r).$$

Given the flat cochains X and Y in R, we now define their *cup product*

$X \smile Y$ as the cochain corresponding to the product of the two forms D_X, D_Y:

$$X \smile Y = \Psi(D_X \vee D_Y). \tag{7}$$

Using (12.3) and (6) gives

$$d(X^r \smile Y^s) = dX^r \smile Y^s + (-1)^r X^r \smile dY^s. \tag{8}$$

The unit 0-cochain I^0 corresponds to the function identically 1. Clearly

$$I^0 \smile X = X \smile I^0 = X. \tag{9}$$

By (5.3), (2) and (3),

$$|X^r \smile Y^s| \leqq \binom{r+s}{r} |X^r| \, |Y|^s, \tag{10}$$

$$|X^r \smile Y^s| \leqq |X^r| \, |Y^s| \qquad \text{if } \; r \text{ or } s \text{ equals } 0, 1, n-1 \text{ or } n. \tag{11}$$

Because of the corresponding properties of the products of forms (I, 6), the product (7) is bilinear and associative, and satisfies

$$X^r \smile Y^s = (-1)^{rs} Y^s \smile X^r. \tag{12}$$

We now find some inequalities on the flat norm of a product, *taking* $R = E$ *for simplicity.* In the general case, *insert the subscript* R; see (VIII, 1.10) and (VIII, 1.21). Since $\binom{r+s}{r} \leqq \binom{r+s+1}{r}$, (V, 4.8) and (10) give

$$\begin{aligned} |X^r \smile Y^s|^\flat &= \sup\{|X \smile Y|, |d(X \smile Y)|\} \\ &\leqq \sup\left\{\binom{r+s}{r}|X|\,|Y|, \binom{r+s+1}{r+1}|dX|\,|Y| + \binom{r+s+1}{r}|X|\,|dY|\right\} \\ &\leqq \binom{r+s+1}{r+1}|dX|\,|Y| + \binom{r+s+1}{r}|X| \sup\{|Y|, |dY|\}, \end{aligned}$$

$$|X^r \smile Y^s|^\flat \leqq \binom{r+s+1}{r+1}|dX|\,|Y| + \binom{r+s+1}{r}|X|\,|Y|^\flat. \tag{13}$$

This gives

$$|X^r \smile Y^s|^\flat \leqq c_{rs}(|X|^\flat\,|Y| + |X|\,|Y|^\flat) \leqq 2c_{rs}|X|^\flat\,|Y|^\flat, \tag{14}$$

$$c_{rs} = \sup\left\{\binom{r+s+1}{r}, \binom{r+s+1}{s}\right\}. \tag{15}$$

For $r = 0$, using (11) gives

$$(16)\qquad |X^0 \smile Y^s|^\flat \leqq |dX|\,|Y| + |X|\,|Y|^\flat \leqq |X|^\flat\,|Y| + |X|\,|Y|^\flat,$$

$$(17)\qquad |X^0 \smile Y^s|^\flat \leqq |X|\,|dY| + |X|^\flat\,|Y| \leqq 2\,|X|^\flat\,|Y|^\flat.$$

EXAMPLE (a). We cannot omit the factor 2 in (17). For, taking $n = 1$, $r = s = 0$, $R =$ unit interval $0 < x < 1$, let

$$D_X(x) = x, \qquad Z = X \smile X.$$

Then, using $\bar{D}_{dZ}(x) = D_{dZ}(x) \cdot e_1$,

$$D_Z(x) = x^2, \qquad \bar{D}_{dZ}(x) = 2x, \qquad |dZ| = 2,$$

and hence, in R,

$$|X|^\flat = 1, \qquad |Z|^\flat = 2 = 2\,|X|^\flat\,|X|^\flat.$$

Now suppose X and Y are sharp. Then they are flat, and the definitions and properties above hold still. If $D_X = \xi$, $D_Y = \eta$, (2) gives

$$|(\xi \vee \eta)(q) - (\xi \vee \eta)(p)|_0 \leqq \binom{r+s}{r}[|\xi(q)|_0\,|\eta(q) - \eta(p)|_0 + |\xi(q) - \xi(p)|_0\,|\eta(p)|_0];$$

therefore

$$(18)\qquad \mathfrak{L}(X^r \smile Y^s) \leqq \binom{r+s}{r}[|X|\,\mathfrak{L}(Y) + \mathfrak{L}(X)\,|Y|].$$

Set $a = \binom{r+s}{r}$, $b = r + s + 1$. Then, by (V, 7.8),

$$\begin{aligned} |X^r \smile Y^s|^\sharp &= \sup \{|X \smile Y|, b\mathfrak{L}(X \smile Y)\} \\ &\leqq \sup \{a\,|X|\,|Y|, ba[|X|\,\mathfrak{L}(Y) + \mathfrak{L}(X)\,|Y|]\} \\ &\leqq ba\mathfrak{L}(X)|Y| + \frac{b}{s+1} a\,|X| \sup \{|Y|, (s+1)\mathfrak{L}(Y)\}, \end{aligned}$$

and since $\dfrac{r+s+1}{s+1}\dbinom{r+s}{r} = \dbinom{r+s+1}{s+1}$ etc.,

$$(19)\quad |X^r \smile Y^s|^\sharp \leqq (r+1)\binom{r+s+1}{r+1}\mathfrak{L}(X)\,|Y| + \binom{r+s+1}{r}|X|\,|Y|^\sharp.$$

This gives

$$(20)\qquad |X^r \smile Y^s|^\sharp \leqq c_{rs}(|X|^\sharp\,|Y| + |X|\,|Y|^\sharp) \leqq 2c_{rs}\,|X|^\sharp\,|Y|^\sharp.$$

For $r = 0$, (18) and (19) give

$$(21)\qquad \mathfrak{L}(X^0 \smile Y^s) \leqq |X|\,\mathfrak{L}(Y) + \mathfrak{L}(X)\,|Y|,$$

$$\begin{aligned} (22)\qquad |X^0 \smile Y^s|^\sharp &\leqq (s+1)\mathfrak{L}(X)\,|Y| + |X|\,|Y|^\sharp \\ &\leqq [|X| + (s+1)\mathfrak{L}(X)]\,|Y|^\sharp \leqq (s+2)\,|X|^\sharp\,|Y|^\sharp. \end{aligned}$$

EXAMPLE (b). We cannot lessen the factor $s + 1$ in (22). To show this, take E^{s+1}, let $D_X(x^1, \cdots) = x^1$, and let $D_Y(p)$ have a single non-zero component, equal to 1. Then in a neighborhood R of the origin,

$$\mathfrak{L}(X) = 1, \quad |Y| = 1, \quad |X \smile Y|^\# = (s+1)\mathfrak{L}(X \smile Y) = s + 1,$$

while for R small enough, $|X|\,|Y|^\#$ is as small as desired.

Recall (Theorem 7D) that for any sharp ϕ and flat X,

$$D_{\phi X}(p) = \phi(p) D_X(p) \text{ a.e.} \tag{23}$$

We show now that the product $Y \smile X$, Y being a 0-cochain, agrees with the products of (VII, 1). That is, if ϕ is any sharp function, and $\bar{\phi}$ is the corresponding 0-cochain ($\bar{\phi} \cdot p = \phi(p)$), then for any r-cochain X, sharp or flat,

$$\bar{\phi} \smile X = X \smile \bar{\phi} = \phi X. \tag{24}$$

Set $W = \bar{\phi} \smile X$. By (7) and (23)

$$D_W = \phi \vee D_X = \phi D_X = D_{\phi X} \quad \text{a.e.,}$$

and (24) follows.

Given ϕ and flat X, recalling (V, 4.11) that $|d\bar{\phi}| = \mathfrak{L}_\phi$ and using (24) and (16) gives

$$|\phi X|^\flat \leqq |\phi|\,|X|^\flat + \mathfrak{L}_\phi |X| \leqq (|\phi| + \mathfrak{L}_\phi)|X|^\flat, \tag{25}$$

which is a strengthening of the second part of (VII, 2.2). The first part of that relation follows from (22).

15. Lebesgue chains. We shall generalize the discussion of (VI, 7). Let $\alpha(p)$ be any measurable summable r-vector valued function in E^n (Euclidean and oriented); then the components $\alpha^\lambda(p)$ are summable. We say α *corresponds* to the flat chain $A = \tilde{\alpha}$ if

$$\int_{E^n} D_X \cdot \alpha = X \cdot A, \qquad \text{all flat } r\text{-cochains } X. \tag{1}$$

(We could use only sharp X, as in (VI, 7.1).) We call $\tilde{\alpha}$ a *Lebesgue chain.*

THEOREM 15A. *The mapping $\alpha \to \tilde{\alpha}$ gives a one-one mapping of equivalence classes into $\mathbf{C}_r^\flat$; the formula $|\tilde{\alpha}| \leftharpoondown \int_{E^n} \langle\alpha\rangle_0$ holds.*

The proof of (VI, Theorem 7A) applies without essential change; see (XI, Theorem 14A).

Note that

$$|\tilde{\alpha} + \tilde{\beta}| = |\tilde{\alpha}| + |\tilde{\beta}| \quad \text{if} \quad \operatorname{car}(\alpha) \cap \operatorname{car}(\beta) = 0; \tag{2}$$

for

$$|\tilde{\alpha} + \tilde{\beta}| = \int_{E^n} \langle\alpha + \beta\rangle_0 = \int_{\operatorname{car}(\alpha)} \langle\alpha\rangle_0 + \int_{\operatorname{car}(\beta)} \langle\beta\rangle_0 = |\tilde{\alpha}| + |\tilde{\beta}|.$$

Recall (App. III, 6) that L^1 is the space of equivalence classes of measurable summable functions, with $\int_{E^n} \langle\phi\rangle$ as norm.

THEOREM 15B. *With E^n Euclidean and oriented, the flat n-chains correspond exactly to the elements Φ of L^1, by*

$$(3) \qquad \Phi \leftrightarrow \tilde{\alpha}, \qquad \alpha(p) = \phi(p)e_{1\cdots n}, \qquad \phi \in \Phi.$$

First, the $\tilde{\alpha}$ are dense in $\mathbf{C}_n^\flat$ (VI, Theorem 7A). Next, since $|\tilde{\alpha}| = \int \langle\alpha_{1\cdots n}\rangle$, the mass, which equals the flat norm (since $r = n$), is the same as the norm in L^1. Since L^1 is complete (App. III, 6), the $\tilde{\alpha}$ give all n-chains A.

For $r = n$, we may write (1) in the form

$$(4) \qquad X\cdot\tilde{\alpha} = \int_{E^n} \bar{D}_X\bar{\alpha}, \quad D_X(p) = \bar{D}_X(p)e^{1\cdots n}, \quad \alpha(p) = \bar{\alpha}(p)e_{1\cdots n}.$$

THEOREM 15C. *For any flat n-chain $A = \tilde{\alpha}$ in E^n, there is a flat n-cochain X with*

$$(5) \qquad |X| = 1, \qquad X\cdot A = |A|.$$

We need merely set $\bar{D}_X(p) = 1$ where $\bar{\alpha}(p) > 0$, $= -1$ where $\bar{\alpha}(p) < 0$, and $= 0$ elsewhere.

Note that (VII, 2.8) holds for any measurable summable α; the proof given there applies.

16. Products of cochains and chains. We take $R = E$ for simplicity. We first define the *cap product* $X \frown A$ of a flat s-cochain X and a Lebesgue $(r+s)$-chain $A = \tilde{\alpha}$ by the formula

$$(1) \qquad X\frown\tilde{\alpha} = \tilde{\beta}, \qquad \beta(p) = D_X(p)\wedge\alpha(p),$$

using the interior product of (I, 7). Since α is summable and D_X is measurable and bounded, β is summable, and hence $\tilde{\beta}$ is a Lebesgue r-chain; see Theorem 15A.

We show that

$$(2) \qquad Y^r\cdot(X^s\frown A^{r+s}) = (Y^r\smile X^s)\cdot A^{r+s},$$

first if $A = \tilde{\alpha}$ is a Lebesgue chain. We use (1), (15.1), (I, 7.1) and (14.7), giving

$$Y\cdot(X\frown\tilde{\alpha}) = \int D_Y\cdot(D_X\wedge\alpha) = \int (D_Y\vee D_X)\cdot\alpha = (Y\smile X)\cdot\tilde{\alpha}.$$

We now prove, always for the moment for $A = \tilde{\alpha}$, with c_{rs} as in (14.15),

$$(3) \qquad |X^s\frown A^{r+s}|^\flat \leqq \left[\binom{r+s+1}{r+1}|X| + \binom{r+s+1}{r}|dX|\right]|A|^\flat,$$

$$(4) \qquad |X^s\frown A^{r+s}|^\flat \leqq 2c_{rs}|X|^\flat|A|^\flat,$$

$$(5) \qquad |X^0\frown A^r|^\flat \leqq (|X| + |dX|)|A|^\flat \leqq 2|X|^\flat|A|^\flat.$$

To prove (3), take any flat r-cochain Y. By (2) and (14.13),

$$| Y\cdot(X\frown A) | = | (Y\smile X)\cdot A | \leqq | X\smile Y |^\flat | A |^\flat$$
$$\leqq \left[\binom{s+r+1}{s+1} | dX | + \binom{s+r+1}{s} | X |\right] | Y |^\flat | A |^\flat;$$

The inequality now follows from (V, 4.3). This inequality gives (4). The last two inequalities follow from (14.16).

We now define $X\frown A$ for any flat X and A. Let $A_1, A_2, \cdots$ be a sequence of continuous chains, $A_i \xrightarrow{\flat} A$ (VI, Theorem 7A); set

$$X\frown A = \lim_{i\to\infty}{}^\flat (X\frown A_i). \tag{6}$$

That the limit exists and is uniquely determined follows at once from (4). The properties given above now extend immediately to flat A.

We prove next

$$| X^s\frown A^{r+s} | \leqq \binom{r+s}{r} | X | | A |, \tag{7}$$

$$| X^s\frown A^{r+s} | \leqq | X | | A | \quad \text{if} \quad s = 0, 1, n-1 \text{ or } n. \tag{8}$$

To prove (7), take any flat r-cochain Y. We have, using (14.10),

$$| Y\cdot(X\frown A) | = | (Y\smile X)\cdot A | \leqq | Y\smile X | | A |$$
$$\leqq \binom{r+s}{r} | Y | |X| | A |;$$

the relation follows from (V, Theorem 16A). To prove (8), use (14.11).

We set

$$X^s\frown A^t = 0 \quad \text{if} \quad s > t. \tag{9}$$

Letting I^0 be the unit 0-cochain, we prove some further properties of the products of flat cochains and chains:

$$X\frown(Y\frown A) = (X\smile Y)\frown A, \tag{10}$$

$$\partial(X^s\frown A^{r+s}) = (-1)^r\, dX\frown A + X\frown\partial A, \tag{11}$$

$$I^0\frown A = A, \qquad I^0\cdot(X^r\frown A^r) = X\cdot A. \tag{12}$$

To prove (10), take any Z of the right dimension. We have, using the associative law for $\smile$,

$$Z\cdot[X\frown(Y\frown A)] = (Z\smile X)\cdot(Y\frown A) = [(Z\smile X)\smile Y]\cdot A$$
$$= [Z\smile(X\smile Y)]\cdot A = Z\cdot[(X\smile Y)\frown A].$$

If $r = 0$, both sides of (11) vanish. To prove (11) for $r > 0$, take any Y^{r-1}, and use (14.8):

$$Y \cdot [\partial(X \frown A)] = dY \cdot (X \frown A) = (dY \smile X) \cdot A$$
$$= [d(Y \smile X) - (-1)^{r-1}(Y \smile dX)] \cdot A = (Y \smile X) \cdot \partial A + (-1)^r (Y \smile dX) \cdot A$$
$$= Y \cdot [X \frown \partial A + (-1)^r \, dX \frown A].$$

Finally, (12) follows from (14.9):

$$Y^s \cdot (I^0 \frown A^s) = (Y^s \smile I^0) \cdot A^s = Y^s \cdot A^s,$$
$$I^0 \cdot (X^r \frown A^r) = (I^0 \smile X^r) \cdot A^r = X^r \cdot A^r.$$

Now suppose X and A are sharp. First suppose $A = \tilde{\alpha}$; then we may define $X \frown A$ as before. We have relations (2), (7) and (8) again, and also

$$(15) \quad |X^s \frown A^{r+s}|^\sharp \leqq \left[\binom{r+s+1}{r+1} |X| + \binom{r+s+1}{r}(s+1)\mathfrak{L}(X)\right] |A|^\sharp,$$

$$(16) \quad |X^s \frown A^{r+s}|^\sharp \leqq 2c_{rs} |X|^\sharp |A|^\sharp,$$

$$(17) \quad |X^0 \frown A^r|^\sharp \leqq [|X| + (r+1)\mathfrak{L}(X)] |A|^\sharp \leqq (r+2) |X|^\sharp |A|^\sharp.$$

To prove (15), use (14.19): For any Y^r,

$$|Y \cdot (X \frown A)| = |(Y \smile X) \cdot A| \leqq |Y \smile X|^\sharp |A|^\sharp$$
$$\leqq \left[(s+1)\binom{s+r+1}{s+1} \mathfrak{L}(X) + \binom{s+r+1}{s} |X|\right] |Y|^\sharp |A|^\sharp,$$

which gives the result. (16) follows from this, and (17) is proved with the help of (14.22).

As in § 14 for the cup product, the definition and above properties of $X \frown A$ for any sharp A now follow. Relations (10) and (12) continue to hold, but not (11) in general, since ∂A need not be defined.

For any sharp ϕ and sharp or flat A, using the notations of (14.24), that relation gives

$$X \cdot \phi A = \phi X \cdot A = (X \smile \bar{\phi}) \cdot A = X \cdot (\bar{\phi} \frown A)$$

for any X; hence

$$(18) \quad \phi A = \bar{\phi} \frown A.$$

Now (5) gives

$$(19) \quad |\phi A|^\flat \leqq (|\phi| + \mathfrak{L}_\phi) |A|^\flat,$$

which strengthens (VII, 1.5). The relation (17) gives (VII, 1.4) again.

17. Products and weak limits. We shall prove

THEOREM 17A. *If the limits on the right hand side exist, then*

$$\text{(1)} \qquad \operatorname*{wkl^\flat}_{i,j\to\infty} (X_i \smile Y_j) = \operatorname*{wkl^\flat}_{i\to\infty} X_i \smile \operatorname*{wkl^\flat}_{j\to\infty} Y_j.$$

Let X, Y be the limits on the right. Say $|X_i|^\flat, |Y_j|^\flat \leqq N$. Because of (14.14), we need merely prove that for any simplex σ of the proper dimension,

$$\text{(2)} \qquad \lim_{i,j\to\infty} [(X_i \smile Y_j)\cdot\sigma] = (X \smile Y)\cdot\sigma.$$

We prove the theorem first in the case that $X = 0$.

Suppose first that X is a 0-cochain. Given σ and $\epsilon > 0$, (V, Lemma 13b) shows that we may choose i_0 so that $|D_{X_i}(p)| \leqq \epsilon' = \epsilon/N|\sigma|$ in a neighborhood U of σ, if $i \geqq i_0$. Considering the cochains in U alone, we have, for $i \geqq i_0$ and for any j, using (14.11) in U,

$$|(X_i \smile Y_j)\cdot\sigma| \leqq |X_i|\,|Y_j|\,|\sigma| \leqq \epsilon' N |\sigma| = \epsilon.$$

Now suppose $X = X^r$, $Y = Y^s$; we shall use induction on r. Given $\epsilon > 0$, choose $\delta > 0$ so that

$$\binom{r+s}{r}\delta N^2,\ \binom{r+s}{r-1}\delta N^2 \leqq \frac{\epsilon}{3|\sigma|}.$$

Cover σ by open spheres $R_1, \cdots, R_m$ of radius $\leqq \delta$. Let p_k be the center of R_k. As in (VII, 10.2), define the $(r-1)$-cochain Z_{ik} in R_k by setting

$$\text{(3)} \qquad Z_{ik}\cdot\tau = X_i\cdot J(p_k, \tau);$$

then if $X'_{ik} = X_i - dZ_{ik}$, we have, using (App. II, 10.3),

$$\text{(4)} \qquad |Z_{ik}| \leqq \delta N, \quad |dZ_{ik}| \leqq (1+\delta)N, \quad |X'_{ik}| \leqq \delta N, \quad |dX'_{ik}| \leqq N.$$

We may cut σ into cells σ_k, so that

$$\sigma = \sum \sigma_k, \quad \sigma_k \subset R_k.$$

Since $\text{wkl}^\flat X_i = 0$, the relations above show that $\text{wkl}^\flat Z_{ik} = 0$ for each k. Hence, by induction, we may find i_0 and j_0 such that, for each k,

$$|(Z_{ik} \smile Y_j)\cdot\partial\sigma_k| < \epsilon/3m \quad \text{if} \quad i \geqq i_0, \quad j \geqq j_0.$$

Now if $i \geqq i_0$, $j \geqq j_0$, (14.8) and (14.10) give

$$|(X_i \smile Y_j)\cdot\sigma| = \left|\sum_k [X'_{ik} \smile Y_j \pm Z_{ik} \smile dY_j + d(Z_{ik} \smile Y_j)]\cdot\sigma_k\right|$$

$$\leqq \sum_k [\,|X'_{ik} \smile Y_j| + |Z_{ik} \smile dY_j|\,]\,|\sigma_k| + \sum_k |(Z_{ik} \smile Y_j)\cdot\partial\sigma_k|$$

$$\leqq \sum_k \left[\binom{r+s}{r}|X'_{ik}|\,|Y_j| + \binom{r+s}{r-1}|Z_{ik}|\,|dY_j|\right]|\sigma_k| + \frac{\epsilon}{3} \leqq \epsilon.$$

We now drop the assumption $X = 0$. Given $\epsilon > 0$, using the case just proved and the hypothesis $\text{wkl}^\flat\, Y_j = Y$ shows that there are numbers i_0 and j_0 such that if $i \geqq i_0$, $j \geqq j_0$, then

$$|[(X_i - X)\smile Y_j]\cdot\sigma| < \epsilon/2,$$
$$|Y_j\cdot(X\frown\sigma) - Y\cdot(X\frown\sigma)| < \epsilon/2.$$

Set $Y_j' = Y_j - Y$. Then

$$|(X\smile Y_j')\cdot\sigma| = |(Y_j'\smile X)\cdot\sigma| = |Y_j'\cdot(X\frown\sigma)| < \epsilon/2,$$

and

$$|(X_i\smile Y_j)\cdot\sigma - (X\smile Y)\cdot\sigma| \leqq |[(X_i - X)\smile Y_j]\cdot\sigma| + |(X\smile Y_j')\cdot\sigma| < \epsilon.$$

As an application of the theorem, we prove:

THEOREM 17B. *Let $X = X^r$ and $Y = Y^s$ be flat in R. Then for any subspace $P = P^t$ of E, with $t \geqq r + s$, using only multivectors lying in P,*

$$D_Z = D_X \vee D_Y \quad \text{a.e. in } P, \text{ if } \quad Z = X\smile Y. \tag{5}$$

For any simplex $\sigma \subset R$ of the correct dimension,

$$(X\smile Y)\cdot\sigma = \int_\sigma D_Z = \int_\sigma D_X \vee D_Y. \tag{6}$$

Note that (5) does not follow directly from (14.7); we know only that $D_Z = D_X \vee D_Y$ a.e. in R.

If we prove (6) for any simplex σ in P, then (5) follows, as in the proof of Theorem 7B.

Choose a sequence of vectors $v_1, v_2, \cdots \to 0$ such that $D_Z = D_X \vee D_Y$ a.e. in each $T_{v_i}P'$, $P' = P(\sigma)$. Let X_i^* be the cochain in part of P' defined by

$$X_i^*\cdot\tau = X\cdot T_{v_i}\tau, \qquad \tau \subset P' \cap T_{-v_i}R.$$

Let $X_{P'}$ be X, considered in $P' \cap R$ only. Define Y_i^*, $Y_{P'}$, similarly. Because of (V, 3.6), it is clear that in any $P' \cap \text{int}_\rho\,(R)$,

$$\text{wkl}^\flat\, X_i^* = X_{P'}, \qquad \text{wkl}^\flat\, Y_i^* = Y_{P'}.$$

By the definition (14.7) of $W = X_{P'}\smile Y_{P'}$, $D_W = D_{X_{P'}} \vee D_{Y_{P'}}$ a.e. in $P' \cap R$, and hence a.e. in σ. Therefore

$$(X_{P'}\smile Y_{P'})\cdot\sigma = \int_\sigma D_{X_{P'}} \vee D_{Y_{P'}} = \int_\sigma D_X \vee D_Y.$$

By (1), this also equals

$$\lim\,[(X_i^*\smile Y_i^*)\cdot\sigma] = \lim \int_{T_{v_i}\sigma} D_X \vee D_Y = \lim \int_{T_{v_i}\sigma} D_Z$$
$$= \lim Z\cdot T_{v_i}\sigma = Z\cdot\sigma,$$

as required.

Note that (5) is equivalent to the relation

$$(X\smile Y)_P = X_P\smile Y_P; \tag{7}$$

this follows also from (X, 11.1), on letting f be the identity mapping of P into E.

18. Characterization of the products. We prove

THEOREM 18A. *The cup product of flat cochains in an open set R is characterized by the following properties:*

(a) *For each X^r, Y^s, $X^r\smile Y^s$ is a flat $(r+s)$-cochain in R; this operation is bilinear.*

(b) *There are numbers a_{rs} such that*

$$|X^r\smile Y^s| \leqq a_{rs}|X^r||Y^s|.$$

(c) *For $r = 0$, $X^0\smile Y^s = D_{X^0}Y^s$, as in* (14.24).

(d) $d(X^r\smile Y^s) = dX^r\smile Y^s + (-1)^r X^r\smile dY^s$.

We know from § 14 that these products exist; we must prove uniqueness. By (c), uniqueness holds for $r = 0$; we use induction on r.

First suppose $X = dZ$ is a coboundary. Then (d), applied to $Z\smile Y$, gives

$$X\smile Y = d(Z\smile Y) - (-1)^{r-1}Z\smile dY;$$

the right hand side is uniquely determined, by induction.

Now take any X. Take any $\sigma = \sigma^{r+s}$. For each sufficiently large integer i, we may choose spheres $R_{i1}, \cdots, R_{im_i}$ of radius $1/2^i$ in R, which cover σ; write $\sigma = \sum_k \sigma_{ik}$, $\sigma_{ik} \subset R_{ik}$. Define Z_{ik}, X'_{ik} as in § 17. Then since $|X'_{ik}| \leqq |dX|/2^i$ (see (17.4)), we have

$$\left|\sum_k (dZ_{ik}\smile Y)\cdot\sigma_{ik} - (X\smile Y)\cdot\sigma\right| = \left|\sum_k [(dZ_{ik} - X)\smile Y]\cdot\sigma_{ik}\right|$$

$$\leqq \sum_k a_{rs}|X'_{ik}||Y||\sigma_{ik}| \leqq a_{rs}|dX||Y||\sigma|/2^i,$$

which $\to 0$ as $i\to\infty$. Therefore

$$(X\smile Y)\cdot\sigma = \lim_{i\to\infty}\sum_k (dZ_{ik}\smile Y)\cdot\sigma_{ik}, \tag{1}$$

and the uniqueness of the right hand side gives that of the left hand side.

Note that since $|Z_{ik}| \leqq |X|/2^i$, we may also write

$$(X\smile Y)\cdot\sigma = \lim_{i\to\infty}\sum_k d(Z_{ik}\smile Y)\cdot\sigma_{ik}. \tag{2}$$

REMARK. We may give a direct proof of the existence of the products, using induction, by means of the formula (2). One proves first the existence of the limit, then (b) and (d), and then finds a bound for $|d(X \smile Y)|$, proving that $X \smile Y$ is a flat cochain; the usual properties of products then follow also. This procedure is rather laborious, and one does not obtain so small values for the a_{rs} as used in (14.10).

X. Lipschitz Mappings

In the applications, a domain of integration very commonly consists of an oriented arc or piece of surface, or more generally, an oriented piece of an r-dimensional manifold in E^m. The first object of this chapter is to represent these as chains. If A is a polyhedral chain, and f is a Lipschitz mapping (II, 4) of the polyhedron $P = \text{spt}(A)$ into E^m, then fA is a "Lipschitz chain" (which is flat, hence also sharp). A domain of the above sort (if compact) is a typical case.

More generally, let f be a Lipschitz mapping of the open set $R \subset E^n$ into the open set $S \subset E^m$. Then a polyhedral chain A in R gives a Lipschitz chain fA in S; this defines a flat chain fA of S for each flat chain A of R, provided f carries the support of A into a set whose closure is in S. For each flat cochain X in S, setting $f^*X \cdot A = X \cdot fA$ for such chains A of R defines a flat cochain X in R. Moreover, flat forms in S go into flat forms in R. The action of mappings on products, and some other miscellaneous topics, are considered.

In the first part of the chapter, we study the Lipschitz chains mentioned above. Take a fine subdivision $\sum \sigma_k$ of P, and let f' be the mapping which agrees with f at the vertices and is affine in each σ_k. We must require the σ_k to be reasonably shaped; see the example of Schwarz mentioned in (IV, 14). Then the $f'\sigma_k$ are in general close to the $f\sigma_k$, both in position and direction, and the polyhedral chain $f'A$ is an approximation to the desired chain fA. We define $fA = \lim^\flat f_i A$, for a sequence of such approximations. To show that $|f_j A - f_i A|^\flat$ is small for large i, j, deform f_i into f_j along line segments; this defines a mapping F of $I \times P$ into E^m (I = unit interval). The image $F(I \times P)$ is not polyhedral in general; but with a simplexwise affine approximation F' to F, we obtain polyhedral chains $F'(I \times A)$, $F'(I \times \partial A)$, of small mass, which may be used to give the desired result.

Next we consider Lipschitz mappings f of open sets (see above). If f maps E^n into E^m, we do not need the considerations of Chapter VIII; but the proofs in the general case are hardly more difficult. For an example of the situation we deal with, let R be the set of all points (r, θ) in the plane (using polar coordinates) such that $0 < \theta < 2\pi$, $r > 0$; set $f(r, \theta) = (r, \theta/2)$. This is not Lipschitz; for the points $(1, \theta)$, $(1, 2\pi - \theta)$, for θ small, are close together, but are mapped into points not close together. But if we use a new metric in R, with distance between two points equal to the lower bound of lengths of curves in R joining them, then with this metric, f is Lipschitz; we say that f is "R-Lipschitz". If this conditions holds, then flat chains A of R go into flat chains fA of S, as noted above. Besides the usual properties of mappings, we give a continuity theorem. The theory of f^*X for flat cochains X in S follows easily.

Let ω be a flat r-form in S. Then there is a corresponding flat r-cochain X in S, this gives a flat cochain $X^* = f^*X$ in R, and this corresponds to a flat form D_{X^*} in R, which we call $f^*\omega$. The desired properties of this f^* hold; in particular, $df^*\omega = f^*d\omega$. In the usual analytical formulation, both ω and f are assumed differentiable. Though one cannot expect the analytical formula for $f^*\omega$ to have meaning for a general flat ω, we prove (Theorem 9B) that if we replace ω by the equivalent flat form D_X, then the formula actually gives $f^*\omega$ a.e.

If f is smooth and its first partial derivatives satisfy a Lipschitz condition, then $f^*\omega$ is sharp whenever ω is sharp, as is easily seen. This is not true for more general f. For example, let f map E^1 into E^1 by setting $f(x) = x^{3/2}$ $(x > 0)$, $f(x) = 0$ $(x \leqq 0)$. Let ω be the unit 1-form in E^1. Then, with the unit vector e in E^1,

$$f^*\omega(x)\cdot e = \omega(f(x))\cdot\nabla f(x, e) = df(x)/dx = (3/2)x^{1/2}, \qquad x > 0,$$

and $f^*\omega(x)\cdot e = 0$ for $x \leqq 0$. Thus $f^*\omega$ is continuous, but does not satisfy a Lipschitz condition.

In § 11 we prove that $f^*(X\smile Y) = f^*X\smile f^*Y$ (the special case $\dim(X) = 0$ was considered in § 10). The proof uses Theorem 9A, and also the theorem on weak limits in (IX, 17). Other properties of the cup and cap products follow easily.

A formula for the norm of a Lipschitz chain is given in § 12; in § 13, we give an improved version of the continuity theorem of § 5.

Remark. Without the assumption that f is Lipschitz, one cannot obtain the results given in this chapter. For example, there is a curve C in the plane E^2 and a smooth function ϕ in E^2 such that the difference $\phi(q) - \phi(p)$ $(p, q \in C)$ cannot be obtained* by integrating derivatives of ϕ along C.

1. Affine approximations to Lipschitz mappings. Let K be a simplicial complex, and let f be a mapping of K into E^m. The *corresponding simplex-wise affine approximation* f' to f is that mapping which is the same as f on each vertex of K, and is affine in each simplex of K (App. I, 12).

* H. Whitney, A function not constant on a connected set of critical points, *Duke Math. Journal, 1* (1935) 514–517.

If $p_1, p_2, \cdots$ are vertices of K, and $\mu_1, \mu_2, \cdots$ are the barycentric coordinates (App. II, 2), then f' is defined by

$$f'\left(\sum \mu_i p_i\right) = \sum \mu_i f(p_i). \tag{1}$$

Recall (App. I, 12.4) that

$$f'(q) - f'(p) = \nabla f'(p^*, q - p) \qquad \text{in each simplex of } K, \tag{2}$$

and $\nabla f'$ is constant in each simplex.

In the case that K itself is a simplex σ, we wish to relate the Lipschitz constant $\mathfrak{L}_{f'}$ (II, 4) to $\mathfrak{L}_f$.

LEMMA 1a. *If f' is the affine approximation to f in the r-simplex σ, then*

$$\mathfrak{L}_{f'} \leqq \mathfrak{L}_f/(r-1)!\Theta(\sigma). \tag{3}$$

REMARK. With a more careful analysis (studying segments in σ of maximum length in each direction), we could replace $(r-1)!$ by (3) by $r!$.

Since f' is affine, we need merely prove

$$|\nabla f'(p^*, u)| \leqq \mathfrak{L}_f |u|/(r-1)!\Theta(\sigma), \qquad p^* \in \text{int}(\sigma),$$

for any vector u in σ. Say

$$\sigma = p_0 \cdots p_r, \qquad u_i = (p_i - p_0)/|p_i - p_0|, \qquad u = \sum a_i u_i.$$

By (2) and (IV, 15.3),

$$|\nabla f'(p^*, u_i)| = |f(p_i) - f(p_0)|/|p_i - p_0| \leqq \mathfrak{L}_f,$$

$$|a_i| \leqq |u|/r!\Theta(\sigma) \qquad (\text{all } i).$$

Hence

$$|\nabla f'(p^*, u)| = \left|\sum a_i \nabla f'(p^*, u_i)\right| \leqq r\mathfrak{L}_f |u|/r!\Theta(\sigma).$$

EXAMPLE. Let σ be the triangle in the (x, y)-plane with vertices at $(0, 0)$, $(1, \epsilon)$, $(2, 0)$. Let f map these vertices into $(0, 0)$, $(1, \epsilon)$, $(1, 0)$. We may extend f over σ so that $\mathfrak{L}_f = 1$ (project points to the right of $x = 1$ into this line). Clearly $\mathfrak{L}_{f'} \geqq 1/2\epsilon$.

By (II, 6.1), (II, 4.1), (I, 12.16) and (II, 4.15), we have the inequality for any Jacobian r-vector

$$|J_f(p)| \leqq \mathfrak{L}_f^r. \tag{4}$$

Applying this to f' and using (3) gives

$$|J_{f'}(p)| \leqq \mathfrak{L}_{f'}^r \leqq \mathfrak{L}_f^r/[(r-1)!\Theta(\sigma)]^r \qquad \text{in } r\text{-simplexes}. \tag{5}$$

2. The approximation on the edges of a simplex. We shall show that if $\nabla_v f$ is nearly constant in most of a simplex, then the vertices are mapped nearly as expected by the values of $\nabla_v f$. By (IX, Theorem 11B), ∇f exists a.e. and is measurable if f is Lipschitz.

LEMMA 2a. *Let f be a Lipschitz mapping of the r-simplex σ into E^n. Let Q be a measurable subset of σ in which ∇f exists, and let ρ be a number such that*

$$(1) \qquad |\sigma - Q| \leqq \rho^r |\sigma|, \qquad 0 < \rho < 1,$$

$$(2) \qquad |\nabla f(q) - \nabla f(p)| \leqq \rho \mathfrak{L}_f \quad \text{if} \quad p, q \in Q.$$

Then for any edge $p_i p_j$ of σ, if $\delta = \operatorname{diam}(\sigma)$,

$$(3) \qquad |f(p_j) - f(p_i) - \nabla f(p^*, p_j - p_i)| < 6\rho\mathfrak{L}_f\delta \quad \text{if} \quad p^* \in Q.$$

Suppose first that $r \geqq 2$. Let

$$\sigma^1 = p_i p_j, \qquad \sigma' = \text{opposite } (r-2)\text{-face}.$$

For any t, $0 \leqq t \leqq 1$, let τ_t be the set of all points

$$p = (1-t)q + tq', \qquad q \in \sigma^1, \qquad q' \in \sigma'.$$

The part of τ_t with a fixed q is an $(r-2)$-simplex similar to σ', of volume $t^{r-2}|\sigma'|$, and the part with q' fixed is a 1-simplex of length $(1-t)|\sigma^1|$; hence

$$|\tau_t| = a(1-t)t^{r-2},$$

a being independent of t. Therefore, if σ_t is the part of σ filled out by the τ_s with $0 \leqq s \leqq t$,

$$|\sigma_t| = b\int_0^t |\tau_s|\, ds = ab\left[\frac{t^{r-1}}{r-1} - \frac{t^r}{r}\right].$$

For $t = 1$, we have $|\sigma| = |\sigma_1| = ab/r(r-1)$; hence

$$(4) \qquad |\sigma_t| = [rt^{r-1} - (r-1)t^r]|\sigma| = [1 + (r-1)(1-t)]t^{r-1}|\sigma|.$$

This gives

$$(5) \qquad |\sigma_t| > t^{r-1}|\sigma| \quad \text{if} \quad t < 1.$$

Combining this with (1) gives, if $H = \sigma - Q$,

$$(6) \qquad |\sigma_\rho \cap H| \leqq |H| < \rho|\sigma_\rho|.$$

Set

$$\sigma^{r-1} = p_i\sigma', \qquad \tau^* = \sigma^{r-1} \cap \sigma_\rho.$$

For each $p \in \tau^*$, the line through p parallel with σ^1 cuts out a segment I_p from σ; these segments fill σ_ρ. Say

$$|I_p \cap H| = \theta(p)|I_p| \qquad (\theta \text{ defined a.e. in } \tau^*).$$

Let τ' be τ^*, projected onto a plane perpendicular to σ^1. Then, using measure in τ' rather than in τ^*,

$$\frac{\int_{\tau^*}\theta(p)\,|\,I_p\,|\,dp}{\int_{\tau^*}\rho\,|\,I_p\,|\,dp} = \frac{|\,H \cap \sigma_\rho\,|}{\rho\,|\,\sigma_\rho\,|} < 1;$$

hence $\theta(p) < \rho$ in a subset of τ^* of positive measure, and we may choose $p' \in \tau^*$ such that

$$(7) \qquad |\,I_{p'} \cap H\,| < \rho\,|\,I_{p'}\,|,$$

and ∇f is defined a.e. in $I_{p'}$.

Say

$$v = p_j - p_i, \qquad I_{p'} = p'p'', \qquad v' = p'' - p';$$

then if $p'_t = p' + tv'$, (II, 1.3) gives, for $p^* \in Q$,

$$|f(p'') - f(p') - \nabla f(p^*, v')\,| \leqq \int_0^1 |\,\nabla f(p'_t, v') - \nabla f(p^*, v')\,|\,dt.$$

Dividing the integral into two parts $\int_1$ and $\int_2$, with those t such that $p'_t \in Q$ and $p'_t \in H$ respectively, (2) gives

$$\int_1 \leqq \rho\mathfrak{L}_f\,|\,v'\,| \leqq \rho\mathfrak{L}_f\delta,$$

while (7) gives

$$\int_2 \leqq 2\mathfrak{L}_f\,|\,v'\,|\,|\,I_{p'} \cap H\,|/|\,I_{p'}\,| < 2\mathfrak{L}_f\delta\rho.$$

Thus

$$(8) \qquad |f(p'') - f(p') - \nabla f(p^*, v')\,| < 3\rho\mathfrak{L}_f\delta, \qquad p^* \in Q.$$

Say $p' = (1 - t')p_i + t'q'$, $q' \in \sigma'$. Then

$$v' = (1 - t')v, \qquad |\,v - v'\,| = t'\,|\,v\,| \leqq \rho\delta.$$

Therefore

$$|\,\nabla f(p^*, v') - \nabla f(p^*, v)\,| \leqq |\,\nabla f(p^*)\,|\,|\,v' - v\,| \leqq \mathfrak{L}_f\rho\delta.$$

Also $|\,p' - p_i\,| \leqq \rho\delta$ etc.; hence

$$|f(p_i) - f(p')\,| \leqq \mathfrak{L}_f\rho\delta, \qquad |f(p_j) - f(p'')\,| \leqq \mathfrak{L}_f\rho\delta.$$

These inequalities give (3).

If $r = 1$, say $\sigma = p'p''$; we may then apply (8).

3. Approximation to the Jacobian. Using Lemma 2a, we can relate the Jacobians (II, 6) J_f, $J_{f'}$:

LEMMA 3a. *With the hypotheses of Lemma 2a, let f' be the affine approximation to f. Then*

$$(1) \qquad |\,J_{f'}(p^*) - J_f(p^*)\,|_0 \leqq 6\rho\mathfrak{L}_f^r/(r-1)!\Theta(\sigma), \qquad p^* \in Q.$$

Using (1.2), set

$$\sigma = p_0 \cdots p_r, \qquad v_i = p_i - p_0, \qquad u_i = \nabla f(p^*, v_i),$$

$$w_i = f(p_i) - f(p_0) = \nabla f'(p^*, v_i).$$

Because of (2.3),

$$|w_i - u_i| < 6\rho\mathfrak{L}_f\delta, \qquad |u_i| \leqq \mathfrak{L}_f\delta, \qquad |w_i| \leqq \mathfrak{L}_f\delta;$$

hence, by (I, 13.9) and the proof of (I, 12.17),

$$|w_1 \vee \cdots \vee w_r - u_1 \vee \cdots \vee u_r|_0 < 6r\rho\mathfrak{L}_f^r\delta^r.$$

The r-direction of σ is, by (III, 1.2),

$$\alpha_0 = v_1 \vee \cdots \vee v_r / r!\,|\sigma|.$$

Hence

$$|J_{f'}(p^*) - J_f(p^*)|_0 = |\nabla f'(p^*, v_1 \vee \cdots \vee v_r) - \nabla f(p^*, v_1 \vee \cdots \vee v_r)|_0 / r!\,|\sigma|$$

$$= |w_1 \vee \cdots \vee w_r - u_1 \vee \cdots \vee u_r|_0 / r!\,|\sigma| < 6\rho\mathfrak{L}_f^r\delta^r/(r-1)!\,|\sigma|,$$

which gives (1).

4. The volume of affine approximations. The definition of r-volume of a "rectifiable r-manifold" may be given for instance through integrating the magnitude of the Jacobian or through finding the limit of the volumes of approximating polyhedra; we show that these definitions are equivalent;* see (7) below.

First note that for simplexes (or convex cells) σ,

(1) $$\{g\sigma\} = |\sigma|\,J_g(p), \quad |g\sigma| = |J_g(p)|\,|\sigma|, \qquad \text{if } g \text{ is affine,}$$

for any $p \in \operatorname{int}(\sigma)$. For, defining the v_i and w_i as in Lemma 3a,

$$r!\{g\sigma\} = w_1 \vee \cdots \vee w_r = \nabla g(p, v_1 \vee \cdots \vee v_r)$$

$$= \nabla g(p, r!\{\sigma\}) = r!\,|\sigma|\,J_g(p).$$

Hence also (the first integral having values in $V_{[r]}$; compare (I, 19.2))

(2) $$\int_\sigma J_g(p)\,dp = \{g\sigma\}, \qquad \int_\sigma |J_g(p)|\,dp = |g\sigma|, \qquad \text{if } g \text{ is affine.}$$

LEMMA 4a. *Let f be a Lipschitz mapping of the oriented r-dimensional polyhedron $P \subset E^s$ into E^n. Then for each $\eta > 0$ and $\epsilon > 0$ there is a $\zeta > 0$ with the following property. Let $\sum \sigma_k$ be any simplicial subdivision of P*

* This is a standard theorem in area theory. See for instance (for $r = 2$, $n = 3$) H. Rademacher, Über partielle und totale Differenzierbarkeit von Funktionen mehrerer Variabeln II, *Math. Annalen*, *81* (1920) 52–63.

of fullness $\geqq \eta$ *and mesh* $\leqq \zeta$, *and let* f' *be the corresponding affine approximation to* f. *Then*

$$\int_P |J_{f'}(p) - J_f(p)|_0 \, dp < \epsilon, \tag{3}$$

$$\sum_k \left| \{f'\sigma_k\} - \int_{\sigma_k} J_f(p)\, dp \right|_0 < \epsilon, \tag{4}$$

$$\left| \{f'P\} - \int_P J_f(p)\, dp \right|_0 < \epsilon, \tag{5}$$

$$\sum_k \left| |f'\sigma_k| - \int_{\sigma_k} |J_f(p)|\, dp \right| < \epsilon, \tag{6}$$

$$\left| |f'(P)| - \int_P |J_f(p)|\, dp \right| < \epsilon. \tag{7}$$

We shall prove (3); the other relations follow at once from this. We may suppose that P is a cell. Choose $\rho < 1$ so that

$$\frac{6\rho\mathfrak{L}_f^r |P|}{(r-1)!\eta} \leqq \frac{\epsilon}{3}, \qquad \frac{2\rho^r\mathfrak{L}_f^r |P|}{[(r-1)!\eta]^r} \leqq \frac{\epsilon}{3}. \tag{8}$$

Set

$$\zeta' = [(r-1)!\eta]^r \rho^r \epsilon / 6\mathfrak{L}_f^r. \tag{9}$$

Since ∇f is measurable (IX, Theorem 11B), we may choose a closed subset Q of P by Lusin's Theorem so that

$$|P - Q| < \zeta', \qquad \nabla f \text{ is continuous in } Q. \tag{10}$$

Choose $\zeta > 0$ so that

$$|\nabla f(q) - \nabla f(p)| \leqq \rho\mathfrak{L}_f \qquad \text{if } p, q \in Q, \ |q - p| \leqq \zeta. \tag{11}$$

Now suppose $P = \sum\sigma_k$ is as stated. Order the σ_k so that, if $H = P - Q$,

$$|\sigma_k \cap H| \leqq \rho^r |\sigma_k| \qquad (k = 1, \cdots, m), \tag{12}$$

and the inequality fails for $k > m$. Set

$$P' = \sigma_1 \cup \cdots \cup \sigma_m, \qquad P'' = \sigma_{m+1} \cup \cdots. \tag{13}$$

For any $k \leqq m$, σ_k satisfies the conditions of Lemma 2a. Hence we may apply (3.1), which, with (8), gives

$$\int_{\sigma_k \cap Q} |J_{f'} - J_f|_0 \leqq \epsilon |\sigma_k| / 3 |P|, \qquad k \leqq m.$$

By (1.5), (1.4) (we may suppose $(r-1)!\eta \leqq 1$, see (IV, 14.3)), (12) and (8),

$$\int_{\sigma_k \cap H} |J_{f'} - J_f|_0 \leqq 2\rho^r |\sigma_k| \mathfrak{L}_f^r / [(r-1)!\eta]^r \leqq \epsilon |\sigma_k| / 3 |P|$$

for $k \leqq m$. Hence

$$\int_{P'} |J_{f'} - J_f|_0 \leqq 2\epsilon/3.$$

The definition of P'', with (10), gives

$$\rho^r \mid P'' \mid = \sum_{k>m} \rho^r \mid \sigma_k \mid < \sum_{k>m} \mid \sigma_k \cap H \mid \leqq \mid H \mid < \zeta'.$$

Hence, by (9),

$$\int_{P''} \mid J_{f'} - J_f \mid_0 < 2(\zeta'/\rho^r)\mathfrak{L}_f^r/[(r-1)!\eta]^r = \epsilon/3.$$

Thus (3) holds.

5. A continuity lemma. We shall give a generalization of (V, 3.6); see also Theorems 6B, 13A, and 7B. If f and g have the same domain of definition Q, we set

(1) $$\delta_{f,g} = \sup \{\mid g(p) - f(p) \mid : \ p \in Q\}.$$

Let f be an affine mapping of the oriented r-simplex σ into E^n. Then $f\sigma$ is an oriented simplex (possibly degenerate), and hence an r-chain, in E^n. If $A = \sum a_i \sigma_i^r$ is a polyhedral chain, and f is affine in each σ_i^r, then $fA = \sum a_i f \sigma_i^r$ is a polyhedral chain in E^n. We shall generalize this in § 6.

Lemma 5a. *Let P be a polyhedron, expressed as a simplicial complex, with simplexes σ_k^s, and let f_0 and f_1 be mappings of P into E^n, affine in each σ_k^s. Suppose*

(3) $$\mathfrak{L}_{f_0} \text{ and } \mathfrak{L}_{f_1} \text{ are} \leqq L \text{ in each } \sigma_k^s.$$

Then for any $A = \sum a_i \sigma_i^r$,

(4) $$\mid f_1 A - f_0 A \mid^\flat \leqq \delta_{f_0,f_1}(L^r \mid A \mid + L^{r-1} \mid \partial A \mid).$$

If the points $F(t, p)$ in (5) lie in the open set S, we may use $\mid f_1 A - f_0 A \mid_S^\flat$ in (4).

Set

(5) $$F(t, p) = (1 - t)f_0(p) + tf_1(p), \qquad 0 \leqq t \leqq 1, \quad p \in P;$$

this is a Lipschitz mapping of the Cartesian product $I \times P$ (I = unit interval) into E^n. Let e_0 be the unit vector along I. Given any σ_k^s, let $e_1, \cdots, e_s$ be an orthonormal system in σ_k^s. Then $e_0, e_1, \cdots, e_s$ is an orthonormal system in $I \times \sigma_k^s$. Take $p \in \operatorname{int}(\sigma_k^s)$. Then writing $\delta = \delta_{f_0,f_1}$, we have, for $i > 0$, $\mid \nabla f_j(p, e_i) \mid \leqq L$, and hence

$$\mid \nabla F(t, p, e_i) \mid = \mid (1 - t)\nabla f_0(p, e_i) + t\nabla f_1(p, e_i) \mid \leqq L, \qquad i > 0.$$

Also

$$\mid \nabla F(t, p, e_0) \mid = \mid f_1(p) - f_0(p) \mid \leqq \delta;$$

hence

(6) $$\mid J_F(t, p) \mid = \mid \nabla F(t, p, e_{0\cdots s}) \mid \leqq \delta L^s \quad \text{in} \quad I \times \sigma_k^s.$$

For some $\eta > 0$, there is an arbitrarily fine subdivision of $I \times P$ of fullness $\geqq \eta$, which gives a subdivision of each $I \times \sigma_k^s$; see (App. II, Lemma 4b) and (App. II, 12). Say $\partial A = \sum b_j \sigma_j^{r-1}$. Given $\epsilon > 0$, set $\epsilon' = \epsilon/(\sum |a_i| + \sum |b_j|)$. Considering the mapping F in each $I \times \sigma_k^s$ for all k and for $s = r$ and $r - 1$, choose $\zeta > 0$ by Lemma 4a small enough for all of these, using ϵ'. Let F' be the corresponding simplexwise affine mapping of $I \times P$. Since $|I \times \sigma_k^s| = |\sigma_k^s|$, (4.7) and (6) give

$$|F'(I \times \sigma_k^s)| \leqq \int_{I \times \sigma_k^s} \delta L^s + \epsilon' = \delta L^s |\sigma_k^s| + \epsilon'.$$

Therefore, using $I \times A = \sum a_i (I \times \sigma_i^r)$ etc.,

$$|F'(I \times A)| \leqq \delta L^r \sum_i |a_i| |\sigma_i^r| + \epsilon' \sum_i |a_i|,$$

$$|F'(I \times \partial A)| \leqq \delta L^{r-1} \sum_j |b_j| |\sigma_j^{r-1}| + \epsilon' \sum_j |b_j|.$$

Since f_0 and f_1 are simplexwise affine, $F'(t, p) = F(t, p)$ for $t = 0, 1$. Hence

$$F'(0 \times A) = f_0 A, \qquad F'(1 \times A) = f_1 A,$$

and (App. II, 13.6) gives

$$(7) \qquad f_1 A - f_0 A = F'(I \times \partial A) + \partial F'(I \times A).$$

These relations give (4); the statement about S is clear.

6. Lipschitz chains. Let A be a polyhedral r-chain, and let f be a Lipschitz mapping of $P = \text{spt}\,(A)$ into E^n. We define the corresponding *Lipschitz chain* fA as follows. Let $\mathfrak{S}_1 P, \mathfrak{S}_2 P, \cdots$ be a full sequence of simplicial subdivisions of P; that is, $\mathfrak{S}_{k+1}$ is a refinement of $\mathfrak{S}_k$, and for some $\eta > 0$, all simplexes are of fullness $\geqq \eta$. Let $f_1, f_2, \cdots$ be the corresponding simplexwise affine mappings of P into E^n. For each k, $f_k A$ is a polyhedral chain in E^n (see § 5). Set

$$(1) \qquad fA = \lim_{k \to \infty}{}^{\flat} f_k A.$$

To show that the limit exists, take any numbers k, l; say $k < l$. For each simplex σ_{li} of $\mathfrak{S}_l P$, both f_k and f_l are affine in σ_{li}; by Lemma 1a,

$$\mathfrak{L}_{f_k}, \mathfrak{L}_{f_l} \leqq L = \mathfrak{L}_f/(r-1)!\eta \qquad \text{in } \sigma_{li}.$$

Hence, by Lemma 5a,

$$|f_l A - f_k A|^{\flat} \leqq \delta_{f_k, f_l}(L^r |A| + L^{r-1} |\partial A|).$$

Clearly $\delta_{f_k, f_l} \to 0$ as $k, l \to \infty$; thus the limit exists.

To show that the limit is unique, consider another sequence $\mathfrak{S}_1' P$ $\mathfrak{S}_2' P, \cdots$ and corresponding mappings $f_1', f_2', \cdots$. Say $\mathfrak{L}_{f_k'} \leqq L'$; set

$L_0 = \sup \{L, L'\}$. Take any k. Using a common simplicial refinement of $\mathfrak{S}_k P$ and $\mathfrak{S}'_k P$ (App. II, Lemma 3b), Lemma 5a shows that

$$| f'_k A - f_k A |^\flat \leqq \delta_{f_k, f'_k}(L_0^r | A | + L_0^{r-1} | \partial A |) \to 0 \quad \text{as} \quad k \to \infty.$$

Suppose f itself is simplexwise affine, in terms of a simplicial subdivision of P. Using a full sequence of refinements of this subdivision, and corresponding mappings f_k, it is clear that $f_k = f$, and hence $f_k A = fA$, for each k. Therefore the new definition of fA coincides with the old when the old applies.

Lemma 6a. *Let f be a Lipschitz mapping of the oriented r-simplex σ into E^n. Let g be an orientation preserving affine mapping of the r-simplex σ' onto σ, and set $f'(p) = f(g(p))$ in σ'. Then $f'\sigma' = f\sigma$.*

Take a full sequence of subdivisions of σ', and let $f'_1, f'_2, \cdots$ correspond. Since g is affine, there is a corresponding full sequence of subdivisions of σ, and corresponding mappings $f_1, f_2, \cdots$. Since $f'_k(p) = f_k(g(p))$ at the vertices of the corresponding subdivision, and these mappings are simplexwise affine, $f'_k(p) = f_k(g(p))$ in σ'; therefore $f'_k\sigma' = f_k(g\sigma') = f_k\sigma$. Letting $k \to \infty$ gives the result.

Because of the lemma, given fA, $P = \text{spt}(A)$, we may take a subdivision of P, replace the (closed) r-cells by disjoint r-cells in E^r, and define a corresponding mapping f' of the union of these cells; clearly f' is Lipschitz. If A' is the corresponding polyhedral chain in E^r, then $f'A' = fA$. We may therefore obtain all Lipschitz r-chains as images of polyhedral chains in E^r.

We show next that

$$|f\sigma| \leqq \int_\sigma |J_f| \leqq \mathfrak{L}_f^r |\sigma|. \tag{2}$$

Take $\mathfrak{S}_1\sigma, \mathfrak{S}_2\sigma, \cdots$, of fullness $\geqq \eta > 0$. Given $\epsilon > 0$, choose $\zeta > 0$ by Lemma 4a. Take any k such that mesh $(\mathfrak{S}_k) < \zeta$. Say $\mathfrak{S}_k\sigma = \sum_i \sigma_{ki}$. Take $p_i \in \sigma_{ki}$. Then with f_k as in the last lemma, (4.2) and (4.3) give

$$|f_k\sigma| \leqq \sum_i |f_k\sigma_{ki}| = \int_\sigma |J_{f_k}| < \int_\sigma |J_f| + \epsilon;$$

letting $k \to \infty$ and using (V, 16.1) gives (2).

Theorem 6A. *For any Lipschitz r-chain fA (hence with J_f defined a.e.),*

$$|fA| \leqq M |A| \quad \text{if} \quad |J_f| \leqq M \quad \text{in spt}(A), \tag{3}$$

$$|fA| \leqq \mathfrak{L}_f^r |A|. \tag{4}$$

This follows at once from (2). See also Theorem 7A.

Theorem 6B. *Let K be a simplicial complex, and let f_0 and f_1 be mappings of K into E^m with $\mathfrak{L}_{f_0}, \mathfrak{L}_{f_1} \leqq L$ in each simplex of K (or in K itself). Then for any simplicial r-chain A of K, the conclusion of Lemma 5a holds.*

For any σ_k^s of K, say ∇f_0 and ∇f_1 are defined and measurable in $Q_k^s \subset \sigma_k^s$, $|\sigma_k^s - Q_k^s|_s = 0$. As in § 5, we find

$$|J_F(t, p)| \leqq \delta L^s, \quad p \in Q_k^s, \quad 0 \leqq t \leqq 1;$$

by (3),

$$|F(I \times \sigma_k^s)| \leqq \delta L^s |I \times \sigma_k^s| = \delta L^s |\sigma_k^s|.$$

Hence $|F(I \times A)| \leqq \delta L^r |A|$, $|F(I \times \partial A)| \leqq \delta L^{r-1} |\partial A|$, and (5.7) with F gives (5.4). Again S may be used.

7. Lipschitz mappings of open sets. Given the open set R in E^n, set

(1) $$\operatorname{dist}_R(p, q) = \inf \{\text{lengths of curves from } p \text{ to } q \text{ in } R\}.$$

We set $\operatorname{dist}_R(p, q) = \infty$ if there is no curve in R joining the points. We could clearly require that the curves be formed of broken line segments. Note that, if pq denotes the line segment from p to q,

(2) $$|q - p| \leqq \operatorname{dist}_R(p, q), \qquad |q - p| = \operatorname{dist}_R(p, q) \quad \text{if } pq \in R.$$

Let f be a continuous mapping of the open set $R \subset E^n$ into the open set $S \subset E^m$. The *R-Lipschitz constant* of f is

(3) $$\mathfrak{L}_{f,R} = \sup \left\{ \frac{\operatorname{dist}_S(f(p), f(q))}{\operatorname{dist}_R(p, q)} \right\},$$

provided this exists; if so, we say f is *R-Lipschitz*. We mention a few elementary facts. The number $\mathfrak{L}_{f,R}$ is independent of S; we could replace S by any neighborhood of $f(R)$ or by E^m. Any curve in R is mapped into a curve at most $\mathfrak{L}_{f,R}$ times its length. We have

(4) $$\operatorname{dist}_S(f(p), f(q)) \leqq \mathfrak{L}_{f,R} |q - p| \quad \text{if} \quad pq \subset R.$$

Hence f is "locally Lipschitz"; in fact, letting $f|Q$ denote the mapping f, considered in the subset Q of R alone,

(5) $$\mathfrak{L}_{f|Q} \leqq \mathfrak{L}_{f,R} \qquad \text{if } Q \subset R \text{ is convex.}$$

It follows that f is Lipschitz in any compact subset of R; therefore, for any polyhedral A in R, fA is defined, as in § 6. If $fA = \lim^\flat f_k A$ as in (6.1), then $f_k A \subset S$ for all sufficiently large k, and $\lim_S^\flat f_k A = fA$, as is apparent from (VIII, Theorem 2B).

We show now how an R-Lipschitz mapping f determines a mapping fA of flat chains A of R, provided A satisfies one condition. We use the notation

(6) $$a^{[r]} = \sup \{a^r, a^{r+1}\}.$$

THEOREM 7A. *Let f be an R-Lipschitz mapping of the open set $R \subset E^n$ into the open set $S \subset E^m$. Then the mapping fB of polyhedral chains B in*

R into Lipschitz chains in S is uniquely extendable to be a linear mapping of flat chains A of R into flat chains fA of S, for all A such that

$$\overline{f(\operatorname{spt}(A))} \subset S, \tag{7}$$

so that the following properties hold ($r = \dim (A)$):

$$f\partial A = \partial f A, \tag{8}$$

$$|fA| \leqq \mathfrak{L}^{r}_{f,R} |A|, \tag{9}$$

$$|fA|^{\flat}_{S} \leqq \mathfrak{L}^{[r]}_{f,R} |A|^{\flat}_{R}, \tag{10}$$

$$\operatorname{spt}(fA) \subset \overline{f(\operatorname{spt}(A))}. \tag{11}$$

Remarks. The condition (7) holds for all compact A, and holds for all A if $\overline{f(R)} \subset S$; simple examples show that it is not true in general. If we weaken the requirement on f, assuming only that it is locally Lipschitz (Lipschitz in a neighborhood of each point, or equivalently, in each compact set), we could still define fA for all compact A in R.

We first prove the above properties for polyhedral chains A; clearly f is linear on these. Take any oriented r-simplex σ in R. Let $\mathfrak{S}_1\sigma, \mathfrak{S}_2\sigma, \cdots$ be a full sequence of subdivisions of σ; this defines also a full sequence of subdivisions of $\partial\sigma$. If f_i is the simplexwise affine mapping of σ defined by f and $\mathfrak{S}_i\sigma$, then, as noted above,

$$\lim{}^{\flat}_{S} f_k\sigma = f\sigma, \qquad \lim{}^{\flat}_{S} f_k\partial\sigma = f\partial\sigma.$$

Since f_k is simplexwise affine, it is clear that $f_k\partial\sigma = \partial f_k\sigma$. Since ∂ is continuous, (8) follows for σ, and hence for polyhedral A.

The inequality (9) for simplexes is an immediate consequence of (6.2) and (5); hence it holds for polyhedral A. The inclusion (11) follows from the definitions of fA and supports.

To prove (10) for polyhedral A, take any $\epsilon > 0$. Choose a polyhedral chain D in R so that

$$|A - \partial D| + |D| < |A|^{\flat}_{R} + \epsilon.$$

Then (8) and (9) give

$$\begin{aligned} |fA|^{\flat}_{S} &\leqq |f(A - \partial D)|^{\flat}_{S} + |\partial f D|^{\flat}_{S} \\ &\leqq \mathfrak{L}^{r}_{f,R} |A - \partial D| + \mathfrak{L}^{r+1}_{f,R} |D| \leqq \mathfrak{L}^{[r]}_{f,R}(|A|^{\flat}_{R} + \epsilon), \end{aligned}$$

which gives (10).

Now take any flat A in R satisfying (7). Then there is a neighborhood U_0 of $Q_0 = \overline{f(\operatorname{spt}(A))}$ with $\bar{U}_0 \subset S$, and a neighborhood U of $Q = \operatorname{spt}(A)$ with $f(U) \subset U_0$. By (VIII, Theorem 3C), there is a sequence $A_1, A_2, \cdots$ of polyhedral chains in U with $\lim^{\flat}_{R} A_i = A$. Applying (10) to $f(A_j - A_i)$

shows that $fA_1, fA_2, \cdots$ is a Cauchy sequence in the flat S-norm; we shall define fA and prove

$$fA = \lim_S^\flat fA_i.$$

Replacing S by E^m shows, first of all, that $fA = \lim^\flat fA_i$ exists as a chain of E^m. Next, choosing a polyhedral chain B_i in U_0 such that $|fA_i - B_i|_S^\flat < 1/i$ (VIII, Theorem 3C) shows that $B_1, B_2, \cdots$ is a Cauchy sequence in the flat S-norm, of polyhedral chains lying in the closed set $\bar{U}_0 \subset S$; now $fA = \lim^\flat B_i$. By definition (VIII, 1), this shows that fA is a flat chain of S and equals $\lim_S^\flat fA_i$. Clearly (10) continues to hold. Now $\operatorname{spt}(fA) \subset \bar{U}_0$, and since U_0 was arbitrary, (11) holds.

Given A again, since $\operatorname{spt}(\partial A) \subset \operatorname{spt}(A)$, ∂A satisfies (7); hence $f\partial A$ is a chain of S. Write $A = \lim_S^\flat A_i$ as above. Then $\partial A = \lim_S^\flat \partial A_i$, and (10) shows that $f\partial A = \lim_S^\flat f\partial A_i$. Hence, using (8) for A_i,

$$f\partial A = \lim_S^\flat \partial fA_i = \partial \lim_S^\flat fA_i = \partial fA,$$

proving (8) in general.

To prove (9) for flat A, we may suppose $|A|$ is finite; then (VIII, Theorem 3C) we may choose the A_i above so that $\lim |A_i| = |A|$. Since (9) holds for the A_i, we have

$$|fA| = |\lim_S^\flat fA_i| \leqq \liminf |fA_i|$$

$$\leqq \mathfrak{L}_{f,R}^r \liminf |A_i| = \mathfrak{L}_{f,R}^r |A|.$$

This completes the proof of the theorem.

We give a continuity theorem.

THEOREM 7B. *With R, S, f as in the last theorem, let A be a flat chain of R satisfying* (7). *Then for any numbers L and $\epsilon > 0$ there is a $\zeta > 0$ such that for any R-Lipschitz mapping g of R into S and any flat chain B of R, if g and B satisfy* (7), *then*

$$(12) \qquad |gB - fA|_S^\flat < \epsilon \quad \text{if} \quad |B - A|_R^\flat < \zeta, \qquad \mathfrak{L}_{g,R} \leqq L, \quad \delta_{f,g} \leqq \zeta.$$

Set $L_0 = \sup\{\mathfrak{L}_{f,R}, L\}$, and choose a polyhedral chain A_1 in R so that

$$|A_1 - A|_R^\flat < \epsilon/4L_0^{[r]}.$$

Set $P = \operatorname{spt}(A_1)$. For some $\eta > 0$, there is a sequence of simplicial subdivisions $\mathfrak{S}_1P, \mathfrak{S}_2P, \cdots$, of fullness $\geqq \eta$. Set $L_1 = L_0/(r-1)!\eta$, and

$$\zeta = \epsilon/4 \sup\{L^{[r]}, L_1^r |A_1| + L_1^{r-1} |\partial A_1|\}.$$

Now take any corresponding g and B. Let f_k and g_k be the simplexwise affine mappings of P corresponding to $\mathfrak{S}_kP$. By Lemma 1a, $\mathfrak{L}_{f_k}$ and $\mathfrak{L}_{g_k}$

are $\leqq L_1$ in each simplex of $\mathfrak{S}_k P$. Also, $\delta_{f_k,g_k} \leqq (1+\lambda_k)\zeta$, where $\lambda_k \to 0$. Hence, by Lemma 5a, for k large enough,

$$|g_k A_1 - f_k A_1|_S^\flat \leqq (1+\lambda_k)\zeta(L_1^r |A_1| + L_1^{r-1}|\partial A_1|) \leqq (1+\lambda_k)\epsilon/4,$$

and letting $k \to \infty$ gives

$$|gA_1 - fA_1|_S^\flat \leqq \epsilon/4.$$

Also, by (10),

$$|fA - fA_1|_S^\flat \leqq \mathfrak{L}_{f,R}^{[r]} |A - A_1|_R^\flat < \epsilon/4,$$

$$|gB - gA_1|_S^\flat \leqq L^{[r]}(|B-A|_R^\flat + |A-A_1|_R^\flat) < \epsilon/2.$$

These relations give (12).

Finally, we prove a transitivity theorem.

THEOREM 7C. *Let R, R' and R'' be open sets in Euclidean spaces E, E', E'' respectively, let f be an R-Lipschitz mapping of R into R', and let g be an R'-Lipschitz mapping of R' into R''. Then gf is an R-Lipschitz mapping of R into R''; also for any flat chain A of R such that*

$$Q = \operatorname{spt}(A), \qquad Q' = \overline{f(Q)} \subset R', \qquad Q'' = \overline{g(Q')} \subset R'',$$

(which holds, for instance, if A is compact), $(gf)A$ is a chain of R'', and

$$(gf)A = g(fA). \tag{13}$$

Set $h = gf$. Suppose first that A is a simplex σ. With a full sequence of subdivisions of σ, let $f_1, f_2, \cdots$ be the corresponding simplexwise affine mappings (into R', if the subdivisions are sufficiently fine). Now $\mathfrak{L}_{f_i} \leqq L$ for some L, and $\lim_{R'}^\flat f_i\sigma = f\sigma$. Hence (10) gives

$$\lim{}_{R''}^\flat g(f_i\sigma) = g(f\sigma).$$

Set $h_i = gf_i$. Applying Lemma 6a to each simplex τ of the ith subdivision gives $h_i\sigma = g(f_i\sigma)$. (If $f_i\tau$ is degenerate, then $h_i\tau = g(f_i\tau) = 0$.) Also $\mathfrak{L}_{h_i} \leqq L\mathfrak{L}_{g,R'}$, since, by (5), $\mathfrak{L}_{g|f_i\tau} \leqq \mathfrak{L}_{g,R'}$ for each τ. Therefore Theorem 6B gives $\lim_{R''}^\flat h_i\sigma = h\sigma$. Hence

$$h\sigma = \lim{}_{R''}^\flat g(f_i\sigma) = g(f\sigma).$$

Therefore (13) holds also for polyhedral A.

Now consider the general case. Since clearly $\mathfrak{L}_{h,R} \leqq \mathfrak{L}_{f,R}\mathfrak{L}_{g,R'}$ and $\overline{h(Q)} \subset Q'' \subset R''$, hA is a flat chain of R''. Choose polyhedral chains A_1, $A_2, \cdots$ in R, with $\lim_R^\flat A_i = A$. Then because of (10),

$$\lim{}_{R'}^\flat fA_i = fA, \quad \lim{}_{R''}^\flat g(fA_i) = g(fA), \quad \lim{}_{R''}^\flat hA_i = hA.$$

Since $g(fA_i) = hA_i$, (13) follows.

8. Lipschitz mappings and flat cochains. We prove

THEOREM 8A. *Let f be an R-Lipschitz mapping of the open set $R \subset E^n$ into the open set $S \subset E^m$. Then for any flat r-cochain X in S, setting*

$$f^* X \cdot A = X \cdot fA \tag{1}$$

for any flat r-chain A of R satisfying (7.7) *defines a flat r-cochain f^*X in R. We have*

$$df^* X = f^*\, dX, \tag{2}$$

$$| f^* X | \leqq \mathfrak{L}^r_{f,R} \, | X |, \tag{3}$$

$$| f^* X |^\flat_R \leqq \mathfrak{L}^{[r]}_{f,R} \, | X |^\flat_S. \tag{4}$$

If also g is an S-Lipschitz mapping into an open set $S' \subset E'$, and Y is a flat cochain in S', then

$$(gf)^* Y = f^*(g^* Y). \tag{5}$$

Take any flat chain A of R satisfying (7.7). Then (1) and (7.10) give

$$| f^* X \cdot A | = | X \cdot fA | \leqq | X |^\flat_S \, | fA |^\flat_S \leqq \mathfrak{L}^{[r]}_{f,R} \, | X |^\flat_S \, | A |^\flat_R;$$

since these chains are dense in the space of flat r-chains of R, (1) defines a flat r-cochain of R, which satisfies (4). Since, for any A satisfying (7.7),

$$df^* X \cdot A = f^* X \cdot \partial A = X \cdot f\, \partial A = X \cdot \partial f A = f^*\, dX \cdot A,$$

(2) holds. Relation (3) follows from (7.9):

$$| f^* X \cdot A | = | X \cdot fA | \leqq | X | \, | fA | \leqq \mathfrak{L}^r_{f,R} \, | X | \, | A |.$$

Finally, given f and g, take any polyhedral A in R. Then

$$(gf)^* Y \cdot A = Y \cdot (gf) A = Y \cdot g(fA) = f^*(g^* Y) \cdot A;$$

hence (5) holds.

We give a theorem on weak limits (see (VIII, 1(g))):

THEOREM 8B. *With f as above, and cochains X and X_i $(i = 1, 2, \cdots)$ in S,*

$$\operatorname*{wkl}^\flat_{R}{}_{i \to \infty} f^* X_i = f^* X \quad \text{if} \quad \operatorname*{wkl}^\flat_{S}{}_{i \to \infty} X_i = X. \tag{6}$$

Take any simplex σ in R. Then

$$\lim (f^* X_i \cdot \sigma) = \lim (X_i \cdot f\sigma) = X \cdot f\sigma = f^* X \cdot \sigma.$$

By (4), the $| f^* X_i |^\flat_R$ are uniformly bounded; hence (6) holds.

9. Lipschitz mappings and flat forms. Let f be a smooth mapping of the open set $R \subset E^n$ into the open set $S \subset E^m$, and let ω be a continuous r-form in S. Then, as in (II, 4), there is a corresponding r-form $f^*\omega$ in R, defined by

$$f^* \omega(p) \cdot \alpha = f^*_p(\omega(q)) \cdot \alpha = \omega(q) \cdot \nabla f(p, \alpha), \qquad q = f(p). \tag{1}$$

Now let f be Lipschitz, and let ω be a flat r-form in S. We cannot expect that (1) may be used to define $f^*\omega$; for instance, f might map R into a subset of S where ω is not defined. We shall define $f^*\omega$ by finding the cochain $X = \Psi\omega$ corresponding to ω (IX, Theorem 7C), forming $X^* = f^*X$ (§ 8), and finding the form $D_{X^*} = \Phi X^*$ corresponding to X^*:

$$f^*\omega = \Phi f^* \Psi \omega. \tag{2}$$

In terms of X, this reads

$$f^* D_X = D_{f^*X}. \tag{3}$$

Theorem 9B below shows that we can obtain a form equivalent to $f^*\omega$ (IX, 7) through the analytic formula (1), provided that we first "improve" ω by replacing it by D_X, $X = \Psi\omega$. Thus one of the important properties of D_X is that it is always defined to a sufficiently large extent to make the analytic formula work. This shows also that the two definitions of $f^*\omega$ agree in the previous situation.

Note that the theorem shows that $D_X(q, \beta)$ exists for certain β, namely, for $\beta = \nabla f(p, \alpha)$; but it does not state that $D_X(q, \beta)$ exists for other β. It might happen that $D_X(q)$, as an r-covector, does not exist for any $q = f(p)$. This is the case, for instance, in the example in the introduction to Chapter IX, if we let f map σ^1 into an interval of the x-axis.

Recall that $D_X(q, \beta)$ is defined by (IX, 4.1) if β is an r-direction, and then by (V, 10.10) for any simple $\beta \neq 0$. Recall also that the existence of $D_X(p)$ implies the linearity of $D_X(p)\cdot\alpha$ in α.

THEOREM 9A. *Let f be an R-Lipschitz mapping of the open set $R \subset E^n$ into the open set $S \subset E^m$, and let X be a flat r-cochain in S. Set $X^* = f^*X$. Then for any $p \in R$ such that f is totally differentiable at p* (IX, 11) *and $D_{X^*}(p)$ exists, $D_X(f(p))$ exists relative to the plane P_p through $f(p)$ in the direction of $\nabla f(p, \alpha(p))$* (IX, 5), *and*

$$D_{X^*}(p) = f_p^* D_X(f(p)). \tag{4}$$

REMARKS. The plane P_p consists of all points $f(p) + \nabla f(p, v)$. The theorem would hold equally well if, in the definition of D_X, we used p-full sequences of simplexes, not necessarily containing p. In place of (8) below, we would use the following relation, coming from (IX, 2.6):

$$\operatorname{diam}(p \cup \sigma) \left| \partial\sigma \right| \leqq (r+1) \left| \sigma \right| / (r-1)!(r!)^{1/r} [\Theta_p(\sigma)]^{1+1/r}. \tag{5}$$

We prove the theorem first in the special case $n = r$. Since f is totally differentiable at p, we may define the affine mapping F of R into E^m by setting

$$F(p') = q + \nabla f(p, p' - p), \qquad q = f(p). \tag{6}$$

We begin by showing that for any $\eta > 0$ and $\epsilon > 0$ there is a $\zeta > 0$ with the following property. For any r-simplex σ,

$$|X\cdot F\sigma - X\cdot f\sigma| \leqq \epsilon |\sigma| \quad \text{if} \quad \Theta(\sigma) \geqq \eta, \quad p \in \sigma \subset U_\zeta(p). \tag{7}$$

There is a sequence of subdivisions of σ of fullness $\geqq \eta' > 0$, for some $\eta' \leqq \eta$. Choose $\epsilon' > 0$ so that

$$|X|_S^\flat \left[L^r + \frac{(r+1)L^{r-1}}{(r-1)!\eta}\right]\epsilon' \leqq \epsilon, \qquad L = \mathfrak{L}_{f,R}.$$

Choose $\zeta \leqq 1/2$ by (c) of (IX, Theorem 11B), using ϵ' and η'. Now take any σ as above. By (IX, 11.1),

$$|f(p') - F(p')| \leqq \epsilon' |p' - p| \leqq \epsilon' \operatorname{diam}(\sigma), \qquad p' \in \sigma.$$

Since $\mathfrak{L}_F \leqq \mathfrak{L}_{f,R}$, Theorem 6B gives

$$|F\sigma - f\sigma|_S^\flat \leqq \delta_{f,F}(L^r|\sigma| + L^{r-1}|\partial\sigma|).$$

By (IX, 2.7),

$$\operatorname{diam}(\sigma)|\partial\sigma| \leqq (r+1)|\sigma|/(r-1)!\eta. \tag{8}$$

Hence, since (in σ) $\delta_{f,F} \leqq \epsilon' \operatorname{diam}(\sigma) \leqq \epsilon'$,

$$|X\cdot(F\sigma - f\sigma)| \leqq |X|_S^\flat \epsilon' \left[L^r + \frac{(r+1)L^{r-1}}{(r-1)!\eta}\right]|\sigma| \leqq \epsilon|\sigma|.$$

Orient $E^r = E^n$ by choosing the r-direction α_0. Suppose first that $J_f(p) = 0$. Then, to prove (4), we need merely show that $D_{X^*}(p, \alpha_0) = 0$. Let $\sigma_1, \sigma_2, \cdots$ be a full p-sequence of r-simplexes; since $J_F(p) = 0$, and hence $F\sigma_i$ is degenerate, $X\cdot F\sigma_i = 0$. Hence (7) gives

$$D_{X^*}(p, \alpha_0) = \lim X^*\cdot\sigma_i/|\sigma_i| = \lim X\cdot f\sigma_i/|\sigma_i| = 0.$$

Now suppose $J_f(p) \neq 0$. Then F is a one-one affine mapping of E^r onto an r-plane P in E^m through q, and has an affine inverse F^{-1} in P. Set

$$b = |J_f(p)|, \qquad b' = 1/b, \qquad \alpha_0' = b'J_f(p);$$

then α_0' is the r-direction of P, properly oriented. Let $\tau_1, \tau_2, \cdots$ be any full q-sequence of simplexes in P, oriented like α_0'; we shall prove that

$$b \lim X\cdot\tau_i/|\tau_i| = D_{X^*}(p, \alpha_0). \tag{9}$$

It will then follow that $D_X(q, J_f(p))$ exists and

$$D_X(q, J_f(p)) = bD_X(q, \alpha_0') = D_{X^*}(p, \alpha_0),$$

which gives (4).

Set $\sigma_i = F^{-1}\tau_i$; these are r-simplexes of E^r, and $|\tau_i| = b|\sigma_i|$, by (4.1). Set $L_1 = \mathfrak{L}_{F^{-1}}$; then

$$\Theta(\sigma_i) = \frac{|\sigma_i|}{\operatorname{diam}^r(\sigma_i)} \geqq \frac{b'|\tau_i|}{L_1^r \operatorname{diam}^r(\tau_i)} = \frac{b'}{L_1^r}\Theta(\tau_i),$$

and hence $\sigma_1, \sigma_2, \cdots$ is a full p-sequence, and

$$\lim X^*\cdot\sigma_i/|\sigma_i| = D_{X^*}(p, \alpha_0).$$

Also, because of (7),

$$\lim\left(b\frac{X\cdot\tau_i}{|\tau_i|} - \frac{X^*\cdot\sigma_i}{|\sigma_i|}\right) = \lim\frac{X\cdot F\sigma_i - X\cdot f\sigma_i}{|\sigma_i|} = 0.$$

These relations give (9), completing the proof for the case $n = r$.

In the general case, the proof above shows that $D_X(q, \nabla f(p, \alpha))$ exists and equals $D_{X^*}(p, \alpha)$ for each simple α; since the second of these terms is linear in α, so is the first.

THEOREM 9B. *Let f and X be as in the last theorem; set $\omega = D_X$ in S. Then* (1) *may be used to define an r-form $f^*\omega$ a.e. in R, and this form is equivalent to that defined by* (2).

This is an immediate consequence of the last theorem; recall that $\omega(q)\cdot\beta$ is needed as a linear function of β only for certain β.

THEOREM 9C. *Let f, X and ω be as in the last theorem. Then with either definition of $f^*\omega$, we have*

$$(10) \qquad df^*\omega = f^*\,d\omega \qquad \text{a.e. in } R.$$

Recall that $d\xi$ is defined in (IX, 12.2). Since this definition uses only the cochain corresponding to ξ, the last theorem shows that the left hand side of (10) does not depend on the definition of $f^*\omega$ used; nor does the right hand side, a.e. With (2), (10) follows at once from (8.2) and (IX, 12).

10. Lipschitz mappings and sharp functions. Given the R-Lipschitz mapping f of R into S, and the function ϕ which is sharp in S, define a corresponding function $\phi^* = f^*\phi$, sharp in R, by

$$(1) \qquad \phi^*(p) = (f^*\phi)(p) = \phi(f(p)).$$

It is clear that

$$(2) \qquad |f^*\phi| \leqq |\phi|, \qquad \mathfrak{L}_{f^*\phi,R} \leqq \mathfrak{L}_{f,R}\mathfrak{L}_{\phi,S}, \qquad |f^*\phi|^\sharp \leqq \mathfrak{L}_{f,R}^{[0]}|\phi|^\sharp.$$

With the notation of (IX, 12), write $\Psi\phi$ for the 0-cochain $\bar{\phi}$ corresponding to ϕ. The above mapping of sharp forms is equivalent to the mapping of the corresponding 0-cochains; that is,

$$(3) \qquad f^*\Psi\phi = \Psi f^*\phi.$$

For

$$f^*\Psi\phi\cdot p = \Psi\phi\cdot f(p) = \phi(f(p)) = \phi^*(p) = \Psi\phi^*\cdot p.$$

If $X = \bar{\phi}$, (3) gives (9.4) in the dimension 0, for all $p \in R$.

For any flat r-cochain X in S, we show that

$$(4) \qquad f^*(\phi X) = (f^*\phi)(f^*X) = \phi^*(f^*X),$$

$$(5) \qquad D_{f^*(\phi X)}(p) = \phi(f(p))D_{f^*X}(p) \qquad \text{if } D_{f^*X}(p) \text{ is defined.}$$

Take any p for which $D_{f^*X}(p)$ is defined, and set $a = \phi(f(p))$. Since $f^*\phi X\cdot\sigma = X\cdot\phi f\sigma$ (VII, 2.1), we find, as in the proof of (IX, 7.6),

$$D_{f^*(\phi X)}(p)\cdot\alpha = \lim \frac{X\cdot\phi f\sigma_i}{|\sigma_i|}, \qquad D_{f^*X}(p)\cdot\alpha = \lim \frac{X\cdot f\sigma_i}{|\sigma_i|},$$

$$|(\phi - a)f\sigma_i| \leqq |\phi - a|\, \mathfrak{L}^r_{f,R}\, |\sigma_i|,$$

$$|\phi f\sigma_i - af\sigma_i| \leqq \epsilon_i \mathfrak{L}_{f,R}\, |\sigma_i| \quad \text{if} \quad |\phi(p) - a| \leqq \epsilon_i \quad \text{in } f(\sigma_i),$$

and letting $i \to \infty$ gives (5). Using this and (IX, 7.6) gives (4).

11. Lipschitz mappings and products. We shall show that the usual formulas of algebraic topology (and of differential forms) carry over.

THEOREM 11A. *Let f be an R-Lipschitz mapping of the open set $R \subset E^n$ into the open set $S \subset E^m$. Then for any flat cochains X and Y in S,*

$$(1) \qquad f^*(X \smile Y) = f^*X \smile f^*Y.$$

Set

$$X^* = f^*X, \qquad Y^* = f^*Y, \qquad Z = X \smile Y,$$
$$Z^* = f^*Z, \qquad Z' = X^* \smile Y^*.$$

First suppose that X and Y are smooth (i.e. D_X and D_Y are). Then, by (IX, 14),

$$D_Z(q) = D_X(q) \vee D_Y(q), \qquad \text{all} \quad q \in S.$$

Let Q be the set of points p of R such that $D_{X^*}(p)$, $D_{Y^*}(p)$ and $D_{Z^*}(p)$ exist, $D_{Z'}(p) = D_{X^*}(p) \vee D_{Y^*}(p)$, and f is totally differentiable at p; then $|R - Q| = 0$. Take any $p \in Q$, and set $q = f(p)$. By Theorem 9A and (I, 10.4),

$$D_{Z^*}(p) = f^*_p(D_Z(q)) = f^*_p(D_X(q) \vee D_Y(q))$$
$$= f^*_p D_X(q) \vee f^*_p D_Y(q) = D_{X^*}(p) \vee D_{Y^*}(p) = D_{Z'}(p).$$

Therefore $D_{Z^*} = D_{Z'}$ a.e. in R, and $Z^* = Z'$.

Now consider the general case. Set $S_k = \mathrm{int}_{1/k}(S)$. Write

$$X = \underset{i\to\infty}{\mathrm{wkl}^\flat_{S_k}}\, X_i, \qquad Y = \underset{i\to\infty}{\mathrm{wkl}^\flat_{S_k}}\, Y_i, \qquad \text{in each } S_k,$$

X_i and Y_i being smooth cochains in S_i, as in (V, 13). Then in any R' such that $f(R') \subset S_k$ for some k, (IX, Theorem 17A), Theorem 8B, and the proof above give

$$f^*(X \smile Y) = f^* \operatorname{wkl}^\flat_{S_k} (X_i \smile Y_i) = \operatorname{wkl}^\flat_{R'} f^*(X_i \smile Y_i)$$
$$= \operatorname{wkl}^\flat_{R'} (f^*X_i \smile f^*Y_i) = \operatorname{wkl}^\flat_{R'} f^*X_i \smile \operatorname{wkl}^\flat_{R'} f^*Y_i = f^*X \smile f^*Y$$

in R', and hence in R.

THEOREM 11B. *With the notations of Theorem* 11A,

$$D_{Z^*}(p) = D_{X^*}(p) \vee D_{Y^*}(p) \qquad \text{a.e. in } R. \tag{2}$$

This follows at once from (1) and the definition (IX, 14.7).

From this formula and Theorem 9A, we can get still further information on the truth of $D_Z = D_X \vee D_Y$:

THEOREM 11C. *With the notations of Theorem* 11A,

(3) $D_Z(f(p)) = D_X(f(p)) \vee D_Y(f(p))$ in the image space of $\nabla f(p)$, a.e. in R.

For we find

$$f_p^* D_Z(f(p)) = f_p^* D_X(f(p)) \vee f_p^* D_Y(f(p)) \qquad \text{a.e. in } R, \tag{4}$$

and this gives (3).

When the first factor is of degree 0, we have a product $\bar{\phi} \smile X = \phi X$; see (IX, 14.24). If we use (2), we find (10.5) a.e. in R.

We end with a formula about the cap product.

THEOREM 11D. *With the notations of Theorem 11A, for any flat chain A of R satisfying* (7.7) *and any flat cochain X in S,*

$$f(f^*X \frown A) = X \frown fA. \tag{5}$$

For let Y be any flat cochain in S of the proper dimension. Then (IX, 16.2) and (1) give

$$Y \cdot f(f^*X \frown A) = f^*Y \cdot (f^*X \frown A) = (f^*Y \smile f^*X) \cdot A$$
$$= f^*(Y \smile X) \cdot A = (Y \smile X) \cdot fA = Y \cdot (X \frown fA).$$

If $X = \bar{\phi}$ is a 0-cochain, (5), (IX, 16.18) and (10.1) give

$$f\phi^* A = \phi f A, \qquad \phi^*(p) = \phi(f(p)). \tag{6}$$

This follows also directly from (10.4) (and vice versa).

12. On the flat norm of Lipschitz chains. Just as the flat norm of a polyhedral chain A was defined by minimizing $|A - \partial D| + |D|$, using polyhedral chains D, so we may find the flat norm of a Lipschitz chain A, using Lipschitz chains D. This is of interest in the study of chains in metric spaces.

THEOREM 12A. *For any Lipschitz chain A in the open set $R \subset E^n$,*

(1) $$|A|^\flat_R = \inf\{|A - \partial D| + |D| : \text{Lipschitz chains } D \subset R\}.$$

Since the inequality $\leqq$ is clear, it is sufficient to show that for any $\epsilon > 0$ there is a Lipschitz chain D in R such that

$$|A - \partial D| + |D| < |A|^\flat_R + \epsilon.$$

Say $A = fA_0$, $A_0 = \sum a_i\sigma_i$. Using a full sequence of subdivisions of the σ_i, we find a number L and a simplexwise affine mapping f' of a subdivision $\sum\tau_j$ of the σ_i such that

$$\mathfrak{L}_{f'} \leqq L, \qquad \delta_{f,f'}(L^r |A_0| + L^{r-l} |\partial A_0|) < \epsilon/2.$$

Let $F(t, p)$ be the deformation of f into f', as in (5.5). Then, as in (5.6), $|J_F| \leqq \delta_{f,f'}L^s$ in the cells $I \times \tau_i^s$. Hence, by (6.3),

$$|F(I \times A_0)| \leqq \delta_{f,f'}L^r |A_0|, \qquad |F(I \times \partial A_0)| \leqq \delta_{f,f'}L^{r-1} |\partial A_0|.$$

Now $D' = F(I \times A_0)$ is a Lipschitz chain, and using a relation like (5.7) gives

$$|f'A_0 - fA_0 - \partial D'| + |D'| < \epsilon/2.$$

This shows, in particular, that $|f'A_0|^\flat_R < |A|^\flat_R + \epsilon/2$. Since $f'A_0$ is polyhedral, there is a polyhedral chain D'' in R such that

$$|f'A_0 - \partial D''| + |D''| < |A|^\flat_R + \epsilon/2.$$

We now have

$$|A - \partial(D'' - D')| + |D'' - D'| < |A|^\flat_R + \epsilon;$$

thus we may use $D = D'' - D'$.

13. Deformations of chains. We wish to generalize Lemma 5a. Given a flat chain A in E^n, we shall define the product $I \times A$ in E^{n+1}, and show that

(1) $$|I \times A|^\flat \leqq 3 |A|^\flat, \qquad |I \times A| = |A|.$$

First suppose $A = \sum a_i\sigma_i$ is polyhedral. Then letting I be the unit segment in the $(n + 1)$-th direction, $I \times \sigma_i$ is a cell in E^{n+1}, and $I \times A = \sum a_i(I \times \sigma_i)$. Given $\epsilon > 0$, choose C and D in E^n so that

$$A = C + \partial D, \qquad |C| + |D| < |A|^\flat + \epsilon.$$

Then $I \times A = (I \times C + 1 \times D - 0 \times D) - \partial(I \times D)$. Since the second part of (1) clearly holds,

$$|I \times C + 1 \times D - 0 \times D| + |I \times D| \leqq |C| + 3 |D|,$$

$$|I \times A|^\flat \leqq 3(|A|^\flat + \epsilon),$$

and the first part of (1) follows for this case.

For a general flat A, let $A = \lim^\flat A_i$, $|A| = \lim |A_i|$, the A_i polyhedral. Then using the first part of (1) shows that $\lim^\flat (I \times A_i)$ exists and defines a flat chain, which we call $I \times A$, uniquely, and the first part of (1) continues to hold; the second part is proved, using $\leqq$.

The remaining inequality may be proved as follows. Given the flat r-cochain X in E^n, we may define the flat $(r+1)$-cochain Y in E^{n+1} by setting

$$D_Y(p) = e^{n+1} \vee D_X(p) \tag{2}$$

(compare (IX, 9)). Then $Y \cdot (I \times A) = X \cdot A$, and $|I \times A| \geqq |A|$ follows with the help of (V, 16.3).

Lemma 13a. *Let A be a flat r-chain of the open set $R \subset E^n$, let f_0 and f_1 be R-Lipschitz mappings of R into E^m, and let F be the mapping of $I \times A$ into E^m defined by the deformation* (5.5) *of f_0 into f_1. Then*

$$|F(I \times A)| \leqq \delta_{f_0, f_1} L^r |A| \quad \text{if} \quad \mathfrak{L}_{f_0,R}, \mathfrak{L}_{f_1,R} \leqq L. \tag{2}$$

Let $A = \lim^\flat_R A_i$, the A_i being polyhedral chains in the closed set $Q \subset R$, with $\lim |A_i| = |A|$ (VIII, Theorem 3C). As in (5.6), $|J_F| \leqq d_{f_0, f_1} L^r$ in the cells of each $I \times A_i$; hence, as in § 5 or (6.3), (2) is true for each A_i. Since $\lim^\flat F(I \times A_i) = F(I \times A)$, (2) holds.

Theorem 13A. *Let f_0 and f_1 be R-Lipschitz mappings of the open set $R \subset E^n$ into the open set $S \subset E^m$, and let A be a flat r-chain of R such that the supports of $f_0 A$, $f_1 A$, and of the deformation chains $F(I \times A)$, $F(I \times \partial A)$ are in S. Then*

$$|f_1 A - f_0 A|^\flat_S \leqq \delta_{f_0, f_1}(L^r |A| + L^{r-1} |\partial A|) \quad \text{if} \quad \mathfrak{L}_{f_0,R}, \mathfrak{L}_{f_1,R} \leqq L. \tag{3}$$

The chains in S are all chains of S (Theorem 7A). Using (5.7) (with F) and (2) gives (3).

Note that (3) expresses a stronger kind of continuity than (7.12). We cannot expect this in general, as we show by an example.

Example. Set $\phi(x) = \sum_{i=1}^{\infty} a_i \sin b_i x$, $0 \leqq x \leqq 1$, with the a_i and b_i chosen so as to give a non-differentiable function; this defines a flat 1-chain $A = \tilde{\phi}$ in E^1. With the unit vector e in E^1, it is easy to see that there is no c such that

$$|T_{he} A - A|^\flat = |T_{he} A - A| \leqq c |h|, \qquad \text{all } h.$$

XI. *Chains and Additive Set Functions*

In this chapter, all chains considered will be sharp and of finite mass. Given the r-chain A, the function $\Lambda_A(D_X) = X \cdot A$ of sharp r-forms D_X (see (V, 10)) is linear, and satisfies a certain continuity condition. The principal theorem of the chapter (Theorem 11A) is that there is a corresponding additive set function γ_A which, through integration, gives the same function of sharp r-forms:

$$X \cdot A = \Lambda_A(D_X) = \int_E D_X \cdot d\gamma_A, \qquad \text{all sharp } X.$$

The values $\gamma_A(Q)$ are r-vectors; the integral is thus a generalization of the usual Lebesgue-Stieltjes integral. Though the integration theory needed is standard, it is not easily available in the form most suitable for our purposes. Hence the first part of the chapter is devoted to setting up the required theory.

The basic properties of additive set functions γ, with values in a finite dimensional Banach space, are given first; the variation $\bar{\gamma}$ of γ is a real valued set function. The definition and elementary properties of the integral $\int_Q \xi \cdot d\gamma$ over Borel sets Q are considered next. As a function of Q, it is additive; we denote it by $\xi \cdot \gamma$. In this notation, the formula $\int \xi \cdot d(\psi \cdot \gamma) = \int (\xi \cdot \psi) \cdot d\gamma$ becomes the associative law: $\xi \cdot (\psi \cdot \gamma) = (\xi \cdot \psi) \cdot \gamma$. In § 5 we show that a set function and its variation may each be expressed as an integral with respect to the other.

The next sections are devoted to proving that a linear functional satisfying certain conditions is given by an integral; we consider first non-negative functionals, then real valued ones, then the general case. We then define the sharp norm $|\gamma|^\#$ of γ, and consider molecular set functions; the properties are similar to those for chains.

The theorem on existence and properties of γ_A follows easily from the theorem on functionals mentioned above; moreover, the sharp norms $|A|^\#$, $|\gamma_A|^\#$ are equal, and the mass of A equals the total variation of γ_A. Through this theorem, analytical properties of A, in particular, of the product ϕA, are easily obtained, generalizing various facts in Chapter VII. The "part A_Q of A in the Borel set Q" may be defined as $\chi_Q A$, χ_Q being the characteristic function of Q (§ 13); the r-vector $\{A_Q\}$ of this chain (VII, 6) is precisely $\gamma_A(Q)$. The relation of continuous chains (VI, 7) to r-vector valued set functions takes on new meaning.

The sharp norm $|\gamma|^\#$ is given an intrinsic characterization in § 15, similar to the characterization of $|A|^\#$ in (V, 8). A direct determination of $|\gamma|^\#$ (for compact γ), similar to the definition of $|A|^\#$ for polyhedral A in (V, 6), is more difficult; it is obtained by "explosions" of γ, breaking it up into pieces and moving them about. Some alternative definitions of $|\gamma|^\#$ are shown not to work.

Since flat chains are sharp, all the results hold for these; but what further can be said we leave completely unanswered. See for instance the problem in § 11. The study of the structure of flat chains is a deep and important problem. See in this connection (VIII, Theorem 5A).

1. On finite dimensional Banach spaces. In the study of set functions whose values are in the finite dimensional Banach space V, we need the two theorems below.

THEOREM 1A. *Given V, there is a number $\eta > 0$ with the following property. Let $v_1, \cdots, v_k$ be any set of elements of V. Then there is a subset $v_{\lambda_1}, \cdots, v_{\lambda_l}$ of these such that*

$$\left|\sum v_{\lambda_j}\right| \geqq \eta \sum |v_i|. \tag{1}$$

Let S be the unit sphere in V (all v with $|v| = 1$). It is covered by a finite set of sets $S_1, \cdots, S_m$ of diameter $\leqq 1/2$. Set $\eta = 1/2m$. Now given $v_1, \cdots, v_k$, we may choose h so that if $v_{\lambda_1}, \cdots, v_{\lambda_l}$ are those for which some positive multiple is in S_h, then

$$\sum |v_{\lambda_j}| \geqq \sum |v_i|/m.$$

Say

$$v_{\lambda_j} = a_{\lambda_j} v'_{\lambda_j}, \qquad v'_{\lambda_j} \in S_h, \qquad a_{\lambda_j} = |v_{\lambda_j}|.$$

Choose $u \in S_h$. Then

$$\begin{aligned}\left|\sum v_{\lambda_j}\right| &= \left|\sum a_{\lambda_j} u + \sum a_{\lambda_j}(v'_{\lambda_j} - u)\right| \\ &\geqq \left|\sum a_{\lambda_j} u\right| - \sum a_{\lambda_j} |v'_{\lambda_j} - u| \geqq \sum a_{\lambda_j} - \tfrac{1}{2}\sum a_{\lambda_j} \\ &= \tfrac{1}{2}\sum |v_{\lambda_j}| \geqq \eta \sum |v_i|.\end{aligned}$$

We show next that from a certain set of non-void closed sets, we may pick a point from each in a reasonable manner. A *Borel function* ϕ is a function such that for each Borel set Q, $\phi^{-1}(Q)$ is a Borel set.

LEMMA 1a. *Let V_1 and V_2 be vector spaces, and let S be a closed subset of $V_1 \times V_2$. For $p \in V_1$, let $S(p)$ be the set of points $q \in V_2$ such that $(p, q) \in S$. Let $S(p)$ be non-void for $p \in P$. Then there is a Borel function $\phi(p)$ $(p \in P)$, with values in V_2, such that $\phi(p) \in S(p)$.*

Let $f(t)$ $(t \geqq 0)$ be a continuous function whose values cover V_2

(a "Peano curve"). Using this to give an ordering of points of V_2, let $\phi(p)$ be the first point of $S(p)$. Equivalently, we may define, for $p \in P$,

$$\tau(p) = \inf \{t\colon f(t) \in S(p)\}, \qquad \phi(p) = f(\tau(p)).$$

The function τ is lower semi-continuous. For suppose $p_i \to p$, $t_i = \tau(p_i) \to t$. Then $q_i = f(t_i) \in S(p_i)$, $q_i \to q = f(t)$, and since $(p_i, q_i) \in S$, we have $(p, q) \in S$. Hence $q = f(t) \in S(p)$, and $\tau(p) \leqq t$. From this it is easily seen that τ is a Borel function; since f is continuous, ϕ is a Borel function.

In a vector space with a scalar product, for each vector $v \neq 0$ there is a unique vector $\phi(v)$ such that $|\phi(v)| = 1$, $\phi(v)\cdot v = |v|$; in fact, $\phi(v) = v/|v|$. In a finite dimensional Banach space V, the following weaker theorem holds.

THEOREM 1B. *Given V, there is a Borel function ϕ defined in V, with values in the conjugate space $\bar{V}$, such that*

$$(2) \qquad |\phi(v)| = 1, \qquad \phi(v)\cdot v = |v| \qquad (v \in V).$$

Let S be the set of all $(v, f) \in V \times \bar{V}$ such that

$$|f| = 1, \qquad f\cdot v = |v|.$$

Then S is closed. Let $S(v)$ be the set of all f such that $(f, v) \in S$. By (App. I, Lemma 8b), $S(v)$ is non-void. The lemma shows that ϕ exists.

REMARK. As noted in the Remark in (I, 13), ϕ is not uniquely defined in general.

2. Vector valued additive set functions. By a *Borel partition* of the Borel set $Q \subset E^n = E$, we mean an expression of Q as the union $\bigcup Q_i$ of a finite or infinite sequence of disjoint Borel sets Q_i.

Let V be a finite dimensional Banach space. By an *additive set function in E with values in V*, we mean a function $\gamma(Q)$ of Borel sets $Q \subset E$, such that $\gamma(Q) \in V$, and

$$(1) \qquad \gamma(Q) = \sum \gamma(Q_i) \text{ for Borel partitions } Q = \bigcup Q_i,$$

the sum always existing in the metric of V.

It is elementary (Saks, pp. 8–9) that

$$(2) \quad \gamma(\lim Q_i) = \lim \gamma(Q_i) \quad \text{if} \quad Q_1 \subset Q_2 \subset \cdots \quad \text{or} \quad Q_1 \supset Q_2 \supset \cdots.$$

The *variation* $\bar{\gamma}$ of γ is the real valued set function defined by

$$(3) \qquad \bar{\gamma}(Q) = \sup \left\{\sum |\gamma(Q_i)| \colon \text{Borel partitions } \bigcup Q_i \text{ of } Q\right\}.$$

Clearly

$$(4) \qquad |\gamma(Q)| \leqq \bar{\gamma}(Q).$$

We could restrict the partitions in (3) to be finite. For if $\bar{\gamma}'$ is the resulting function, clearly $\bar{\gamma}'(Q) \leqq \bar{\gamma}(Q)$. To prove the converse, take any Borel partition $\bigcup Q_i$ of Q. Then for each m,

$$\sum_{i=1}^{m} | \gamma(Q_i) | \leqq \sum_{i=1}^{m} | \gamma(Q_i) | + | \gamma(\bigcup_{i=m+1}^{\infty} Q_i) | \leqq \bar{\gamma}'(Q),$$

and letting $m \to \infty$ shows that $\bar{\gamma}(Q) \leqq \bar{\gamma}'(Q)$.

We must prove that $\bar{\gamma}(Q)$ is finite. Suppose not. Then (compare Saks, p. 10) we shall choose Borel sets $Q_1 \supset Q_2 \supset \cdots$ such that

$$\bar{\gamma}(Q_i) = \infty, \qquad | \gamma(Q_i) | \geqq i - 1. \tag{5}$$

Set $Q_1 = Q$. Suppose $Q_1, \cdots, Q_k$ have been found. Since $\bar{\gamma}(Q_k) = \infty$, we may choose a finite Borel partition $Q_k = \bigcup_i Q_{ki}$ such that

$$\sum_i | \gamma(Q_{ki}) | \geqq [| \gamma(Q_k) | + k]/\eta,$$

with η as in Theorem 1A. There is a subset $Q_{k\lambda_j}$ of the Q_{ki} such that

$$| \gamma(\bigcup_j Q_{k\lambda_j}) | = \left| \sum_j \gamma(Q_{k\lambda_j}) \right| \geqq | \gamma(Q_k) | + k.$$

Set $Q' = \bigcup_j Q_{k\lambda_j}$. If $\bar{\gamma}(Q') = \infty$, we may set $Q_{k+1} = Q'$. If not, then set $Q_{k+1} = Q_k - Q'$. Then clearly $\bar{\gamma}(Q_{k+1}) = \infty$ (see the proof of additivity below), and $| \gamma(Q_{k+1}) | \geqq | \gamma(Q') | - | \gamma(Q_k) | \geqq k$.

Set $Q^* = \lim Q_k$. Then, by (2),

$$| \gamma(Q^*) | = | \lim \gamma(Q_k) | = \lim | \gamma(Q_k) | = \infty,$$

a contradiction; hence $\bar{\gamma}(Q)$ is finite.

We now prove that $\bar{\gamma}$ satisfies (1); this will show that $\bar{\gamma}$ is a (finite) Borel measure. Take any Borel Q, and any Borel partition $Q = \bigcup Q_i$. First we prove $\bar{\gamma}(Q) \leqq \sum \bar{\gamma}(Q_i)$. Let $Q = \bigcup Q'_j$ be any Borel partition of Q. Set $Q_{ij} = Q_i \cap Q'_j$. Then $\bigcup_j Q_{ij}$ is a Borel partition of Q_i, and hence $\sum_j | \gamma(Q_{ij}) | \leqq \bar{\gamma}(Q_i)$. Therefore

$$\sum_j | \gamma(Q'_j) | = \sum_j \left| \sum_i \gamma(Q_{ij}) \right| \leqq \sum_{i,j} | \gamma(Q_{ij}) | \leqq \sum_i \bar{\gamma}(Q_i),$$

as required. To prove the reverse inequality, given $\epsilon > 0$, choose Borel partitions $Q_i = \bigcup_j Q_{ij}$ so that

$$\sum_j \left| \gamma(Q_{ij}) \right| > \bar{\gamma}(Q_i) - \epsilon/2^i.$$

Now $\bigcup_{i,j} Q_{ij}$ is a Borel partition of Q, and $\sum_{i,j} | \gamma(Q_{ij}) | > \sum_i \bar{\gamma}(Q_i) - \epsilon$, which gives the result.

Lemma 2a. *Given any Borel set Q and $\epsilon > 0$, there is a compact set $P \subset Q$ and an open set $R \supset Q$ such that for any Borel Q',*

$$| \gamma(Q') - \gamma(Q) | < \epsilon \quad \text{if} \quad P \subset Q' \subset R. \tag{6}$$

We may choose P and R (compare Saks, p. 69; Halmos, p. 183) so that $\bar{\gamma}(Q-P)$ and $\bar{\gamma}(R-Q)$ are $< \epsilon/2$. Then

$$|\gamma(Q') - \gamma(Q)| = |\gamma(Q'-Q) - \gamma(Q-Q')|$$
$$\leqq \bar{\gamma}(Q'-Q) + \bar{\gamma}(Q-Q') \leqq \bar{\gamma}(R-P) < \epsilon.$$

The *total value* $[\gamma]$ and *total variation* $|\gamma|$ of γ are defined by

(7) $$[\gamma] = \gamma(E), \qquad |\gamma| = [\bar{\gamma}] = \bar{\gamma}(E).$$

If $\gamma_1, \cdots, \gamma_m$ are additive set functions with values in V, we may define the additive set function

(8) $$\gamma = \sum a_i\gamma_i\colon\ \gamma(Q) = \sum a_i\gamma_i(Q).$$

The additive set functions now form a linear space, and $[\sum a_i\gamma_i] = \sum a_i[\gamma_i]$.

3. Vector valued integration. Let V, V' and V'' be finite dimensional Banach spaces, and let $v' \cdot v = v''$ be a bilinear multiplication of the pair V', V into V'', such that

(1) $$|v' \cdot v| \leqq |v'|\,|v|.$$

The two cases we shall use most are:

(a) $V' =$ reals and $V'' = V$; $a \cdot v = av$.

(b) $V' =$ conjugate space $\bar{V}$ of V, and $V'' =$ reals. In particular, with the vector space $W = V(E)$ (App. I, 10), use $V = W_{[r]}$, $V' = W^{[r]}$, and let $v' \cdot v$ be the multiplication of (I, 2.2); the norms used are the mass and comass. See (I, 13.4).

Let γ be an additive set function in E with values in V. We shall define the integral $\int_R \xi \cdot d\gamma$, with values in V'', ξ being any bounded Borel function defined at least in the Borel set $R \subset E$, with values in V'.

First suppose ξ is a *Borel step function*; that is, there is a finite Borel partition $R = Q_1 \cup \cdots \cup Q_m$, and there are elements $\xi_1, \cdots, \xi_m$ of V', such that $\xi(p) = \xi_i$ for $p \in Q_i$. In this case, we define

(2) $$\int_R \xi \cdot d\gamma = \sum_i \xi_i \cdot \gamma(Q_i).$$

Now consider the general case. For each $\epsilon > 0$ we shall define a set $\Phi_\epsilon \subset V''$, as follows. An element v'' of V'' is in Φ_ϵ provided the following is true. There is a Borel set $S \subset V'$ containing all values $\xi(p)$, and a Borel partition $S = S_1 \cup \cdots \cup S_k$, each S_i being of diameter $< \epsilon$; there is an element $\xi_i \in S_i$; and

(3) $$v'' = \sum_i \xi_i \cdot \gamma(Q_i), \qquad Q_i = \xi^{-1}(S_i) \cap R.$$

Note that we may choose $p_i \in Q_i$ (if $Q_i \neq 0$), and use $\xi_i = \xi(p_i)$.

We prove

$$\text{diam } (\Phi_\epsilon) \leqq 2\epsilon\bar{\gamma}(R). \tag{4}$$

For take two elements of Φ_ϵ, defined by $(S_1, \cdots, S_k, \xi_1, \cdots, \xi_k)$ and $(S'_1, \cdots, S'_l, \xi'_1, \cdots, \xi'_l)$ respectively. Set

$$S_{ij} = S_i \cap S'_j, \qquad Q_{ij} = \xi^{-1}(S_{ij}) \cap R.$$

and choose $\xi_{ij} \in S_{ij}$ if $S_{ij} \neq 0$. Let $\sum'_{i,j}$ denote the sum over all (i, j) such that S_{ij} is non-void. Since $\bigcup_j Q_{ij}$ is a Borel partition of Q_i (omitting (i, j) if $S_{ij} = 0$), and $|\xi_{ij} - \xi_i| \leqq \epsilon$, we find

$$\left|\sum_{i,j}{}' \xi_{ij}\cdot\gamma(Q_{ij}) - \sum_i \xi_i\cdot\gamma(Q_i)\right| = \left|\sum_{i,j}{}'(\xi_{ij} - \xi_i)\cdot\gamma(Q_{ij})\right|$$
$$\leqq \sum_{i,j}{}' |\xi_{ij} - \xi_i| \, |\gamma(Q_{ij})| \leqq \epsilon\bar{\gamma}(R).$$

A similar inequality holds with the S'_j and ξ'_j, and (4) follows.

Since $\Phi_\epsilon \neq 0$, and $\Phi_{\epsilon'} \subset \Phi_\epsilon$ if $\epsilon' < \epsilon$, there is a unique element of V'' which is in the closure of every Φ_ϵ; this shall be $\int_R \xi\cdot d\gamma$. Thus the integral is the limit of approximating sums of the form (3).

The elementary properties of integrals are easily proved; for instance,

$$\left|\int_Q \xi\cdot d\gamma\right| \leqq \int_Q \langle\xi\rangle \, d\bar{\gamma} \leqq |\xi| \, \bar{\gamma}(Q) \leqq |\xi| \, |\gamma|, \tag{5}$$

$$\int_Q \xi\cdot d\gamma = \xi_0\cdot\gamma(Q) \quad \text{if} \quad \xi(p) = \xi_0 \text{ in } Q, \tag{6}$$

$$\int_Q \xi\cdot d\gamma = \sum \int_{Qi} \xi\cdot d\gamma \qquad \text{if } \bigcup Q_i \text{ is a Borel partition of } Q. \tag{7}$$

From these properties we see at once to what degree the expression (3) approximates to the integral: For Borel partitions $Q = \bigcup Q_i$,

$$\left|\sum_i \xi_i\cdot\gamma(Q_i) - \int_Q \xi\cdot d\gamma\right| \leqq \epsilon\bar{\gamma}(Q) \quad \text{if} \quad |\xi(p) - \xi_i| \leqq \epsilon \quad \text{in } Q_i. \tag{8}$$

One could of course define the integral by introducing coordinate systems into the Banach spaces, thus reducing the definition to the real valued case.

Given the set function γ and the bounded Borel function ξ, we define a set function $\xi\cdot\gamma$, by

$$(\xi\cdot\gamma)(Q) = \int_Q \xi\cdot d\gamma; \quad \text{then} \quad [\xi\cdot\gamma] = \int_E \xi\cdot d\gamma. \tag{9}$$

Now $\xi\cdot\gamma$ and $[\xi\cdot\gamma]$ are bilinear functions of ξ and γ.

For any Borel partition $\bigcup Q_i$ of Q, (5) and (7) give

$$\sum |(\xi\cdot\gamma)(Q_i)| \leqq \sum \int_{Q_i} \langle\xi\rangle \, d\bar{\gamma} = \int_Q \langle\xi\rangle \, d\bar{\gamma};$$

hence we have the inequalities on variation

$$\overline{\gamma\cdot\xi}(Q) \leqq \int_Q \langle\xi\rangle \, d\bar{\gamma}, \qquad |\xi\cdot\gamma| \leqq \int_E \langle\xi\rangle \, d\bar{\gamma} = [\langle\xi\rangle\cdot\bar{\gamma}]. \tag{10}$$

These need not be equalities, for instance if (e_1, e_2) is a base in V, with dual base (e^1, e^2), and the values of γ and of ξ are in the directions of e_1 and e^2 respectively. But see (4.2) below.

The set function γ defines a semi-norm in the space of bounded Borel functions ξ:

$$|\xi|_\gamma = |\xi\cdot\gamma|. \tag{11}$$

One may define γ-measurable functions, like Lebesgue measurable functions; we do not need this, except for measures γ: $\gamma(Q) \geqq 0$.

LEMMA 3a. *Given γ, the bounded Borel function ξ, and $\epsilon > 0$, there is a Borel step function η such that*

$$|\eta - \xi|_\gamma \leqq \int_E \langle \eta - \xi \rangle \, d\bar{\gamma} < \epsilon, \qquad |\eta| \leqq |\xi|; \tag{12}$$

there is also a sharp function η with these properties.

For Borel step functions, this is an immediate consequence of the definition of the integral. We shall find a sharp function η, assuming ξ is a Borel step function; compare Halmos, pp. 241–242.

Say $E = Q_1 \cup \cdots \cup Q_m$, $\xi(p) = \xi_i$ in Q_i. Choose a compact set $Q'_i \subset Q_i$ such that

$$\bar{\gamma}(Q_i - Q'_i) < \epsilon/mN, \qquad N = \sup\{|\xi_i|\}.$$

For some $\zeta > 0$, the sets $U_\zeta(Q'_i)$ are disjoint. Let ϕ_i be a real valued sharp function with $0 \leqq \phi_i(p) \leqq 1$, such that $\phi_i = 1$ in Q'_i and $\phi_i = 0$ outside $U_\zeta(Q'_i)$ (App. III, Lemma 1a). Set

$$\eta_i(p) = \phi_i(p)\xi_i, \qquad \eta(p) = \sum \eta_i(p).$$

Then η is sharp, and since $\eta(p) = \xi_i$ in Q'_i and $|\eta(p)| \leqq N$, we find

$$\int_E \langle \eta - \xi \rangle \, d\bar{\gamma} \leqq \sum \int_{Q_i - Q'_i} N d\bar{\gamma} \leqq N \sum \bar{\gamma}(Q_i - Q'_i) < \epsilon.$$

REMARK. We cannot in general obtain η from ξ by smoothing as in (App. III, 3), unless $\bar{\gamma}$ is absolutely continuous with respect to Lebesgue measure.

4. Point functions times set functions. Given the set function γ and the real bounded Borel function ϕ, define the set function $\phi\gamma = \phi\cdot\gamma$ by (3.9). In particular, if χ_Q is the characteristic function of Q ($= 1$ in Q and $= 0$ in $E - Q$),

$$\chi_Q\gamma(P) = \gamma(Q \cap P), \qquad [\chi_Q\gamma] = \gamma(Q). \tag{1}$$

We give a formula and an inequality for the variation of $\phi\gamma$:

$$\overline{\phi\gamma} = \langle \phi \rangle \bar{\gamma}, \qquad \overline{\phi\gamma}(Q) \leqq |\phi| \, \bar{\gamma}(Q), \quad \phi \text{ real valued.} \tag{2}$$

That $\leqq$ holds in the first relation (for any Q) was proved in (3.10). To prove the reverse inequality, take any $\epsilon > 0$. We may choose a Borel partition $Q = \bigcup Q_i$ so that $\sum |\gamma(Q_i)| \geqq \bar{\gamma}(Q) - \epsilon$, and so that for some $p_i \in Q_i$, $|\phi(p) - \phi(p_i)| \leqq \epsilon$ in Q_i (each i). Say

$$|\gamma(Q_i)| = \bar{\gamma}(Q_i) - \eta_i;$$

then $\sum \eta_i \leqq \epsilon$. Using (3.8) gives

$$|\phi\gamma(Q_i)| - \phi(p_i)\gamma(Q_i)| \leqq \epsilon\bar{\gamma}(Q_i),$$

$$||\phi(p_i)|\,\bar{\gamma}(Q_i) - \langle\phi\rangle\bar{\gamma}(Q_i)| \leqq \epsilon\bar{\gamma}(Q_i),$$

$$||\phi(p_i)\gamma(Q_i)| - |\phi(p_i)|\,\bar{\gamma}(Q_i)| = |\phi(p_i)|\,\eta_i,$$

and hence

$$\overline{\phi\gamma}(Q) \geqq \sum |\phi\gamma(Q_i)| \geqq \sum [\langle\phi\rangle\bar{\gamma}(Q_i) - 2\epsilon\bar{\gamma}(Q_i) - |\phi|\,\eta_i]$$
$$\geqq \langle\phi\rangle\bar{\gamma}(Q) - 2\epsilon\bar{\gamma}(Q) - \epsilon\,|\phi|\,,$$

giving the required inequality. The second part of (2) follows from this.

We prove (compare Saks, p. 37; Halmos, p. 134)

(3) $$\int_Q \xi\cdot d(\phi\gamma) = \int_Q \phi\xi\cdot d\gamma; \quad \text{thus} \quad \xi\cdot\phi\gamma = \phi\xi\cdot\gamma.$$

Given $\epsilon > 0$, choose a partition $Q = \bigcup Q_i$ and points $p_i \in Q_i$ so that

$$|\xi(p) - \xi(p_i)|, \qquad |\phi(p) - \phi(p_i)| \leqq \epsilon \quad \text{in } Q_i, \text{ each } i.$$

Then $|\phi(p)\xi(p) - \phi(p_i)\xi(p_i)| \leqq (|\phi| + |\xi|)\epsilon$ in Q_i, and

$$\left|\int_Q \xi\cdot d(\phi\gamma) - \sum \xi(p_i)\cdot\phi\gamma(Q_i)\right| \leqq \epsilon\overline{\phi\gamma}(Q) \leqq \epsilon\,|\phi|\,\bar{\gamma}(Q),$$

$$\left|\int_Q \phi\xi\cdot d\gamma - \sum \phi(p_i)\xi(p_i)\cdot\gamma(Q_i)\right| \leqq \epsilon(|\phi| + |\xi|)\bar{\gamma}(Q),$$

$$\left|\sum \xi(p_i)\cdot[\phi\gamma(Q_i) - \phi(p_i)\gamma(Q_i)]\right|$$
$$\leqq |\xi| \sum \int_{Q_i} \langle\phi - \phi(p_i)\rangle d\bar{\gamma} \leqq \epsilon\,|\xi|\,\bar{\gamma}(Q);$$

hence

$$\left|\int_Q \xi\cdot d(\phi\gamma) - \int_Q \phi\xi\cdot d\gamma\right| \leqq 2\epsilon(|\phi| + |\xi|)\bar{\gamma}(Q),$$

giving (3).

Remark. (3) clearly holds in a more general case: ξ, ψ and γ have values in vector spaces V_1, V_2 and V_3 respectively (ξ and ψ being Borel point functions, and γ, an additive set function), and $(v_1\cdot v_2)\cdot v_3$ and $v_1\cdot(v_2\cdot v_3)$ are defined and are equal. Then $\xi\cdot(\psi\cdot\gamma) = (\xi\cdot\psi)\cdot\gamma$.

We shall give some relations involving limits. Write

(4) $$\lim \gamma_i = \gamma \quad \text{if} \quad \lim \gamma_i(Q) = \gamma(Q), \qquad \text{all Borel } Q,$$

(5) $$\lim \uparrow \phi_i = \phi, \quad \text{or} \quad \phi_i \uparrow \phi, \quad \text{if} \quad \phi_1 \leqq \phi_2 \leqq \cdots, \quad \lim \phi_i = \phi,$$

and similarly for $\phi_i \downarrow \phi$.

If $(\phi - \phi_i) \downarrow 0$, then

$$|\phi\gamma(Q) - \phi_i\gamma(Q)| = \left|\int_Q (\phi - \phi_i)\, d\gamma\right| \leqq \int_Q \langle \phi - \phi_i \rangle\, d\bar{\gamma} \to 0$$

(compare Saks, p. 27; Halmos, p. 112); hence

(6) $$\lim \phi_i\gamma = \phi\gamma \quad \text{if} \quad \phi_i \uparrow \phi \quad \text{or} \quad \phi_i \downarrow \phi.$$

Hence also, for real bounded Borel ϕ_i and ϕ, using (2),

(7) $$\overline{\phi\gamma} = \phi\bar{\gamma} = \sum \phi_i\bar{\gamma} = \sum \overline{\phi_i\gamma} \quad \text{if} \quad \sum \phi_i = \phi,\ \phi_i \geqq 0.$$

In particular,

(8) $$|\gamma| = \sum |\chi_{Q_i}\gamma|, \qquad \text{Borel partitions } \bigcup Q_i \text{ of } E.$$

For bounded Borel functions ξ_i, ξ, ϕ_i, ϕ, and a number N,

(9) $$\lim \xi_i \cdot \gamma = \xi \cdot \gamma \quad \text{if} \quad \xi_i \to \xi, \qquad |\xi_i| \leqq N \quad (\text{all } i),$$

(10) $$\lim \phi_i\gamma = \phi\gamma \quad \text{if} \quad \phi_i \to \phi, \qquad |\phi_i| \leqq N \quad (\text{all } i);$$

for instance, if $\eta_i = \xi - \xi_i$, $|(\eta_i \cdot \gamma)(Q)| \leqq (\langle \eta_i \rangle \bar{\gamma})(Q) \to 0$ (Saks, p. 29; Halmos, p. 110).

5. Relation between a set function and its variation. We wish to express each of γ, $\bar{\gamma}$ as an integral with respect to the other. For measures μ, write $\xi = \eta$ a.e. (μ) if the set Q of points where $\xi(p) \neq \eta(p)$ satisfies $\mu(Q) = 0$.

THEOREM 5A. *Given γ as in § 2, there is a Borel function $\Gamma(p)$ with values in V, such that*

(1) $$\gamma(Q) = \int_Q \Gamma\, d\bar{\gamma}, \quad \text{all Borel } Q: \qquad \gamma = \Gamma\bar{\gamma},$$

(2) $$|\Gamma(p)| = 1 \qquad \text{in } E.$$

The function Γ is unique up to $\bar{\gamma}$-equivalence.

Choose a base $(e_1, \cdots, e_m)$ in V, and write

$$\gamma(Q) = \sum \gamma^i(Q) e_i;$$

then the γ^i are real valued additive set functions, and clearly, for some N,

$$|\gamma^i(Q)| \leqq N |\gamma(Q)| \leqq N\bar{\gamma}(Q), \qquad \text{all } i.$$

Therefore, by the Radon-Nikodym Theorem (Saks, p. 36; Halmos, p. 128), $\bar{\gamma}$-measurable functions Γ^i exist, satisfying (1), with γ^i. There is a Borel function $\Gamma_0^i = \Gamma^i$ a.e. $(\bar{\gamma})$ (Saks, pp. 75–76). Then $\Gamma'(p) = \sum \Gamma_0^i(p)e_i$ satisfies (1). The usual proof of uniqueness shows that Γ' is unique as stated.

Suppose that $|\Gamma'(p)| \geqq 1$ a.e. $(\bar{\gamma})$ is false. Then for some $\epsilon > 0$ there is a Borel set Q such that $\bar{\gamma}(Q) > 0$, and $|\Gamma'(p)| \leqq 1 - \epsilon$ in Q. Now for any Borel partition $Q = \bigcup Q_i$,

$$\sum |\gamma(Q_i)| \leqq \sum \left| \int_{Q_i} (1 - \epsilon)\, d\bar{\gamma} \right| = (1 - \epsilon)\bar{\gamma}(Q),$$

contradicting the definition of $\bar{\gamma}(Q)$.

Suppose finally that $|\Gamma'(p)| \leqq 1$ a.e. $(\bar{\gamma})$ is false. Then there is an $\epsilon > 0$, a Borel subset S of V, and an element $v \in S$, such that

$$|v| \geqq 1 + \epsilon, \qquad \operatorname{diam}(S) \leqq \epsilon/2,$$
$$Q = \Gamma'^{-1}(S), \qquad \bar{\gamma}(Q) > 0.$$

Now $|\Gamma'(p) - v| \leqq \epsilon/2$ for $p \in Q$, and hence

$$|\gamma(Q)| \geqq \left| \int_Q v\, d\bar{\gamma}(p) \right| - \left| \int_Q [\Gamma'(p) - v]\, d\bar{\gamma}(p) \right|$$
$$\geqq (1 + \epsilon)\bar{\gamma}(Q) - (\epsilon/2)\bar{\gamma}(Q) > \bar{\gamma}(Q),$$

again a contradiction. Hence $\Gamma' = 1$ a.e. $(\bar{\gamma})$, and we may replace Γ' by Γ, with $|\Gamma(p)| = 1$ in E.

In the second theorem, we cannot apply the Radon-Nikodym Theorem in the usual form, and we no longer have uniqueness (see the Remark following Theorem 1B).

THEOREM 5B. *With γ as before, there is a Borel function f in E, with values in the conjugate space $\bar{V}$ of V, such that*

$$(3) \qquad |f(p)| = 1, \quad \int_Q f \cdot d\gamma = \bar{\gamma}(Q), \quad \text{all Borel } Q: \qquad \bar{\gamma} = f \cdot \gamma.$$

Define ϕ by Theorem 1B, and set

$$(4) \qquad f(p) = \phi(\Gamma(p)).$$

Then $f(p) \cdot \Gamma(p) = |\Gamma(p)| = 1$, and (1) and (4.3) with the following Remark give

$$f \cdot \gamma = f \cdot (\Gamma \cdot \bar{\gamma}) = (f \cdot \Gamma) \cdot \bar{\gamma} = \bar{\gamma}.$$

We give a further formula for $\bar{\gamma}$:

THEOREM 5C. *With γ as before, using Borel point functions f with values in $\bar{V}$,*

$$(5) \qquad \bar{\gamma}(Q) = \sup \left\{ \int_Q f \cdot d\gamma : |f| \leqq 1 \right\}.$$

We may require f to be a Borel step function or a sharp function in E. If Q is open, we may require f to be a sharp function, vanishing outside Q.

The inequality $\geqq$ is clear in each case, while $\leqq$ in (5) follows from (3). That we may obtain $\leqq$, using Borel step functions or sharp functions, follows from Lemma 3a. To prove the last statement (with $\leqq$), let Q be open, and let $\epsilon > 0$ be given. Choose g_0 (Theorem 5B) so that $|g_0| = 1$ and $\int g_0 \cdot d\gamma = \bar{\gamma}(Q)$. Choose a sharp g_1 by Lemma 3a so that $|g_1| \leqq 1$ and $\int_E \langle g_1 - g_0 \rangle \, d\bar{\gamma} < \epsilon/2$. Choose a compact set $Q' \subset Q$ so that $\bar{\gamma}(Q - Q') < \epsilon/2$ (Lemma 2a). Choose a real valued sharp function ϕ in E so that $0 \leqq \phi(p) \leqq 1$, $\phi = 1$ in Q' and $\phi = 0$ in $E - Q$ (App. III, Lemma 1a), and set $f = \phi g_1$. It is easy to see that $\int_Q f \cdot d\gamma > \bar{\gamma}(Q) - \epsilon$.

6. On positive linear functionals. We wish to show that a certain linear function Λ of a certain class of functions ϕ is given by integrating ϕ with respect to a Carathéodory measure. This is a standard theorem in Lebesgue theory. Let L^+ denote the set of real valued non-negative sharp functions in E.

LEMMA 6a. *Let Λ be a real valued non-negative function defined in L^+, such that*

$$\Lambda(a\phi + b\psi) = a\Lambda(\phi) + b\Lambda(\psi) \quad \text{if} \quad a, b \geqq 0; \quad \phi, \psi \in L^+; \tag{1}$$

$$\lim \Lambda(\phi_i) = 0 \quad \text{if} \quad \phi_i \downarrow 0. \tag{2}$$

Then there is a uniquely defined Carathéodory measure μ in E such that

$$\Lambda(\phi) = \int \phi \, d\mu, \qquad \text{all } \phi \in L^+. \tag{3}$$

Note that every such measure μ gives rise to such a Λ; see (4.6).

Take $a = b = 1$ in (1). If we take $\phi = \psi = 0$, we find $\Lambda(0) = 0$. If we use ϕ, $\psi - \phi$ in place of ϕ, ψ, we find

$$\Lambda(\phi) \leqq \Lambda(\psi) \quad \text{if} \quad \phi(p) \leqq \psi(p). \tag{4}$$

Using ϕ_i, $\phi - \phi_i$ in place of ϕ, ψ, gives

$$\Lambda(\phi) = \lim \Lambda(\phi_i) \quad \text{if} \quad \phi_i \uparrow \phi, \qquad \phi_i, \phi \in L^+. \tag{5}$$

Let L^* be the set of all bounded functions ϕ expressible as $\lim \uparrow \phi_i$, $\phi_i \in L^+$. Set

$$\Lambda(\phi) = \lim \Lambda(\phi_i) \quad \text{if} \quad \phi_i \uparrow \phi, \quad \phi_i \in L^+, \quad \phi \in L^*. \tag{6}$$

Since $\Lambda(\phi_i) \leqq \Lambda(\bar{\phi})$ ($\bar{\phi}(p) = |\phi|$, all p), $\Lambda(\phi)$ is finite. It is independent of the sequence chosen. For let $\phi_i' \uparrow \phi$ also. Set $\psi_{ij}(p) = \inf \{\phi_i(p), \phi_j'(p)\}$. For each j, $\psi_{ij} \uparrow \phi_j'$; hence

$$\lim_{i \to \infty} \Lambda(\phi_i) \geqq \lim_{i \to \infty} \Lambda(\psi_{ij}) = \Lambda(\phi_j'), \qquad \text{all } j,$$

and hence $\lim \Lambda(\phi_i) \geqq \lim \Lambda(\phi_j')$. The reverse inequality follows similarly.

Clearly (1) and (4) hold now with $\phi, \psi \in L^*$.

Give any open set R, its characteristic function χ_R is in L^*. (We may set $\phi_i(p) = \inf \{i \operatorname{dist}(p, E - R), 1\}$; then $\phi_i \uparrow \chi_R$). Set

$$\mu(R) = \Lambda(\chi_R). \tag{7}$$

For any set Q, the outer measure $\mu^*(Q)$ is defined by

$$\mu^*(Q) = \inf \{\mu(R): Q \subset R, R \text{ open}\}. \tag{8}$$

Clearly $\mu^*(Q_1) \leqq \mu^*(Q_2)$ if $Q_1 \subset Q_2$, and $\mu^*(R) = \mu(R)$ if R is open. We must prove (Saks, Chapter II; Halmos, p. 48 and Chapter X)

$$\mu^*\left(\bigcup Q_i\right) \leqq \sum \mu^*(Q_i), \tag{9}$$

$$\mu^*(Q_1 \cup Q_2) = \mu^*(Q_1) + \mu^*(Q_2) \quad \text{if} \quad \operatorname{dist}(Q_1, Q_2) > 0. \tag{10}$$

If we prove these for open sets, they follow easily for the general case. Suppose $R = \bigcup_{i=1}^{\infty} R_i$. Let $\chi_{R_i} = \lim \uparrow_{j\to\infty} \psi_{ij}$, $\psi_{ij} \in L^+$. Set

$$\phi_j = \inf \{\psi_{1j} + \cdots + \psi_{jj}, 1\}.$$

Then $\phi_j \uparrow \chi_R$, so that $\mu(R) = \lim \Lambda(\phi_j)$. Also

$$\Lambda(\phi_j) \leqq \Lambda(\psi_{1j} + \cdots + \psi_{jj}) = \Lambda(\psi_{1j}) + \cdots + \Lambda(\psi_{jj}) \leqq \sum_{i=1}^{\infty} \mu(R_i),$$

and (9) follows for open sets. To prove (10) for open sets (we need merely $R_1 \cap R_2 = 0$), let $\chi_{R_i} = \lim \uparrow_{j\to\infty} \psi_{ij}$, and set $\phi_j = \psi_{1j} + \psi_{2j}$. Then if $R = R_1 \cup R_2$, $\chi_R = \lim \uparrow \phi_i$, and

$$\mu(R) = \lim \Lambda(\phi_i) = \lim [\Lambda(\phi_1) + \Lambda(\phi_2)] = \mu(R_1) + \mu(R_2).$$

Thus μ^* is an outer Carathéodory measure. We shall write $\mu(Q) = \mu^*(Q)$ for μ-measurable sets, in particular, for Borel sets Q.

To prove (3), take any $\epsilon > 0$. Let R_i be the set of points p such that $\phi(p) > i\epsilon$ $(i = 0, 1, \cdots)$; then R_i is open, and for some m, R_m is void. Set

$$\psi(p) = \begin{cases} (i+1)\epsilon, & p \in R_i - R_{i+1}, \\ 0, & p \in E - R_0; \end{cases}$$

then

$$\psi(p) = \epsilon \sum_{i=0}^{m} \chi_{R_i}(p), \qquad \psi(p) - \epsilon < \phi(p) \leqq \psi(p).$$

Because of (7), $\Lambda(\chi_{R_i}) = \int \chi_{R_i}\, d\mu$; hence $\Lambda(\psi) = \int \psi\, d\mu$. Also

$$\Lambda(\psi) \leqq \Lambda(\phi) + \epsilon\mu(E) \leqq \Lambda(\psi) + \epsilon\mu(E),$$

and the same inequalities hold for the corresponding integrals. This shows that

$$\left| \Lambda(\phi) - \int \phi \, d\mu \right| \leqq \epsilon\mu(E),$$

and (3) follows.

Using (3), it is clear that μ must be as constructed; hence μ is unique.

7. On bounded linear functionals. We give a lemma like Lemma 6a, using the space $\mathbf{C}^{\#0}$ of all real sharp functions, and a different continuity hypothesis. The two hypotheses are related by the following lemma.

LEMMA 7a. *Let $\phi_i \downarrow 0$ (the ϕ_i continuous); then $\phi_i \to 0$ u.c.s. (uniformly in compact sets).*

Given the compact set Q and $\epsilon > 0$, let Q_i be the set of points $p \in Q$ such that $\phi_i(p) \geqq \epsilon$. Then $\bigcap Q_i = 0$, and since the Q_i are compact, some Q_{i_0} is void. Now $\phi_i(p) < \epsilon$ in Q for $i \geqq i_0$.

Given the real valued function Λ in $\mathbf{C}^{\#0}$, define

$$(1) \qquad | \Lambda | = \sup \{ | \Lambda(\phi) | : \ \phi \text{ sharp}, \quad | \phi | \leqq 1 \};$$

Λ is *bounded* if this is finite.

LEMMA 7b. *Let Λ be a linear function in $\mathbf{C}^{\#0}$ with the following property. Given the $\phi_i \in \mathbf{C}^{\#0}$ and N,*

$$(2) \qquad \lim \Lambda(\phi_i) = 0 \quad \text{if} \quad | \phi_i | \leqq N, \qquad \phi_i \to 0 \text{ u.c.s.}$$

Then there is a unique additive set function γ in E such that

$$(3) \qquad \Lambda(\phi) = \int_E \phi \, d\gamma, \qquad \phi \in \mathbf{C}^{\#0}; \qquad | \Lambda | = | \gamma |.$$

First we show that Λ is bounded. If not, then there is a sequence $\phi_i, \phi_2, \cdots$ such that $| \phi_i | \leqq 1$ and $| \Lambda(\phi_i) | \geqq i$. Set $\psi_i = \phi_i/i$; then $\psi_i \to 0$ uniformly, and hence $\Lambda(\psi_i) \to 0$; but $| \Lambda(\psi_i) | \geqq 1$, a contradiction.

Define Λ_1 and Λ_2 in L^+ (§ 6) by

$$(4) \qquad \Lambda_1(\phi) = \sup \{ \Lambda(\psi) : \quad \psi \text{ sharp}, \quad 0 \leqq \psi \leqq \phi \},$$

$$(5) \qquad \Lambda_2(\phi) = \Lambda_1(\phi) - \Lambda(\phi).$$

Taking $\psi = 0$ in (4) shows that $\Lambda_1(\phi) \geqq 0$. For any ψ in (4), $| \Lambda(\psi) | \leqq | \Lambda | | \psi | \leqq | \Lambda | | \phi |$; hence $\Lambda_1(\phi)$ is finite, and $\Lambda_2(\phi)$ is also. Taking $\psi = \phi$ shows that $\Lambda_1(\phi) \geqq \Lambda(\phi)$, $\Lambda_2(\phi) \geqq 0$.

To show that Λ_1 (and hence Λ_2) is linear in L^+, note first that $\Lambda_1(a\phi) = a\Lambda_1(\phi)$ $(a \geqq 0)$. Now take any ϕ_1, ϕ_2, $\phi = \phi_1 + \phi_2$ in L^+. Given $\epsilon > 0$, choose ψ_i so that $0 \leqq \psi_i \leqq \phi_i$ and $\Lambda(\psi_i) > \Lambda_1(\phi_i) - \epsilon/2$ $(i = 1, 2)$. Then $0 \leqq \psi_1 + \psi_2 \leqq \phi$, and hence

$$\Lambda_1(\phi) \geqq \Lambda(\psi_1 + \psi_2) = \Lambda(\psi_1) + \Lambda(\psi_2) > \Lambda_1(\phi_1) + \Lambda_1(\phi_2) - \epsilon.$$

Conversely, choosing ψ so that $0 \leqq \psi \leqq \phi$ and $\Lambda(\psi) > \Lambda_1(\phi) - \epsilon$, set $\psi_1 = \inf \{\phi_1, \psi\}$, $\psi_2 = \psi - \psi_1$; then we see easily that $0 \leqq \psi_i \leqq \phi_i$ $(i = 1, 2)$, and hence

$$\Lambda_1(\phi_1) + \Lambda_1(\phi_2) \geqq \Lambda(\psi_1) + \Lambda(\psi_2) = \Lambda(\psi) > \Lambda_1(\phi) - \epsilon,$$

which proves $\Lambda_1(\phi) = \Lambda_1(\phi_1) + \Lambda_1(\phi_2)$.

Suppose $\phi_i \downarrow 0$. Choose ψ_i so that $0 \leqq \psi_i \leqq \phi_i$, $\Lambda(\psi_i) > \Lambda_1(\phi_i) - 1/2^i$. By Lemma 7a, $\phi_i \to 0$ u.c.s.; hence $\psi_i \to 0$ u.c.s. Also $|\psi_i| \leqq |\phi_1|$. Therefore $\Lambda(\psi_i) \to 0$, and hence $\Lambda_1(\phi_i) \to 0$. This gives also $\Lambda_2(\phi_i) \to 0$.

By Lemma 6a, there are Carathéodory measures μ_1, μ_2 such that $\Lambda_i(\phi) = \int \phi \, d\mu_i$, $\phi \in L^+$. Set

$$\gamma = \mu_1 - \mu_2: \quad \gamma(Q) = \mu_1(Q) - \mu_2(Q). \tag{6}$$

Then

$$\int \phi \, d\gamma = \int \phi \, d\mu_1 - \int \phi \, d\mu_2 = \Lambda_1(\phi) - \Lambda_2(\phi) = \Lambda(\phi), \qquad \phi \in L^+.$$

Given any $\phi \in \mathbf{C}^{\sharp 0}$, set $\phi_1 = \sup \{\phi, 0\}$, $\phi_2 = \phi_1 - \phi$; then ϕ_1 and ϕ_2 are in L^+, and

$$\Lambda(\phi) = \Lambda(\phi_1) - \Lambda(\phi_2) = \int \phi_1 \, d\gamma - \int \phi_2 \, d\gamma = \int \phi \, d\gamma,$$

proving the first part of (3). For the other equality and uniqueness, see Theorem 8A.

8. Linear functions of sharp r-forms. The theorem below will be an immediate extension of Lemma 7b. It is the basis of part of Theorem 11A. We use mass of r-vectors and comass of r-forms (II, 3.2) as norms.

THEOREM 8A. *Let Λ be a real valued linear function of sharp r-forms in E, such that*

$$\lim \Lambda(\omega_i) = 0 \quad \text{if} \quad |\omega_i|_0 \leqq N, \quad \omega_i \to 0 \text{ u.c.s.} \tag{1}$$

Then there is a unique r-vector valued additive set function γ such that

$$\Lambda(\omega) = \int_E \omega \cdot d\gamma = [\omega \cdot \gamma], \qquad \text{all sharp } \omega; \tag{2}$$

we have

$$|\Lambda| = |\gamma|. \tag{3}$$

Note that every γ gives rise to a Λ satisfying (1) (without the "u.c.s."); see (4.9).

Let $(e_1, \cdots, e_n)$ be a base in $V(E)$. For each $\lambda = (\lambda_1, \cdots, \lambda_r)$, set

$$\Lambda^\lambda(\phi) = \Lambda(\phi e^\lambda), \qquad \text{sharp real valued } \phi. \tag{4}$$

By Lemma 7b there is a real valued additive set function γ^λ such that

$$\Lambda(\phi e^\lambda) = \int \phi \, d\gamma^\lambda.$$

Set

$$\gamma = \sum_{(\lambda)} \gamma^\lambda e_\lambda\colon \quad \gamma(Q) = \sum_{(\lambda)} \gamma^\lambda(Q) e_\lambda, \qquad \text{Borel } Q.$$

The definition of the integral in § 3 as the limit of a sequence of sums, each sum using a Borel partition of E, shows that

$$\int_E \omega \cdot d\gamma = \lim \sum_i \omega(p_i) \cdot \gamma(Q_i)$$

$$= \lim \sum_i \sum_{(\lambda)} \omega_\lambda(p_i) \gamma^\lambda(Q_i) = \sum_{(\lambda)} \int_E \omega_\lambda \, d\gamma^\lambda;$$

hence

$$\Lambda(\omega) = \sum_{(\lambda)} \Lambda(\omega_\lambda e^\lambda) = \sum_{(\lambda)} \int_E \omega_\lambda \, d\gamma^\lambda = \int_E \omega \cdot d\gamma.$$

To prove uniqueness, suppose also $\Lambda(\omega) = \int_E \omega \cdot d\gamma'$, $\gamma' \neq \gamma$. Set $\gamma_1 = \gamma' - \gamma$; then $|\gamma_1| > 0$. By Theorem 5C, there is a sharp ω such that

$$\int_E \omega \cdot d\gamma_1 > |\gamma_1|/2, \qquad \int_E \omega \cdot d\gamma' \neq \int_E \omega \cdot d\gamma,$$

a contradiction.

The relation (3) follows at once from (5.5) and (2).

We mention a theorem that can be used in place of Theorem 8A in the proof of Theorem 11A below. Let K_0 be the normed linear space of compact sharp r-forms ω in E (we could use compact continuous forms), with the comass $|\omega|_0 = \sup \{|\omega(p)|_0\}$ as norm. The completion K of K_0 is easily seen to be the Banach space of continuous r-forms which "approach 0 at infinity"; that is, given $\omega \in K$ and $\epsilon > 0$, there is a compact set Q such that $|\omega(p)|_0 < \epsilon$ in $E - Q$. Let $\bar{K}_0 = \bar{K}$ denote the conjugate space of K_0 (or of K).

THEOREM 8B. *With the above notations, $\bar{K}$ is the Banach space of r-vector valued additive set functions in E, with total variation as norm. More explicitly, given Λ in $\bar{K}$, there is a unique γ such that*

$$\Lambda(\omega) = \int_E \omega \cdot d\gamma, \qquad \text{all } \omega \in K; \tag{5}$$

also, $|\Lambda| = |\gamma|$. Conversely, each such γ gives rise to such a Λ.

The theorem reduces at once to the real variable case. Using a decomposition as in (7.5), we may apply Theorem D, p. 247, of Halmos to each of Λ_1, Λ_2.

9. The sharp norm of r-vector valued set functions. Define $|\omega|^\#$ for r-forms as in (V, 10.2). If γ is r-vector valued, its *sharp norm* is

$$(1) \qquad |\gamma|^\# = \sup \left\{ \int_E \omega \cdot d\gamma : \quad \omega \text{ sharp}, \quad |\omega|^\# \leqq 1 \right\}.$$

That $|\gamma|^\# > 0$ if $\gamma \neq 0$ follows from Theorem 5C.

If $\omega = \omega_0$ is constant, then, by (3.6), $[\omega \cdot \gamma] = \omega_0 \cdot [\gamma]$. Hence, by (I, 13.6) and Theorem 5C,

$$(2) \qquad |[\gamma]|_0 \leqq |\gamma|^\# \leqq |\gamma|.$$

The *support* spt (γ) of γ is the set of points p such that for each neighborhood U of p there is a Borel set $Q \subset U$ with $\gamma(Q) \neq 0$; equivalently, E — spt (γ) is the largest open set R such that $\bar{\gamma}(R) = 0$. Write $\gamma \subset Q$ if spt $(\gamma) \subset Q$. We say γ is *compact* if spt (γ) is compact. We prove

$$(3) \qquad |\gamma|^\# \leqq \rho |\gamma|/(r+1) + |[\gamma]|_0 \quad \text{if} \quad \gamma \subset \bar{U}_\rho(p).$$

For take any $\epsilon > 0$. Choose a sharp ω such that

$$|\omega|^\# \leqq 1, \qquad [\omega \cdot \gamma] > |\gamma|^\# - \epsilon.$$

Set $\omega_0 = \omega(p)$. Then, as in the proof of (VII, 7.2),

$$(4) \qquad \begin{aligned} |[\omega \cdot \gamma]| &\leqq \left| \int_{U_\rho(p)} (\omega - \omega_0) \cdot d\gamma \right| + \left| \int_E \omega_0 \cdot d\gamma \right| \\ &\leqq \rho \mathfrak{L}_0(\omega) |\gamma| + |\omega|_0 \cdot |[\gamma]|_0, \end{aligned}$$

which gives (3).

Note that

$$(5) \qquad \text{spt}(\bar{\gamma}) = \text{spt}(\gamma), \qquad \text{spt}(\phi\gamma) = \text{spt}(\phi) \cap \text{spt}(\gamma).$$

The second relation is easily proved if $\phi \geqq 0$, $\gamma = \bar{\gamma}$; in the general case, use the first relation and (4.2).

10. Molecular set functions. We say the set function γ is *atomic*, and is *at* p, if spt $(\gamma) = p$. Then $[\gamma] = \gamma(p)$. If $\gamma(p) = \alpha$, we write $\gamma = \gamma_{p,\alpha}$. A *molecular set function* is a finite sum of atomic set functions. Note that for any bounded Borel ω,

$$(1) \qquad [\omega \cdot \gamma_{p,\alpha}] = \omega(p) \cdot \alpha.$$

LEMMA 10a. *The molecular set functions are dense in the space of r-vector valued additive set functions, in the sharp norm.*

Take any γ and $\epsilon > 0$. Let Q be a compact set such that $\bar{\gamma}(E - Q) < \epsilon/2$ (Lemma 2a). Let $\bigcup Q_i$ be a finite Borel partition of Q such that diam $(Q_i) \leqq \rho = (r+1)\epsilon/4|\gamma|$. Choose $p_i \in Q_i$, and set $Q_0 = E - Q$, and

$$\alpha_i = \gamma(Q_i), \qquad \gamma_i = \gamma_{p_i,\alpha_i} \quad (i \geqq 1), \qquad \gamma'_i = \chi_{Q_i}\gamma \quad (i \geqq 0).$$

Then (4.1) and (9.2) give, for $i \geqq 1$,

$$[\gamma_i] = \alpha_i = [\gamma_i'], \qquad |\gamma_i| = |\alpha_i|_0 \leqq |\gamma_i'|.$$

By (4.8), $\sum_{i\geqq 0}|\gamma_i'| = |\gamma|$. Hence, by (9.2) and (9.3),

$$\Big|\gamma - \sum_{i\geqq 1}\gamma_i\Big|^\# \leqq |\gamma_0'|^\# + \sum_{i\geqq 1}|\gamma_i' - \gamma_i|^\# \leqq |\gamma_0'| + \sum_{i\geqq 1}\rho\cdot 2|\gamma_i'|/(r+1)$$
$$< \epsilon/2 + 2\rho|\gamma|/(r+1) = \epsilon,$$

as required.

REMARK. The molecular set function has the property

$$(2) \qquad \Big|\sum_{i\geqq 1}\gamma_i\Big| \leqq \sum|\gamma_i'| \leqq |\gamma|.$$

For any molecular set function, the definition of total variation shows that

$$(3) \qquad \Big|\sum\gamma_{p_i,\alpha_i}\Big| = \sum|\gamma_{p_i,\alpha_i}| = \sum|\alpha_i|_0 \quad \text{if the } p_i \text{ are distinct.}$$

11. Sharp chains and set functions. Consider the following linear spaces:

$\mathfrak{M}_r$, consisting of all sharp r-chains in E of finite mass, with mass or sharp norm as norm;

$\mathbf{M}_r$, consisting of all r-vector valued additive set functions in E, with total variation or sharp norm as norm.

We shall show that these are isomorphic, preserving both norms.

Say $A \in \mathfrak{M}_r$ and $\gamma \in \mathbf{M}_r$ *correspond* (compare (VI, 7)) if, for every sharp r-cochain X in E,

$$(1) \qquad X\cdot A = \int_E D_X\cdot d\gamma = [D_X\cdot\gamma].$$

THEOREM 11A. *The above correspondence $A \to \gamma = \gamma_A$ is a one-one linear mapping of $\mathfrak{M}_r$ onto $\mathbf{M}_r$, such that*

$$(2) \qquad [\gamma_A] = \{A\}, \qquad |\gamma_A|^\# = |A|^\#, \qquad |\gamma_A| = |A|.$$

To each A corresponds a linear function Λ_A of sharp r-forms, defined by $\Lambda_A(D_X) = X\cdot A$. Suppose $X_1, X_2, \cdots$ is a sequence such that $|D_{X_i}|_0 \leqq N$ for some N, and $D_{X_i} \to 0$ u.c.s. We shall show that $\lim \Lambda_A(D_{X_i}) = 0$. Given $\epsilon > 0$, choose a real compact sharp function ϕ (VII, Theorem 4A) such that

$$|\phi A - A| < \epsilon/2N.$$

Set $Q = \operatorname{spt}(\phi)$. Now $\phi D_{X_i} \to 0$ uniformly, and (V, 10.4) and (VI, 8.4) show that

$$|\phi X_i| = |D_{\phi X_i}|_0 = |\phi D_{X_i}|_0 \to 0.$$

Hence we may choose i_0 so that

$$|\phi_{X_i}| < \epsilon/2|A|, \qquad i \geqq i_0.$$

Now take any $i \geqq i_0$. Using (VII, 2.1) gives

$$\begin{aligned} |\Lambda_A(D_{X_i})| = |X_i \cdot A| &\leqq |X_i \cdot (A - \phi A)| + |\phi X_i \cdot A| \\ &\leqq N \cdot \epsilon / 2N + (\epsilon / 2 |A|) |A| = \epsilon, \end{aligned}$$

proving $\Lambda_A(D_{X_i}) \to 0$. Hence, by Theorem 8A, there is a unique $\gamma = \gamma_A$ such that (1) holds. Clearly the correspondence is linear.

Take any r-covector ω_0; let X be the cochain such that $D_X(p) = \omega_0$ (all p). By (VII, 6.3) and (3.6),

$$\omega_0 \cdot \{A\} = X \cdot A = \int_E \omega_0 \cdot d\gamma_A = \omega_0 \cdot [\gamma_A];$$

it follows that $[\gamma_A] = \{A\}$. By the formulas (V, 16.3) for $|A|$ (the sharp case) and (5.5) for $|\gamma|$, with (V, 10.4) and (1), $|\gamma_A| = |A|$. By (V, 4.3) (the sharp case), (9.1), (V, 10.4) and (1), $|\gamma_A|^\sharp = |A|^\sharp$.

Suppose $\gamma_A = 0$, i.e. $\gamma_A(Q) = 0$, all Borel Q. Then for any X, $X \cdot A = [D_X \cdot \gamma_A] = 0$, and hence $A = 0$. Hence the correspondence is one-one.

Suppose A is the chain at p such that $\{A\} = \alpha$ (VII, 7), and $\gamma = \gamma_{p,\alpha}$ (§ 10). Then (VII, 7.1) and (10.1) give

$$X \cdot A = D_X(p) \cdot \alpha = [D_X \cdot \gamma_{p,\alpha}]$$

for all sharp X; hence $\gamma_A = \gamma_{p,\alpha}$, which shows that the molecular chains correspond to the molecular set functions.

To show that the correspondence is onto $\mathbf{M}_r$, take any $\gamma \in \mathbf{M}_r$. For each integer k, let γ_k be a molecular set function such that $|\gamma - \gamma_k|^\sharp < 1/2^k$ (Lemma 10a). Say $\gamma_{A_k} = \gamma_k$. Now

$$|A_j - A_k|^\sharp = |\gamma_j - \gamma_k|^\sharp \to 0,$$

and hence $A_1, A_2, \cdots$ is a Cauchy sequence in the sharp norm, and has a limit A. By (V, Theorem 16B), (2) and (10.2),

$$|A| \leqq \liminf |A_k| = \liminf |\gamma_k| \leqq |\gamma|,$$

which is finite. Also

$$|\gamma_A - \gamma_k|^\sharp = |A - A_k|^\sharp \to 0,$$

hence $|\gamma_A - \gamma|^\sharp = 0$ and $\gamma_A = \gamma$, as required. This completes the proof.

Examples. For $r = 0$, $\gamma_A(Q)$ is a real number. If $A = \sum a_i p_i$, $\gamma_A(Q)$ is the sum of those a_i for which $p_i \in Q$; $[\gamma_A] = \{A\} = \sum a_i$. If A is the 1-chain defined by an oriented arc from p to q (X, 6), then $[\gamma_A]$ is the vector $q - p$; if Q is a Borel set containing just the subarc $p'q'$, then $\gamma_A(Q) = q' - p'$. (Compare § 13.)

Problem. Let A be a flat chain of finite mass. Then A is sharp (V, Theorem 14B), and hence $X \cdot A = \int_E D_X \cdot d\gamma_A$, all sharp X. Does this relation hold also for flat X?

We end by proving

(3) $$\operatorname{spt}(\gamma_A) = \operatorname{spt}(A).$$

Suppose $p \in \operatorname{spt}(\gamma_A)$. Take any neighborhood U of p. Say $\gamma_A(Q) \neq 0$, $Q \subset U$. Then $\bar{\gamma}_A(U) \geqq \bar{\gamma}_A(Q) > 0$, and by Theorem 5C, there is a sharp function ω, vanishing outside U, such that $[\omega \cdot \gamma_A] \neq 0$. Say $D_X = \omega$; then $X \cdot A \neq 0$, proving $p \in \operatorname{spt}(A)$. If, conversely, p is not in $\operatorname{spt}(\gamma_A)$, then there is a neighborhood U of p such that $\gamma_A(Q) = 0$ for all $Q \subset U$. Now $[\omega \cdot \gamma_A] = 0$ if $\operatorname{spt}(\omega) \subset U$, hence $X \cdot A = 0$ if $\operatorname{spt}(X) \subset U$, and p is not in $\operatorname{spt}(A)$.

12. Bounded Borel functions times chains. If A is a chain of finite mass and ϕ is a real bounded Borel function, let ϕA be the chain corresponding to $\phi\gamma_A$ (Theorem 11A); thus

(1) $$\gamma_{\phi A} = \phi\gamma_A, \qquad \text{bounded Borel } \phi.$$

If ϕ is sharp, then for any sharp X, (1), (4.3) and (VI, 8.4) give

$$X \cdot \phi A = [D_X \cdot \phi\gamma_A] = [\phi D_X \cdot \gamma_A] = [D_{\phi X} \cdot \gamma_A] = \phi X \cdot A;$$

using (VII, 2.1) shows that the definition of ϕA agrees with that in (VII, 1) in this case.

If we define

(2) $$\omega \cdot A = \int_E \omega \cdot d\gamma_A = [\omega \cdot \gamma_A]$$

for the bounded Borel r-form ω, then the proof above shows that, for real bounded Borel ϕ,

(3) $$\phi\omega \cdot A = \omega \cdot \phi A.$$

A particular case of (2) is $D_X \cdot A = X \cdot A$ (X sharp).

The function ϕA is linear in both variables. By (4.3) and (1),

(4) $$\phi(\psi A) = (\phi\psi)A.$$

By (11.2), (1) and (4.2),

(5) $$|\phi A| = |\langle\phi\rangle A| = |\phi\gamma_A| = \int_E \langle\phi\rangle\, d\bar{\gamma}_A.$$

By (11.2) and (1),

(6) $$\{\phi A\} = [\phi\gamma_A] = \int_E \phi\, d\gamma_A.$$

We prove, for bounded Borel ϕ_i, ϕ, ω_i, ω,

(7) $$\lim |\phi A - \phi_i A| = 0 \quad \text{if} \quad \phi_i \uparrow \phi \quad \text{or} \quad \phi_i \downarrow \phi,$$

(8) $$\lim |\phi A - \phi_i A| = 0 \quad \text{if} \quad \phi_i \to 0, \qquad |\phi_i| \leqq N,$$

(9) $$\lim \omega_i \cdot A = \omega \cdot A \qquad \text{if} \quad \omega_i \to \omega, \qquad |\omega_i|_0 \leqq N.$$

These follow from (4.6), (4.9), (4.10) and (3.10), using (4.2). Thus

$$|(\phi - \phi_i)A| = |(\phi - \phi_i)\gamma_A| = \langle \phi - \phi_i \rangle \bar{\gamma}_A(E) \to 0,$$

$$|(\omega - \omega_i)\cdot A| \leqq |(\omega - \omega_i)\cdot \gamma_A| \leqq \langle \omega - \omega_i \rangle_0 \bar{\gamma}_A(E) \to 0.$$

Note that (7) and (8) give, for bounded Borel ϕ_i, ϕ,

(10) $\phi_i A \xrightarrow{\#} \phi A$ if $\phi_i \uparrow \phi$ or $\phi_i \downarrow \phi$ or $\phi_i \to \phi$, $|\phi_i| \leqq N$.

Generalizing (VII, 1.17) (if $|A|$ is finite), we prove, for bounded Borel ϕ_i, ϕ,

(11) $$\sum |\phi_i A| = |\phi A| \quad \text{if} \quad \phi_i \geqq 0, \qquad \sum \phi_i = \phi.$$

For finite sums, this follows from (5); for infinite sums, apply (7).

13. The part of a chain in a Borel set. Let A be of finite mass, and let Q be a Borel set, with characteristic function χ_Q. We define the *part of A in Q* by

(1) $$A_Q = \chi_Q A.$$

By (12.10),

(2) $$A_Q = \sum A_{Q_i} \quad \text{if} \quad \bigcup Q_i \text{ is a Borel partition of } Q.$$

Thus $\Phi_A(Q) = A_Q$ is an additive set function, whose values are sharp chains of finite mass. We prove

(3) $$\{A_Q\} = \gamma_A(Q);$$

for (4.1), (12.1) and (11.2) give

$$\gamma_A(Q) = [\chi_Q \gamma_A] = [\gamma_{\chi_Q A}] = \{\chi_Q A\}.$$

Because of (12.4),

(4) $$\phi(A_Q) = \phi(\chi_Q A) = \chi_Q(\phi A) = (\phi A)_Q.$$

Setting $\phi = \chi_Q$ in (12.5) gives

(5) $$|A_Q| = \bar{\gamma}_A(Q).$$

Hence

(6) $$|A_Q| = \sum |A_{Q_i}|, \qquad \text{Borel partitions } \bigcup Q_i \text{ of } Q.$$

More generally, by (12.4) and (12.5),

(7) $$|(\phi A)_Q| = |(\chi_Q \phi)A| = \int_Q \langle \phi \rangle \, d\bar{\gamma}_A.$$

By (3), (VII, 6.2) and (5),

(8) $$|\gamma_A(Q)|_0 \leqq |A_Q|^\# \leqq |A_Q| = \bar{\gamma}_A(Q).$$

Note that (using the notation in (3.11))

$$(9) \qquad |\phi|_A = |\phi|_{\bar{\gamma}_A} = |\phi A| = \int_E \langle\phi\rangle \, d\bar{\gamma}_A$$

gives a semi-norm in the space of bounded Borel functions ϕ.

We give a theorem similar to (VII, Theorem 8C); the latter could be derived from the present theorem (if $|A|$ is finite).

THEOREM 13A. *For each chain A of finite mass and $\epsilon > 0$ there is a Borel partition $E = \bigcup Q_i$ such that*

$$(10) \qquad \sum |A_{Q_i}|^\sharp \geqq \sum |\{A_{Q_i}\}|_0 > |A| - \epsilon.$$

By definition of $|\gamma_A|$, we may choose the Q_i so that $\sum|\gamma_A(Q_i)|_0 > |\gamma_A| - \epsilon$; using (3), (8) and (11.2) gives (10).

14. Chains and point functions. Let A be an r-chain of finite mass. By Theorems 5A and 5B, there are Borel functions $\Gamma_A(p)$ and $\xi_A(p)$, whose values are r-vectors and r-covectors respectively, such that

$$(1) \qquad \gamma_A(Q) = \int_Q \Gamma_A \, d\bar{\gamma}_A, \qquad |\Gamma_A(p)|_0 = 1,$$

$$(2) \qquad \bar{\gamma}_A(Q) = \int_Q \xi_A \cdot d\gamma_A, \qquad |\xi_A(p)|_0 = 1.$$

The function Γ_A is uniquely determined a.e. ($\bar{\gamma}_A$), but ξ_A need not be.

Applying the Remark following (4.3) gives, for sharp X,

$$(3) \qquad D_X \cdot \gamma_A = D_X \cdot (\Gamma_A \cdot \bar{\gamma}_A) = (D_X \cdot \Gamma_A)\bar{\gamma}_A;$$

hence

$$(4) \qquad X \cdot A = [D_X \cdot \gamma_A] = [(D_X \cdot \Gamma_A)\bar{\gamma}_A] = \int_E (D_X \cdot \Gamma_A) \, d\bar{\gamma}_A.$$

EXAMPLES. Let A be the 1-chain formed by a smooth oriented arc (see (X, 6)). Then we may let $\Gamma_A(p)$ be the tangent vector at each point p of the arc, and let $\Gamma_A(p) = 0$ elsewhere. (Γ_A is arbitrary outside the arc, since $\bar{\gamma}_A = 0$ there.) If A is formed by a portion of a smooth oriented r-manifold M, then $\Gamma_A(p)$ ($p \in M$) is the r-direction of the oriented tangent plane at p.

We now discuss the possibility of representing $\gamma_A(Q)$ in the form $\int_Q \alpha(p) \, dp$, α summable. Let μ denote Lebesgue measure; then this integral may be written as

$$\int_Q \alpha = \int_Q \alpha \, d\mu = (\alpha\mu)(Q).$$

Note that the proof of (4.2) gives

$$(5) \qquad \overline{\alpha\mu} = \langle\alpha\rangle_0 \mu, \qquad |\alpha\mu| = \int_E \langle\alpha\rangle_0 \, d\mu = \int_E |\alpha(p)|_0 \, dp.$$

THEOREM 14A. *The mapping* $\alpha \to \tilde{\alpha}$ *of* (VI, *Theorem* 7A) *exists with Lebesgue measurable summable* α *in place of continuous summable* α; *we have*

$$\gamma_{\tilde{\alpha}} = \alpha\mu. \tag{6}$$

First, $\tilde{\alpha}$ is defined for continuous summable α, by the theorem quoted. Given the Lebesgue measurable summable α, choose, for each integer i, a continuous summable α_i such that (App. III, 6)

$$|(\alpha - \alpha_i)\mu| < 1/2^i.$$

Then, as in the proof of (VI, Theorem 7A), $A = \lim^\flat \tilde{\alpha}_i$ exists, $X \cdot A = \int_E D_X \cdot \alpha \, d\mu$ for sharp X, and hence $\gamma_A = \alpha\mu$.

By (11.2) and (5),

$$|A| = |\gamma_A| = |\alpha\mu| = \int_E \langle\alpha\rangle_0 \, d\mu. \tag{7}$$

Note that the mapping $\alpha \to \tilde{\alpha}$ is one-one only in the sense that $\tilde{\alpha}' = \tilde{\alpha}$ implies $\alpha' = \alpha$ a.e.

We prove

THEOREM 14B. *Given the chain* A *of finite mass, there is a Lebesgue measurable summable* r*-vector valued function* α *such that* $A = \tilde{\alpha}$:

$$X \cdot A = \int_E D_X \cdot \alpha \, d\mu, \qquad \text{all sharp } X, \tag{8}$$

if and only if γ_A *is absolutely continuous with respect to Lebesgue measure.*

If $A = \tilde{\alpha}$, then for any sharp X, (11.1), (8) and (4.3) give $[D_X \cdot (\gamma_A - \alpha\mu)] = 0$; by Theorem 5C, $\gamma_A = \alpha\mu$, which is absolutely continuous. Conversely, if γ_A is absolutely continuous, then by the Radon-Nikodym Theorem, we may write $\gamma_A = \alpha\mu$; now $|\alpha\mu| = |\gamma_A| = |A|$, and hence α is summable; also, by (4.3), $X \cdot A = [D_X \cdot \gamma_A] = [(D_X \cdot \alpha)\mu]$, and hence $A = \tilde{\alpha}$.

15. Characterization of the sharp norm. Given the set function γ and the vector v, define the set function $T_v\gamma$, the *translation* of γ by v, by

$$T_v\gamma(Q) = \gamma(T_{-v}Q), \tag{1}$$

T_uQ denoting the set of points $p + u$, $p \in Q$.

THEOREM 15A. *The sharp norm* $|\gamma|^\sharp$ ($\gamma \in \mathbf{M}_r$) *is the supremum* $|\gamma|^\sharp_s$ *of semi-norms* $|\gamma|'$ *satisfying*

(a) $|\gamma|' \leqq |\gamma|$,

(b) $|T_v\gamma - \gamma|' \leqq |v|\,|\gamma|/(r+1)$,

(c) *for each point* p *and* $\epsilon > 0$ *there is a* $\zeta > 0$ *such that*

$$|\gamma|' \leqq \epsilon|\gamma| \quad \text{if} \quad \gamma \subset U_\zeta(p) \quad \text{and} \quad [\gamma] = 0. \tag{2}$$

By (App. I, Lemma 15b), $|\;|^\sharp_s$ is a semi-norm in $\mathbf{M}_r$.

We show first that the sharp norm satisfies the conditions; this will prove that $|\gamma|^\sharp \leqq |\gamma|^\sharp_s$, and hence that $|\ |^\sharp_s$ is a norm. Properties (a) and (c) follow from (9.2) and (9.3) respectively. To prove (b), take any sharp ω such that $|\omega|^\sharp \leqq 1$. If $\omega_v(p) = \omega(p - v)$, then

$$(3) \qquad |[\omega\cdot(T_v\gamma - \gamma)]| = \left|\int_E (\omega_v - \omega)\cdot d\gamma\right| \leqq \mathfrak{L}_0(\omega)\,|v|\,|\gamma|.$$

By (V, 10.2), $\mathfrak{L}_0(\omega) \leqq 1/(r + 1)$; hence (9.1) gives (b).

Let diam (γ) denote diam (spt (γ)). We say the set functions $\gamma_1, \cdots, \gamma_m$ form a *pure ζ-partition* of γ if there is a finite Borel partition $\bigcup Q_i$ of spt (γ) such that

$$(4) \qquad \gamma_i = \chi_{Q_i}\gamma \quad \left(\text{hence } \sum \gamma_i = \gamma\right), \qquad \text{diam } (\gamma_i) \leqq \zeta.$$

If γ is compact, this exists, for each $\zeta > 0$ (take diam $(Q_i) \leqq \zeta$).

To prove that $|\gamma|^\sharp_s \leqq |\gamma|^\sharp$, we start by showing that the semi-norm $|\ |^\sharp_s$ satisfies (9.3). Take any p, $\rho > 0$, and $\gamma \subset \bar{U}_\rho(p)$. Let $|\ |'$ be any semi-norm satisfying (a), (b), (c). Given any $\epsilon > 0$, choose $\zeta > 0$ by (c). Let $\gamma_1, \cdots, \gamma_m$ be a pure ζ-partition of γ; we may choose vectors $v_1, \cdots, v_m$, $|v_i| \leqq \rho$, such that $\gamma' = \sum T_{v_i}\gamma_i \subset \bar{U}_\zeta p)$. By (b) and (4.8),

$$|\gamma' - \gamma|' \leqq \sum |v_i|\,|\gamma_i|/(r + 1) \leqq \rho\,|\gamma|/(r + 1).$$

Set $\alpha = [\gamma]$, $\beta = \gamma_{p,\alpha}$ (§ 10). Then $[\beta - \gamma'] = 0$, $|\beta| = |\alpha|_0$, and (c) gives

$$|\beta - \gamma'|' \leqq \epsilon\,|\beta - \gamma'| \leqq \epsilon(|[\gamma]|_0 + |\gamma|).$$

Hence, if $N = 2|\gamma|$,

$$|\gamma|' \leqq |\gamma - \beta|' + |\beta| \leqq \rho\,|\gamma|/(r + 1) + N\epsilon + |[\gamma]|_0,$$

and (9.3) follows for $|\ |'$, and hence for $|\ |^\sharp_s$.

Next, by what has just been proved, we may apply the proof of Lemma 10a, showing that the molecular set functions are dense in the norm $|\ |^\sharp_s$. We shall show that the polyhedral set functions γ_A (A polyhedral) are dense also. It is sufficient to show that for any atomic set function $\gamma = \gamma_{p,\alpha}$ and $\epsilon > 0$ there is a polyhedral chain A such that $|\gamma - \gamma_A|^\sharp_s < \epsilon$.

By definition of $|\alpha|_0$ (I, 13.1), there are simple r-vectors $\alpha_1, \cdots, \alpha_m$ such that

$$\alpha = \sum \alpha_i, \qquad \sum |\alpha_i| < |\alpha|_0 + 1.$$

Choose ρ so that $\rho(2|\alpha|_0 + 1)/(r + 1) < \epsilon$. Choose simplexes σ_i in $U_\rho(p)$ and numbers a_i such that

$$\{a_i\sigma_i\} = \alpha_i, \qquad |a_i\sigma_i| = |\alpha_i|.$$

Set $A = \sum a_i\sigma_i$. Now

$$\{A\} = \sum \alpha_i = \alpha, \qquad |A| < |\alpha|_0 + 1.$$

Also

$$|\gamma - \gamma_A| \leqq |\gamma| + |A| < 2|\alpha|_0 + 1,$$

and (9.3) for $|\;|_s^\#$ gives (using (11.3))

$$|\gamma - \gamma_A|_s^\# \leqq \rho|\gamma - \gamma_A|/(r+1) < \epsilon,$$

as required.

Next we show that $|\gamma_A|_s^\# \leqq |\gamma_A|^\#$ for A polyhedral. Take any norm $|\;|'$ satisfying (a), (b) and (c). Set $|A|' = |\gamma_A|'$ for A polyhedral. Using (11.2) and the fact that $[\gamma_{\partial\sigma}] = \{\partial\sigma\} = 0$, $\gamma_{T_vA} = T_v\gamma_A$, we see that the norm $|A|'$ satisfies the conditions of (V, Theorem 8B); see (V, 8.8). Hence $|\gamma_A|' = |A|' \leqq |A|^\# = |\gamma_A|^\#$, proving the statement.

We know now that $|\gamma_A|_s^\# = |A|^\# = |\gamma_A|^\#$ for polyhedral chains A. Take any γ. For each integer i, there is a polyhedral chain A_i such that $|\gamma - \gamma_{A_i}|_s^\# < 1/2^i$ (see above). Now

$$|\gamma|_s^\# = \lim|\gamma_{A_i}|_s^\# = \lim|\gamma_{A_i}|^\# = |\gamma|^\#,$$

since $|\gamma - \gamma_{A_i}|^\# \leqq |\gamma - \gamma_{A_i}|_s^\# \to 0$. This completes the proof.

16. Expression for the sharp norm. We give an expression for the sharp norm of any compact $\gamma \in \mathbf{M}_r$. For a general γ, given $\epsilon > 0$, we can choose a compact Q such that $\bar{\gamma}(E - Q) < \epsilon$, and hence $|\gamma - \gamma_Q|^\# \leqq |\gamma - \gamma_Q| < \epsilon$ if $\gamma_Q = \chi_Q\gamma$; thus a limiting process will give the sharp norm of any γ.

Let $\gamma_1, \cdots, \gamma_m$ be compact. A *ρ-explosion* $\mathscr{E}$ of this set consists of the following:

(A) pure partitions (§ 15) $\sum_j\gamma_{ij}$ of each γ_i;
(B) a regrouping of the set of all γ_{ij} into a set of set functions β_{kl};
(C) a set of vectors v_{kl} such that, for some set of points p_k,

$$T_{v_{kl}}\beta_{kl} \subset \bar{U}_\rho(p_k) \qquad \text{(all } k, l).$$

(We could assume simply $\sum_l T_{v_{kl}}\beta_{kl} \subset \bar{U}_\rho(p)$, all k.)

Given the explosion $\mathscr{E}$, set

$$N(\gamma_1, \cdots, \gamma_m; \mathscr{E}) = \sum_{k,l} \frac{|v_{kl}|\,|\beta_{kl}|}{r+1} + \sum_k \left|\left[\sum_l \beta_{kl}\right]\right|_0. \tag{1}$$

Thus we push the γ_{ij} around, so that they lie in a set of sets of small diameter; we add the "amount of push" to the sum of the masses of the r-vectors of the corresponding set functions. The analogy with (V, 6.1) is clear.

Define

(2) $$N_\rho(\gamma_1, \cdots, \gamma_m) = \inf \{N(\gamma_1, \cdots, \gamma_m; \mathscr{E}): \quad \rho\text{-explosions } \mathscr{E}\}.$$

Since any ρ'-explosion is a ρ-explosion if $\rho' < \rho$, we have

(3) $$N_{\rho'}(\gamma_1, \cdots, \gamma_m) \geqq N_\rho(\gamma_1, \cdots, \gamma_m) \quad \text{if} \quad \rho' < \rho.$$

Hence we may define

(4) $$N(\gamma_1, \cdots, \gamma_m) = \lim_{\rho \to 0} N_\rho(\gamma_1, \cdots, \gamma_m).$$

Finally, for any compact γ, set

(5) $$|\gamma|_{\mathrm{e}}^{\#} = \inf \{N(\gamma_1, \cdots, \gamma_m): \quad \gamma = \sum \gamma_i, \quad \text{the } \gamma_i \text{ compact}\}.$$

Clearly

$$N_\rho(\gamma_1, \cdots, \gamma_1', \cdots) \leqq N_\rho(\gamma_1, \cdots) + N_\rho(\gamma_1', \cdots);$$

hence we see that $|\ |_{\mathrm{e}}^{\#}$ is a semi-norm. We prove

(6) $$|\gamma|^{\#} = |\gamma|_{\mathrm{e}}^{\#} \quad \text{if } \gamma \text{ is compact.}$$

We prove first that $|\gamma|_{\mathrm{e}}^{\#} \leqq |\gamma|^{\#}$ (γ compact). By Theorem 15A, it is sufficient to show that the semi-norm $|\ |_{\mathrm{e}}^{\#}$ satisfies (a), (b) and (c) of that theorem.

To prove (a), it is sufficient to show that $N_\rho(\gamma) \leqq |\gamma|$ for all ρ. Take a pure ρ-partition $\sum \gamma_i$ of γ; use no regrouping, and let the $v_{ij} = 0$. For the resulting ρ-explosion $\mathscr{E}$, we have (using (4.8))

$$N(\gamma; \mathscr{E}) = \sum_i |[\gamma_i]|_0 \leqq \sum_i |\gamma_i| = |\gamma|,$$

giving the result.

To prove (b), given γ and v, take any $\rho > 0$. Let $\gamma_1, \cdots, \gamma_m$ be a pure ρ-partition of $-\gamma$. Set

$$\gamma_{m+i} = T_v \gamma_i, \qquad v_i = v, \qquad v_{m+i} = 0 \qquad (i = 1, \cdots, m).$$

Group γ_i with γ_{m+i} for each i; this gives a ρ-explosion $\mathscr{E}$ of the pair $-\gamma$, $T_v\gamma$. Also, by (1),

$$N(-\gamma, T_v\gamma; \mathscr{E}) = \sum_{i=1}^{m} \frac{|v||\gamma_i|}{r+1} = \frac{|v||\gamma|}{r+1}.$$

Hence $N_\rho(-\gamma, T_v\gamma) \leqq |v||\gamma|/(r+1)$, and the same inequality holds for $N(-\gamma, T_v\gamma)$ and for $|T_v\gamma - \gamma|_{\mathrm{e}}^{\#}$, as required.

To prove (c), given p and $\epsilon > 0$, set $\zeta = (r+1)\epsilon$. Now take any $\gamma \subset \bar{U}_\zeta(p)$ such that $[\gamma] = 0$. Take any $\rho > 0$. Choose a pure partition

$\sum\gamma_i$ of γ, and points p_i, such that $\gamma_i \subset \bar{U}_\rho(p_i)$. Set $v_i = p - p_i$; then $|v_i| \leqq \zeta$, and $T_{v_i}\gamma_i \subset \bar{U}_\rho(p)$. Also $[\sum T_{v_i}\gamma_i] = [\gamma] = 0$. Grouping all the γ_i together, we have a ρ-explosion $\mathscr{E}$ of γ, such that

$$N(\gamma; \mathscr{E}) = \sum |v_i| |\gamma_i|/(r+1) \leqq \zeta |\gamma|/(r+1);$$

hence $N_\rho(\gamma) \leqq \epsilon |\gamma|$, and (c) follows.

We now prove that $|\gamma|^\# \leqq |\gamma|^\#_e$. By (9.1), it is sufficient to show that

(7) $$|[\omega\cdot\gamma]| \leqq |\gamma|^\#_e \quad \text{if} \quad |\omega|^\# \leqq 1.$$

Take any $\epsilon > 0$. Write

$$\gamma = \gamma_1 + \cdots + \gamma_m, \qquad N(\gamma_1, \cdots, \gamma_m) < |\gamma|^\#_e + \epsilon,$$

the γ_i being compact. Take any $\rho > 0$. Since $N_\rho(\gamma_1, \cdots) \leqq N(\gamma_1, \cdots)$, there is a ρ-explosion $\mathscr{E}$ such that

$$N(\gamma_1, \cdots, \gamma_m; \mathscr{E}) < |\gamma|^\#_e + \epsilon.$$

With the notations above, using (15.3), (9.4), (1) and (4.8) gives

$$\begin{aligned}
|[\omega\cdot\gamma]| &= \Big|\sum_{k,l}[\omega\cdot(\beta_{kl} - T_{v_{kl}}\beta_{kl})] + \sum_k \Big[\omega\cdot\sum_l T_{v_{kl}}\beta_{kl}\Big]\Big| \\
&\leqq \mathfrak{L}_0(\omega)\sum_{k,l}|v_{kl}||\beta_{kl}| \\
&\quad + \sum_k\Big[\rho\mathfrak{L}_0(\omega)\Big|\sum_l T_{v_{kl}}\beta_{kl}\Big| + |\omega|_0\Big|\Big[\sum_l T_{v_{kl}}\beta_{kl}\Big]\Big|_0\Big] \\
&\leqq \sum_{k,l}\frac{|v_{kl}||\beta_{kl}|}{r+1} + \frac{\rho}{r+1}\sum_{k,l}|\beta_{kl}| + \sum_k\Big|\Big[\sum_l \beta_{kl}\Big]\Big|_0 \\
&< |\gamma|^\#_e + \epsilon + \rho\sum_i |\gamma_i|/(r+1).
\end{aligned}$$

Since ρ is arbitrary, $|[\omega\cdot\gamma]| \leqq |\gamma|^\#_e + \epsilon$, and (6) follows, completing the proof.

17. Other expressions for the norm. We shall say the expression $\gamma = \sum\gamma_i$ of γ is a *Borel partition* of γ if there are Borel functions $\phi_1, \cdots, \phi_m$ in E^n such that

(1) $$0 \leqq \phi_i(p) \leqq 1, \qquad \sum_i \phi_i(p) = 1, \qquad \gamma_i = \phi_i\gamma.$$

We say the expression $\gamma = \sum\gamma_i$ is a *Lipschitz partition* if there are functions as above which are Lipschitz.

THEOREM 17A. *In the definition of $|\gamma|^\#_e$ (γ compact) in § 16, the pure partitions may be replaced by either Borel partitions or Lipschitz partitions.*

Let $|\gamma|_B^\sharp$ denote $|\gamma|_e^\sharp$, except that Borel partitions are used in place of pure partitions. Since pure partitions are Borel partitions, $|\gamma|_B^\sharp \leqq |\gamma|_e^\sharp$. Next, the proof of (16.7), using (4.7), shows that $|[\omega\cdot\gamma]| \leqq |\gamma|_B^\sharp$ for all sharp ω such that $|\omega|^\sharp \leqq 1$; hence $|\gamma|^\sharp \leqq |\gamma|_B^\sharp$. Since $|\gamma|_e^\sharp = |\gamma|^\sharp$, we have $|\gamma|_B^\sharp = |\gamma|_e^\sharp$.

Define $|\gamma|_L^\sharp$, using Lipschitz partitions; then clearly $|\gamma|_L^\sharp \geqq |\gamma|_B^\sharp$. The proof that $|\gamma|_L^\sharp \leqq |\gamma|^\sharp$ follows exactly the proof in § 16 that $|\gamma|_e^\sharp \leqq |\gamma|^\sharp$. Thus $|\gamma|_L^\sharp = |\gamma|_e^\sharp$.

In the definition of $|\gamma|_e^\sharp$, *we cannot use*

$$(2) \qquad |\gamma|_?^\sharp = \lim_{\rho\to 0}\Big[\inf\Big\{N_\rho(\gamma_1, \cdots)\colon \quad \sum \gamma_i = \gamma, \text{ the } \gamma_i \text{ compact}\Big\}\Big].$$

For, let σ be an $(r+1)$-simplex, and take any $\rho > 0$. If we write $\sigma = \sum\sigma_i$, diam $(\sigma_i) < \rho$, and use $\gamma = \gamma_{\partial\sigma}$, $\gamma_i = \gamma_{\partial\sigma_i}$, then considering each $\gamma_{\partial\sigma_i}$ separately and letting the $v_i = 0$ shows that $N_\rho(\gamma_1, \cdots) = 0$, and $|\gamma_{\partial\sigma}|_?^\sharp = 0$.

It is also not true in general that $|\gamma|_e^\sharp = N(\gamma)$. For, set $\gamma = \gamma_{A_0}$, with A_0 defined as follows. In E^2, let A and B be two perpendicular oriented line segments of length $\lambda = 0.1$, with an end in common. Let A' and B' be formed by translating A and B by the vector v of length $h = 1.8$. Set

$$A_0 = A + B - A' - B'.$$

Let C and C' be the corresponding hypotenuses, and D and D' the triangles thus formed. Now

$$\begin{aligned}|A_0|^\sharp &= |\partial D - \partial D' + (T_v C - C)|^\sharp \\ &\leqq |D| + |D'| + |v|\,|C|/2 = \lambda^2 + 2^{1/2}h\lambda/2 < 2^{1/2}\lambda.\end{aligned}$$

We shall prove that $N(\gamma_{A_0}) \geqq 2^{1/2}\lambda$; in fact, if $\rho = 0.1$, then $N_\rho(\gamma_{A_0}) \geqq 2^{1/2}\lambda$. Take any pure partition of γ_{A_0}; we may clearly partition further, giving partitions of $\gamma_{A'}$, $\gamma_{B'}$. This gives partitions of A, B, A', B'. Working with these, translations are grouped together into polyhedral chains $L_1, \cdots, L_s$, of diameter $<\rho$; we must study (16.1). Each L_i is formed by translating parts A_i, B_i, A_i', B_i' of A, B, A', B' respectively. Let h_i be the minimum length of vectors used for the pieces of A_i and B_i, and h_i', for the pieces of A_i' and B_i'. Clearly

$$h_i + h_i' > h - 2\rho - 2^{1/2}\lambda > 1.45 > 2^{1/2}.$$

(If L_i contains no parts of A_i' or B_i' for instance, we may set $h_i' = 1.45$ in what follows.) Set

$$2a_i = h_i\,|A_i| + h_i'\,|A_i'| + 2^{1/2}\,\big||A_i| - |A_i'|\big|,$$

$$2b_i = h_i\,|B_i| + h_i'\,|B_i'| + 2^{1/2}\,\big||B_i| - |B_i'|\big|;$$

we shall show that the right hand side of (16.1) is at least

$$a = \sum (a_i + b_i). \tag{3}$$

The algebraic inequality $(a - b)^2 \geqq 0$ gives $a^2 + b^2 \geqq (a + b)^2/2$; setting

$$\alpha_i = \left| \, | A_i | - | A'_i | \, \right|, \qquad \beta_i = \left| \, | B_i | - | B'_i | \, \right|,$$

we have therefore

$$| \{A_i + B_i - A'_i - B'_i\} |_0 = (\alpha_i^2 + \beta_i^2)^{1/2} \geqq 2^{-1/2}(\alpha_i + \beta_i),$$

and the statement follows.

Note next that using positive numbers, if $h + h' \geqq k$ and $\alpha \geqq \alpha'$, then $h\alpha + h'\alpha' \geqq k\alpha'$; hence, for positive numbers,

$$h\alpha + h'\alpha' + k \, | \alpha - \alpha' | \geqq k \sup \{\alpha, \alpha'\} \quad \text{if} \quad h + h' \geqq k.$$

Applying this gives

$$2a_i \geqq 2^{1/2} \sup \{ | A_i |, | A'_i | \};$$

similarly for $2b_i$. Hence

$$\begin{aligned} a &\geqq (2^{1/2}/2) \sum_i [\sup \{ | A_i |, | A'_i | \} + \sup \{ | B_i |, | B'_i | \}] \\ &\geqq (2^{1/2}/4) \sum_i (| A_i | + | A'_i | + | B_i | + | B'_i |) = 2^{1/2}\lambda. \end{aligned}$$

This proves $N(\gamma_{A_0}) \geqq 2^{1/2}\lambda > | A_0 |^\# = | \gamma_{A_0} |^\#$, as stated.

APPENDICES

Appendix I

Vector and linear spaces

Linear spaces are encountered throughout this book; in particular, the set of domains of integration (chains) and the set of integrands (cochains) form linear spaces (in fact, Banach spaces). The most important finite dimensional vector spaces which occur are the vector space $V = V(E^n)$ (§ 10) associated with Euclidean space E^n, and the spaces $V_{[r]}$ and $V^{[r]}$ of multivectors and multicovectors formed from V (Chapter I). We review here the basic properties which are assumed in the body of the book.

The treatment of affine spaces (§ 10) is somewhat novel; an axiomatic approach is given, based on the theory of vector spaces.

We recall a few common terms. A transformation f of a set S into a set S' is *one-one* if $f(p) = f(q)$ implies $p = q$; f is *onto* if the image $f(S)$ of S is the whole of S', i.e. each $p' \in S'$ is $f(p)$ for some $p \in S$. If an operation $p \circ q$ is defined in both S and S' (for instance, addition or multiplication), f is a *homomorphism* if $f(p \circ q) = f(p) \circ f(q)$. If S' is a group, the *kernel* of the homomorphism f is the set of elements of S going into the identity element of S'. A transformation of one algebraic system into another of the same nature is an *isomorphism* if it is one-one, and all the operations are preserved. A transformation of one metric space into another is an *isometry* if it preserves distances (hence it is one-one).

We use the symbol 0 equally well for the number zero or for the identity element in an additive group (for instance, the vector 0). Commonly λ, μ etc. denote ordered sets of integers; thus, $\lambda = (\lambda_1, \cdots, \lambda_r)$. The sum $\sum_\lambda$ denotes the sum over all sets λ, while

$$\sum_{(\lambda)} \text{ is the sum, with the restriction } \lambda_1 < \cdots < \lambda_r.$$

Certain numerical functions are useful:

$$\epsilon_\lambda = \epsilon_{\lambda_1 \cdots \lambda_n}, \qquad \delta_\lambda^\mu = \delta_{\lambda_1 \cdots \lambda_n}^{\mu_1 \cdots \mu_n};$$

ϵ_λ is 1 if $(\lambda_1, \cdots, \lambda_n)$ is an even permutation of $(1, \cdots, n)$, is -1 if the permutation is odd, and is 0 if the λ_i are not all distinct; δ_λ^μ is 1 if the λ_i and the μ_i are each distinct and one of these sets is an even permutation of the other, is -1 if the permutation is odd, and is 0 in all other cases.

Let $\hat{\imath}$ mean that this symbol is to be omitted. Thus

$$(p_1, \cdots, \hat{p}_i \cdots, p_n) = (p_1, \cdots, \hat{\imath}, \cdots, p_n)$$
$$= (p_1, \cdots, p_{i-1}, p_{i+1}, \cdots, p_n).$$

We write $G \approx G'$ if the groups or vector spaces G, G' are isomorphic.

1. Vector spaces. A *linear space* V is a set of elements, which may be called points, or vectors, which form an Abelian group under the operation of addition, and which may be multiplied by scalars (real numbers in this book), so that $a(u + v) = au + av$, $(a + b)u = au + bu$, $(ab)v = a(bv)$, $1v = v$. We assume known the theory of linear dependence. The *dimension* $\dim (V)$ of V is the maximum number of independent elements of V; V is a *vector space* if its dimension is finite. There is then a *base* $e_1, \cdots, e_n$ $(n = \dim (V))$ in V, such that any vector v of V may be written uniquely in the form $v = \sum v^i e_i$; the numbers v^i are the *components of* v relative to this base.

A set H of vectors of V *spans* the subspace H^* of V, consisting of all vectors v which can be written in the form $\sum a_i u_i$, the u_i in H.

Suppose $e_1, \cdots, e_n$ and $e'_1, \cdots, e'_n$ are bases in V. Then we may write

(1) $$e_i = \sum_j a_i^j e'_j, \qquad e'_i = \sum_j a'^j_i e_j.$$

Combining these gives $e_i = \sum_j a_i^j \sum_k a'^k_j e_k$ etc.; hence

(2) $$\sum_j a_i^j a'^k_j = \sum_j a'^j_i a_j^k = \delta_i^k,$$

i.e. The matrices $\| a_i^j \|$, $\| a'^j_i \|$ are inverses. Writing

$$v = \sum_i v'^i e'_i = \sum_j v^j e_j = \sum_j v^j \sum_i a_j^i e'$$

gives the law of transformation of components of a vector v:

(3) $$v'^i = \sum_j a_j^i v^j, \qquad v^i = \sum_j a'^i_j v'^j.$$

A vector space V (but not a linear space in general) has a *natural topology*: open and closed sets, also limit points of sequences, are defined, with the usual properties. To show this, choose a base $e_1, \cdots, e_n$, and use this to define a metric, as in § 9 below; different choices of bases give different metrics in general, but the same topology.

An important example of a vector space is *arithmetic n-space* $\mathfrak{A}^n$, whose elements are ordered sets $(a_1, \cdots, a_n)$ of n real numbers, with the obvious definition of addition and multiplication by scalars. It has a *natural base*:

(4) $$\bar{e}_1 = (1, 0, \cdots, 0), \bar{e}_2 = (0, 1, \ldots, 0), \cdots, \bar{e}_n = (0, 0, \cdots, n).$$

Using this base gives the natural metric of $\mathfrak{A}^n$ (§ 9).

2. Linear transformations. A transformation ϕ of a linear space V into another one W is *linear* if

$$\phi(u + v) = \phi u + \phi v, \qquad \phi(au) = a\phi u. \tag{1}$$

Clearly ϕ is one-one if and only if $\phi u = 0$ implies $u = 0$.

The set of linear transformations of V into W forms a linear space $L(V, W)$, under the definition

$$(\phi + \psi)v = \phi v + \psi v, \qquad (a\phi)v = a(\phi v). \tag{2}$$

If ϕ and ψ are linear transformations of V into W and of W into X respectively, then $\psi \circ \phi = \psi\phi$, defined by

$$(\psi\phi)(v) = \psi(\phi v), \tag{3}$$

is a linear transformation of V into X.

3. Conjugate spaces. Letting $\mathfrak{A} = \mathfrak{A}^1$ denote the real numbers, the *conjugate space* of the vector space V is the vector space

$$\bar{V} = L(V, \mathfrak{A}); \tag{1}$$

thus the elements of $\bar{V}$ are the real valued linear functions f in V. (For the case of normed linear spaces, see § 8 below.) We call the elements of $\bar{V}$ *covectors* of V.

If $e_1, \cdots, e_n$ is a base in V, the *dual base* $e^1, \cdots, e^n$ in $\bar{V}$ is the set of elements in $\bar{V}$ defined by

$$e^i(e_j) = \delta^i_j. \tag{2}$$

Since any element of $\bar{V}$ is defined by naming its values on the e_i, we see easily that the e^i actually form a base in $\bar{V}$. Hence also $\dim(\bar{V}) = \dim(V)$, and therefore V and $\bar{V}$ are isomorphic (but the isomorphism depends on the base chosen).

The *components* of the covector $f = \sum f_i e^i$ relative to the above base are the numbers f_i. Working out $e^i(v)$ and $f(e_i)$ and using (2) gives

$$v^i = e^i(v), \qquad f_i = f(e_i). \tag{3}$$

Also

$$f(v) = \sum f_i v^i. \tag{4}$$

For each vector v of V, set

$$\Phi_v(f) = f(v), \qquad f \in \bar{V}; \tag{5}$$

then Φ_v is a linear function in $\bar{V}$, i.e. an element of the conjugate $\bar{\bar{V}}$ of $\bar{V}$. Clearly Φ is a linear transformation of V into $\bar{\bar{V}}$.

LEMMA 3a. *The above* Φ *is an isomorphism of the vector space* V *onto* $\bar{\bar{V}}$.

REMARK. This may fail for Banach spaces; see § 14.

To show that Φ is one-one, suppose $v \neq 0$. There are elements $e_2, \cdots, e_n$ of V such that $v, e_2, \cdots, e_n$ is a base. There is an element f of $\bar{V}$ such that $f(v) = 1$, $f(e_i) = 0$; then $\Phi_v(f) = f(v) \neq 0$, proving $\Phi_v \neq 0$. Hence also $\dim(\Phi(V)) = \dim(V) = \dim(\bar{V}) = \dim(\bar{\bar{V}})$; therefore Φ is onto.

Take two bases in V as in § 1, and let $e^1, \cdots$ and $e'^1, \cdots$ be the dual bases. Using (1.1) and (1.2) gives

$$\Big(\sum_k a^i_k e^k\Big)(e'_j) = \Big(\sum_k a^i_k e^k\Big)\Big(\sum_l a'^l_j e_l\Big) = \sum_{k,l} a'^l_j a^i_k\, \delta^k_l = \delta^i_j;$$

a similar relation holds with e'^k, e_j. Hence

$$e'^i = \sum_j a^i_j e^j\,, \qquad e^i = \sum_j a'^i_j e'^j. \tag{6}$$

As in (1.3), we find the law of transformation of components of a covector:

$$f'_i = \sum_j a'^j_i f_j, \qquad f_i = \sum_j a^j_i f'_j. \tag{7}$$

Let ϕ be a linear transformation of the vector space V into the vector space W. Then there is a *dual* (or *conjugate*, or *adjoint*) linear transformation ϕ^* of $\overline{W}$ into $\bar{V}$, defined by

$$(\phi^* f)\,(v) = f(\phi v), \qquad f \in \overline{W},\ v \in V. \tag{8}$$

Each ϕ gives a ϕ^*; thus we have a transformation of $L(V, W)$ into $L(\overline{W}, \bar{V})$, which is clearly linear:

$$(\phi_1 + \phi_2)^* = \phi_1^* + \phi_2^*, \qquad (a\phi)^* = a\phi^*. \tag{9}$$

Also, if ϕ and ψ are linear transformations of V into W and of W into X respectively, then

$$(\psi\phi)^* = \phi^*\psi^*. \tag{10}$$

For, given $f \in \bar{X}$ and $v \in V$,

$$((\psi\phi)^* f)(v) = f((\psi\phi)v) = f(\psi(\phi v))$$

$$= (\psi^* f)\,(\phi v) = (\phi^*(\psi^* f))(v) = ((\phi^*\psi^*)f)(v).$$

4. Direct sums, complements. If X and Y are sets, their *Cartesian product* $X \times Y$ is the set of all pairs (x, y) or $x \times y$, with $x \in X$, $y \in Y$. If

V and W are linear spaces, their Cartesian product can be turned into a linear space $V \oplus W$, the *direct sum* of V and W, by the definitions

$$(1) \qquad (v_1, w_1) + (v_2, w_2) = (v_1 + v_2, w_1 + w_2), a(v, w) = (av, aw).$$

Similarly we may define $V_1 \oplus \cdots \oplus V_r$. Clearly $\dim (V \oplus W) = \dim (V) + \dim (W)$ for vector spaces. Note that $\mathfrak{A}^n \approx \mathfrak{A} \oplus \cdots \oplus \mathfrak{A}$ (n summands).

Let V_1 and V_2 be subspaces of the linear space V. If every element of V can be written uniquely in the form $v = v_1 + v_2$ $(v_1 \in V_1, v_2 \in V_2)$, we say V_1 and V_2 are *complementary* subspaces of V. Then clearly $V \approx V_1 \oplus V_2$. The linear transformation π defined by

$$(2) \qquad \pi(v_1 + v_2) = v_1 \qquad (v_1 \in V_1, v_2 \in V_2)$$

leaves all vectors of V_1 fixed and sends all vectors of V_2 into 0; it is the *projection* of V onto V_1, along V_2.

5. Quotient spaces. Let V_1 be a subspace of the linear space V. Write

$$(1) \qquad v \equiv w \bmod V_1 \quad \text{if} \quad w - v \in V_1.$$

This relation is reflexive, symmetric and transitive; hence the elements of V fall into equivalence classes with respect to this relation. Let C_v be the class containing v. Setting

$$(2) \qquad C_v + C_w = C_{v+w}, \qquad aC_v = C_{av},$$

turns these classes into the elements of a linear space, the *quotient*, *factor*, or *difference* space of V over V_1; we shall call it V *mod* V_1.

If V_1 has a complement V_2 in V, then each class C_v contains exactly one element πv in V_2; setting $\Phi(C_v) = \pi v$ defines an isomorphism of V mod V_1 onto V_2.

6. Pairing of linear spaces. The results of this section will be used in the next one. We say the linear spaces V, W are *paired* if a bilinear multiplication $w \cdot v$ is defined, with real numbers as values. For any subspace V_1 of V, the *annihilator* ann (V_1) is the subspace of W consisting of all w such that $w \cdot v = 0$ for $v \in V_1$; similarly for ann (W_1).

LEMMA 6a. *Let H and H^* be paired vector spaces, and suppose* ann (H) *and* ann (H^*) *contain only* 0 *in H^* and H respectively. Then setting $[\Phi(h^*)](h) = h^* \cdot h$ gives an isomorphism Φ of H^* onto $\bar{H}$.*

Clearly Φ is a linear transformation of H^* into $\bar{H}$. If $\Phi(h^*) = 0$, then $h^* \in$ ann (H); hence $h^* = 0$, and Φ is one-one; also $\dim (H^*) \leqq \dim (\bar{H})$. By the same reasoning,

$$\dim (\bar{H}) = \dim (H) \leqq \dim (\overline{H^*}) = \dim (H^*);$$

hence Φ is onto.

7. Abstract homology. We use notations similar to those in (App. II, 8). Given a vector space $\mathbf{C}$ and a linear transformation ∂ of $\mathbf{C}$ into itself such that $\partial\partial A = 0$, we shall define corresponding "homology" and "cohomology" spaces $\mathbf{H}$ and $\mathbf{H}^*$. Write $X \cdot A$ for $X(A)$ ($A \in \mathbf{C}$, $X \in \overline{\mathbf{C}}$).

Let $\mathbf{Z}$ and $\mathbf{B}$ be the kernel and image of ∂ respectively, let $d = \partial^*$ be the dual of ∂, mapping $\overline{\mathbf{C}}$ into $\overline{\mathbf{C}}$, and let $\mathbf{Z}^*$ and $\mathbf{B}^*$ be the kernel and image of d respectively. We prove

(1) $\mathbf{Z}^* = \operatorname{ann}(\mathbf{B})$, (2) $\mathbf{Z} = \operatorname{ann}(\mathbf{B}^*)$,

(3) $\mathbf{B}^* = \operatorname{ann}(\mathbf{Z})$, (4) $\mathbf{B} = \operatorname{ann}(\mathbf{Z}^*)$.

That $X \in \mathbf{Z}^*$ means $dX = 0$, i.e. $dX \cdot A = X \cdot \partial A = 0$ (all $A \in \mathbf{C}$), i.e. $X \in \operatorname{ann}(\mathbf{B})$. Hence also $\mathbf{B} \subset \operatorname{ann}(\mathbf{Z}^*)$. Given $A \in \operatorname{ann}(\mathbf{Z}^*)$, setting $\phi(dX) = X \cdot A$ gives clearly a linear function ϕ in $\mathbf{B}^*$, which has a linear extension ϕ_1 over $\overline{\mathbf{C}}$. Since $\overline{\overline{\mathbf{C}}} \approx \mathbf{C}$, there is an element $B \in \mathbf{C}$ with $Y \cdot B = \phi_1(Y)$, all $Y \in \overline{\mathbf{C}}$. Now

$$X \cdot \partial B = dX \cdot B = \phi_1(dX) = \phi(dX) = X \cdot A, \qquad \text{all } X \in \overline{\mathbf{C}},$$

proving $A = \partial B$, $A \in \mathbf{B}$. Thus (1) and (4) are proved. Interchanging the roles of $\mathbf{C}$ and $\overline{\mathbf{C}}$ and using the isomorphism $\overline{\overline{\mathbf{C}}} \approx \mathbf{C}$ gives the other two relations.

Since $\partial\partial A = 0$, $\mathbf{B} \subset \mathbf{Z}$. Also

$$ddX \cdot A = dX \cdot \partial A = X \cdot \partial\partial A = 0$$

(all $A \in \mathbf{C}$); hence $ddX = 0$, and $\mathbf{B}^* \in \mathbf{Z}^*$. Therefore we may define

(5) $$\mathbf{H} = \mathbf{Z} \bmod \mathbf{B}, \qquad \mathbf{H}^* = \mathbf{Z}^* \bmod \mathbf{B}^*$$

Define a pairing between $\mathbf{H}$ and $\mathbf{H}^*$ by

(6) $$h^* \cdot h = X \cdot A \quad \text{if} \quad X \in h^*, \qquad A \in h.$$

From (1) and (2) we see at once that this is well defined. Suppose $h^* \cdot h = 0$, all $h \in \mathbf{H}$. Take $X \in h^*$; then $X \cdot A = 0$, all $A \in Z$, and by (3), $X \in \mathbf{B}^*$, and $h^* = 0$; hence $\operatorname{ann}(\mathbf{H}) = 0$. Similarly $\operatorname{ann}(\mathbf{H}^*) = 0$. Therefore, by Lemma 6a, the pairing defines isomorphisms

(7) $$\mathbf{H}^* \approx \overline{\mathbf{H}}, \qquad \mathbf{H} \approx \overline{\mathbf{H}^*}.$$

8. Normed linear spaces. Let V be a linear space. Any real valued function $|v|$ in V with the properties

(1) $$|v + w| \leqq |v| + |w|,$$

(2) $$|av| = |a||v|,$$

we call a *semi-norm* in V.

LEMMA 8a. *For any semi-norm,*

$$|0| = 0, \tag{3}$$

$$|v| \geqq 0. \tag{4}$$

The first relation follows on setting $a = 0$ in (2), and the second from $2|v| = |v| + |-v| \geqq |v - v| = 0$.

The semi-norm $|v|$ is a *norm* provided

$$|v| \neq 0 \quad \text{if} \quad v \neq 0. \tag{5}$$

If this holds, we may define the *distance* between two vectors v and w by

$$\text{dist}(v, w) = |w - v|; \tag{6}$$

this turns V into a metric space. (The triangle inequality follows at once from (1).) We now suppose that V is normed.

A real bounded linear function f in V is a real valued linear function such that

$$|f| = \sup\{f(v)\colon\ v \in V, |v| = 1\} \tag{7}$$

is finite. Then $|f|$ is the smallest number such that

$$|f(v)| \leqq |f|\,|v|, \qquad \text{all } v \in V. \tag{8}$$

Because of (8), *f is continuous in V.*

LEMMA 8b. *For any $v \in V$ there is a bounded linear function f in V such that*

$$|f| = 1, \qquad f(v) = |v|. \tag{9}$$

This is an immediate consequence of the Hahn-Banach extension theorem; see Banach, p. 27.

REMARK. In general, f is not unique; see the end of (I, 13).

The set of bounded linear functions in V forms a linear space, in which clearly (1) and (2) hold; (5) is an immediate consequence of the lemma. Hence these functions form a normed linear space, the *conjugate space* $\bar{V}$ of V.

LEMMA 8c. *With the above norms,*

$$|v| = \sup\{f(v)\colon\ f \in \bar{V}, |f| = 1\}. \tag{10}$$

Let $|v|'$ denote the right hand side. Because of (8), $|v|' \leqq |v|$; but by Lemma 8b, $|v|' \geqq |v|$.

Let V be a normed vector space. Then any linear function f in V is bounded; hence V, considered simply as a vector space, is exactly the conjugate space of V as in § 3.

LEMMA 8d. *Given the normed linear space V, let Φ be the linear transformation of V into $\bar{\bar{V}}$ defined by* (3.5). *Then Φ is an isomorphism into.*

Because of Lemma 8b, the proof of Lemma 3a applies.

Suppose V and $\bar{V}$ are conjugate as in § 3. Let $|v|$ and $|f|$ be norms in these spaces. We call these norms *conjugate* if (7) and (10) hold. By Lemma 8c, it is sufficient to prove (7). We now show that it is also sufficient to prove (10).

LEMMA 8e. *Let V and $\bar{V}$ be conjugate vector spaces, and let them be normed. If* (10) *holds, then so does* (7), *and the norms are conjugate.*

The norm $|f|$ in $\bar{V}$ determines a norm in $\bar{\bar{V}}$; with Φ defined as above, the norm of any element Φ_v of $\bar{\bar{V}}$ is, by (10),

$$(11) \quad |\Phi_v| = \sup\{\Phi_v(f): |f| = 1\} = \sup\{f(v): |f| = 1\} = |v|.$$

Therefore, applying Lemma 8c to the pair $\bar{V}$, $\bar{\bar{V}}$,

$$|f| = \sup\{\Phi_v(f): |\Phi_v| = 1\} = \sup\{f(v): |v| = 1\}$$

as required.

9. Euclidean linear spaces. The linear space V is *Euclidean* if a *scalar product* $u \cdot v$, with real values, is defined, such that:

(1) The product is bilinear and symmetric.

$$(2) \quad v \cdot v > 0 \text{ if } v \neq 0.$$

Define *norm* and *distance* by

$$(3) \quad |v| = (v \cdot v)^{1/2}, \qquad \operatorname{dist}(v, w) = |w - v|.$$

We prove the Schwarz inequality:

$$(4) \quad |u \cdot v| \leqq |u|\,|v|.$$

We may suppose $|v| = 1$. Setting $a = u \cdot v$, we find

$$0 \leqq |u - av|^2 = |u|^2 - 2a(u \cdot v) + a^2 = |u|^2 - (u \cdot v)^2,$$

giving (4). (Another proof may be found in (I, 12.8).)

Using (4) gives

$$|u + v|^2 = |u|^2 + 2(u \cdot v) + |v|^2 \leqq (|u| + |v|)^2,$$

proving (8.1). Hence we see that $|v|$ is actually a norm in V, and therefore V is metric.

The vectors u, v are *orthogonal* if $u \cdot v = 0$; if $u \neq 0$, $v \neq 0$, we call them *perpendicular*. The sets P, Q of vectors are *orthogonal* if $u \cdot v = 0$ for all $u \in P$, $v \in Q$. The set of vectors $v_1, \cdots, v_r$ is an *orthonormal set* if $v_i \cdot v_j = \delta_i^j$; thus they are mutually orthogonal unit vectors.

Let $e_1, \cdots, e_n$ be an orthonormal base in V (supposed of finite dimension). Then writing vectors in terms of their components (§ 1), we find

$$u \cdot v = \Big(\sum_i u^i e_i\Big) \cdot \Big(\sum_j v^j e_j\Big) = \sum_i u^i v^i, \tag{5}$$

$$|v| = \Big|\sum_i v^i e_i\Big| = \Big[\sum_i (v^i)^2\Big]^{1/2}. \tag{6}$$

Note that the base $\bar{e}_1, \cdots, \bar{e}_n$ in $\mathfrak{A}^n$ (§ 1) is orthonormal.

Given a vector space V, we may choose a base $e_1, \cdots, e_n$ in it, and define scalar products by (5), thus turning V into a Euclidean vector space.

LEMMA 9a. *The mapping ϕ of the Euclidean vector space V into $\bar{V}$ defined by $\phi_u(v) = u \cdot v$ is an isometry onto; setting $\phi_u \cdot \phi_v = u \cdot v$ makes $\bar{V}$ Euclidean, with the same norm as in* (8.7).

Since the scalar product is bilinear, $\phi_u \in \bar{V}$, and ϕ is linear. If $u \neq 0$, then $\phi_u(u) = |u|^2 \neq 0$, $\phi_u \neq 0$; hence ϕ is one-one. Since $\dim(\bar{V}) = \dim(V)$, ϕ is onto. The norm in $\bar{V}$ is

$$|\phi_u| = \sup\{\phi_u(v): |v| = 1\} = \sup\{u \cdot v: |v| = 1\}.$$

By (4), $|\phi_u| \leqq |u|$. If $u \neq 0$, setting $v = u/|u|$ shows that $|\phi_u| \geqq |u|$. Hence $|\phi_u| = |u|$, and (8.6) shows that ϕ is an isometry.

Note that if $v \neq 0$, then setting $u = v/|v|$, $f = \phi_u$ satisfies (8.9) and is the only element of $\bar{V}$ with this property.

The scalar product $\phi_u \cdot \phi_v = u \cdot v$ satisfies (1). Also $(\phi_u \cdot \phi_u)^{1/2} = (u \cdot u)^{1/2} = |u| = |\phi_u|$. Finally, if $\phi_u \neq 0$, then $u \neq 0$, $|u| \neq 0$, and $|\phi_u| \neq 0$, proving (2).

Let $e_1, \cdots, e_n$ be an orthonormal base in V. Since $e_i \cdot e_j = \delta_i^j$, the set $\phi_{e_1}, \cdots, \phi_{e_n}$ is the dual base.

Since $\phi_u(e_i) = u \cdot e_i = \phi_u \cdot \phi_{e_i} = e^i \cdot \phi_u$, $f(e_i) = e^i \cdot f$. Also $e^i(v) = \phi_{e_i}(v) = e_i \cdot v$. Hence, by (3.3),

$$v^i = e^i(v) = e_i \cdot v, \quad f_i = f(e_i) = e^i \cdot f, \quad \text{with an orthonormal base.} \tag{7}$$

Also (always with an orthonormal base)

$$u \cdot v = \sum_i u^i v^i, \qquad f \cdot g = \sum_i f_i g_i. \tag{8}$$

Since $\phi_v = \sum v^i \phi_{e_i} = \sum v^i e^i$, we have

$$f_i = v^i \quad \text{if} \quad f = \phi_v. \tag{9}$$

10. Affine spaces. An affine space is, roughly speaking, a vector space, but without a particular vector being chosen as zero. It may be defined as follows.

An *affine space* E, of dimension n, is a system composed of a set of points p, a vector space $V = V(E)$ of dimension n, and operations $p + v$ of V on E, called *translations*, with the following properties.

$$(p + u) + v = p + (u + v). \tag{1}$$

$$p + 0 = p. \tag{2}$$

$$p + v \neq p \quad \text{if} \quad v \neq 0. \tag{3}$$

$$\text{For each } p \text{ and } q \text{ there is a } v \text{ such that } p + v = q. \tag{4}$$

In other words, V is a simply transitive group of transformations on E. We prove:

$$\text{If } p + u = p + v, \quad \text{then} \quad u = v. \tag{5}$$

For

$$\begin{aligned} p = p + 0 = p + (u + (-u)) &= (p + u) + (-u) \\ &= (p + v) + (-u) = p + (v - u), \end{aligned}$$

and (3) gives $v - u = 0$.

Because of (5), the vector v in (4) is unique; we call it $q - p$.

If $p + v = q$, then $q + (-v) = p + (v + (-v)) = p$; hence the following relations are equivalent:

$$p + v = q, \quad v = q - p, \quad q + (-v) = p, \quad -v = p - q. \tag{6}$$

Since, by (1) and (6),

$$\begin{aligned} p + [(p' - p) + (p'' - p')] &= [p + (p' - p)] + (p'' - p') \\ &= p' + (p'' - p') = p'', \end{aligned}$$

using (6) again gives

$$p'' - p = (p'' - p') + (p' - p). \tag{7}$$

Choose a fixed point $O \in E$, which may be called the "origin", and set $\phi(v) = O + v$ $(v \in V)$. This sets up a one-one correspondence between the vectors of V and the points of E.

Example. Let V^{n+1} be a vector space, and V^n a subspace, of the dimensions shown. Choose a vector $O \in V^{n+1}$ not in V^n, and let E be the set of all vectors of the form $p = O + v$ $(v \in V^n)$. Then E is an affine space, with V^n as the associated vector space.

Any vector in V^{n+1} may be written in the form

$$tp + v \qquad (t \in \mathfrak{A}, \quad p \in E, \quad v \in V^n), \tag{8}$$

with t uniquely defined; also

$$tp + v = tp' + v' \quad \text{if and only if} \quad v' - v = t(p - p'). \tag{9}$$

We now show that all affine spaces are of the above character. Given E and $V = V(E)$, define V' to be the set of all expressions of the form (8), subject to the rule (9). Clearly the relation (9) is reflexive, symmetric and transitive; hence V' is a well defined set of elements. Let tp denote $tp + 0$.

To turn V' into a vector space, choose $O \in E$. Note that (9) gives

$$tp + v = tO + v', \qquad v' = t(p - O) + v. \tag{10}$$

Thus any element of V' may be written in the form $tO + v'$; this expression is unique, by (9). Set

$$a(tO + v) = (at)O + av, \tag{11}$$

$$(tO + v) + (t'O + v') = (t + t')O + (v + v'). \tag{12}$$

Take any $O' \in E$; set $w = O' - O$. Then these relations and (9) and (10) give

$$a(tO') = a(tO + tw) = (at)O + atw = (at)O',$$

$$tO' + t'O' = (tO + tw) + (t'O + t'w) = (t + t')O + (tw + t'w)$$

$$= (t + t')O + (t + t')w = (t + t')O',$$

showing that the operations (11) and (12) are independent of the choice of O.

Define the direct sum V^* (§ 4) and the one-one mapping of V^* onto V' by

$$V^* = \mathfrak{A} \oplus V, \qquad \psi(t, v) = tO + v. \tag{13}$$

Since V^* is a vector space and the operations are preserved by ψ, V' is proved to be a vector space.

Setting $\phi(p) = 1p$ puts E into V', bringing the situation back to that of the example.

The subset E' of E is an *affine subspace* of E if there is a vector subspace V' of $V(E)$ such that the operations $p' + v'$ ($p' \in E'$, $v' \in V'$) make E' affine, with $V' = V(E')$.

11. Barycentric coordinates. We define certain "linear combinations" of points of the affine space E; that in (1) is a point of E, and that in (2) is a vector of $V = V(E)$. With a fixed $O \in E$, set (for any k)

$$\sum_{i=0}^{k} a_i p_i = O + \sum_{i=0}^{k} a_i(p_i - O) \quad \text{if} \quad \sum_{i=0}^{k} a_i = 1, \tag{1}$$

$$\sum_{i=0}^{k} a_i p_i = \sum_{i=0}^{k} a_i(p_i - O) \qquad \text{if} \quad \sum_{i=0}^{k} a_i = 0. \tag{2}$$

Define V' as in § 10. The above have meaning in V', and are true relations; hence, considering E as imbedded in V', the right hand sides (which have direct meaning in E and V) are independent of the choice of O. (A direct proof is easy to give.)

The point p in (1) is the center of mass of a set of masses, in amount a_i at p_i.

The points $p_0, \cdots, p_k$ of E are *dependent* if they are contained in an affine subspace of E of dimension $< k$; otherwise, they are *independent.*

If the p_i are independent, different sets a_i in (1) determine different points p; the a_i are the *barycentric coordinates* of p in terms of the p_i. In particular, the point

$$(1 - t)p + tq = p + t(q - p) \tag{3}$$

runs from p to q as t runs from 0 to 1.

For another example, let p_0, p_1, p_2 be the vertices of a triangle. Operating in V', we see that the point two thirds of the way from p_0 to the mid point of p_1p_2 is

$$q = \tfrac{1}{3}p_0 + \tfrac{2}{3}(\tfrac{1}{2}p_1 + \tfrac{1}{2}p_2) = \tfrac{1}{3}p_0 + \tfrac{1}{3}p_1 + \tfrac{1}{3}p_2.$$

The symmetry of the expression shows that the medians of the triangle intersect at q.

Any non-void subset Q of E *spans* a subspace E' of E, namely, the smallest subspace of E containing Q. If $p_0, \cdots, p_k$ is a maximal independent set of points in Q, then E' consists of all the points in (1).

There is clearly a *natural topology* in E, just as in $V(E)$ (see § 1).

12. Affine mappings. Let E and E' be affine spaces, let Q be a subset of E, and let f be a transformation of Q into E'. We say f is *affine* if

$$f\left(\sum a_i p_i\right) = \sum a_i f(p_i) \qquad \left(\sum a_i = 1\right), \tag{1}$$

provided each p_i and $\sum a_i p_i$ are in Q.

Suppose f is affine in Q. Then it is continuous (hence it is a mapping). Let $p_0, \cdots, p_k$ be a maximal set of independent points of Q. Set

$$F(p_i - p_0) = f(p_i) - f(p_0); \tag{2}$$

this may be extended to define a linear transformation F of the vector space $V(Q)$ spanned by all vectors $q - p_0$, and hence by all vectors $q - p$ $(p, q \in Q)$ into $V(E')$. We show that

$$f(q) - f(p) = F(q - p), \qquad p, q \in Q. \tag{3}$$

For we may write $p = \sum a_i p_i$, $q = \sum b_i p_i$; then

$$f(q) - f(p) = \sum (b_i - a_i) f(p_i) = \sum (b_i - a_i)[f(p_i) - f(p_0)]$$

$$= \sum (b_i - a_i) F(p_i - p_0) = F\left[\sum (b_i - a_i)(p_i - p_0)\right] = F(q - p).$$

It follows that

(4) $f(q) - f(p) = \nabla f(p^*, q - p), \qquad p, q \in Q, \quad p^* \in \operatorname{int}(Q),$

and $\nabla f(p, v)$ (v fixed) *is constant in* int (Q).

Conversely, suppose there exists a linear transformation F of $V(Q)$ into $V(E')$ such that (3) holds. Then f is affine in Q. For if $p_0, \cdots, p_k$ and $p = \sum a_i p_i$ are in Q, then

$$\sum a_i f(p_i) - f(p) = \sum a_i [f(p_i) - f(p)] = \sum a_i F(p_i - p)$$
$$= F\left[\sum a_i(p_i - p)\right] = F\left(\sum a_i p_i - p\right) = 0.$$

An *affine coordinate system* χ in the affine space $E = E^n$ is an affine transformation of $\mathfrak{A}^n$ onto E. With the notations of § 1, set

(5) $O = \chi(0), \qquad e_i = \nabla\chi(0, \bar{e}_i).$

Then for any $x = (x^1, \cdots, x^n)$ in $\mathfrak{A}^n$, (4) gives

(6) $\chi(x) = O + \nabla\chi\left(0, \sum x^i \bar{e}_i\right) = O + \sum x^i e_i;$

the x^i are called the *coordinates* of $\chi(x)$.

13. Euclidean spaces. Let E be an affine space such that $V = V(E)$ is Euclidean (§ 9). Then we say E is a *Euclidean space*. Now $|q - p|$ is defined for $p, q \in E$, and is the *distance* from p to q.

An *orthonormal coordinate system* χ in $E = E^n$ is an affine coordinate system such that the corresponding mapping $\nabla\chi(x)$ (which is independent of x) of $V(\mathfrak{A}^n) = \mathfrak{A}^n$ into $V(E)$ preserves distances. An elementary argument shows that it preserves scalar products also; hence, with the notations of (12.5),

(1) $e_i \cdot e_j = \bar{e}_i \cdot \bar{e}_j = \delta_i^j.$

14. Banach spaces. A *Banach space* is a normed linear space V (§ 8) which is complete. That is, if $v_1, v_2, \cdots$ is a sequence in V such that $|v_j - v_i| \to 0$ as $i, j \to \infty$, then there is an element v of V such that $v_i \to v$.

If V_0 is a normed linear space, we may complete it, forming a Banach space V, as follows. Call two Cauchy sequences $(v_1, v_2, \cdots)$ and $(v_1', v_2', \cdots)$ equivalent if $\lim |v_i' - v_i| = 0$; the elements of V are the equivalence classes of Cauchy sequences in V_0. Let the norm of the equivalence class of $(v_1, v_2, \cdots)$ be $\lim |v_i|$. Imbed V_0 in V by setting $\phi(v) =$ equivalence class of $(v, v, \cdots)$. Now V is complete, and V_0 is a dense subset of V.

The conjugate space $\bar{V}$ of a normed linear space V (§ 8) is complete, and hence is a Banach space. For let $f_1, f_2, \cdots$ be a Cauchy sequence in $\bar{V}$. For any $v \in V$, since $|f_j(v) - f_i(v)| \leqq |f_j - f_i|\,|v| \to 0$, the sequence

$f_1(v), f_2(v), \cdots$ is Cauchy, and has a limit, which we call $f(v)$. Clearly f is a bounded linear function in V, and $\lim |f_i - f| = 0$.

Lemma 14a. *Let V_0 be a normed linear space, with completion V. Then any real bounded linear function f in V_0 is uniquely extendable to be a bounded linear function in V. This defines a one-one isometric linear mapping of the conjugate space $\bar{V}_0$ onto $\bar{V}$. The norm in $\bar{V}$ may be determined by using V_0 alone:*

$$(1) \qquad |f| = \sup \{f(v): \ v \in V_0, |v| = 1\}, \qquad f \in \bar{V}.$$

Let f be bounded and linear in V_0. For each $v = \lim v_i \in \dot{V}$ (the v_i in V_0), set $f(v) = \lim f(v_i)$; this is uniquely defined, and is clearly bounded and linear in V. The rest of the lemma follows easily.

A metric space is *separable* if it contains a sequence of points $p_1, p_2, \cdots$ which is dense in the space. Any vector space is separable. (With an affine coordinate system, the set of vectors whose components are all rational is denumerable.)

The Banach space V is *reflexive* if the mapping Φ of V into $\bar{\bar{V}}$ of Lemma 8d is onto. The conjugate space of a non-separable space is non-separable; hence if V is separable and $\bar{V}$ is not, then V is not reflexive. This occurs in (V, 18).

15. Semi-conjugate spaces. Let V be a linear space with a semi-norm $|v|$ (§ 8). Let V^* be the subspace of V containing those v with $|v| = 0$, and set $V' = V \bmod V^*$ (§ 5). If $v_1 - v_2 \in V^*$, then $|v_1| = |v_2|$; hence a norm is determined in V'. We may define bounded linear functions f in V as before, and use (8.7) to define $|f|$. These functions form what might be called the *semi-conjugate space* $\bar{V}$ of V. For any $f \in \bar{V}$, $|f(v)| \leqq 0$ and hence $f(v) = 0$ for all $v \in V^*$; therefore f defines a function f' in V', which is clearly linear and bounded. Setting $|f'| = |f|$ defines $|f'|$ uniquely. Clearly this definition is that defining the norm in the conjugate space $\bar{V}'$ of V'. We thus have a natural linear transformation of $\bar{V}$ into $\bar{V}'$.

Lemma 15a. *The above transformation is an isomorphism of $\bar{V}$ onto $\bar{V}'$. If the semi-norm in V is a norm, this transformation is the identity transformation of $\bar{V}$ onto $\bar{V}$.*

The proof is simple.

The following lemma will be used in (V, 8).

Lemma 15b. *For each element h of a set H, let $|v|_h$ be a semi-norm. Then*

$$(1) \qquad |v| = \sup \{|v|_h: \ h \in H\},$$

if finite for all $v \in V$, is a semi-norm.

Since $|v + w|_h \leqq |v|_h + |w|_h \leqq |v| + |w|$, all $h \in H$, (8.1) holds. We prove $|av| \leqq |a||v|$ similarly; then also, if $a \neq 0$ and $b = 1/a$, $|b(av)| \leqq |b||av|$, hence $|av| \geqq |a||v|$, and (8.2) follows.

Appendix II

Geometric and Topological Preliminaries

A good portion of the concepts and results in this appendix, in particular, Sections 1–3, 5–7, and 10–13, are used freely in the body of the book (starting with Chapter III). Section 4 is used in Chapter VII, Sections 8 and 9 in IV C and the end of VII, and Sections 14 through 16, in IV B. Proofs of some elementary facts are sketched or omitted; other proofs are given in detail. Though some small parts of the book clearly have algebraic topology as a foundation, the few sections of this appendix devoted to the subject give a sufficient background. In a few places, proofs are given with the help of results in early chapters.

The elements of point set theory are assumed. The following definitions and notations will be used. The sets will lie in a space S, generally an affine or Euclidean space or a linear space. Write $Q_1 \subset Q_2$ if Q_1 is a subset of Q_2; $p \in Q$ if p is an element of Q. Write $Q_1 \cap Q_2$, $Q_1 \cup Q_2$, $Q_1 - Q_2$, for the intersection, the union, and the difference (set of points in the first set but not in the second), of Q_1 and Q_2 respectively; $\bigcup_i Q_i$ and $\bigcap_i Q_i$ mean the union and the intersection of the Q_i respectively. Write $\bar{Q}$ for the closure of Q, fro $(Q) = \bar{Q} \cap \overline{S - Q}$ for the frontier of Q, and int $(Q) = Q -$ fro (Q) for the interior of Q. For cells σ, int (σ) is defined with reference to the plane of σ; see § 1.

A mapping f of Q into S' is a continuous function with domain Q whose range $f(Q)$ lies in S'. The set $f^{-1}(Q')$ is the set of all points p such that $f(p) \in Q'$. We suppose the reader is familiar with the basic properties of mappings and of compact sets. Recall that a subset of Euclidean space is compact if and only if it is bounded and closed.

In a metric space, dist (p, q) is the distance from p to q; this is $|q - p|$ if the space is Euclidean. The distance from the set P to the set Q is dist $(P, Q) = \inf \{\text{dist}\,(p, q)\}$ (notation in (App. III)) for $p \in P$, $q \in Q$. The diameter diam (Q) of Q is $\sup \{\text{dist}\,(p, q)\}$ for $p, q \in Q$. The ζ-neighborhood of Q is $U_\zeta(Q)$, the set of all p such that dist $(p, Q) < \zeta$; let $\bar{U}_\zeta(Q)$ denote its closure. A set of the form $U_\zeta(p)$ or $\bar{U}_\zeta(p)$ is called an *open ball* or *closed ball* respectively. Let $\text{int}_\zeta\,(Q)$ denote the set of all p such that $\bar{U}_\zeta(p) \subset \text{int}\,(Q)$; thus $\text{int}_\zeta(Q) = S - \bar{U}_\zeta(S - Q)$.

A *homotopy* of f_0 into f_1 is a set of mappings f_t $(0 \leqq t \leqq 1)$ such that $F(t \times p) = f_t(p)$ is continuous in $I \times S$ (I being the unit interval); f_0 and f_1 are *homotopic* if this homotopy exists.

1. Cells, simplexes. A *closed half space* in an affine space E of dimension n is the set of points which lie on a given side of an affine subspace P of E of dimension $n - 1$, together with P itself. A *convex polyhedral cell* σ in E, or *cell* for short, is a non-void bounded (closed) subset of E expressible as the intersection of a finite set of closed half spaces. The *plane* $P(\sigma)$ of σ is the smallest affine subspace containing σ; the *dimension* $\dim(\sigma)$ of σ is the dimension of $P(\sigma)$. If this is r, we call the cell an *r-cell*, and denote it by σ^r. Note that if σ_1 and σ_2 are cells, so is $\sigma_1 \cap \sigma_2$, if non-void.

The frontier fro (σ) of σ, considered as a subset of $P(\sigma)$, is the *boundary* $\partial\sigma$ of σ; if σ is oriented, then $\partial\sigma$ is the set of points lying on the chain which is called $\partial\sigma$ in § 7 below. The *interior* int (σ) is $\sigma - \partial\sigma$. If $\dim(\sigma_1) = \dim(\sigma_2)$, say σ_1 and σ_2 are *non-overlapping* if int $(\sigma_1) \cap$ int $(\sigma_2) = 0$.

We may express $\partial\sigma$ as the union of a finite set of $(n - 1)$-cells, the $(n - 1)$-*faces* of σ. Each of these has faces, etc.; the set of all these we call the *proper faces* of σ. We consider σ as an *improper face* of itself.

The *vertices* of σ are the faces of dimension 0; they are the points of σ which are interior to no segment lying in σ. If $p_0, \cdots, p_r$ are the vertices of σ, then σ is the smallest convex set which contains these points; the points of σ are expressible in the form

$$p = \mu_0 p_0 + \cdots + \mu_r p_r, \qquad \text{each } \mu_i \geqq 0, \quad \sum \mu_i = 1; \tag{1}$$

see (App. I, 11). Allowing the μ_i to be negative gives all points of $P(\sigma)$. The *edges* of σ are the 1-faces of σ.

A *simplex* σ in E is a set of points expressible in the form (1), the p_i being independent (App. I, 11). Then $\dim(\sigma) = r$, and the expression (1) is unique. The μ_i are the *barycentric coordinates* of p in terms of the p_i. We write $\sigma = p_0 \cdots p_r$. Then the simplexes $p_{\lambda_0} \cdots p_{\lambda_k}$ are the faces of σ, and the p_i are its vertices.

If σ is a cell, its *center* is the center of mass of its vertices; if $p_0, \cdots, p_k$ are the vertices, then the center is

$$p_\sigma = a p_0 + \cdots + a p_k, \qquad a = 1/(k+1). \tag{2}$$

Let f be an affine mapping (App. I, 12) of the r-simplex $\sigma = p_0 \cdots p_r$ into an affine space E'; then $f(\sigma)$ is an r-simplex in E', if the $q_i = f(p_i)$ are independent; this happens if and only if $P(f(\sigma))$ is of the same dimension r as σ, or, if and only if, considering f in int (σ) only, the Jacobian J_f is $\neq 0$ there (II, 6). If the q_i are dependent, we say f is *degenerate* in σ, or, $f(\sigma)$ is *degenerate*. For more general cells σ, we say $f(\sigma)$ is *degenerate* if $J_f = 0$ in int (σ).

If the simplex σ is in Euclidean space, it is easy to see that diam (σ) is the length of the longest edge of σ.

2. Polyhedra, complexes. A *polyhedron* Q in an affine space E is a (closed) point set which is expressible as the union of a finite set of cells; its *dimension* dim (Q) is the largest of the dimensions of the cells. If there is an expression in which all the cells are of the same dimension as E, we call Q a *polyhedral region* in E.

One could consider polyhedra abstractly, without reference to a containing space; we do not need to do this.

A *complex* K is a finite set S of cells, with the following properties. If $\sigma \in S$, then each face of σ is the union of cells of S. No int (σ) intersects cells of lower dimension of S. Each intersection $\sigma \cap \sigma'$ is the union of cells of S.

We shall use only complexes in which each face of a cell σ of S is itself a cell of S; thus we shall not subdivide the proper faces of σ into smaller cells.

The union of the cells of K form a polyhedron, which we also call K. Note that each point of K is in exactly one of the sets int (σ), $\sigma \in S$.

A *simplicial complex* K is a complex whose cells are all simplexes, each face of a simplex of K being a simplex of K. If $p_1, p_2, \cdots$ are the vertices of K, then each simplex of K is of the form $p_{\lambda_0} \cdots p_{\lambda_r}$. Each point of K may be written uniquely in the form

$$(1) \qquad p = \sum \mu_i(p) p_i, \qquad \text{each } \mu_i(p) \geqq 0, \quad \sum \mu_i(p) = 1,$$

with the condition that if $p \in p_{\lambda_0} \cdots p_{\lambda_r}$, then $\mu_j(p) = 0$ for $j \neq \lambda_0, \cdots, \lambda_r$. The $\mu_j(p)$ are continuous functions of p in K; they are the *barycentric coordinates* of p in K.

The *star* St (σ) of σ in K is the point set consisting of all int (σ') such that σ is a face of σ'; it is an open set in the space consisting of the points of K. The *closed star* $\overline{\text{St}}$ (σ) is the closure in K of St (σ); the *star boundary* ∂St (σ) is $\overline{\text{St}}$ (σ) — St (σ).

Note that in a simplicial complex K, $p \in$ St $(p_{\lambda_0} \cdots p_{\lambda_r})$ if and only if $\mu(p_{\lambda_i}) > 0$ for each i.

3. Subdivisions. If the polyhedron Q is the set of points of a complex K, we say K is a *subdivision* of Q. If K is a simplicial complex, we say K is a *simplicial subdivision* or *triangulation* of Q. Occasionally we allow K to be an infinite complex, for instance if, in place of Q, we use E^n or an open subset of E^n. Since Q is in an affine space, barycentric coordinates in σ are well defined, for each cell σ (App. I, 11). Through smooth mappings, we may run into "cells" $\sigma' = f(\sigma)$ in an affine space E' which are "curved" in E'; thus we may find a "curvilinear" subdivision of a subset of E'; see for instance (III, 7). In this case, barycentric coordinates in σ do not carry over to barycentric coordinates in E'.

LEMMA 3a. *Any polyhedron has a subdivision.*

This can be shown by elementary arguments.

Suppose K and K' are subdivisions of P, with the property that each cell of K' is contained in a cell of K. Then we say K' is a *refinement* of K. Then each cell of K is a union of cells of K'.

The *regular subdivision* K' of a complex K is defined as follows. For each cell σ of K, its center p_σ (see § 1) is a vertex of K'; in particular, each vertex of K is a vertex of K'. For each increasing sequence $\sigma_0 \subset \sigma_1 \subset \cdots \subset \sigma_r$ of cells of K, there is a corresponding simplex $\tau = p_{\sigma_0} p_{\sigma_1} \cdots p_{\sigma_r}$ of K'. Thus K' is simplicial.

Clearly each τ as above is in σ_r. If we show that each point p of K is in int (τ) for a unique simplex τ of K', this will prove that K' is a refinement of K. Say $p \in \text{int}\,(\sigma)$. If $p = p_\sigma$, then $p \in \text{int}\,(p_\sigma) = p_\sigma$. Suppose that $p \neq p_\sigma$. Then there is a unique segment $p_\sigma p'$ containing p, with $p' \in \partial\sigma$. Using induction, we see easily that there is a unique $\tau' \subset \partial\sigma$ with $p' \in \text{int}\,(\tau')$, and that if $\tau' = p_{\sigma_0} \cdots p_{\sigma_k}$, then $p \in \text{int}\,(\tau)$ with $\tau = p_{\sigma_0} \cdots p_{\sigma_k} p_\sigma$, and τ is unique.

LEMMA 3b. *Any two subdivisions K_1, K_2 of a polyhedron P have a common simplicial refinement.*

Let K be the complex whose cells are the non-void intersections $\sigma_1 \cap \sigma_2$ (σ_1 in K_1, σ_2 in K_2). The regular subdivision K' of K has the required properties.

If the complex K is in a metric space, then the numbers diam (σ) for cells σ of K are defined; the largest of these is the *mesh* of K.

LEMMA 3c. *Any complex K_0 has a sequence of consecutive simplicial refinements K_1, K_2, $\cdots$ with mesh $(K_i) \to 0$.*

We may let K_i be the regular subdivision of K_{i-1}. The last statement of the lemma is obvious if, instead, we let K_i be the standard refinement of K_{i-1} $(i \geq 2)$; see the next section.

Let C be an n-cube; with its faces, it forms a complex K. Let $v_1, \cdots, v_n$ be the edge vectors of C. Let K' be the regular subdivision of K. Then for each n-simplex τ of K', its last vertex is the center of C, its next to the last vertex is the center of one of the $2n$ $(n-1)$-faces C' of C, the preceding vertex is the center of one of the $2(n-1)$ $(n-2)$-faces C'' of C', etc. Hence there are $2^n n!$ n-simplexes in K', and each has $\pm\frac{1}{2}v_1, \cdots, \pm\frac{1}{2}v_n$ as a set of edge vectors.

4. Standard subdivisions. If K_i is the regular subdivision of K_{i-1} $(i = 1, 2, \cdots)$, we get simplexes of worse and worse shapes, as is seen at once in the case that K_0 is a 2-simplex; the numbers $\Theta(\sigma)$ (IV, 14) have no positive lower bound. We shall give a manner of subdivision which corrects this†; it will be useful in Chapter IX.

Let $\sigma = p_0 \cdots p_r$ be a simplex, with its vertices given in the order shown; we construct its *standard subdivision* $\mathfrak{S}\sigma$ as follows. Set

(1) $$p_{ij} = \tfrac{1}{2}p_i + \tfrac{1}{2}p_j, \qquad i \leqq j;$$

in particular, $p_{ii} = p_i$. These are the vertices of $\mathfrak{S}\sigma$. Define a partial ordering among these vertices by setting

(2) $$p_{ij} \leqq p_{kl} \quad \text{if} \quad k \leqq i \text{ and } j \leqq l.$$

For instance, $p_{22} < p_{23} < p_{13}$, while p_{13} and p_{24} are unrelated. The simplexes of $\mathfrak{S}\sigma$ are all those formed from the p_{ij} which are in increasing order. For example, if $r = 2$, the 2-simplexes are

$$p_0p_{01}p_{02}, \quad p_1p_{01}p_{02}, \quad p_1p_{12}p_{02}, \quad p_2p_{12}p_{02},$$

and if $r = 3$, the 3-simplexes are

$$p_0p_{01}p_{02}p_{03}, \quad p_1p_{01}p_{02}p_{03}, \quad p_1p_{12}p_{02}p_{03}, \quad p_2p_{12}p_{02}p_{03},$$

$$p_1p_{12}p_{13}p_{03}, \quad p_2p_{12}p_{13}p_{03}, \quad p_2p_{23}p_{13}p_{03}, \quad p_3p_{23}p_{13}p_{03}.$$

There are 2^r r-simplexes in $\mathfrak{S}\sigma$, which may be found conveniently as follows. The last vertex is p_{0r}. The next to the last is either $p_{0,r-1}$ or p_{1r}. In general, having chosen a vertex p_{ij}, the preceding one is either $p_{i+1,j}$ or $p_{i,j-1}$. Clearly the interiors of these simplexes are disjoint; it is not hard to see that they actually form a simplicial complex, which is a subdivision of σ.

Set

(3) $$v_i = p_i - p_{i-1} \qquad (i = 1, \cdots, r);$$

using (1) shows that

(4) $$p_{i,j+1} - p_{ij} = \tfrac{1}{2}v_{j+1}, \qquad p_{i-1,j} - p_{ij} = -\tfrac{1}{2}v_i.$$

If K is a simplicial complex, and we order its vertices in some fixed fashion, then $\mathfrak{S}_1K$ may be formed by subdividing each of its simplexes as above. Now each simplex of $\mathfrak{S}_1K$ has its vertices ordered, and we may subdivide again, forming $\mathfrak{S}_2K$, etc. Thus we form the *sequence of standard subdivisions* of K, relative to the given order of the vertices.

In the following lemmas, we use the notations of (III, 1) and (IV, 14).

LEMMA 4a. *If $\sigma = p_0 \cdots p_r$, $v_i = p_i - p_{i-1}$, then each r-simplex of $\mathfrak{S}_k\sigma$ has $\pm v_1/2^k, \cdots, \pm v_r/2^k$ as edge vectors, and has volume $|\sigma|/2^{kr}$.*

This follows from (4) and (III, 1.3).

LEMMA 4b. *Given the simplicial complex K with ordered vertices, at most a finite number of shapes occur among the simplexes of the $\mathfrak{S}_kK$. For some $\eta > 0$, $\Theta(\tau) \geqq \eta$ for all simplexes τ of all $\mathfrak{S}_kK$.*

† A similar type of subdivision was given by H. Freudenthal, *Annals of Math.* *43* (1942), 580–582.

The first statement follows from the last lemma, and the second statement follows from the first.

The following lemma is needed in (IX, 8).

LEMMA 4c. *Let τ be an $(r-1)$-simplex of $\mathfrak{S}\sigma$, $\sigma = p_0 \cdots p_r$, which does not lie in $\partial\sigma$. Then the plane P of τ contains p_{0r} but no other point of p_0p_r.*

This being trivial for $r = 1$, we use induction. First, τ contains p_{0r}; for if not, then all its vertices would lie either in $\sigma_0 = p_1 \cdots p_r$ or in $\sigma_r = p_0 \cdots p_{r-1}$, which would give $\tau \subset \partial\sigma$. Now the vertices of τ other than p_{0r} form a simplex τ'. Either $\tau' \subset \sigma_0$ or $\tau' \subset \sigma_r$, say the former. Suppose $\tau' \subset \partial\sigma_0$; say $\tau' \subset p_1 \cdots \hat{p}_j \cdots p_r$. If $j < r$, then $\tau \subset p_0 \cdots \hat{p}_j \cdots p_r \subset \partial\sigma$, a contradiction; hence $\tau' \subset p_1 \cdots p_{r-1}$, and the plane P' of τ' does not contain p_r. If τ' is not in $\partial\sigma_0$, then P' does not contain p_r, by induction. Since P contains P' and also a point p_{0r} not in the plane of σ_0, $P \cap \sigma_0 = P' \cap \sigma_0$, and P does not contain p_r; hence the last statement of the lemma follows.

5. Orientation. As seen in (I, 8), an affine space E of dimension n $(n \geqq 1)$ has two orientations; an orientation is given by the choice of an ordered set $(v_1, \cdots, v_n)$ of n independent vectors. For $n = 0$, E is a single point; we do not give it an orientation.

A cell σ is *oriented* by orienting its plane $P(\sigma)$. If $P(\sigma)$ has already been oriented, we say σ is oriented *like* $P(\sigma)$, or σ lies *positively* in $P(\sigma)$, if the two orientations of $P(\sigma)$ agree. We may do the same for a polyhedral region in an oriented affine space.

By the *oriented simplex* $\sigma = p_0 \cdots p_r$, we mean the simplex σ, oriented by the set of vectors $(p_1 - p_0, \cdots, p_r - p_0)$, or equivalently, $(p_1 - p_0, \cdots, p_r - p_{r-1})$.

LEMMA 5a. *For any permutation $(\lambda_0, \cdots, \lambda_r)$ of $(0, \cdots, r)$, $p_{\lambda_0} \cdots p_{\lambda_r}$ has the same or opposite orientation as $p_0 \cdots p_r$ according as the permutation is even or odd.*

The proof is elementary.

We now consider the orientation of the $(n-1)$-faces of an oriented n-cell σ. If $n = 1$, then σ is a 1-simplex pq; we say q is in $\partial\sigma$ *positively*, and p, *negatively*. Now suppose $n \geqq 2$, and σ' is an $(n-1)$-face. We may choose $(v_1, \cdots, v_n)$ orienting σ^n, so that $v_2, \cdots, v_n$ lie in σ'. Choose $p \in \operatorname{int}(\sigma')$; then either no point $p + tv_1$ is in σ for $t > 0$, or all such points are for $t > 0$ sufficiently small; we say v_1 points *out of* σ or *into* σ at σ' in these two cases respectively. In the first case, we say the orientation of σ' defined by $(v_2, \cdots, v_n)$ *agrees* with that of σ, or, σ' thus oriented lies *positively* in $\partial\sigma$; in the second case, the opposite is true.

In particular, suppose $\sigma = p_0p_1 \cdots p_r$ and $\sigma' = p_1 \cdots p_r$, oriented as shown. Set $v_i = p_i - p_{i-1}$. Then $(v_1, \cdots, v_r)$ orients σ, $(v_2, \cdots, v_r)$

orients σ', and v_1 points out of σ at σ'; hence σ' lies positively in $\partial\sigma$. Using Lemma 5a, we see easily that $p_0 \cdots \hat{p}_i \cdots p_r$ lies positively or negatively in $\partial\sigma$ according as i is even or odd.

Suppose R is an oriented open set in E^n, with fro (R) consisting in part of a smooth $(n - 1)$-manifold σ; near a given point p, say σ is given by the equation $x_1 = f(x_2, \cdots, x_n)$, using rectangular coordinates, f being continuously differentiable. We may "orient" σ near p by orienting the tangent plane there; the definition of σ being "positively" in ∂R is clear.

6. Chains and cochains. Let K be a complex, and let its cells be oriented. An (algebraic) *r-chain* of K is an expression of the form $A = \sum a_i\sigma_i^r$, the a_i being real numbers. (We could let the a_i be elements of an Abelian group.) We write dim $(A) = r$. Two chains are *added* by adding corresponding coefficients; to *multiply* a chain by a real number, multiply each coefficient by that number. Thus the set of r-chains becomes a linear space $\mathbf{C}_r = \mathbf{C}_r(K)$. The dimension of this space is clearly the number of cells of K of dimension r.

We allow ourselves to put in or drop out a term 0σ with zero coefficient; also, we may write σ in place of 1σ. Now the oriented r-cells themselves become chains; they form a base in $\mathbf{C}_r$.

If we let $-\sigma$ denote the cell σ oppositely oriented, we may identify $(-a)(-\sigma)$ with $a\sigma$.

In the case $r = 0$, if $p_1, p_2, \cdots$ are the vertices of K, then any 0-chain is $\sum a_i p_i$; questions of orientation do not arise.

An (algebraic) *r-cochain X of K* is an element of the conjugate space $\mathbf{C}^r = \overline{\mathbf{C}}_r$ of $\mathbf{C}_r$. We write $X\cdot A$ in place of $X(A)$. Set $X_i = X\cdot\sigma_i^r$. If we let σ_i^r denote not only an oriented simplex or a chain, but also the cochain defined by $\sigma_i^r\cdot\sigma_j^r = \delta_i^j$, then the σ_i^r form also a base for the r-cochains; for any X as above, clearly $X = \sum X_i\sigma_i^r$. Even if we occasionally use this notation, we keep distinct the meaning of chains and cochains. Note that

$$X\cdot A = \sum_i X_i\sigma_i^r \cdot \sum_j a_j\sigma_j^r = \sum_i X_i a_i. \tag{1}$$

The *unit* 0-cochain I^0 is $\sum p_i$, summed over all vertices; thus for any vertex p_i, $I^0\cdot p_i = 1$. Now

$$I^0\cdot\sum a_i p_i = \sum a_i, \tag{2}$$

often called the "Kronecker index" of the 0-chain $\sum a_i p_i$.

The definition of *polyhedral r-chains* in an affine space E is given in (V, 1); see also (III, 2). We may clearly consider polyhedral chains in a polyhedron. Each r-chain of a complex K clearly corresponds to a well defined polyhedral r-chain in K.

7. Boundary and coboundary. The *boundary* ∂A of an r-chain A is an $(r-1)$-chain, defined as follows. For $r = 0$, set $\partial A = 0$. Now take any r-cell σ. If $r = 1$, then $\sigma = pq$; set $\partial\sigma = q - p$. If $r \geqq 2$, let $\partial\sigma$ be the sum of the $(r-1)$-faces of σ, oriented to agree with σ (§ 5). Set $\partial \sum a_i \sigma_i^r = \sum a_i \partial \sigma_i^r$. For a simplex, we have (see § 5)

$$\partial(p_0 \cdots p_r) = \sum_{i=0}^{r} (-1)^i p_0 \cdots \hat{p}_i \cdots p_r. \tag{1}$$

Now the operation ∂ is a linear mapping of $\mathbf{C}_r$ into $\mathbf{C}_{r-1}$, for each $r \geqq 1$.

Each oriented $(r-2)$-face σ' of an r-cell σ is a face of two $(r-1)$-faces σ_1, σ_2 of σ (if $r \geqq 2$); if σ_1 and σ_2 are oriented to agree with σ, it is easily seen that the orientation of σ' agrees with one of these and disagrees with the other. It follows that $\partial\partial\sigma = 0$, and hence

$$\partial\partial A = 0, \qquad \text{all chains } A. \tag{2}$$

These considerations apply to polyhedral chains also.

The *coboundary* dX of the r-cochain X in K is defined by

$$dX \cdot A = X \cdot \partial A; \tag{3}$$

this is an $(r+1)$-cochain of K. Because of (2), we have

$$ddX = 0, \qquad \text{all cochains } X. \tag{4}$$

Note that, in a simplicial complex, by (1) and (3),

$$d(p_0 \cdots p_r) = \sum_k{}^{*} p_k p_0 \cdots p_r, \tag{5}$$

summing over all k such that $p_k p_0 \cdots p_r$ is a simplex of K.

Let Q be the oriented parallelepiped of (III, 11); we shall prove (III, 11.2). Since the orientation of Q is given by the ordered set $(v_1, \cdots, v_n)$, that of $(-1)^{i-1} Q$ is given by $(v_i, v_1, \cdots, \hat{v}_i, \cdots, v_n)$. Since $(v_1, \cdots, \hat{v}_i, \cdots, v_n)$ orients A_i^+ and v_i points out of Q at A_i^+, A_i^+ lies positively in $\partial[(-1)^{i-1}Q]$, and $(-1)^{i-1}A_i^+$ lies positively in ∂Q. Clearly $-(-1)^{i-1}A_i^-$ lies positively in ∂Q, and the relation follows. (We may also use (12.2) below and induction.)

8. Homology and cohomology. A chain A of K is a *cycle* if $\partial A = 0$; a *boundary* if $A = \partial B$ for some B. Define cocycles and coboundaries similarly. Let $\mathbf{Z}_r, \mathbf{B}_r, \mathbf{Z}^r, \mathbf{B}^r$ denote the linear spaces of cycles, boundaries, cocycles and coboundaries respectively, of dimension r. Because of (7.2) and (7.4), $\mathbf{B}_r \subset \mathbf{Z}_r$ and $\mathbf{B}^r \subset \mathbf{Z}^r$; hence we may define

$$\mathbf{H}_r = \mathbf{Z}_r \bmod \mathbf{B}_r, \qquad \mathbf{H}^r = \mathbf{Z}^r \bmod \mathbf{B}^r. \tag{1}$$

These are the rth *homology* and *cohomology spaces* of K (with real coefficients) respectively.

The cycles A and B of K are *homologous*, written $A \sim B$, if $A - B$ is a boundary; then A and B determine the same element of $\mathbf{H}_r$. Similarly for *cohomologous* cocycles, written $X \backsim Y$.

Let $\mathbf{C}$ be the direct sum of the spaces $\mathbf{C}_0, \mathbf{C}_1, \cdots$ (App. I, 4); this is the linear space formed by all chains of all dimensions of K. Let $\mathbf{C}^*$ be the direct sum of the $\mathbf{C}^i$, i.e. the linear space formed by all cochains; this may clearly be considered as the conjugate space of $\mathbf{C}$. We have subspaces $\mathbf{Z}$ and $\mathbf{B}$ of $\mathbf{C}$, and $\mathbf{Z}^*$ and $\mathbf{B}^*$ of $\mathbf{C}^*$; defining $\mathbf{H}$ and $\mathbf{H}^*$ as in (App. I, 7.5), it is clear that these are the direct sums

$$(2) \qquad \mathbf{H} = \mathbf{H}_0 \oplus \mathbf{H}_1 \oplus \cdots, \qquad \mathbf{H}^* = \mathbf{H}^0 \oplus \mathbf{H}^1 \oplus \cdots$$

The pairing of (App. I, 7.6) between $\mathbf{H}$ and $\mathbf{H}^*$ defines a pairing between $\mathbf{H}_r$ and $\mathbf{H}^r$ for each r, and because of (App. I, 7.7) we have

$$(3) \qquad \mathbf{H}^r \approx \overline{\mathbf{H}}_r, \qquad \mathbf{H}_r \approx \overline{\mathbf{H}}^r.$$

Hence also $\mathbf{H}_r$ and $\mathbf{H}^r$ are of the same dimension, and are therefore isomorphic; but there is in general no natural manner of setting up this isomorphism.

It is shown in treatises on algebraic topology that the spaces $\mathbf{H}_r$ and $\mathbf{H}^r$ depend on the polyhedron of K only, not on the particular subdivision; the dimension of $\mathbf{H}_r$ is the rth "Betti number" of K. (A proof is given in (VII, 12).)

9. Products in a complex. We discuss briefly pairings among the cohomology and homology spaces of a complex K.

There exists a bilinear operation $X^r \smile Y^s = Z^{r+s}$, called a *cup product* of the cochains X^r and Y^s, with the following properties.

(a) If σ^{r+s} appears with a non-zero coefficient in $\sigma^r \smile \sigma^s$, then σ^r and σ^s are faces of σ^{r+s}.

(b) $d(X^r \smile Y^s) = dX^r \smile Y^s + (-1)^r X^r \smile dY^s$.

(c) $I^0 \smile X^r = X^r \smile I^0 = X^r$.

Because of (b), this defines a bilinear operation between $\mathbf{H}^r$ and $\mathbf{H}^s$, with values in $\mathbf{H}^{r+s}$; this is also called the cup product. It may be shown† that this operation is uniquely determined, independently of the choice of product in the spaces $\mathbf{C}^k$. In this manner $\mathbf{H}^*$ becomes a ring, with the cohomology class of I^0 as unit element.

Given the cup product, we define the *cap product* $X^r \frown A^{r+s} = B^s$ of a cochain and a chain, the result being a chain, by the definition

$$(1) \qquad Y^s \cdot (X^r \frown A^{r+s}) = (Y^s \smile X^r) \cdot A^{r+s}$$

† See for instance H. Whitney, On products in a complex, *Annals of Math.* 39 (1938), 397–432; S. Lefschetz, *Algebraic Topology*, Am. Math. Soc., New York, 1942, Chapter V.

for all s-cochains Y^s. This defines a bilinear operation between $\mathbf{H}^r$ and $\mathbf{H}_{r+s}$, with values in $\mathbf{H}_s$, also uniquely determined. Now $\mathbf{H}^*$ is a ring of operators on $\mathbf{H}$.

10. Joins. The *join* $J(P, Q)$ of two point sets P, Q in an affine space E is the set of all points on all segments pq, $p \in P$, $q \in Q$. We shall use especially a corresponding algebraic operation.

Let $\sigma = p_0 \cdots p_r$ be an oriented r-simplex in E, and let p be a point not in the plane $P(\sigma)$. The *join* of p with σ is then

$$(1) \qquad J(p, \sigma) = p\sigma = pp_0 \cdots p_r;$$

this is an oriented $(r + 1)$-simplex in E. We could define similarly $J(\sigma^r, \sigma^s) = \sigma^{r+s+1}$, if this is a simplex. If $p \in P(\sigma)$, we set $J(p, \sigma) = 0$. Define

$$(2) \qquad J\left(p, \sum a_i\sigma_i\right) = \sum a_i J(p, \sigma_i);$$

this is a polyhedral chain in E. We prove

$$(3) \qquad \partial J(p, A) = A - J(p, \partial A) \quad \text{if} \quad \dim(A) \geqq 1,$$

$$(4) \qquad \partial J(p, A) = A - (I^0 \cdot A)p \quad \text{if} \quad \dim(A) = 0.$$

We may suppose A is a simplex $\sigma = p_0 \cdots p_r$. Suppose first $p\sigma$ is non-degenerate. Then for $r \geqq 1$, (3) follows at once from (1) and (7.1), while for $r = 0$,

$$\partial J(p, p_0) = \partial(pp_0) = p_0 - p = p_0 - (I^0 \cdot p_0)p.$$

If $p \in P(\sigma)$, we must prove that the right hand side of (3), which we call B, is 0. (This is clear for (4).) Now B is a polyhedral r-chain in the r-dimensional space $P(\sigma)$, and using induction shows that $\partial B = 0$. From the definition of ∂B (V, 1) it is clear that $B = 0$.

LEMMA 10a. *Let A be a polyhedral cycle in the convex subset Q of affine space E; if* $\dim(A) = 0$, *suppose* $I^0 \cdot A = 0$. *Then* $A = \partial B$, *B in Q.*

Choose $p \in Q$, and set $B = J(p, A)$. That $\partial B = A$ follows at once from (3) or (4).

11. Subdivisions of chains. Let K be a complex, and let K' be a refinement of K. For each oriented r-cell σ of K, let $\mathfrak{S}\sigma$ be the chain in K' composed of the r-cells of K' in σ, oriented like σ. Set $\mathfrak{S}\sum a_i\sigma_i = \sum a_i\mathfrak{S}\sigma_i$. We prove

$$(1) \qquad \partial\mathfrak{S}A = \mathfrak{S}\,\partial A.$$

We may suppose $A = \sigma$. Say $\partial\sigma = \sum\sigma_i$, $\mathfrak{S}\sigma = \sum\sigma'_j$, $\mathfrak{S}\sigma_i = \sum\sigma'_{ij}$. We must prove $\sum_i \partial\sigma'_i = \sum_{i,j}\sigma'_{ij}$. Each cell of dimension $r - 1$ interior to σ

appears in some $\partial\sigma'_k$ and in some $\partial\sigma'_l$, with opposite signs; hence it drops out of $\sum_i \partial\sigma'_i$. The remaining cells make up exactly $\sum_{i,j}\sigma'_{ij}$.

Suppose K' is the regular subdivision of K. We may define $J(p, \tau)$ if p and τ are both in some cell σ of K. It is easily seen that

$$\mathfrak{S}\sigma = J(p_\sigma, \mathfrak{S}\partial\sigma), \tag{2}$$

p_σ being the center of σ; we may use this to prove (1) in this case.

REMARK. If we consider the chain A of K as a polyhedral chain in K, then $\mathfrak{S}A$ is the same polyhedral chain in K; see (V, 1).

12. Cartesian products of cells. Let σ^r and σ^s be cells, in affine spaces P_1, P_2. The Cartesian product $P_1 \times P_2$ (App. 1, 4) is clearly an affine space P, and the subset $\sigma^r \times \sigma^s$ is a cell in P.

Suppose σ^r and σ^s are oriented, by $(v_1, \cdots, v_r)$ and $(w_1, \cdots, w_s)$ respectively. Then we orient $\sigma^r \times \sigma^s$ by $(v_1, \cdots, v_r, w_1, \cdots, w_s)$. If $\sum a_i\sigma_i^r$ and $\sum b_j\sigma_j^s$ are polyhedral chains in P_1 and P_2 respectively, their *Cartesian product* is the polyhedral chain $\sum a_i b_j(\sigma_i^r \times \sigma_j^s)$ in P.

We shall prove the boundary formula (1); the case $r = 2$ is used in (V, 9), but elsewhere, we use only the case $r = 1$.

Say $\partial\sigma^r = \sum\sigma_i^{r-1}$, $\partial\sigma^s = \sum\sigma_j^{s-1}$, as polyhedral chains. It is easy to see that the $(r + s - 1)$-faces of $\sigma^r \times \sigma^s$ are the cells $\sigma_i^{r-1} \times \sigma^s$ and $\sigma^r \times \sigma_j^{s-1}$. Hence, to prove

$$\partial(\sigma^r \times \sigma^s) = \partial\sigma^r \times \sigma^s + (-1)^r\sigma^r \times \partial\sigma^s, \tag{1}$$

we need merely show that each $(r + s - 1)$-face has the correct coefficient. Consider for example the face $\sigma^r \times \sigma_j^{s-1}$. We may suppose $(w_2, \cdots, w_s)$ orients σ_j^{s-1}, and w_1 points out of σ^s at σ_j^{s-1}. Now $(v_1, \cdots, v_r, w_2, \cdots, w_s)$ orients $\sigma^r \times \sigma_j^{s-1}$, while w_1 points out of $\sigma^r \times \sigma^s$ at $\sigma^r \times \sigma_j^{s-1}$; since $(w_1, v_1, \cdots, v_r, w_2, \cdots, w_s)$ orients $(-1)^r\sigma^r \times \sigma^s$, $\sigma^r \times \sigma_j^{s-1}$ lies positively in the boundary of this cell, as required.

Let I be the unit interval $0 \leqq t \leqq 1$ in $\mathfrak{A}$; this is an oriented 1-cell. Then (1) gives

$$\partial(I \times A) = 1 \times A - 0 \times A - I \times \partial A. \tag{2}$$

Let $\sigma = p_0 \cdots p_r$, the vertices ordered as shown; we give a corresponding simplicial subdivision of $I \times \sigma$. Set $q_i = 0 \times p_i$, $q'_i = 1 \times p_i$, and

$$\mathfrak{S}_0(I \times \sigma) = \sum_{i=0}^{r} (-1)^i\tau_i, \qquad \tau_i = q_0 \cdots q_i q'_i \cdots q'_r. \tag{3}$$

It is elementary to show that the τ_i are non-overlapping and fill out σ. We shall show that the orientations are correct. Set

$$v_i = p_i - p_{i-1} = q_i - q_{i-1} = q'_i - q'_{i-1}.$$

Let e be the unit vector in $\mathfrak{A}$. Then τ_i is oriented by the set $(v_1, \cdots, v_i, e, v_{i+1}, \cdots, v_r)$, and hence $(-1)^i\tau_i$ is oriented by the set $(e, v_1, \cdots, v_r)$, which also orients $I \times \sigma$.

Using the same definition of the subdivision of the cells of $I \times \partial\sigma$, a simple calculation shows that

$$\partial\mathfrak{S}_0(I \times \sigma) = 1 \times \sigma - 0 \times \sigma - \mathfrak{S}_0(I \times \partial\sigma). \tag{4}$$

13. Mappings of complexes. Let f be a mapping of the complex K (i.e. of the corresponding polyhedron) into an affine space E. We say f is *cellwise affine* if f, considered in each cell alone, is affine. If f maps K into another complex K', we may use the same definition, provided that each cell of K is mapped into some cell of K'. If K is simplicial, we speak of a *simplexwise affine* mapping. If both K and K' are simplicial, and f is simplexwise affine and maps each vertex of K into a vertex of K', then f is *simplicial*. Then f is determined by its values on the vertices p_i of K, and, using (2.1),

$$f\left(\sum\mu_i(p)p_i\right) = \sum\mu_i(p)f(p_i) \qquad \left(\mu_i(p) \geqq 0, \quad \sum\mu_i(p) = 1\right). \tag{1}$$

Now for each simplex σ of K, if f maps the vertices of σ into distinct vertices of K', then $f(\sigma)$ is a simplex of K', while in the contrary case, f is degenerate in σ.

Let f be a cellwise affine mapping of the complex K into E (or into a polyhedron P). For each oriented cell σ of K, oriented by $(v_1, \cdots, v_r)$, $f(\sigma)$ is a cell, which we orient by

$$(\nabla f(p, v_1), \cdots, \nabla f(p, v_r)), \tag{2}$$

for any $p \in \sigma$; these vectors are independent if $J_f \neq 0$ in σ. If this is so, we let $f\sigma$ denote the cell thus oriented; in the contrary case, set $f\sigma = 0$. In particular, if $\sigma = p_0 \cdots p_r$ and $f(p_i) = q_i$, then $f\sigma = q_0 \cdots q_r$. Setting $f\sum a_i\sigma_i = \sum a_i f_i$, f now becomes a transformation of polyhedral chains in K into polyhedral chains in E (or in P). In particular, if f is a simplicial mapping of K into K', and A is a chain of K, then fA may be considered as a chain of K'.

For any cellwise affine mapping f and chain A, we have

$$\partial fA = f\partial A. \tag{3}$$

It is sufficient to prove this in case A is an oriented cell σ. This is obvious if $J_f \neq 0$ in int (σ); otherwise, $f\sigma = 0$, and that $f\partial\sigma = 0$ follows as in the proof of (10.3).

Let f be a cellwise affine mapping of K into the affine space E, and let v be a vector in E. Setting

$$f_v(p) = f(p) + v, \qquad \mathscr{D}_v(t \times p) = f(p) + tv, \tag{4}$$

defines mappings f_v of K into E and $\mathscr{D}_v$ of $I\times K$ into E. The latter defines a *deformation* or *homotopy* of f into f_v, as t runs from 0 to 1. These define transformations f_v and $\mathscr{D}_v$ of chains A and $I\times A$ respectively into polyhedral chains in E. Since $\mathscr{D}_v(0\times A)=fA$ and $\mathscr{D}_v(1\times A)=f_vA$, using (3) and (12.2) gives

$$\partial\mathscr{D}_v(I\times A)=f_vA-fA-\mathscr{D}_v(I\times\partial A). \tag{5}$$

More generally, let $\mathfrak{S}(I\times K)$ be a subdivision of $I\times K$, and let F be a cellwise affine mapping of this complex into E (or into a polyhedron P). If $f_i(p)=F(i\times p)$ $(i=0,1)$, then

$$\partial F(I\times A)=f_1A-f_0A-F(I\times\partial A), \tag{6}$$

the chains again being polyhedral.

14. Some properties of planes. We prove some lemmas that will be used in (IV, B); here, E^m denotes Euclidean m-space.

Let P and P' be planes in E^m, of any dimensions; let π_Pv be the orthogonal projection into P of the vector v. The *independence* of P and P' we define to be

$$\operatorname{ind}(P,P')=\inf\{|v-\pi_Pv|:\ v\in P',|v|=1\}. \tag{1}$$

This depends only on the directions of the planes (I, 12), not on their positions. Clearly the independence is 0 if and only if the planes have a non-zero vector in common, and is 1 if and only if the planes are orthogonal.

Since the set of unit vectors is compact, we may choose a unit vector v in P' such that $|\pi_Pv-v|=\operatorname{ind}(P,P')$. Supposing $\pi_Pv\neq 0$, set $u=\pi_Pv/|\pi_Pv|$. Let L be a line containing v; then $|\pi_Lu-u|=|\pi_Pv-v|$, and hence

$$|\pi_P u-u|\leqq|\pi_Lu-u|=\operatorname{ind}(P,P'),$$

proving $\operatorname{ind}(P',P)\leqq\operatorname{ind}(P,P')$ if P and P' are not orthogonal. This works both ways; hence

$$\operatorname{ind}(P',P)=\operatorname{ind}(P,P'). \tag{2}$$

Recall that $P(\sigma)$ is the plane of σ.

LEMMA 14a. *Let σ be an s-cell and let P be an n-plane in E^m, such that*

$$s+n\geqq m,\qquad \operatorname{dist}(P,\sigma)<\operatorname{dist}(P,\partial\sigma). \tag{3}$$

Then $s+n=m$, P intersects σ in a single point, and

$$\operatorname{ind}(P,P(\sigma))>\operatorname{dist}(P,\partial\sigma)/\operatorname{diam}(\sigma). \tag{4}$$

Let d, d' denote the distances in (3). Choose p and q so that

$$p\in\sigma,\qquad q\in P,\qquad |q-p|=d.$$

(Actually, $p = q$.) Suppose there were a vector $v \neq 0$ common to σ and P. Then for some a, $p + av \in \partial\sigma$ and $q + av \in P$, contradicting $d' > d$. Hence there is no such v, which shows that $s + n = m$. Hence also the vectors of P together with those of σ span those of E^m, and we can write

$$q - p = u_1 + u_2, \qquad u_1 \in \sigma, \qquad u_2 \in P.$$

Set $p^* = p + u_1 = q - u_2$; then $p^* \in P(\sigma) \cap P$. If p^* were not in σ, then

$$p' = p^* - au_1 \in \partial\sigma \quad \text{for some } a, \qquad 0 < a \leqq 1;$$

setting $q' = p^* + au_2$ gives $d' \leqq |q' - p'| \leqq d$, again a contradiction. Hence $p^* \in \sigma$. Since σ and P have no common vector $\neq 0$, there is no other point in $P \cap \sigma$.

Now take any unit vector v in $P(\sigma)$. Choose $a > 0$ so that $p' = p^* + av \in \partial\sigma$, and set $q' = p^* + a\pi_P v$. Then

$$d' \leqq |q' - p'| = a\,|\pi_P v - v| < \operatorname{diam}(\sigma)\,|\pi_P v - v|,$$

and (4) follows.

Lemma 14b. *Let P^* be a plane in E, let P be a plane in P^*, let Q be a closed set in P^*, let p^* be a point of E not in Q, and let Q^* be the join $J(p^*, Q)$. Then*

(5) $$\operatorname{dist}(Q^*, P) \geqq \operatorname{dist}(Q, P)\operatorname{dist}(p^*, P^*)/\operatorname{diam}(Q^*).$$

Suppose this relation, which we write in the form $c \geqq ab/d$, were false. Let Q^{**} be the set of all points on all rays which start at a point of Q and pass through p^*. Then we may choose $p' \in Q^{**}$ and $q' \in P$ so that

$$|q' - p'| = \operatorname{dist}(Q^{**}, P) \leqq c < ab/d.$$

Say p' is on the ray from p'' through p^* ($p'' \in Q$). Since $Q \subset P^*$, $b \leqq d$; hence $c < ab/d \leqq a$, and $p' \neq p''$. Therefore $p'q'$ is perpendicular to $p''p'$. Let L be the line through p'' and q', and let q^* be the nearest point of L to p^*; then q^*p^* is perpendicular to L. Since q' is nearer to p' than to p'', q^* and q' are on the same side of p'' in L. Now the triangles $p''p'q'$ and $p''q^*p^*$ are similar, and hence

$$|p' - q'| = |q' - p''|\,|q^* - p^*|/|p^* - p''| \geqq ab/d,$$

a contradiction, proving the lemma.

15. Mappings of n-pseudomanifolds into n-space. This section will be used in (IV, B). We shall say a simplicial complex K is an *oriented n-pseudomanifold* if K is of dimension n, each simplex is a face of an n-simplex, each $(n - 1)$-simplex is a face of either one or two n-simplexes, each n-simplex σ_i^n is oriented, and $\partial \sum \sigma_i^n$ contains only those $(n - 1)$-simplexes on just one n-simplex. (Generally one assumes also connectedness in some sense.) We shall let ∂K denote the set of $(n - 1)$-simplexes just mentioned, or the set of points on these simplexes.

We call the mapping f of K into oriented E^n *simplexwise positive* if for each σ_i^n, f is smooth and one-one in σ_i^n (the partial derivatives being continuous on the boundary), and $\bar{J}_f > 0$ there (II, 6). For any complex L, let L^k, or $(L)^k$, denote the subcomplex containing all cells of L of dimension $\leqq k$. With f in K as above, any point q of $E^n - f(K^{n-1})$ is in the image of a certain number h of n-simplexes of K; we say q is *covered h times*. (If some Jacobians were negative, we would consider the algebraic number of times q was covered.)

The proof of the following lemma could be somewhat shortened if we used theorems from algebraic topology.

Lemma 15a. *Let f be simplexwise positive in K. Then for any connected open subset R of $E^n - f(\partial K)$, any two points of R not in $f(K^{n-1})$ are covered the same number of times. If this number is* 1, *then f, considered in the open subset $R' = f^{-1}(R)$ of K only, is one-one onto R.*

By the inverse function theorem (II, Theorem 7A), $f(\text{int}\,(\sigma_i^n))$ is open, for each σ_i^n. We show first that for any σ^{n-1} not in ∂K, $f(\text{St}\,(\sigma^{n-1}))$ (see § 2) is open. Let σ_1^n and σ_2^n be the n-simplexes of K with σ^{n-1} as face. Take any $p \in \text{int}\,(\sigma^{n-1})$; we need merely show that $f(\text{St}\,(\sigma^{n-1}))$ covers a neighborhood of $f(p)$. Since f is one-one in each σ_i^n, there is a neighborhood U of $f(p)$ not touching $f(\partial\,\text{St}\,(\sigma^{n-1}))$. Since f is smooth in σ^{n-1}, with Jacobian $(n - 1)$-vector $\neq 0$, we may choose U so that $f(\sigma^{n-1})$ cuts it into two connected parts, U_1 and U_2. (The proof is elementary.) Let pp_i be a segment in σ_i^n, $p_i \in \text{int}\,(\sigma_i^n)$, mapping into an arc A_i in U $(i = 1, 2)$. If we orient σ^{n-1}, it is in $\partial\sigma_1^n$ and $\partial\sigma_2^n$ with opposite signs; since $\bar{J}_f > 0$ in each σ_i^n, the tangent vectors to A_1 and to A_2 at $f(p)$ are on opposite sides of $f(\sigma^{n-1})$ (see (13.2)); hence we may suppose $f(p_1) \in U_1$, $f(p_2) \in U_2$. Now suppose there were a point $q \in U - f(\sigma^{n-1})$ not in $f(\text{St}\,(\sigma^{n-1}))$. There is an arc A in $U - f(\sigma^{n-1})$ joining q to either $f(p_1)$ or $f(p_2)$. There is a first point q^* in A which is in $f(\overline{\text{St}}\,(\sigma^{n-1}))$; by the choice of U and A, $q^* \in U_j' = f(\text{int}\,(\sigma_j^n))$ for $j = 1$ or 2. But U_j' is open, contradicting the definition of q^*, and the statement is proved.

Suppose the first conclusion of the lemma were false. Then we may express $R - f(K^{n-1})$ as the union of two disjoint sets R_1 and R_2, such that for some h, each point of R_1 is covered h times and each point of R_2 is covered a different number of times. We may choose an arc A from a point of R_1 to a point of R_2, lying in $R - f(K^{n-2})$, which crosses from R_1 to R_2 at a point q; then $q \in f(\sigma^{n-1})$ for some σ^{n-1}. Let $\sigma_1^{n-1}, \cdots, \sigma_k^{n-1}$ be the $(n - 1)$-simplexes of K whose images contain q; say σ_i^{n-1} is a face of the n-simplexes σ_i, σ_i'. Since f is one-one in the n-simplexes, we see at once that these n-simplexes are distinct. By the proof above, there is a neighborhood U of q such that for each i, each point of $U - f(K^{n-1})$ is in just one of $f(\sigma_i), f(\sigma_i')$. We may suppose U touches no $f(\sigma_j^{n-1})$ for any other j; then any other $f(\sigma_l^n)$ containing q contains U. Hence all points of $U - f(K^{n-1})$ are covered the same number of times, contradicting the choice of q.

Next we show that for any simplex σ^k of K, $f(\text{St}\,(\sigma^k))$ is open. Given $p \in \text{int}\,(\sigma^k)$, we must show that $f(\text{St}\,(\sigma^k))$ covers a neighborhood U of $q = f(p)$. We may suppose U is connected and does not touch $f(\partial\,\text{St}\,(\sigma^k))$. Now $L = \overline{\text{St}}(\sigma^k)$ is an oriented n-pseudomanifold, and the proof above shows that all points of U not in $f(L^{n-1})$ are covered the same number N of times by n-simplexes of L. Since some points near q are covered, $N \geqq 1$; hence all points of U are in $f(L)$.

We now prove the last part of the lemma. Since the number of times points are covered is 1, f maps R' onto R. Now suppose $f(p_1) = f(p_2) = q$, $p_1 \neq p_2$. Say $p_i \in \text{int}\,(\sigma_i)$ (dimension of σ_i unspecified). Since f is one-one in all simplexes, $\text{St}\,(\sigma_1) \cap \text{St}\,(\sigma_2) = 0$. By the proof above, $f(\text{St}\,(\sigma_i))$ covers all points of some neighborhood U_i of q not in $f(K^{n-1})$ a number of times $N_i > 0$, for $i = 1, 2$. But this shows that f, in K, covers points of $U = U_1 \cap U_2$ at least twice, a contradiction, completing the proof of the lemma.

16. Distortion of triangulations of E^m. We show that if certain triangulations of E^m are slightly distorted, they remain triangulations.

Lemma 16a. *Given the integer m, there is a number $\rho^* > 0$ with the following property. Let K_0 be a subdivision of E^m into cubes of side length h, and let K be the regular subdivision of K_0, with vertices $p_0, p_1, \cdots$. For each i, let p_i' be a point with*

$$|p_i' - p_i| \leqq \rho^* h. \tag{1}$$

Let f be the simplexwise affine mapping of K into E^m defined by $f(p_i) = p_i'$. Then f is one-one in K onto E^m, and the simplexes $f(\sigma)$ form a simplicial subdivision of E^m.

We may clearly suppose $h = 1$. Let $\sigma = p_0 \cdots p_m$ be an m-simplex of K_0, with center p_σ; set $c = \text{dist}\,(p_\sigma, \partial\sigma)$. We may choose $\rho^* < c/2$ so that if (1) holds, then $p_0' \cdots p_m'$ is a non-degenerate simplex, oriented like σ (see (IV, Lemma 14c)).

Now let the p_i' and f be given. Orient E^m, and orient the m-simplexes like E^m. Clearly $f(p_\sigma)$ is covered by $f(\sigma)$ only. The proof of Lemma 15a applies, with ∂K void, even though K is infinite, showing that f is one-one onto. The last statement is a consequence of this.

Appendix III

Analytical Preliminaries

We present here some results of analysis that may not be well known to the reader, and are of importance in different parts of the book. The special functions constructed in § 1 are used in § 2; partitions of unity described there appear for instance in §§ 8, 10, 15 and 18 of Chapter III and also in (IV, 27), (VII, 3) and (VIII, 2). A very useful tool is the approximation to a continuous or measurable function by a smooth function; this is discussed in §§ 3 and 4. Some topics from Lebesgue theory are considered in §§ 5 and 6.

A function ϕ is *k-smooth* in an open set R if it is continuous and has continuous partial derivatives through the order k in R (using an affine coordinate system). Then "0-smooth" and "continuous" are the same. We say ϕ is *smooth* if it is 1-smooth. Say ϕ is *cellwise continuous* in E^n if, for some subdivision of E^n (App. II, 3), ϕ is defined and uniformly continuous in the interior of each n-cell; similarly if ϕ is defined in a polyhedron. The definition of *cellwise constant* is similar. Two cellwise continuous functions are considered the same if they differ only in a polyhedral set of dimension $< n$. Say ϕ (assumed continuous or cellwise continuous or measurable) is *summable* in R if $\int_R |\phi(p)|\,dp$ is finite. Say ϕ is *locally summable* in R if each $p \in R$ is in a neighborhood U such that ϕ is summable in U. The *carrier* car (ϕ) of ϕ is the set Q of points where $\phi \neq 0$; the *support* spt (ϕ) of ϕ is $\bar{Q}$.

Write sup $\{a, b\}$ for the larger of the two numbers a, b; write

$$\sup\,\{\phi(p)\colon\ C(p)\} = \sup_{C(p)}\,\{\phi(p)\}$$

for the least upper bound of the numbers $\phi(p)$, using only those p such that the condition $C(p)$ holds. Use inf $\{\}$ similarly for the greatest lower bound.

If ϕ is a real function, or if ϕ has values in a Banach space, let $\langle\phi\rangle$ be the function whose value at p is $|\phi(p)|$; thus $\langle\phi\rangle(p) = |\phi(p)|$. Define

$$|\phi|_R = \sup\,\{|\phi(p)|:\ p \in R\}, \qquad |\phi| = \sup\,\{|\phi(p)|\}.$$

If $|\ |_0$ is used for the norm in the Banach space, as in (I, 13), we use the notation $\langle\phi\rangle_0$, $|\phi|_0$.

Let $|Q|$ denote the volume, or more generally, the Lebesgue measure, of the set $Q \subset E^n$. See also § 5.

1. Existence of certain functions. We prove the existence of real valued functions with certain properties; for instance, they equal one in a given closed set and equal zero in another set. A function is *sharp* if it is bounded and satisfies a Lipschitz condition (II, 4).

LEMMA 1a. *Let Q and Q' be closed sets in E^n whose distance apart is $\zeta > 0$. Then there is a real sharp function ϕ in E^n which equals 1 in Q and equals 0 in Q'.*

A formula for ϕ is given in (V, 12.3), with σ replaced by Q.

We now define some real functions which are ∞-smooth (i.e. all partial derivatives exist). Set

$$\Psi_0(t) = 1/t(1-t), \qquad \Phi_0(t) = e^{-\Psi_0(t)}, \quad \text{if} \quad 0 < t < 1, \tag{1}$$

and $\Phi_0(t) = 0$ for other values of t. Then each derivative of Φ_0 approaches 0 as $t \to 0$ or $t \to 1$; hence Φ_0 is ∞-smooth for all t. Next, set

$$a = \int_0^1 \Phi_0(t)dt, \qquad \Phi_1(t) = \frac{1}{a}\int_{-\infty}^{t} \Phi_0(s)\, ds. \tag{2}$$

then Φ_1 is ∞-smooth, and

$$\begin{gathered} \Phi_1(t) = 0 \quad (t \leqq 0), \qquad \Phi_1(t) = 1 \quad (t \geqq 1), \\ 0 < \Phi_1(t) < 1 \qquad (0 < t < 1). \end{gathered} \tag{3}$$

Suppose $0 < t_0 < t_1$. Using Φ_1, we may clearly construct an ∞-smooth function $\Phi^*(t)$ such that $\Phi^*(t) = 1$ for $|t| \leqq t_0$, $\Phi^*(t) = 0$ for $|t| \geqq t_1$, and $0 < \Phi^*(t) < 1$ elsewhere.

LEMMA 1b. *Let $Q \subset Q'$ be concentric closed cubes or balls in E^n. Then there is an ∞-smooth real function $\Phi(p)$ in E^n such that*

$$\begin{gathered} \Phi(p) = 1 \text{ in } Q, \qquad \Phi(p) = 0 \text{ in } E^n - Q', \\ 0 < \Phi(p) < 1 \quad \text{in} \quad \operatorname{int}(Q') - Q. \end{gathered} \tag{4}$$

If Q and Q' are balls of radii t_0 and t_1 respectively, with center p_0, we may set $\Phi(p) = \Phi^*(|p - p_0|)$. If they are cubes, defined say by $|x^i| \leqq t_k$ $(i = 1, \cdots, n)$, for $k = 0, 1$, we may set $\Phi(p) = \Phi^*(x^1) \cdots \Phi^*(x^n)$.

LEMMA 1c. *Let Q_1 and Q_2 be disjoint closed sets in E^n. Then there is an ∞-smooth real function Φ in E^n such that*

$$\begin{gathered} \Phi(p) = 1 \quad \text{in} \quad Q_1, \qquad \Phi(p) = 0 \quad \text{in} \quad Q_2, \\ 0 \leqq \Phi(p) \leqq 1 \quad \text{in} \quad E^n. \end{gathered} \tag{5}$$

Let $P_1, P_2, \cdots$ be closed balls not touching Q_2, whose interiors cover Q_1, such that each point of E^n is in a neighborhood which touches but a finite number of the P_i. (If Q_1 is bounded, we need only a finite number of P_i). Let Φ_i' be an ∞-smooth function in E^n which is >0 in int (P_i) and $= 0$ in $E^n - P_i$ (Lemma 1b). Set $\Phi'(p) = \sum \Phi_i'(p)$. Let U be a neighborhood of Q_1 such that $\Phi'(p) > 0$ in U. Define $\Phi''(p)$ in the same fashion as $\Phi'(p)$ was defined, using $E^n - U$ and Q_1 in place of Q_1 and Q_2. Set

$$\Phi(p) = \Phi'(p)/[\Phi'(p) + \Phi''(p)];$$

clearly the required properties hold.

2. Partitions of unity. We say a set of real valued functions $\phi_1, \phi_2, \cdots$ in E^n forms a *partition of unity* in E^n if

$$(1) \qquad 0 \leqq \phi_i(p) \leqq 1 \qquad (\text{all } p, i), \qquad \sum \phi_i(p) = 1 \quad \text{in } E^n.$$

Commonly ϕ_i is required to vanish outside of a given open set U_i. A construction of a partition of unity is given in (III, 8). For a corresponding construction in a smooth manifold, see (III, 10).

We shall give a particular construction here.

LEMMA 2a. *Let $Q \subset E^n$ be compact, let $U_1, \cdots, U_m$ be open, and suppose $Q \subset \bigcup U_i$. Then there are ∞-smooth functions $\phi_0, \phi_1, \cdots, \phi_m$ such that*

(a) $0 \leqq \phi_i(p) \leqq 1 \qquad (i = 0, 1, \cdots, m)$,

(b) $\phi_0(p) = 0$ *in a neighborhood of* Q,

(c) $\phi_i(p) = 0$ in $E^n - U_i \qquad (i = 1, \cdots, m)$,

(d) $\phi_0(p) + \phi_1(p) + \cdots + \phi_m(p) = 1$ *in* E^n.

For some $\zeta > 0$, we may set (see notations in App. II)

$$U_i' = \text{int}_\zeta (U_i), \qquad i = 1, \cdots, m,$$

and will have $\bar{U}_i' \subset U_i$, $Q \subset \bigcup U_i'$. (If the last condition failed for $\zeta = 1/k$, for each k, take $q_k \in Q - \bigcup U_i'$, using $\zeta = 1/k$; there is a limiting point q of the sequence $q_1, q_2, \cdots$ in Q, and we quickly find a contradiction.) For $i = 1, \cdots, m$, let Φ_i be an ∞-smooth function in E^n which is >0 in U_i' and $= 0$ in $E^n - U_i$ (Lemma 1c). For some $\epsilon > 0$, $\bar{U}_\epsilon(Q) \subset \bigcup U_i'$. Let Φ_0 be an ∞-smooth function which is >0 in $E^n - \bigcup U_i'$ and $= 0$ in $\bar{U}_\epsilon(Q)$. Then $\sum_{j \geqq 0} \Phi_j(p) > 0$ in E^n, and we may use

$$\phi_i(p) = \Phi_i(p)/[\Phi_0(p) + \cdots + \Phi_m(p)], \qquad i = 0, 1, \cdots, m.$$

3. Smoothing functions by taking averages. The simplest way to smooth a function ϕ in E^n is to use $\bar{\phi}_\rho(p)$, which equals the average of the values of ϕ in a ball $U_\rho(p)$ about p. If $|U_\rho(p)| = a_n\rho^n$, then (the first integral being in $V = V(E^n)$)

$$(1) \qquad \bar{\phi}_\rho(p) = \frac{1}{a_n\rho^n} \int_{U_\rho(0)} \phi(p + v)\, dv = \frac{1}{a_n\rho^n} \int_{U_\rho(p)} \phi(q)\, dq.$$

If ϕ is continuous, then $\bar{\phi}_\rho$ is smooth. We wish to define ϕ_ρ, which will be ∞-smooth. We will suppose that ϕ is continuous or cellwise continuous, or more generally, Lebesgue measurable and locally summable. It may have real values, or values in a Banach space. It may be defined in E^n, or in an open set $R \subset E^n$; in the latter case, ϕ_ρ is defined in $\mathrm{int}_\rho\,(R)$.

Let $\bar{\kappa}_\rho(t)$ be an ∞-smooth real function, defined for $t \geqq 0$, which is monotone decreasing, is constant in some interval $(0, t_0)$, and equals 0 for $t \geqq \rho$. (It may be defined as in § 1.) Set $\kappa_\rho(v) = \bar{\kappa}_\rho(\,|\,v\,|\,)$ in $V = V(E^n)$. Then κ_ρ is ∞-smooth, and (multiplying $\bar{\kappa}_\rho$ by some constant c_ρ if necessary)

$$\kappa_\rho(v) = 0 \quad (|\,v\,| \geqq \rho), \quad \int_V \kappa_\rho(v)\,dv = 1. \tag{2}$$

We define the *ρ-average* ϕ_ρ of ϕ by

$$\phi_\rho(p) = \int_{U_\rho(0)} \kappa_\rho(v)\phi(p+v)\,dv = \int_R \kappa_\rho(q-p)\phi(q)\,dq, \quad p \in \mathrm{int}_\rho\,(R). \tag{3}$$

We will have occasion to use another notation also:

$$\kappa_i'(v) = \kappa_{1/i}(v), \qquad A_i\phi(p) = \phi_{1/i}(p) \qquad (i = 1, 2, \cdots). \tag{4}$$

LEMMA 3a. *Let ϕ be continuous in R. Then*

$$\lim_{\rho \to 0} \phi_\rho(p) = \phi(p), \qquad p \in R, \tag{5}$$

and this holds uniformly in any compact set $Q \subset R$.

Because of (2),

$$\phi_\rho(p) - \phi(p) = \int_{U_\rho(0)} \kappa_\rho(v)[\phi(p+v) - \phi(p)]\,dv, \quad p \in \mathrm{int}_\rho\,(R). \tag{6}$$

Take ρ_0 such that $U_{\rho_0}(Q) \subset R$ and $|\phi(p+v) - \phi(p)| < \epsilon$ if $p \in Q$, $|\,v\,| < \rho_0$; then (6) gives $|\,\phi_\rho(p) - \phi(p)\,| < \epsilon$ if $\rho \leqq \rho_0$, $p \in Q$.

LEMMA 3b. *If ϕ is measurable and locally summable, then*

$$\lim_{\rho \to 0} \phi_\rho(p) = \phi(p) \qquad \text{a.e.} \quad \text{in } R. \tag{7}$$

To show this, we first note that since $\kappa_\rho(v)$ depends on $|\,v\,|$ alone, $\phi_\rho(p)$ is expressible in terms of $\bar{\phi}_\eta(p)$, $0 \leqq \eta \leqq \rho$. To find such an expression, set

$$\nu_\rho(h) = \inf\,\{t\colon\ \bar{\kappa}_\rho(t) \leqq h\}; \tag{8}$$

this is defined for $0 < h \leqq b = \bar{\kappa}_\rho(0)$; ν_ρ is the inverse of $\bar{\kappa}_\rho$ in any interval where $\bar{\kappa}_\rho$ is decreasing. Take the graph of the function κ_ρ; a section at the height h $(0 < h < b)$ cuts the graph in an open ball of radius $t = \nu_\rho(h)$, which is of n-volume $a_n t^n$. Hence, clearly,

$$\phi_\rho(p) = \int_0^b a_n[\nu_\rho(h)]^n \bar{\phi}_{\nu_\rho(h)}(p)\,dh = \int_0^\rho a_n t^n \left[-\frac{d}{dt}\bar{\kappa}_\rho\,(t)\right] \bar{\phi}_t(p)\,dt. \tag{9}$$

This may be verified analytically as follows. Set

$$\bar{\kappa}_{\rho h}(t) = \sup \{\bar{\kappa}_\rho(t) - h, 0\}, \tag{10}$$

$$\phi_{\rho h}(p) = \int_V \bar{\kappa}_{\rho h}(|v|)\phi(p + v)\, dv. \tag{11}$$

Then $\partial\bar{\kappa}_{\rho h}(t)/\partial h = -1$ if $t < \nu_\rho(h)$ and $= 0$ if $t > \nu_\rho(h)$, provided that $\nu_\rho(h)$ is a point of increase of κ_ρ, and

$$\frac{\partial \phi_{\rho h}(p)}{\partial h} = \int_V \frac{\partial \bar{\kappa}_{\rho h}(|v|)}{\partial h} \phi(p + v)\, dv = -\int_{U_{\nu(h)}(0)} \phi(p + v)\, dv$$

for such h; since $\phi_{\rho 0}(p) = \phi_\rho(p)$, $\phi_{\rho b}(p) = 0$, we have

$$\phi_\rho(p) = -\int_0^b \frac{\partial \phi_{\rho h}(p)}{\partial h}\, dh = \int_0^b \int_{U_{\nu(h)}(0)} \phi(p + v)\, dv\, dh,$$

giving (9). Note that taking $\phi(p) = 1$ (all p) in (9) gives

$$\int_0^b a_n[v_\rho(h)]^n\, dh = \int_0^\rho a_n t^n \left[-\frac{d\bar{\kappa}_\rho}{dt}\right] dt = 1. \tag{12}$$

It is well known (see § 5) that $\lim_{\rho\to 0} \bar{\phi}_\rho(p) = \phi(p)$ a.e. in R. We show that (7) holds whenever this holds. Given $\epsilon > 0$, choose ρ_0 so that $|\bar{\phi}_t(p) - \phi(p)| < \epsilon$ if $t \leqq \rho_0$. Then (9) and (12) give

$$|\phi_\rho(p) - \phi(p)| = \left| \int_0^\rho a_n t^n \left[-\frac{d\bar{\kappa}_\rho}{dt}\right] [\bar{\phi}_t(p) - \phi(p)]\, dt \right| < \epsilon, \quad \rho \leqq \rho_0.$$

LEMMA 3c. *$\phi_\rho(p)$ is ∞-smooth in* $\mathrm{int}_\rho\,(R)$.

We first show that, with a base $e_1, \cdots, e_n$ in V,

$$\nabla_{e_i}\phi_\rho(p) = -\int_R \nabla_{e_i}\kappa_\rho(q - p)\phi(q)\, dq. \tag{13}$$

(See also (4.1).) We have

$$\left| \frac{\phi_\rho(p + te_i) - \phi_\rho(p)}{t} + \int_R \nabla_{e_i}\kappa_\rho(q - p)\phi(q)\, dq \right|$$
$$\leqq \int_{U_\rho(p)} \left| \frac{\kappa_\rho(q - p - te_i) - \kappa_\rho(q - p)}{t} + \nabla_{e_i}\kappa_\rho(q - p) \right| |\phi(q)|\, dq,$$

and (13) follows from the fact that the first factor in the last integrand approaches 0 uniformly (we may use (II, Lemma 2c)).

Since κ_ρ is ∞-smooth, we may repeat this process, and the lemma follows.

4. The Weierstrass approximation theorem. We prove

LEMMA 4a. *Let ϕ be m-smooth ($m \geqq 0$) in the open set $R \subset E^n$. Then for any compact set $Q \subset R$ and any $\epsilon > 0$ there is a function ϕ' which is*

∞-smooth in E^n, such that the function $\psi(p) = \phi'(p) - \phi(p)$, and all its derivatives of order $\leqq m$ (in a fixed coordinate system), are $< \epsilon$ in Q.

REMARKS. We suppose ϕ is real valued. If ϕ has values in a Euclidean space E', we may introduce an orthonormal coordinate system into E', and apply the lemma to each coordinate separately, thus obtaining it for ϕ. For $m = 0$, ϕ is assumed merely to be continuous. We may phrase the conclusion in the form "ϕ' approximates to ϕ in Q, together with partial derivatives of order $\leqq m$, with an error $< \epsilon$", or, "ϕ' approximates (ϕ, Q, m, ϵ)."

Choose $\rho_0 > 0$ so that $U_{3\rho_0}(Q) \subset R$. Take a cubical subdivision of E^n, with cubes of diameter ρ_0, and let P be the set of all points of cubes touching $U_{\rho_0}(Q)$; then $P \subset R$. Set $\phi^*(p) = \phi(p)$ in P and $\phi^*(p) = 0$ in $E^n - P$; then $\phi^* = \phi$ in $U_{\rho_0}(Q)$. For any $\rho \leqq \rho_0$, ϕ^*_ρ is defined as in § 3, and is ∞-smooth in E^n. For $m = 0$, we may use $\phi' = \phi^*_\rho$ for a sufficiently small ρ, as follows directly from Lemma 3a. For $m = 1$, use integration by parts in (3.13) to give

$$(1) \qquad \nabla_{e_i}\phi_\rho(p) = \int_R \kappa_\rho(q - p)\, \nabla_{e_i}\phi(q)\, dq \qquad (\phi \text{ smooth}),$$

that is, $(\nabla_{e_i}\phi)_\rho(p) = \nabla_{e_i}\phi_\rho(p)$. Applying Lemma 3a to each of ϕ, $\partial\phi/\partial x^1$, $\partial\phi/\partial x^2, \cdots$ shows that we may use $\phi' = \phi^*_\rho$ again. Repeating the process gives the lemma for general m.

5. Lebesgue theory. In the last chapters of the book, we assume a familiarity with the Lebesgue integral. We describe here a few terms and facts of special importance.

We write "a.e." for "almost everywhere" (everywhere except in a set of measure 0). In Chapters IX and X, the term "measurable" means "Lebesgue measurable"; we let $|Q|$ denote the measure of the set Q. However, we often consider an s-plane P^s in the space E^n; if we say "ϕ is measurable in P^s", we mean that considering ϕ in P^s alone, ϕ is measurable there. Also, if $Q \subset P^s$, we let $|Q|_s$ denote the Lebesgue measure of Q, using measure in the space P^s. If ϕ is a real measurable function defined in the measurable set Q, we let $|\phi|$ denote its essential supremum:

$$(1) \qquad |\phi| = \text{ess sup}\,\{\phi\} = \inf\,\{\sup\,\{|\phi(p)|:\ p \in Q'\}:\ |Q - Q'| = 0\};$$

this agrees with the former definition if ϕ is continuous or cellwise continuous. The definition is the same if ϕ has values in a normed linear space. If ϕ has values in $V_{[r]}$ or $V^{[r]}$, and we use mass or comass as norm (I, 13), we write $|\phi|_0$ in place of $|\phi|$.

The theorem on term by term integration is important: Suppose ψ is measurable and summable in the measurable set R, ϕ_i $(i = 1, 2, \cdots)$

and ϕ are measurable in R, $0 \leqq \phi_i(p) \leqq \psi(p)$ in R, and $\lim \phi_i(p) = \phi(p)$ a.e. in R. Then

$$\lim_{i \to \infty} \int_R \phi_i = \int_R \phi.$$

We recall some facts from the theory of derivation (see Saks, Chapter IV). For a bounded measurable set $Q \subset E^n$, set

$$\Theta(Q) = |Q|/[\text{diam}(Q)]^n \tag{2}$$

(see (IV, 14), (IX, 2)). Let Φ be a countably additive real set function in the measurable set $R \subset E^n$, such that for some N,

$$|\Phi(Q)| \leqq N|Q|, \qquad \text{all measurable } Q \subset R; \tag{3}$$

or more generally, assume that Φ is absolutely continuous. For any $p \in R$, take any sequence $Q_1, Q_2, \cdots$ of measurable sets in R which contain p, such that $\Theta(Q_i) \geqq \eta$ (all i) for some $\eta > 0$ and diam $(Q_i) \to 0$; set

$$D_\Phi(p) = \lim \Phi(Q_i)/|Q_i|, \tag{4}$$

if this limit exists and is independent of the sequence chosen. Then D_Φ exists a.e. in R and is measurable, and

$$\Phi(Q) = \int_Q D_\Phi, \qquad \text{measurable } Q \subset R. \tag{5}$$

We could use sequences $Q_1, Q_2, \cdots$ which do not contain p, provided that, letting Q_i' denote Q_i together with the point p, we have $\Theta(Q_i') \geqq \eta$ (all i) for some $\eta > 0$; for $|Q_i'| = |Q_i|$. We could restrict the type of sets Q_i used; for instance, we could require them to be simplexes, as in (IX, 4.1). Then D_Φ will be defined at possibly more points than the former D_Φ.

One form of Fubini's Theorem may be stated as follows. Let ϕ be a real measurable and summable function in the measurable set $R \subset E^n$. Let P^s and P^{n-s} be planes in E^n with just one point in common (hence they span E^n). For each $q \in P^{n-s}$, let $P^s(q)$ be the plane through q parallel to P^s. Then for almost all $q \in P^{n-s}$, ϕ is measurable and summable in $P^s(q) \cap R$, and

$$\int_R \phi(p)\, dp = \int_{P^{n-s}} \int_{P^s(q) \cap R} \phi(p)\, dp\, dq.$$

Also, if $Q \subset R$ is such that $|R - Q| = 0$, then for almost all $q \in P^{n-s}$, $|P^s(q) \cap R - Q|_s = 0$.

We shall have occasion to use the following extension of Fubini's Theorem. Let P^{n-s} and P^{s-1} be planes without common point (hence they span E^n). For each $q \in P^{n-s}$, let $P^s(q)$ be the plane spanned by P^{s-1} and q. (The $P^s(q)$ cover almost all of E^n.) Let ϕ and Q be as before. Then for almost all $q \in P^{n-s}$, ϕ is measurable in $P^s(q) \cap R$ and

$| P^s(q) \cap R - Q |_s = 0$. (The formula for $\int_R \phi$ holds if suitably modified.) This may be seen as follows. Any p_0 not in P^{s-1}, but in some $P^s(q_0)$, is in a neighborhood U covered by the $P^s(q)$; there is a smooth homeomorphism of U onto an open set U', such that the $P^s(q) \cap U$ go into parts of parallel planes cutting U'. The function ϕ and the set Q go into a function ϕ' and a set Q' in U'. By Fubini's Theorem, for almost all of these parallel planes P', ϕ' is measurable in $P' \cap U'$ and $| P' \cap U' - Q' |_s = 0$. Hence, for some set H_U in P^{n-s} of $(n - s)$-measure 0, ϕ is measurable in $P^s(q) \cap U$ and $| P^s(q) \cap U - Q |_s = 0$ for $q \in P^{n-s} - H_U$. The union of the $P^s(q) - P^{s-1}$ is covered by a denumerable set of such neighborhoods U_i; let H be the union of the H_{U_i}. Then $| H |_{n-s} = 0$, and each $P^s(q)$ $(q \in P^{n-s} - H)$ has the desired property.

6. The space L^1. Consider the real measurable summable functions ϕ in E^n; two of these are *equivalent* if they are equal a.e. The equivalence classes Φ are the points of L^1; the norm of Φ is $\int_{E^n} \langle \phi \rangle$, for any $\phi \in \Phi$. Since L^1 is complete (see Halmos, p. 175, or E. J. McShane, *Integration*, Princeton University Press, 1951, pp. 182–184), it is a Banach space.

Let S be the subspace of L^1 formed by the continuous summable functions. Then S is dense in L^1. This may be seen as follows. Take any $\phi \in \Phi \in L^1$. Given $\epsilon > 0$, we may choose a function ψ which is bounded and equals 0 outside a bounded set R, such that $\int \langle \phi - \psi \rangle < \epsilon/2$. Form the averages $A_i\psi$ $(i = 1, 2, \cdots)$ as in § 3. Since $A_i\psi \to \psi$ a.e. (Lemma 3b), $\langle A_i\psi - \psi \rangle \to 0$ a.e. Also for some fixed summable ψ', $\langle A_i\psi - \psi \rangle < \psi'$ in E^n, for all i. Hence $\lim \int \langle A_i\psi - \psi \rangle = 0$, and we may choose i_0 such that $\int \langle A_{i_0}\psi - \psi \rangle < \epsilon/2$. Now $A_{i_0}\psi$ is continuous, and $\int \langle A_{i_0}\psi - \phi \rangle < \epsilon$. In this connection, see also (XI, Lemma 3a).

We remark that the operation in L^1 defined by a translation in E^n $(T_v\phi(p) = \phi(p - v))$ is continuous. A more general theorem follows from (IX, Theorem 15B) and (X, Theorem 7B).

Index of Symbols

We list first symbols defined in the appendices.

Appendix I

$\sum\limits_{(\lambda)}$, ϵ_λ, δ^μ_λ 341
$\hat{\imath}$, $\approx$, dim (V), e_i, $\mathfrak{A}^n$, $\bar{e}_i$ 342
$L(V, W)$, $\mathfrak{A}$, e^i 343
$\bar{V}$ 343, 347
ϕ^*, $X \times Y$, $x \times y$ 344
$V \oplus W$, V mod W, ann (V_1) 345
$|v|$ 346, 348
dist (v, w) 347, 348
$u \cdot v$ 348
$|\phi|$ 349
$V(E)$, $p + v$ 350
$\Sigma a_i p_i$ 351
E^n 353

Appendix II

$\subset$, $\in$, $\cap$, $\cup$, $\bigcup$, $\bigcap$, $Q_1 - Q_2$, $\bar{Q}$, int (Q), $f^{-1}(Q)$, dist (p, q), dist (P, Q), diam (Q), $U_\zeta(Q)$, $\bar{U}_\zeta(Q)$, int$_\zeta$ (Q) 355
$P(\sigma)$, dim (σ), int (σ) 356
$\partial\sigma$ 356, 362
$p_0 \cdots p_r$ 356, 360
dim (Q), St (σ), $\overline{\text{St}}$ (σ), ∂St (σ) 357
$\Sigma a_i \sigma_i^r$, dim (A), $\mathbf{C}_r$, $\mathbf{C}^r$, I^0 361
∂A 362, 153, 155
dX 362, 157
$\mathbf{H}_r$, $\mathbf{H}^r$, $\mathbf{H}$, $\mathbf{H}^*$ 362, 363
$X \smile Y$, $X \frown A$ 363, 278, 281
$J(P, Q)$, $J(p, A)$, $J(\sigma^r, \sigma^s)$ 364
$\sigma^r \times \sigma^s$, I 365
$\mathscr{D}_v(t \times p)$ 366
ind (P, P') 367

Appendix III

car (ϕ), spt (ϕ), sup { }, inf { }, $\langle\phi\rangle$, $|\phi|_R$, $\langle\phi\rangle_0$ 371
$|\phi|$, $|\phi|_0$ 371, 376, 51
$|Q|$ 372, 376, 86
κ_i', κ_ρ, $A_i\phi$ 374
a.e., $|Q|_s$, ess sup 376
$\Theta(Q)$ 377, 125, 256
L^1 378

Introduction

$-\sigma$ 4
$X \cdot \sigma$ 4
$X \cdot A$, $\partial\sigma$ 5, 156
∂A 5, 153, 155
$|\sigma|$ 6, 81
$|v|$ 6
$\{\sigma\}$ 8, 80
$D_X(p)$ 8, 170, 262
$v_1 \vee v_2$, $v_1 \vee \cdots \vee v_r$ 10, 36
e_i, v^i 11, 342
$e_{\lambda_1 \cdots \lambda_r}$, $\alpha^{\lambda_1 \cdots \lambda_r}$ 11, 38, 40
$V_{[r]}$ 11, 36
$\bar{V}$ 12, 343, 347
e^i, f_i 12, 343
$e^{\lambda_1 \cdots \lambda_r}$, $\xi_{\lambda_1 \cdots \lambda_r}$ 12, 38, 40
$V^{[r]}$ 12, 37
$f \vee g$, $f^1 \vee \cdots \vee f^r$ 13, 37
α_0, ω_0 14, 66, 67
$u \times v$ 14
∇f, $\nabla_v f(p) = \nabla f(p, v)$ 15, 58, 59
$f^*\omega$ 16, 62
$J_f(p)$, $\bar{J}_f(p)$ 16, 66, 67

$\tilde{J}_f(p)$ 17

∂M 21, 108

dX 21, 28, 157

$d\omega$ 21, 70, 104

∇ 23

$|X|$, $|X|^\flat$, $|A|$, $|A|^\flat$ 27, 153, 156

$|X|^\sharp$, $|A|^\sharp$ 28, 159, 160

Chapter I

$u\vee v$, $v_1\vee\cdots\vee v_r$, $V_{[r]}$ 36

$f\vee g$, $f^1\vee\cdots\vee f^r$, $V^{[r]}$, $\omega\cdot\alpha$ 37

e_λ, e^λ, α^λ, ω_λ 38, 40

$L^r_{\text{alt}}(V)$ 39

$\alpha\vee\beta$, $\omega\vee\xi$ 41

$\omega\wedge\alpha$ 42

$\rho(R, R')$, $\tilde{\rho}(\tilde{R}, \tilde{R}')$ 45

ϕ^* 46

$\mathscr{D}\alpha$, $\mathscr{D}'\omega$ 47

$\alpha\cdot\beta$, $\omega\cdot\xi$ 48, 49

$|\phi|$ 51

$|\alpha|_0$, $|\omega|_0$ 52

$\cos\theta$, $|W_2 - W_1|$ 56

Chapter II

$\nabla_v f(p) = \nabla f(p, v)$ 58

$\nabla f(p)$ 59

$|\nabla f(p)|$, $|\nabla f|_Q$, $|\nabla f|$ 60

$gf = g\circ f$, ω_0, $\bar{\omega}(p)$ 61

$|\omega|$, $|\omega|_0$, $\nabla f(p, \alpha)$, f_p^*, $f^*\omega$, $\omega\vee\xi$ 62

$\mathfrak{L}_f = \mathfrak{L}(f)$, $\mathfrak{L}_{f|Q}$, $e_i(p)$ 63

$e_\lambda(p)$, $e^i(p) = dx^i$, $\omega_\lambda(p)$ 64

α_0, $J_f(p)$ 66

$\bar{J}_f(p)$, ω_0 67

$d\omega$ 70, 77, 104, 106, 273

Chapter III

$\{\sigma\}$ 80, 8

$|\sigma|$, $|\sigma|_r$ 81

$\Sigma a_i\sigma_i^r$ 81, 152

$\{A\}$ 82, 220

$\omega\circ A$ 82

$\int_\sigma\omega$, $\int_A\omega$ 83

$|A|$, $\int_{\bar{\sigma}}\bar{\omega}$ 84

$\int_R\omega$ 85

$Q_1\subset Q_2\subset\cdots\to R$, $|R|$ 86

spt (ω) 89

$f\sigma$, $\int_{f\sigma}\omega$ 92, 107

$\int_M\omega$ 93

$d\omega$ 104, 106, 70, 77, 273

u.c.s. 105

∂M, $\partial_0 M$ 108

Chapter IV

L_f 113

$\Theta(\sigma)$ 125, 256

I^0, $\mathbf{1}$ 139

$\mathbf{H}^r_\mu$, $\mathbf{H}^*_\mu$, $\mathbf{h}\smile\mathbf{h}'$ 142

$X\smile Y$, γ_f 143

Chapter V

$\Sigma a_i\sigma_i$, $A(p)$ 152

$\mathbf{C}_r^{\text{pol}}$, spt (A) 153

∂A 153, 155

$|A|$ 153, 179, 232, 241

$|A|^\flat$, $\mathbf{C}_r^\flat$ 154

T_vA, $\mathscr{D}_vA$ 155

$X\cdot A$, $\mathbf{C}^{\flat r}$, $|X|^\flat$ 156

$|X|$ 156, 233

dX 157

$|A|^\sharp$, $\mathbf{C}_r^\sharp$ 159

$|X|^{\#}$, $\mathbf{C}^{\# r}$ 160
$\mathfrak{L}_X = \mathfrak{L}(X)$ 161
$\mathfrak{L}_0(\omega)$, $|\omega|^{\#}$ 167
$D_X(p, \alpha)$ 167, 258
$D_X(p)$ 170, 262

$\phi_{\sigma,\zeta}(p)$ 174
$\lim^{\flat}$, $\lim^{\#}$, $\xrightarrow{\flat}$, $\xrightarrow{\#}$, $\text{wkl}^{\flat}$ 175
X_η 176
$|A|^{\flat}_{\rho}$, $|A|^{\#}_{\rho}$, $|X|^{\flat}_{\rho}$, $|X|^{\#}_{\rho}$, $|A|^{\odot}_{\rho}$, $|X|^{\odot}_{\rho}$, $\mathbf{C}^{\flat}_{\rho,r}$, $\mathbf{C}^{\#}_{\rho,r}$, $\mathbf{C}^{\flat r}_{\rho}$, $\mathbf{C}^{\# r}_{\rho}$ 178

Chapter VI

$\bar{t}$ 186, 191
$\tilde{\phi}$ 187, 199
$|\gamma|$ 190
$\bar{\gamma}(a, b)$ 191
$\gamma^{\odot}$ 192

spt (ψ), A_T 194
γ_A, I^0 196
$\mathbf{M}$, $|\gamma|^{\#}$, $\mathbf{C}$ 198
$\tilde{\alpha}$ 199, 280
$d^*\alpha$ 204

Chapter VII

$|\ |^{\odot}$, $N^{(r)}_{\phi}$ 208
ϕA 208, 328
ϕX 212
spt (A), spt (X) 213, 237
$\gamma_\eta(t)$ 214

$\{A\}$ 220, 82
$P(\sigma)$, $X(\sigma)$ 225
α/β 227
$\mathbf{H}^{\flat}$, $\mathbf{H}^{\flat r}$, $\mathbf{H}^*$ 229

Chapter VIII

$\mathscr{D}_v(Q)$, $|A|^{\flat}_{R}$ 232
$\mathbf{C}^{\flat}_{r}(R)$, $\mathbf{C}^{\flat r}(R)$, $\lim^{\flat}_{R}$, $|X|^{\flat}_{R}$ 233
$|A|^{\#}_{R}$, $\mathbf{C}^{\# r}(R)$, $\mathfrak{L}_{X,R}$, $\mathfrak{L}_{\phi,R}$ 234
$\mathfrak{L}_0(\omega, R)$, $|\omega|^{\#}_{R}$, $\text{wkl}^{\odot}_{R}$ 235
$N^{(r)}_{\phi,R}$ 236

spt (A), spt (X) 237, 213
$|A|^{\flat}_{R,\rho}$, $|A|^{\#}_{R,\rho}$ 238
$|X|^{\flat}_{R,\rho}$, $|X|^{\#}_{R,\rho}$ 239
$|A|^{T}_{R,\rho}$ 248

Chapter IX

$\Theta(\sigma)$, $\Theta_p(\sigma)$ 256
$P(\sigma)$, $\alpha(\sigma)$, $\alpha(P)$, $P(p, \alpha)$, $D_X(p, \alpha)$ 258
$|H|_s$ 260
$D_X(p)$ 262
$\bar{d}\omega$ 266
$\tilde{d}_\mu\omega$ 268

$\Psi\omega$, $\Phi X = D_X$ 272
$d\omega$ 273
$X \smile Y$, I^0 278, 363
$\bar{\phi}$, $\tilde{\alpha}$ 280
$X \frown A$ 281, 363

Chapter X

$\delta_{f,g}$ 295
fA 296, 300

$\text{dist}_R\,(p, q)$, $\mathfrak{L}_{f,R}$, $f|Q$, $a^{[r]}$ 298
$f^*X, f^*\omega$ 302

Chapter XI

$\bar{\gamma}$ 312
$[\gamma]$, $|\gamma|$, $\int_R \xi \cdot d\gamma$ 314
$\xi \cdot \gamma$ 315
$|\xi|_\gamma$, χ_Q 316
$\lim \gamma_i$, $\uparrow$, $\downarrow$ 318
L^+, L^* 320
u.c.s. 322

$|\gamma|^{\#}$, spt (γ), $\gamma_{p,\alpha}$ 325
$\mathfrak{M}_r$, $\mathbf{M}_r$, γ_A 326
ϕA 328, 208
A_Q 329
$\Gamma_A(p)$, $\xi_A(p)$ 330
$\tilde{\alpha}$, $T_v\gamma$ 331
diam (γ) 332

Index of Terms